T0379547

The Routledge Companion to Planning in the Global South

The Routledge Companion to Planning in the Global South offers an edited collection on planning in parts of the world that, more often than not, are unrecognised or unmarked in mainstream planning texts. In doing so, its intention is not to fill a 'gap' that leaves this 'mainstream' unquestioned but to re-theorise planning from a deep understanding of 'place' as well as a commitment to recognise the diverse modes of practice that come within it.

The chapters thus take the form not of generalised, 'universal' analyses and prescriptions, but instead are critical and located reflections in thinking about how to plan, act and intervene in highly complex city, regional and national contexts. Chapter authors in this Companion are not all planners, or are planners of very different kinds, and this diversity ensures a rich variety of insights, primarily based on cases, to emphasise the complexity of the world in which planning is expected to happen.

The book is divided into a framing Introduction followed by five sections: planning and the state; economy and economic actors; new drivers of urban change; landscapes of citizenship; and planning pedagogy. This volume will be of interest to all those wanting to explore the complexities of planning practice and the need for new theories of knowledge from which to draw insight to face the challenges of the twenty-first century.

Gautam Bhan is Lead, Academics and Research, at the Indian Institute for Human Settlements, Bangalore. He holds a PhD from the University of California, Berkeley, and is most recently the author of *In the Public's Interest: Evictions, Citizenship and Inequality in Contemporary Delhi* (University of Georgia Press, 2016).

Smita Srinivas is an economic development scholar with a PhD from MIT. She is Visiting Professor, Economics Department and the Innovation, Knowledge, Development Centre (IKD), Open University, UK; Visiting Senior Fellow, International Development Department, London School of Economics and Political Science (LSE); Honorary Professor, Indian Council for Research on International Economic Relations (ICRIER). Her last book *Market Menagerie* (Stanford University Press, 2012) won the EAEPE 2015 Myrdal Prize.

Vanessa Watson is Professor of City Planning at the University of Cape Town (South Africa) and is a Fellow of this University. She holds degrees, including a PhD, from South African universities and the Architectural Association of London and is on the executive committee of the African Centre for Cities.

The Routledge Companion to Planning in the Global South

Edited by Gautam Bhan, Smita Srinivas and Vanessa Watson

LONDON AND NEW YORK

First published 2018
by Routledge
2 Park Square, Milton Park, Abingdon, Oxon OX14 4RN

and by Routledge
711 Third Avenue, New York, NY 10017

Routledge is an imprint of the Taylor & Francis Group, an informa business

© 2018 selection and editorial matter, Gautam Bhan, Smita Srinivas and Vanessa Watson; individual chapters, the contributors

The right of Gautam Bhan, Smita Srinivas and Vanessa Watson to be identified as the authors of the editorial material, and of the authors for their individual chapters, has been asserted in accordance with sections 77 and 78 of the Copyright, Designs and Patents Act 1988.

All rights reserved. No part of this book may be reprinted or reproduced or utilised in any form or by any electronic, mechanical, or other means, now known or hereafter invented, including photocopying and recording, or in any information storage or retrieval system, without permission in writing from the publishers.

Trademark notice: Product or corporate names may be trademarks or registered trademarks, and are used only for identification and explanation without intent to infringe.

British Library Cataloguing-in-Publication Data
A catalogue record for this book is available from the British Library

Library of Congress Cataloging-in-Publication Data
Names: Bhan, Gautam, editor. | Srinivas, Smita, editor. |
Watson, Vanessa, editor.
Title: The Routledge companion to planning in the Global South /
edited by Gautam Bhan, Smita Srinivas and Vanessa Watson.
Description: Abingdon, Oxon ; New York, NY : Routledge, 2018. |
Includes bibliographical references and index.
Identifiers: LCCN 2017012167 | ISBN 9781138932814
(hardback : alk. paper) | ISBN 9781315678993 (ebook)
Subjects: LCSH: Regional planning—Developing countries. |
City planning—Developing countries. | Economic development—
Developing countries. | Urban policy—Developing countries.
Classification: LCC HT395.D44 R68 2018 | DDC 307.1/209724—dc23
LC record available at https://lccn.loc.gov/2017012167

ISBN: 978-1-138-93281-4 (hbk)
ISBN: 978-1-315-67899-3 (ebk)

Typeset in Bembo
by Swales & Willis Ltd, Exeter, Devon, UK

Printed in the United Kingdom
by Henry Ling Limited

Contents

Figures

Tables

Contributors

Adriana Allen is Professor of Development Planning and Urban Sustainability at The Bartlett Development Planning Unit, University College London, where she leads the Research Cluster on Environmental Justice, Urbanisation and Resilience. Originally trained as an urban and regional planner in her native country Argentina, she specialised over the years in the fields of urban environmental planning and political ecology. She has 30 years of international experience in research and consultancy undertakings in over 20 countries in Asia, Africa and Latin America. Both as an academic and practitioner, her work focuses on investigating and promoting transformative links between socio-environmental change, justice and sustainability in urban and peri-urban contexts.

Isabelle Anguelovski is ICREA Research Professor at the Universitat Autònoma de Barcelona Spain. Her research examines the extent to which urban plans and policy decisions contribute to more just, resilient, healthy and sustainable cities, and how community groups in distressed neighbourhoods contest the existence, creation or exacerbation of environmental inequities as a result of urban (re)development processes and policies.

Ragui Assaad is Professor at the Humphrey School of Public Affairs at the University of Minnesota, where he chairs the Global Policy area and co-chairs the Master of Development Practice programme. His current research focuses on youth and gender issues in the labour market in the Arab World.

Clive Barnett is Professor of Geography and Social Theory at the University of Exeter, UK. He is author most recently of *The Priority of Injustice: Locating Democracy in Critical Theory* (University of Georgia Press, 2017).

Jane Battersby is a senior researcher at the African Centre for Cities, UCT, and is the research co-ordinator of the ESRC/DFID-funded Consuming Urban Poverty project. An urban geographer by background, her work focuses on urban food security and food systems.

Jigar Bhatt is a PhD candidate at Columbia University and received his Master's degree in City Planning from MIT. His research focuses on issues of political economy and sociology of expertise. He also has several years of professional experience in infrastructure planning and programme evaluation, mostly in southern Africa.

Chris Bonner was the Director of the Organisation and Representation Programme of the WIEGO (Women in Informal Employment: Globalising and Organising) Network (2004–2016), and is currently programme advisor. She helped build a capacity of informal workers organisations and facilitated transnational networking among organisations of domestic workers, home-based workers, street vendors and waste pickers. She has contributed to research and writing on organising in the informal economy. Prior to this she was a trade unionist and during 1997–2003, the founding Director of DITSELA Workers Education Institute in Johannesburg.

Mercy Brown-Luthango holds a DPhil in Sociology from Stellenbosch University. She has 16 years research experience and has worked on a wide array of research projects, all with a strong focus on inclusive and equitable development, participatory processes and the improvement of development outcomes for poor, marginalised groups.

Françoise Carré is Research Director of the Center for Social Policy at the University of Massachusetts Boston J.W. McCormack Graduate School of Policy and Global Studies (USA) and research coordinator for the WIEGO Network. She specialises in applied labour economics, industrial relations and comparative employment policy, as well as studies of community based organisations. She has written extensively about temporary, short-term and informal work in the US and in international perspective and about low-wage employment. She holds a PhD in Urban and Regional Studies from the Massachusetts Institute of Technology.

Fernando M. Carrión is an architect at Universidad Central del Ecuador, holds a Master's degree in Regional and Urban Development from El Colegio de México and is a doctoral candidate at the University of Buenos Aires. He was former Director of CIUDAD, Planning Director of the Municipality of Quito, General Coordinator of RED CIUDADES for Latin America, Director of FLACSO–Ecuador, Council for the Metropolitan District of Quito, Coordinator of the Workgroup 'The right to the city' of CLACSO and President of the Latin-American Organization of Historical Centers (OLACCHI). He specialises in urban planning and development, decentralisation, historical centres, housing, citizen security and football sociology. Currently he is a researcher at FLACSO–Ecuador.

Jia-Ching Chen is an urban, development and environmental studies researcher and Assistant Professor in the Department of Global Studies at the University of California, Santa Barbara. Currently, his interests are in China's role in shaping the global green economy and the spread of Chinese planning expertise through its international development activities. He also has professional experience in social movements and organised labour. Dr. Chen received his PhD from the University of California, Berkeley, in City and Regional Planning with a designated emphasis in Global Metropolitan Studies and outside fields in Geography and Anthropology.

Martha Alter Chen is a Lecturer in Public Policy at the Harvard Kennedy School, Affiliated Professor at the Harvard Graduate School of Design and International Coordinator of the WIEGO Network. Her expertise is on the intersection of work, gender and poverty with a special focus on the informal economy. She has many years of resident experience in Bangladesh and India.

Eric Chu is a Lecturer in Planning and Human Geography in the School of Geography, Earth and Environmental Sciences at the University of Birmingham, UK. His research focuses on the politics of governing climate change adaptation, resilience and development in cities.

Rhonda Douglas holds a Master's degree in Management for the Voluntary Sector from McGill University and has over two decades of experience in resource mobilisation and organisational development in the voluntary sector.

Edesio Fernandes is a Brazilian-British legal scholar, international consultant and author who has specialised in the legal dimensions of urban development processes, land and housing rights, informal settlements and urban and metropolitan governance. He is a member of DPU Associates and of the Teaching Faculty of the Lincoln Institute of Land Policy, has been associated with Tulane University, NYU Wagner and several Brazilian universities, and has regularly worked for UN-Habitat, World Bank and several governments and NGOs.

Mônica A. Haddad is an Associate Professor of Urban and Regional Planning at Iowa State University. She is the Director of the Graduate Certificate in Geographic Information Systems. She received her PhD and Master's degrees in Urban and Regional Planning from the University of Illinois at Urbana-Champaign. Her research agenda focuses on four main issues, in the context of Brazilian cities: (a) social justice, (b) urban infrastructure provision, (c) poverty alleviation and (d) environmental planning.

Patrick Heller is the Lyn Crost Professor of Social Sciences and Professor of Sociology and International Studies at Brown University. His main area of research is the comparative study of social inequality and democratic deepening. His most recent books include *Bootstrapping Democracy: Transforming Local Governance and Civil Society in Brazil* (Stanford University Press, 2011) and *Deliberation and Development: Rethinking the Role of Voice and Collective Action in Unequal Societies* (The World Bank Group, 2015). He has published articles on urbanisation, comparative democracy, social movements, development policy, civil society and state transformation. His most recent project, Cities of Delhi, conducted in collaboration with the Centre for Policy Research, explores the dynamics of governance and social exclusion in India's capital.

Amlanjyoti Goswami is Head, Law and Regulation, and Faculty at the Indian Institute for Human Settlements, Bangalore. He co-leads the development of the law and governance concentration for the proposed IIHS Masters in Urban Practice (MUP) programme. Amlan works on law, policy and governance, including research in land acquisition, decentralisation and knowledge epistemologies. He is interested in conceptual questions of urban theory, legality and its relationships with space, politics and culture. He holds degrees in law and sociology from Harvard University and the University of Delhi.

Rita Lambert is an architect and urban development planner originally from Ethiopia. She is currently a Teaching Fellow at The Bartlett Development Planning Unit, UCL and a co-investigator on several research projects in Africa and Latin America. Her current research focuses on the relationship between planning and spatial knowledge production, manipulation and circulation, as well as the development of tools that can be adopted by ordinary citizens to navigate institutional barriers and expand the room for manoeuvre towards socio-environmentally just urbanisation.

Colin McFarlane is Professor in Urban Geography at Durham University. His research examines the experience and politics of informal settlements, and has included work in several cities, especially Mumbai and, more recently, Kampala. He is interested in the ways in which cities are composed, lived and politicised on the margins of global urbanism, reflected in books and papers on urban informality, infrastructure, learning, densities and smart cities.

Maureen Mackintosh is Professor of Economics at the Open University and Director of the Innovation, Knowledge and Development (IKD) research centre. She is a development economist, specialising in the economics of markets for health care and medicines, with particular reference to Africa. She has collaborated with Paula Tibandebage in research and writing for over 20 years, most recently on a project on 'Industrial productivity and health system performance', funded by the UK Economic and Social Research Council. Her most recent book is *Making Medicines in Africa: The Political Economy of Industrializing for Local Health* (Palgrave Macmillan, 2016; co-edited with G. Banda, P. Tibandebage and W. Wamae).

Anjali Mahendra is the Lead for Global Research at the World Resources Institute's (WRI) Ross Centre for Sustainable Cities. Her work straddles research and practice in transport policy and related economic, climate and urban development issues, focusing on cross-national comparisons. She has a PhD in Urban and Regional Planning and Master's degrees in Transportation and City Planning from MIT, and was trained as an architect at the School of Planning and Architecture, New Delhi.

Eduardo Marques holds a PhD in Social Sciences (IFCH/Unicamp) and is a Livre-docente Professor at the Department of Political Science (DCP) and a researcher at the Center for Metropolitan Studies (CEM), University of São Paulo. Currently, he is president of the Research Committee 21 – Urban and regional development of the International Sociological Association (ISA). He is the author most recently of *São Paulo in the Twenty-First Century: Spaces, Heterogeneities, Inequalities* (Routledge, 2016).

Sudeshna Mitra is faculty at the Indian Institute for Human Settlements. Her research looks at urban land and real estate, from perspectives of planning, governance, economic transitions, value propositions and property regimes. She has taught at the School of Planning and Architecture, Cornell University and Cornell in Washington. She has also worked as a consultant on urban and economic infrastructure projects in India and South East Asia.

Nancy Odendaal is an Associate Professor in City and Regional Planning at the University of Cape Town. Her research focuses on three interconnected areas of enquiry: infrastructure development, specifically technology innovation, and socio-spatial change in cities, planning education and the theoretical and empirical parameters of case study research. Previously, Nancy was based at the African Centre for Cities, where she coordinated the expansion of the Association of African Planning Schools (AAPS), and managed an Africa-wide project on curricula reform. She is currently chair of the Global Planning Education Association Network, and incoming chair of AAPS.

Dele Olowu is a retired professor and an international consultant based in the Netherlands. He is currently a Fellow of the University of Leiden, African Studies Center and co-Director of Global Peace Compact. He was Professor of Public Administration and Local Government studies at Obafemi Awolowo University, Ile-Ife in Nigeria, where he taught for 20 years. He also taught public policy and management at the Universities of Addis Ababa, Ethiopia, University of Namibia, Windhoek, and at the Institute of Social Studies, Erasmus University, Rotterdam in the Netherlands. In addition, he had worked for the United Nations Economic Commission for Africa (UNECA), the African Development Bank (AfDB) and the World Bank. He has published widely on African governance and public administration including intergovernmental relations, leadership, regional, local and indigenous governments.

Susan Parnell is Professor of Geography at the University of Cape Town and is closely associated with the African Centre for Cities. Recent co-edited books include *Climate Change at the City Scale* (Routledge, 2012), *Routledge Handbook on Cities of the Global South* (Routledge, 2014) and *Africa's Urban Revolution* (Zed Books, 2014).

Alessandro Petti is a Professor of Architecture and Social Justice at the Royal Institute of Art in Stockholm and a Loeb Fellow at Harvard University Graduate School of Design. He is an architect who combines theoretical research with an architectural, artistic and pedagogical practice engaged in the struggle for justice and equality. In 2012 with Sandi Hilal he founded 'Campus in Camps', an experimental educational programme in the Dheisheh refugee camp in Bethlehem (www.campusincamps.ps). With Sandi Hilal and Eyal Weizman, Petti created DAAR (Decolonizing Architecture Art Residency) in Beit Sahour, Palestine, an architectural studio and residency programme that has gathered architects, artists, activists, urbanists, filmmakers and curators to work collectively on politics and architecture (www.decolonizing.ps). They published together *Architecture After Revolution*, revisiting today's struggles for justice and equality not only from the historical perspective of revolution, but also as a continued struggle for decolonisation.

Aromar Revi is Director of the Indian Institute for Human Settlements (IIHS), India's prospective independent National University for Research and Innovation addressing its challenges of urbanisation, through an integrated programme of education, research, practice and training. He is an international practitioner, consultant, researcher and educator with over 30 years of inter-disciplinary experience in public policy and governance, the political economy of reform, development, technology, sustainability and human settlements. He is a member of the Leadership Council of the UN Sustainable Development Solutions Network (SDSN), and co-chair of its urban thematic group, which led the global campaign for an urban Sustainable Development Goal (SDG).

Elena Reyes holds Master's degrees in International Cooperation and Urban Development (Technische Universität Darmstadt and Università Tor Vergata a Roma) and Sustainable Development and Design of the City (Tecnológico de Monterrey). Her research interests are urban planning and informal settlements. She co-authored *The Handbook for Sustainable Housing Design in Mexico* (Editorial Trillas, 2013).

Debra Roberts is Acting Head of the Sustainable and Resilient City Initiatives Unit and Chief Resilience Officer at eThekwini Municipality and Honorary Professor in the School of Life Sciences at the University of KwaZulu-Natal (South Africa). She also serves as co-chair of Working Group II of the Intergovernmental Panel on Climate Change (IPCC) Sixth Assessment Report.

Qianqi Shen got her PhD in Planning and Public Policy from Bloustein School of Planning and Public Policy at Rutgers University and Master's degree in Science in Urban Planning from Columbia University. Currently, she is doing field research in China, focusing on intergovernmental negotiations in land development and governance.

AbdouMaliq Simone is an urbanist with particular interest in emerging forms of collective life across cities of the so-called global South. He is presently Research Professor at the Max Planck Institute for the Study of Religious and Ethnic Diversity, Visiting Professor of Sociology, Goldsmiths College, University of London and Visiting Professor of Urban Studies at the African Centre for Cities, University of Cape Town.

Caroline Skinner is a Senior Researcher at the African Centre for Cities at the University of Cape Town and Urban Research Director for the global research and policy network WIEGO. For nearly two decades, Skinner's work has interrogated the nature of the urban informal economy with a focus on informing advocacy processes and livelihood-centred policy and planning responses.

Paul Smoke is Professor of Public Finance and Planning and Director of International Programs at the New York University's Robert F. Wagner Graduate School of Public Service. His research and policy interests include the political economy of public sector reform, especially decentralisation, intergovernmental fiscal relations and service delivery, as well as urban development planning. He has published in numerous journals and authored or edited several books on decentralisation. Smoke has worked in many countries, especially in Africa and Asia, and with a wide range of international development partners.

Paula Tibandebage is a former Principal Researcher with REPOA, a non-government policy research institute in Tanzania. She has over 20 years' research experience, specialising in issues of social protection and social services provisioning including health and education. Recent publications include: *Making Medicines in Africa: The Political Economy of Industrializing for Local Health* (Palgrave Macmillan, 2016; co-edited with M. Mackintosh, G. Banda and W. Wamae); and "Can managers empower nurse-midwives to improve maternal health care? A comparison of two resource-poor hospitals in Tanzania" (*International Journal of Health Planning and Management*, 2015), which reports findings from her Wellcome Trust-supported project on Ethics, Payments and Maternal Survival in Tanzania.

Ivan Turok is Executive Director at the Human Sciences Research Council in South Africa. He is Editor-in-Chief of the journal *Regional Studies* and Honorary Professor at the University of Glasgow. He is also Chairman of the City Planning Commission for Durban, and an adviser to the United Nations, OECD and national governments. He has authored over 120 academic publications on city and regional development, urban transformation and national urban policies.

Alexandra Velasco studied Applied Ecology and Environmental Communications at Universidad San Francisco de Quito, holds an MBA from Instituto de Altos Estudios Nacionales, and a Master's degree in Infrastructure Planning from the University of Stuttgart. She has worked as President of Biciacción, an NGO in Quito, as Metropolitan Director for Sports and Recreation in the Municipality of Quito, and she is the former Director for Non-motorised Transport, and CEO of consultancy firm MÓVERE. She is author of more than 50 articles related to transport and mobility in newspapers and digital media. Currently she works in the German Development Cooperation.

Tanja Winkler is an Associate Professor in the School of Architecture, Planning and Geomatics at the University of Cape Town, South Africa. Her research interests include community-led planning interventions, exploring the nature and meaning of ethical values for planning theory and practice, and planning education through engaged scholarship.

Christopher Yap is a PhD candidate and participatory video maker based at the Center for Agroecology, Water and Resilience, Coventry University. Born in the UK, his research explores the political and spatial significance of community-managed gardens in London and Seville, with the aim of building solidarity between citizen collectives working in different contexts but facing similar challenges. Christopher worked previously at The Bartlett Development Planning Unit, conducting action-research in cities in the global North and South.

Oren Yiftachel researches and teaches political geography and urban planning at Ben-Gurion University of the Negev, Beersheba. He is currently the Head of the Department for Multi-disciplinary Studies, holding the Chair of Urban Studies. Over the last two decades he has been active in civil society organisations, working with dispossessed communities in Israel/Palestine.

Acknowledgements

First and foremost, we want to thank Saskia Greyling, based in the Environmental and Geographical Science department at the University of Cape Town, for project managing and editing this book. Her attention to detail, her awesome competence and her unfailing good humour in sometimes difficult circumstances ensured that this book was finally completed. Vanessa Watson thanks the South Africa National Research Fund for the resources to support Saskia.

We would like to thank our Routledge team, including editorial assistant, Egle Zigaite, production editor Sophie Watson and production manager Caroline Watson, for all their help with this book. We also thank the following for permission to use their material: Nick Anim, Campus in Camps/Anna Sara, cLIMA sin Riesgo, DAAR, Economic Research Forum, Luise Fischer, Rita Lambert, Oxford University Press, REPOA, Routledge and Eva Vivalt.

Introduction

Gautam Bhan, Smita Srinivas and Vanessa Watson

Planning, across the many disagreements of what it may mean and hold within it, has long held one thread across them: the shared assumption that it is both possible and desirable to initiate and direct processes of change towards goals such as social and spatial justice and sustainability. We see planning as broadly the domains that include public policy design, institutional design, plans and plan-making, spatial and territorial planning, public administration and public capacity building. What has organised our differences within this book is to deliberately reveal the different types of contestation that remain. These include the terms of such direction of social change, the theories of knowledge that must underlie them, and the modes and processes that emerge across diverse disciplinary homes, professional praxis and institutional locations. This is a valuable – if often frustrating – tradition of debate and dissent that we believe is one of the great strengths of the field of planning, and one that this book both inherits and speaks to. Rarely are books available, it seems to us, that make these contestations evident within a field of professional engagement, or reveal the different institutional mechanisms of engagement that help build many co-existing, workable theories 'from here'. We seek a two-fold thrust in the book to showcase not only how the field has arrived here in time and multiple locations, but also where we go from here.

On planning in the contemporary moment

The first is how to assess planning at a time of profound global change. Writing this introduction in February of 2017 is to write surrounded by what feels like seismic shifts – in the world balance of political power; in the stability of what have come to be known as liberal democracies; in the complexity of how 'national' developmental paradigms straddle an increasingly re-arranged and polycentric 'global'; in the environmental health of our planet; in the emergence of new technologies; and in the way societies and individuals perceive and project themselves, their identities and their shared or more often highly antagonistic values. Growth, jobs and trade, seemingly on the agenda of all politics these days, offer both a chance for a progressive and regressive set of plans and for a more or less connected, and shared, world. The growth and prosperity the last century has brought is undeniable in both its impact on human life as well as its continuing fractures, omissions and exclusions. What certainly seems to have shifted is the complexity of the negotiations that define our present, from scale to priorities, means to ends. What does it mean to initiate and direct change – to plan – in this moment? How must we conceive of and re-imagine 'planning' to respond to changing contexts and challenges?

As editors, we argue through this book that this complexity makes the practices of planning more rather than less critical. Across scales from neighbourhoods to supra-national bodies, we are not alone in this assertion. As the United Nations adopts a new set of Sustainable Development Goals and a New Urban Agenda (NUA) is put forth, what finds central place is not just a focus on key normative goals on equity, sustainability and the enhancement of urban economies, but also a focus on planning as a central tool and mode for achieving these. The NUA is but one example of an acknowledgement that achieving these ambitious aims will require governments to readdress the way we plan, finance, develop, govern and manage cities, requiring strong urban policies, stronger institutions of governance and reinvigorated long-term and integrated urban and territorial planning and design.

What this re-framing looks like is a central pre-occupation of this book. Planning systems have been shaped significantly by the contexts in which they have emerged, even while many of the regions covered in this book have planning systems strongly influenced by their colonial histories. Professionals who call themselves planners can be found in government at all levels, in non-governmental organisations and civil society movements, international development agencies and in the private sector. They come from a wide variety of disciplinary approaches. Some deal with spatial frameworks and regulations, others deal with policy development in various sectors of government. We see this presence of planners in multiple organisational and institutional environments as a clear strength of moving forward to reinforce plans and their processes, rather than focus only on their dissipation or illegitimacy.

We have therefore responded to this diversity in two ways. One is to deliberately remain open on the issue of 'what is planning', seeking not a forced, pithy consensus but insisting only on the centrality of the question. These are editorial choices as we assembled the book and drew in authors who have their own complex engagements in theorising development patterns. As we asked our authors to talk about 'planning', it soon became apparent that there remains a great diversity across the globe in how the terms 'planning' and 'planners' are understood. As a result, we asked authors, where possible, to explain how they interpret the terms. We also note that a number of our authors would not classify themselves as planners of any kind, but are clearly engaged in analysis of plans: we asked them to contribute due to their understanding of the larger forces that shape national, urban or regional change in relation to the global South and their responses to our requests to define 'planning' reflect their views from 'outside' of the field. In our view, this sort of deliberate engagement with 'process' rather than professional or disciplinary labels permits a book of this kind to actively seek out contention and complex perspectives to planning. In our part introductions, we reflect on the positions they articulated in response, leaving them as juxtapositions in order for the readers to understand what planning practice looks like today.

Second, we chose to root ourselves in actually existing practices of planning in all their forms and locations. We thus sought authors who could write from practice and experience as well as those writing from theory and the world of ideas, and – ideally – those who manage to do both. In doing so we have sought to not let a compartmentalised job market within and across universities, sectors/industry and disciplines become the basis of a narrow and reductive theory–practice divide, or to let any one definition of 'planning' dominate our text. A focus on practice is immensely useful in this regard because the substantive domains of planning are diverse and peculiar to different epistemological and methods traditions. Mobility plans and economic development may often be closely linked, yet professionals in these domains come with very different types of training and thinking about the world of relevant ideas or their practical scales of intervention. The same is true of those who work on housing and those who come from traditions of environmental planning. They go together, and yet they may not.

Much has been written about the disconnect between theory and practice in planning and the gradual disengagement of some theory from the work of practitioners, leaving the latter to wonder if their work is at all theoretically informed, and whether theorists 'know' what they are speaking about. This is especially important to a book concerned with parts of the world where everyday living can be a struggle. From a position that practice benefits from concepts and ideas (and critique) and that theory can more usefully draw on the dynamics of context and 'place', rather than placeless abstractions, we hope that this book can help unsettle and re-constitute this 'divide' into a much more productive knowledge relation that seeks to theorise from and for practice, and sees practice as a mode of knowledge generation and not just 'implementation' or 'application' of already formed ideas. We do this not to argue that there can be a one-to-one relationship between single types of theorising and types of practice. This too, is a false understanding. Looking at any planning domain – for example, something as vital as the environment – makes this immediately clear. What constitutes appropriate theories for environmental gains is unclear beyond the contexts in which they are applied, and within the field, 'environmental planning' must attempt its own integration across many fields of planning.

On place, and the global South

The second debate that we wish to enter is one on place. As editors, we would argue that in this respect too, this book is somewhat different to many of the volumes that emerge on planning. Its geography and deep engagement with substantive domains of planning help us selectively forefront cities, regions (including rural dilemmas) and countries, which have not generally co-habited single planning publications. As such, it can be seen as counterbalance to volumes covering economically advanced regions and to those that claim to be international yet leave out or disregard those parts of the world covered in this book. At the same time, our substantive domain emphasis as well as our early hopes for the book created engagements with authors to clarify their selective use of theories and empirical work when they argue for why one place is different from another. We have tried to tease these out in the chapters and hope these also serve for teaching purposes to make clear that both planning institutions and planning as institutional change are context-specific. We hope this will generate for the reader, in quite diverse ways, a sense of why and how place matters.

Asking questions about planning to ourselves as editors and to our authors within such a curated geography marks our concern about planning ideas that are highly abstract, context-less, place-less, sometimes circulating as global 'best practice' models. We have all known places where such models have been imposed where they are not only entirely inappropriate to a particular context but also do irreparable harm and damage to places where they land. As editors we ask ourselves the question: what kind of planning processes should we consider 'from here' i.e. from the unique nature of particular places? Is it possible for one physical place to learn from another? What would this mean in terms of substantive domains of planning and the required institutional scaffolding that goes with it? Public plans invariably have multiple institutional homes and organisation partners, often outside the state. How do we then generalise from the differences within each location as well as the multiplicity of patterns – or the absence thereof – when we read across them? How can speaking from a new set of perhaps conceptual geographies not just give further testimony or case studies that fill 'gaps' in a notional 'canon' but produce new knowledge on its own terms?

One way to organise such knowledge is to speak of it as 'Southern'. We recognise that reference to the South and 'Southernness' as an epistemological concept has been emerging across a number of disciplines in the last decade, and indeed the geographies of this book reflect some of this emphasis. However, it is important to stress that this marker or framing remains open and contested even between us. As editors, we have collectively been careful not to simply equate the concept of global South with a geographical South (see individual editor contributions below). Beyond this, while we share our agreement on the importance of theorising from place, we disagree on whether the global South is the most appropriate way. Because of our substantive engagements, disciplines, and domain's histories and methods, the term means different things for the three of us. We see it as a perspective, a theoretical orientation, or a provocation to debate difference. In this book, each of us engages with this acknowledgement differently, from embracing it to arguing for its abandonment.

Again, it bears repeating that we have chosen to hold and highlight these differences rather than to try and reach a forced and artificial consensus on our respective views. In our teaching experiences, and many professional engagements, we have repeatedly found that people take theories as universal in large part because they see neither the art nor science of theorising emerge from particular histories and locations, nor from individuals themselves. In a lighter vein, this form of organising an introduction is also a reflection of one way of thinking about planning praxis and its recent debates on consensus, participation and difference, attributes we find to be particularly strong in professional domains. This may be because in these domains, orthodoxy is always insecure, invariably shown up by empirical context. Worldly relevance holds uneven value across and within disciplines. To make these differences productive, however, we also reflect on their roots, locating them in our own relationships to place, disciplinary training and professional lives. We have lived, worked and studied in different parts of the world – can our own personal biographies have something to offer in terms of the way in which we understand geographies of knowledge? In order to both take on the critical self-reflexivity we demanded of our authors, and to further open the debates we are inviting our readers to enter, we use the section below in the somewhat unusual task of articulating our views on planning and ourselves in individually written sections.

Locating ourselves

Gautam Bhan

In this section, I lay out three dislocations that brought me to this book. I see this book as marking and responding to a moment within urban and planning theory and practice of what can variously be called as opening, correction, crisis, rebellion or transversal engagement. Since a good crisis, as is often said, is a terrible thing to waste, I describe in the following words my own engagement with this theoretical opening. I do so in a slightly particular choice of form: part biography, part conversation with theorists, words and ideas that have helped me navigate the dislocations I describe and locate – rather than lose – myself within them. First, I lay out an understanding of the South and of Southern inquiry. Then, I argue that the space of such inquiry represents an opportunity to re-think an object called 'planning', precisely because such re-thinking is demanded by the (dis)locations the book inhabits in its chapters. Finally, using my own relationship to this tension, I argue for the need to imagine modes of practice that emerge from a new theory of knowledge.

The South as an ethos of inquiry

My first encounter with the idea that place matters – and matters deeply – in the act of theorising was in writing about my own city, New Delhi (Bhan 2016). I had not expected such an encounter for I had not thought of myself as dislocated in any way. I was a native, writing about the city he knew best. In writing the story of the evictions of bastis – income-poor informal settlements often misrecognised as 'slums'– that had scarred the landscape of my home, I realised I was mistaken. In a long tradition of theorists within the formal academy, I sought to write of the basti from outside it. I sought to explain evictions, to describe them in order to contain them and find ways to respond to them. I could not, for a long time, listen to them. Slowly, this shifted. I found myself reversing my location. I sought to write from the basti, instead of about it. I sought to write about my own locations: the academy and planning theory, the authority and institutions of planning practice as well as of myself. I did so because that was what theorising from the basti compelled me to do. As I wrote then: "to look at the basti from within planning theory or practice is to tell a tale of exception. To look at planning from the basti is to tell a tale of the fiction of the rule" (Bhan 2016: 34–35).

This dislocation for me was the first realisation – more affectively then than theoretically – of what I would later think of as theorising from the South. This origin, however, marked for me then as it does now that such theorising was first an ethical space, or perhaps more accurately, an ethos of inquiry. Within this ethos, a set of questions are not just asked but prioritised. These questions come precisely from an act of locating – in the basti, in this case – and looking from rather than looking at. The more formal project of theorising from the South expands from this instinct, this moment of dislocation. It argues that the basti is not an incidental but a typical location. It is a periphery. My sense of the periphery is shaped by the work of people like Teresa Caldeira and James Holston as a conceptual rather than physical location that not just enables certain inquiries but insists upon them (Caldeira and Holston 2005). Peripheries move across scale, space and time: the peripheries of the world economic and political system both historically and today; peripheries within cities themselves; peripheries of geographies of authoritative knowledge. Seen as a periphery, the global South is then a relational geography. It is not just a collection of previously underdeveloped countries or the boundaries of the post-colonial world but a dynamic and changing set of locations. I see the global South then as a 'project' (Urban Poverty and Inequality Collective 2015) rather than a place, an ex-centric location that allows a different "angle of vision in telling the history of the ongoing global present" (Comaroff and Comaroff 2015).

The empirics of similar-but-not-same

Yet when we speak of 'Southern urban theory', I am convinced that this relationality is embedded, at least in this current historical conjuncture, within a broader empirical specificity. This specificity is not fixed, it is hard to tangibly bind, but it is nevertheless pivotal in marking the current manifestation of the 'periphery' at different scales, in different spaces and across time. It is this specificity that allows the idea of Southern urban theory to be both relational yet specific, and to speak – both accurately and inaccurately – of a set of locations at once as 'cities of the global South'. Let me take an example, borrowing again the words of Teresa Caldeira. Writing about "peripheral urbanization", she defines it as modes of the production of space that "(a) operate with a specific temporality and agency, (b) engage transversally with official logics, (c) generate new modes of politics, and (d) create highly unequal and heterogeneous cities" (Caldeira 2016: 1). In her article, she draws upon examples from São Paulo, Istanbul, Santiago,

Mexico City and New Delhi. She argues that writing from all of these locations reminds us that "peripheral urbanization is remarkably pervasive, occurring in many cities of the South, regardless of their different histories of urbanization and political specificities" (p. 2). This does not mean that peripheral urbanisation plays out the same way in these cities, or that the forms of "unequal and heterogenous cities" look the same. It does not mean that all cities of the South exhibit it. Yet, it does mean that from a set of specific locations, it is possible to develop a "model" that "articulates general features while remaining open and provisional to account for the ways in which the modes of operation it characterizes vary and constantly transform" (p. 3).

Other writers of the South – sometimes boldly, at others times hesitantly – also refer to such shared empirical contexts. If for Caldeira, it is the mode of production of space, for Susan Parnell and Edgar Pieterse, it is the shared fates of "large, fairly well-resourced places that nevertheless have very large concentrations of chronically poor people who are institutionally excluded from the government support structures that are necessary for their well-being" (Parnell and Pieterse 2010: 147). Quickly, in the article, this amorphous set of cities become more specific in their shared historical geographies. They are "embryonic post-colonial local state structures" (p. 148) or "post-colonial contexts where local and provincial governments are rather belated constructions, with limited fiscal and human capacity and with incomplete administrative systems at their disposal" (p. 150). When AbdouMaliq Simone writes of "people as infrastructure", his description moves quickly between inner-city Johannesburg – the location of that first dislocation, the first compelled inquiry – to argue that "*African cities* are characterized by incessantly flexible, mobile, and provisional intersections of residents that operate without clearly delineated notions of how the city is to be inhabited and used" (2004: 407; emphasis added).

My own education was rooted in the discovery – which I felt like an intuition as much as a lesson I was taught – of the limits, nature and bounds of shared empiricism. Living and studying in the US, that sense that what I was taught didn't 'fit' never left me in all the years I spent there. The sense that the city outside my window wasn't what I understood as 'the city' never quietened. It is this that made me in my own work repeatedly argue against the reduction of the specificity of Southern locations to differences in degree and not kind, to hold ground against the kind of argument that states that between "Paris and Palestine, London and Rio, Johannesburg and New York", there are merely "differences of degree not substance" (Merrifield 2014: 29) in a new (but still singular) urban question. I do so not to argue about the degree of similarity versus difference in a regression-model view of planetary urbanisation. I do so to hold onto – perhaps only for now, provisionally – the fact that a project of thinking from the South continues to provoke different questions and inquiries from its locations, the surest evidence that something distinct holds such an assemblage together.

Towards a Southern practice

For the second dislocation, I return to New Delhi. The ethos that motivated that work was one that sought not just to understand, but to somehow engage. Evictions had angered, hurt and horrified me. I went to the city's (this time quite literal) peripheries at the time seeking something: an outlet, some absolution, some answers. Mostly, I sought ways to move, to imagine a theory of practice in hindsight through understanding how we had gotten to a place where evictions of the homes of the city's poorest residents were possible in the world's largest democracy. My search led me first to the Courts and, via them, back to the city's master plan, a document that until that day I had never had any reason to encounter. Again, the encounter was not incidental, but, in some senses, typical. In a city marked by peripheral urbanisation, where the basti was built in tension with planning and the law, it was perhaps inevitable that I and basti residents would find ourselves face to face with both. The basti did not have to lead me to planning, and yet the basti could only have led me to planning.

In its interface with the city, planning represents a privileged site to assess the translation of an ethos of inquiry into the political fields of practice, intervention and engagement. The project to think from place must be inevitably tied to the production of theories of practice that then follow. This too must be part of the ethos of Southern inquiry and it must be a dislocation about the ethics of knowledge production in addition to its geographies. If we are motivated by conditions that are, at least in part, marked by peripheries, then we must ask how a new body of thought and language gives us new ways of moving, new modes of practice as well as theories of knowledge from which they can emerge.

From 'here' in the global South, what does this new body of thought mean for modes of practice? The Southern city is a privileged site within which to ask this question. For each Southern provocation, the challenge to conceive a mode of practice is inescapable. If peripheral urbanisation, for example, is a model pervasive to many cities of the South, then we must ask: how must master plans respond to the auto-constructed neighbourhood built in 'transversal engagement' with law and plans? Should they post-facto regularise, undermining their own logics? How must planning imagine itself in a non-linear, incremental temporality – are its tools prepared to retrofit rather than prepare? If informality is not a moment of transition as part of modernisation but a long-term end state, then how do we structure urban economic development or livelihood promotion policies in cities 'here'? If deep differences (Watson 2006) mark the social geographies of certain cities, then how must we re-structure 'participation'? If James Ferguson is right that new forms of post-welfare economic life create not just the desire for empowerment but the need to make "declarations of dependence" (Ferguson 2013), then how we must we construct welfare regimes? If the South is indeed marked by local and provincial governments that are "belated constructions", as Parnell and Pieterse (2010) argue, then what actors are capable of delivering this welfare?

The provocations above are deliberately wide-ranging, pulling planning from any comfortable location only in spatial governance into every aspect of urban life from the economy to culture, political forms to engineering systems. This then is the third dislocation that I wish to insist on – to challenge the reduction of planning to the particular instantiations of its professional life, certain disciplinary locations, and narrow traditions that root it only to the built environment. As the questions charged by Southern locations push us, they must also dislocate our understandings of what planning is. I seek in my own practice – whether in projects or policy, in writing or through teaching – to perform this dislocation. Doing so for me is to see as much to scatter as to cohere; to learn from the auto-constructed city to pick apart, re-assemble, grow slowly, find and shape meaning as one goes along. I wish to, in my own way, auto-construct an object called 'planning' from some of its peripheries as well as from within its centre, to challenge official logics to geographies of authoritative knowledge, and to create new politics and knowledge on the way.

Smita Srinivas

Planning comprises many actors: scholarly institutional analysts, public policy and public administration practitioners and draws from many disciplines – in short, '*planners*'. This is a complex and dilemma-ridden professional domain, filled with different appetites for intervention, or theory-action frameworks. Planning is an eternal human activity, and some form of representative public planning has existed over time in every place. The contention is by whom, for what, and how actors/agents can argue for these in the public benefit. This last, of public benefit, is easier said than done for reasons I discuss further in this section. These processes connect the individual's inner world and their outer actions and responsibilities, and shape their assessment of progress over time.

My own sphere of economic development involves public planning processes especially in industrial transformation. I focus on the role of the state, and whether public and private firms can be dynamic and offer public benefit. These benefits can be diverse: revenues and jobs, learning institutions and skills, relevant products, investment strategies and social protections. My book *Market Menagerie* (Srinivas 2012) focused on governing public plans for economic transformation and industrial organisation and drew on the history and comparative development of the health industry. The book was deliberately integrative, providing explanations and heuristics to study the tremendous impact of the health industry, but also its disconnectedness from health plans, service delivery and deeply uneven accessibility. In related work on the same industry, I analyse problem-solving capabilities versus the wider notion of state capacity (Srinivas 2015). There is no given single path to plans or policy analysis, but these fields are at the heart of the economy and public life.

My early training was in mathematics and physics, then economics, with a PhD at MIT focused on microeconomics, public finance, technology plans and industrial organisation, at what was possibly the strongest planning and public policy programme in the US at the time. At MIT and later at Harvard's Kennedy School in science, technology and public policy, I was fortunate to work with extremely thoughtful, irascible, (famous in cases), economics and development scholars, many with long histories of engagement in developing countries. At Columbia University I designed, re-designed and taught economics and several planning/policy courses for those going into professional MS and PhDs. At Harvard, at MIT and at HEI, or in India with a range of organisational efforts over the years, I worked closely with students seeking 'real-world' translation, invariably unsatisfied with their disciplinary focus alone. These students today form a large and novel professional community worldwide, often in deeply contested development contexts and harsh professional realities.

My own interest in institutional change has originated in two tracks: one from training in physics and mathematics, drawing on philosophy and methods, technique and measurement, and human curiosity faced with the beauty of natural patterns. At best, these areas situate us in wider questions of our disciplines', and our own, relevance. In applied matters they can also point us to self-reflection about the physical world and improvements in material reality. My second, related, path was in this vein: economics training focused on economic development, the analysis and study of micro-economics and industrial organisation, economic history and impact. Economics (and its wider political economy history) provided me ways of thinking about technological change, markets and non-market institutions, investment and exchange, and the dilemmas of governance of social choices. Economic development has sometimes explicitly been concerned with different methods of moral philosophy and priorities in technological advance and knowledge institutions.

From my two paths of training have come many points of convergence and questions. Can technological changes be planned, and if so, through what means and to what ends? My current research covers types of knowledge transformations and technological learning taking place across spatial and institutional sites: from prosthetics to water-pollution drones, from health diagnostics, vaccines and food processing techniques to energy systems and waste processing. These technological shifts involve new or recognised knowledge, industrial policies, urban and regional investment and growth plans, technical standards and human task formation, laboratory and experimental methods, physical and organisational improvements in essential industries and increasingly inter-linked areas of manufacturing, agriculture and services. Often these plans make strong philosophical assumptions about market structure or regulatory and public ethics. In this world of climate, poverty or health concerns, a concern for institutional design would re-steer governance and incentives towards industries that are safe, humane and location- and planet-friendly. Even if the world switched to solar and wind farms, we will still need such economic plans and public processes that ensure transparent governance and public benefit.

Economic development plans require everyone's visions, but also involve specialist and domain-specific process. Opportunities in national, and sub-national contexts, with private industry, and at grassroots with diverse organisations, has made me distrustful of 'North–South' divisions or too easy 'Southern' solidarities. We need a more particular framing of history, discipline, context and thematic questions to ask under what conditions development actually occurs. My work with several organisations has only helped further clarify the uneasy history of what economic development plans did or did not, could or could not do, and how to think of the evolution of nation-states and their planning procedures and impact. Over time, I have developed some preliminary insights at best into the interlinked processes of economic and social policy and the challenges of scaling up. These perspectives and curiosity about the value-based arguments about economic change have influenced my work from then to now (Lund and Srinivas 2000; Papaioannou and Srinivas 2016; Srinivas 2010, 2014, 2015, 2017; Srinivas and Sutz 2008).

Professional engagement has to draw more critically on scholarship through actual deployment. Scholarship alone, devoid of the experiential aspect of problem-framing and problem-engagement, is somewhat empty. Because of my early training, the wars in ideology and methods within the social sciences called only from a distance even when I turned to economics. This allowed me an intellectually freeing, omnivorous approach to methods. Technological learning analysis and frameworks of industrial plans have preoccupied me because of their enormous impact on the daily lives and practical philosophies of people. This body of research has therefore included both highly localised technical product studies, skills and work issues (and immersive planning workshops run through the Technological Change Lab), as well as wider questions of how we describe the evolution and institutions of demand and 'meso' level approaches to innovation and structural change (see Srinivas 2017). At the practical level, informal work, health care, boundaries for firms and systems, technological learning and industrial policy, taxonomies of innovation and investment strategies – these have become the pieces of a continuing economic puzzle. I have become increasingly critical of the narrow training most economists receive and the equally poor economics training that many public policy and planning schools dole out. Strong interdisciplinary work and multi-disciplinary teams require more disciplinary expertise, not less, to ask what progress and progressive disciplinary and social change might entail.

We might say that to plan is to always have to answer questions such as: 'plan what?', 'for whom?', 'through what instruments?' and 'which motivations?', and to specify timeline, actors, politics and public processes. In economics of the older variety, all plans were located in political philosophy and moral life. As advances in evolutionary perspectives, of the technical roots of economic growth, and the behavioural 'revolution' have come about, not surprisingly there is far greater interest in a pluralist economics and its influences on participation, of agents and interests, of methodological individualism, and the 'soft' norms and 'hard' rules of institutional change. These norms for professional life cannot be left to theory if we ask about motivations and instruments. Planners often publicly self-identify with a progressive politics, to focus on inequality, uneven service delivery, governance norms and social justice. But the words mean little it seems to me, when personal integrity is absent or trust and long-term group relations are absent. Whatever one's disciplinary background, we come to our fields with an implicit or explicit sense of values and personal ethics. I was extraordinarily fortunate to grow up with stories of multi-generational attempts to make India a good place to be, and to witness from a very early age how individuals can embody exceptional standards for trust, community, public service and ethics. I have seen the far-reaching impacts of these actions and sacrifices on the quality of 'Indian' governance, of course many disappointing outcomes too, but equally exceptional examples of individuals and organisations extending assistance to those worse off. Several of these examples exist within government. Consequently, the multi-generational effects of individual actions must be assessed both in terms of aspirational models for those who observe this hard-earned integrity with its attendant sacrifices, as well as for those who have enjoyed its substantial benefits in improvements to their lives. Norms grow from these actions perhaps more strongly than the narrower domain of electoral politics.

It is therefore worth calling out the obvious, that plans require human beings to work with integrity and self-reflection in public service. Yet also that academics, government professionals, CEOs or senior managers of private firms, elected or other public leaders, act daily in ways inconsistent with the world they claim to wish to see. These questions of uncertainty, norms, ethical agency and institutional variety in public processes and the state (which both the proverbial Left and Right have increasingly come to distrust) have pushed me to view industrial organisation increasingly from the vantage point of evolutionary economics and its provocations. This is to better imbue economic transformation with philosophical nuances of individual actions, uncertainty, time and open-ended outcomes, and to situate the growth dynamics of firms and regulation with their well-documented evolutionary characteristics. Because technological change too is characterised by these facets of uncertainty and dynamism, the study of norms and institutional design raises the ante for planning. How best might we think of iterative, non-determinist processes and transparent procedural design, and consider how some economic priorities may have to be defended over others in a public, responsible, arena beyond electoral democracy?

Worryingly, it seems to me, an increasing number of public policy/planning scholars are theorists alone, perhaps even heralding the demise of the relevance of these fields. Practice, after all, makes the moral philosophical imperatives stark and personally uneasy. This can allow theory and professional engagements to remain more honest and agile. Of course, some planners usefully deploy these existential dilemmas to refine their field and to seek consensual ground with other stakeholders. Yet, there is a further question of how to identify and build on personal integrity to separate the political rhetoric of social transformation and economic opportunity from self-serving personal behaviour. We can also use these dilemmas to refine the theoretical assumptions of the disciplines we trained with. This question of deeply personalising the social goals and ethics of planning may sit uncomfortably with individuals, professional associations (even those with codes of ethics) and certainly in disciplines. For scholars who analyse plans and critique them but rarely act to intervene or shape the outcome, planning then remains an object of enquiry but rarely one in which they are protagonists (see my statement about easy critique in Chapter 10, this volume). In fact, their training in some instances, such as anthropology or some strands of economics, actively cautions against intervention.

Many leading economic development scholars were institutionalists driven by professional involvements and personal stories: Hirschman knows and says much about how to make plans and processes better because of his professional engagements and undoubtedly his own story of forced emigration and personal sacrifices. The same is true in different ways of Arthur Lewis and Alice Amsden whose framing of economic development, though dated in some respects, is remarkably prescient about its challenges. Sophisticated institutional challenges about planning processes (and also on technological change) have come from Marc Toole, Mike Piore, Judith

Tendler and Ha-Joon Chang, from Gandhi and several Gandhians, and from wide-ranging planning and policy scholars looking critically at economics such as Maureen Mackintosh, Judith Sutz, Elliott Sclar, Rodrigo Arocena, Joanna Chataway, Carlota Perez or Richard Nelson, to name a few. Furthermore, the dominance of Western and post-Enlightenment European influences is too one-sided to study the economy. This transfer of models and the credibility of theories and their plans deserve greater scrutiny in a world where English, French, the national dominance of languages such as Hindi or Spanish, the Internet and selective cultural histories in education are arguably over-represented or read exclusively in political language as opposed to personal action. My own interests have increasingly found fertile cross-cultural pollination in geographic and civilisational approaches, some ancient and modern in economics. For example, both Greek and Vedanta 'schools' inform approaches to 'late' industrial development. Likewise, there are new intersections of theories on technological learning of Penrose with Coase, of Sen with Schumpeter, of Lewis with others, that offer up framing perspectives on development plans and what they might best represent.

Planning, with policy analysis as a subset, has itself become an intra- and inter-social science phenomenon, and has meant specialised debates rejecting the old, tired, variety of Cold War rhetoric of centralised planning versus Hayek. Technological change also requires a 'soft' determinism in analysis and any approach to public plans whether they are urban or rural, or at different levels or departments (Srinivas 2012). Plans are increasingly discussions of learning and knowledge systems. These reflect measurement and action debates of inequality, national targeting of sectors, industrial policy strategies and 'good' and 'bad' protectionism. In some respects these wider linkages have also made it much more obvious how much these frameworks call attention to, or conversely, sidestep individual choice and action, which are urgent and essential facets of personal integrity and service whether these be found in bureaucracies that are essential for learning feedback, or in private firms investing in technology platforms.

Given these complexities, I point out (in Chapter 10) that the idea of a shared global South or Southern identity should be taken up with considerable caution. Knowledge strategies, skills profiles and the growth patterns of firms are shaped by, but also affect their urban, rural or regional environments in unexpected ways. Accountability of governments and the regulatory design that allows dynamic economic activity and opportunity to flourish, remain the important questions of both citizenship and global economic opportunity. The end of the Cold War has also reduced the influence of 'less-' developed countries to practically nothing in geopolitical venues, and bypassed the others who have few other types of economic leverage to display, making their domestic planning mandate more difficult. International trade, environmental concerns and development aid are also converging in inopportune ways. These give more credece to a customised 'planning from here', but at the same time make evident the absence of state capacity and problem-solving capabilities.

A pivotal, shared, normative vision of a one 'planning' or of a notion of a shared 'South' as a progressive endeavour is therefore more complex to disaggregate and its moral high ground difficult to occupy for long. It may even mark the demise of a 'planning theory' because planners' spheres of action and scales of influence often contradict or conflict with each other. Despite these contradictions, my optimism about professional engagement in economic development has also grown as the discipline of economics has become more pluralist, and the study of technological change, and contributions in evolutionary and behavioural economics have been usefully incorporated into professional strategies. New forms of multilateralism and distributed power also raise useful challenges to the nation-state in economic development. If planners have to play this awkward bridge in ensuring that economic plans connect between different levels of government and push to new models of governance, they will have to act as selective agents of change. This requires them to identify with specific investments, and sector and regional priorities. These are moral and difficult value-propositions, not value-neutral arenas of industrial dynamics (see Papaioannou and Srinivas 2016). It is hard to dodge the question of winners and losers, or difficult compromises in any planning intervention. Nevertheless, 'planning from here' in economic development terms inevitably requires us understanding the impacts of plans 'for here', but often made elsewhere.

This willingness to intervene and to make explicit value propositions in plans and policies is certainly part of the task of professionals and not of armchair disciplinary theorists. Planning involvements require the stomach for intervention as a moral act in public service, and require getting off the proverbial disciplinary fence (or armchair). No utopian disciplinary world of plans and policies exists where scholars are not also protagonists. I suspect this acknowledgement may push us to a more honest struggle for personal integrity and public benefit.

Vanessa Watson

My interest in this book arose out of my ongoing concern that much planning research, publication and policy development is produced through conceptual frameworks and world-views strongly influenced by a geographical bias to the advanced economic regions of the world (the global North). This bias may be shaped by individual scholar location, training and institutional home, but a growing literature (see Connell 2014; Graham 2015; Paasi 2005) also documents the geography of academic knowledge influenced by factors such as research funding, conference location and access, publication companies, journal editorial boards and reviewers, and importantly, the dominance of the English language as the medium for intellectual development and exchange. While the location of knowledge production may be less of an issue in some fields, the discipline and profession of city and regional planning, for me, cannot escape the importance of the socio-spatial and environmental context in which knowledge about it emerges and in which it is applied and practical ideas are put to work. A concentration of planning knowledge from territories usually classified as the global North is particularly problematic when this work fails to specify the contextual informants on which it is based (such as Western liberal democracy or well-resourced institutions) and further (and as is mostly the case in social sciences) claims its findings and policy applications to be relevant in all parts of the world. Connell (2014) offers a more cynical view of this. She argues that an understanding of the global "political economy of knowledge production" shows how knowledge produced in Southern regions is controlled and marginalised by dominant theoretical production processes, reinforcing the assumption that planning ideas from the global North have superior value and can be used beyond their source region as 'placeless' generalisations.

As a result of this kind of bias it is hard to find cities or regions in the global South that at some point have not been subjected to planning concepts and models imported from the global North, usually imposed on territories as part of a process of colonisation and retained thereafter as a way of protecting land values and interests of a new elite. Decontextualised 'best practice' transfers continue, from global North to South but now as well between Southern regions carrying labels such as world-class cities, and eco or smart cities. I see their impacts as almost always falling far short of predictions and usually with negative outcomes in some or other form (Watson 2014a).

I believe that this kind of placeless generalisation of knowledge and application raises important issues in the planning field. This is because planning outcomes so often have implications for the spatial and sectoral allocation of resources, and hence they have societal distributive impacts. They are also likely to spatially shape public and private sector urban investments as well as the distribution of land and building controls with direct cost and exclusionary implications. Planning processes and outcomes therefore have to be highly sensitive to social, economic and environmental dynamics in any context and the kind of impact they can have on human lives and futures. I therefore see myself as a planner who believes that understanding space (equally time, or history) is critical for planning to make a difference. At the same time, I distance myself from the label 'spatial planner', which rather often implies an abstract, Euclidean and decontextualised view of the world.

I have come to this personal position largely through my work over a number of decades on planning practice, theory and education on the African continent. In the early days of my career as a planning teacher, researcher and anti-apartheid activist in South Africa, it was impossible to ignore the ways in which the state used spatial planning ideas from the global North to justify the planning of apartheid cities as international 'best practice'. Garden cities, neighbourhood units, Radburn and Milton Keynes layouts, and satellite cities were used to argue that the removal of people of colour from their homes in urban areas declared 'whites only' was in their best interests, and they were being relocated to improve their quality of life. The devastating personal and societal effects of these forced removals to peripheral and sterile new housing estates, and the wider impact of racially divided cities, continue to be felt today, over 20 years after the end of apartheid. Income divides still closely mirror older racial divides giving rise to cities that are more unequal than almost anywhere else in the world.

In more recent years I have been involved in a long-term project to shift planning education and an understanding of planning on the African continent through the Association of African Planning Schools. Most African countries have inherited colonial planning legislation that continues (mostly unchanged) to shape current national planning law and planning curricula. Global North planning ideas (and more recently hyper-modernist city planning and architectural ideas from cities such as Dubai or Singapore), rooted in assumptions that physical, social and institutional contexts are not very different from the regions of origin of these ideas, continue to block meaningful and effective management of African cities and regions. This view of planning, rooted in top-down master-planning, still shapes planning curricula in some parts of the continent. City planning through physical master plans, land-use zoning and building controls is highly disconnected from the reality of most African cities where rapid growth, poverty and informality is the norm. But while such laws continue to protect the land values of the urban elite and can be used as a useful tool to evict and banish political opposition they are unlikely to change. An approach to planning that directly addresses African urban issues (recognising their huge diversity), rather than the current obsession with creating 'world class cities', is clearly needed.

I am therefore inspired by a literature that argues for a perspective on planning 'from the global South' (Watson 2014b). The work of Oren Yiftachel, Ananya Roy, Faranak Miraftab, Libby Porter and Gautam Bhan, to name a few, has been an important informant. In the field of urban studies heated debates on Southern urbanism have laid useful ground for me to develop my own Southern perspective in planning, as well as its potentials and limitations. The title of this book nonetheless remains contentious. Is there a global South planning, and what is the global South anyway? My own position is that there is not, and cannot be, planning theory/practice for the global South separate and different from planning theory/practice for the global North. Global South is not a geographical entity. To claim this would, I believe, be to set up a false binary across regions of the world that are highly interconnected, diverse and consist of multiple and changing cores and peripheries at various scales. My position is that a perspective 'from the South' can open new questions in other and different parts of the world. This is to counter what I suggest is a dominance of questions raised by experience and ideas in the global North, giving planning a constrained, limited and parochial set of ideas to use in the vastly different contexts beyond the North, and perhaps posing new questions and perspectives for the global North as well.

However, a perspective from the South needs to be firmly located in an appreciation of place (or context). This implies that the values and objectives of planning in that place are always surfaced, that concepts from other parts of the world are tested (not simply applied) in context and new ideas (not 'best practices') can feed back to the growing and diverse international 'pot' of planning theories and concepts. The value of case study research is important here, and this book offers a wealth of new and productive case material. The complexity and diversity of urbanisms and urban processes that are emerging from in-depth case study work in Southern contexts is significant. It shows that the very different processes and factors that produce cities and regions defy the possibility of capture in a single universal theoretical model of urbanism, or simple planning responses. Understanding new and previously un-researched cases and practices through a single case or in comparative ways can, I suggest, raise new propositions for testing elsewhere and over time can (inductively) build up new meso-level theorising relevant for perhaps certain parts of the world or certain issues. A degree of generalisation beyond the single case is an essential step in the development of ideas and knowledge, but equally important is knowing where and when they do not apply.

At the same time, an appreciation of the importance of place and context (of planning 'from here') also has to be connected to a wider and global frame. A number of chapters in the book offer this larger continental or cross-continental perspective. Differences caused by past and ongoing global relations of dominance and dependence are important. I suggest that the definition used by Dados and Connell (2012: 13) comes close to the mark when they argue that the term global South is far more than a geographical south: "It references an entire history of colonialism, neo-imperialism, and differential economic and social change through which large inequalities in living standards, life expectancy and access to resources are maintained".

The process of building planning knowledge and ideas from place, in and from the global South, is in its infancy. It runs counter to a long-standing dominance of ideas that gloss over the depth and importance of difference, and to theoretical traditions in planning supported by powerful Northern knowledge-producing institutions and systems. It also runs counter to the growing dominance of private sector and developer-led visions of future cities, shaped by hyper-modernist aesthetics. For me, this book is a further stake in the ground to signify that an alternative is possible.

The chapters

This part of the introduction describes the five thematic sections within the book. In doing so, it traces two simultaneous registers of argument. The first is the curation of the sections themselves and the chapters within them. This curation represents the multiplicity within planning, particularly in its intellectual and thematic dimensions, disciplinary allegiances, professional practices and institutional locations. The second is then within each section where chapters hold the core debates of this book – on planning and the global South – within and across thematics.

In the sections that follow, we thus describe the individual and collective contribution of the chapters within and across the sections, teasing out the diverse approaches to 'planning' as well as agreements and disagreements on its relationship to the 'global South'. Our aim, as is evident in the introduction thus far, is again to open up, make uncertain, problematise and challenge, with the belief that such opening takes our praxis richly forward.

Part I: Planning and/as the state

The opening section of this book takes on the configurations and meanings of planning as a state apparatus and action. How do the diverse practices of planning locate themselves within the different state forms, scales and institutions? Is this diversity shaped or determined by a Southern location?

The section starts at the supra-national scale. In their chapter on the New Urban Agenda, Clive Barnett and Susan Parnell argue that global-level agendas are key to understanding national planning practice, especially in rapidly urbanising contexts like Africa and Asia. Their analysis of the new agenda argues for the emergence of "a process of implementing and monitoring priorities for housing provision, infrastructure investment and land development that have significant potential for reconfiguring the role and understanding of local government in general and the more specific role of spatial planning" (Chapter 1, this volume).

If this focus on the local government indeed is to be significant, then several chapters in this section play perfect foil to assess how they will come into this role. Patrick Heller sets up a direct comparative frame looking at what he calls the 'local state' within a clearly Southern geography: the mega-cities of India, South Africa and Brazil. How does this local state, he asks, direct growth and development 'here'? He argues that there are deep variations within three contexts that are often described collectively as well as singularly, and that it is these variations that "[highlight] the political nature of the institutional conditions for effective planning" (Chapter 2, this volume). Heller instead points us to thinking about the local state's bureaucratic, fiscal and coordination capacities as the key site to assess what he calls the efficacy and effects of urban governance whether for growth or for inclusion.

Looking at the growth of the peri-urban areas of Kolkata and Hyderabad, Sudeshna Mitra takes a slightly different entry point. She argues that frames of looking at planning in the South have focused too much on notions of 'failure'. This division between ideas of planned/unplanned, she argues, misses the ways in which planning practice is evolving, "particularly ways in which there is a break from static imaginations of territorial state power and absolute spatial control, embedded in planning instruments such as master plans" (Chapter 6, this volume).

Mitra and Heller are looking at what they both describe as deeply dynamic and rapidly changing urban contexts. 'Planning' for them is thus inscribed in the politics of how institutions of the state – at different scales – move and attempt to direct this change towards particular ends. Yet writing from several African urban contexts, Dele Olowu (Chapter 4) argues that the central question is one of whether the 'local state' has the capacity to play such an institutional role. Olowu asserts that urbanisation in several parts of Africa have seen decentralisation and democratisation move in opposite directions, effectively incapacitating the local state. Looking at cases from Nigeria, Rwanda and Ethiopia in particular, Olowu reminds us of a particular historical geography where the distribution of funds, functions and functionaries remains a site of deep political contestation across the scales of governance in ways that critically impact planning.

Eduardo Marques also offers a different challenge for state capacity to plan: data. Drawing from the experience of the paucity of housing data in the Brazilian planning and policy-making process, he argues that, particularly in Southern cities, the central concern of data is not just availability and accuracy. Instead, data practices that seek to make the urban legible and visible face spatial patterns that are "very distant from the normative ideas implicitly embedded in technical standards and models" (Chapter 5, this volume). The relative dominance and heterogeneity of precarity in Southern cities, he argues, represents a particular challenge though the exact form of this challenge varies widely within the 'South'.

Also in Brazil, Edesio Fernandes looks at one of the most discussed contemporary innovations in urban planning: the country's 2001 City Statute that sought to give, as he describes it, "legal support to urban reform" that would institutionalise participation, the social function of property and inclusive planning. Fernandes argues that the statute represented a renewed life for urban and spatial planning within a new governance framework but that it has failed to take this opportunity. Instead, older socio-spatial segregation processes have consolidated. What lies behind this? Fernandes argues that one of the core failings lies in the narrow understanding of planning itself, its separation from institutional, legal and governance frameworks as well as from public management. With this separation, he says, "urban planners and urban managers remain, and have seemingly become increasingly, hostages to exclusionary land and property markets that they have created and fomented in the first place" (Chapter 3, this volume).

Part II: Economy and economic actors

The chapters in this section of the Companion set a high bar for planning as public policy, institutional reform and public administrative capabilities. The authors use plans as a way to describe domestic choices within national economies to engage with global transformations. They provide contrasting views on whether it is national economic priorities and outcomes rather than urbanisation itself that should be the focus of future development goals. There is a surprising degree of agreement that the developing world is vastly heterogeneous. Not surprisingly, the chapters diverge in how important 'Southern' vocabulary is to the debate on plans in large part because their economic units of analysis are different as are the effects by which they measure progress. Rather than a minimalist view or a retreat in the face of immense challenges, the authors highlight the need for new ambition and expectations of public plans and institutional design.

The chapters often repeat the critical national development context in which institutional prerequisites for state action and sub-national outcomes have to be assessed. For Turok (Chapter 7), this contingent institutional design lies in land and infrastructure links; for Smoke (Chapter 12), in the actual deployment of fiscal instruments and powers; for Srinivas (Chapter 10), with state-led industrial policies and associated labour, trade and welfare choices; and Shen (Chapter 8) builds on this contingency of state-led trade and industry in spatially de-limited zones. In tandem, chapters such as Assaad's (Chapter 9) and Skinner and Watson's (Chapter 11) emphasise the local context of work, through measurement, access, land use rights and civic amenities.

Is promoting urbanisation the object of planning, or should plans act as directing processes that guide and mediate national development planning processes to improvements in incomes and amenities and the quality of life? Paul Smoke (Chapter 12) draws on the fiscal thread of governance and its political economy, showcasing the common missing element of attention to development finance. He demonstrates the low fiscal and administrative state capacity that has accompanied decentralisation and devolution mandates. This planning mismatch leaves a substantial precariousness to how urbanisation and rural transformation is funded and monitored: not least of which include budget shortfalls, unfunded mandates, lack of transparency, and inability of local governments to set their own plan priorities.

Qianqi Shen (Chapter 8) reminds us of nation-state-led plans, decentralisation and sub-national capacity again by demonstrating the diversity of multiple 'China stories'. Focusing on SEZs, she counters critics of spatially de-limited zones by revealing the variety of actors and politics in plan-making and execution in the largest manufacturing and urbanisation re-steering the world has yet seen. She argues that economic and physical-spatial planners are increasingly forced to subsume their vision to the power of bureaucrats who frame intergovernmental negotiations in ways that are politically amenable. Shen also pulls on the thread of planning and democracy, suggesting that planning processes seen through state and decentralisation would do better to plan for both the divergence of process and outcome.

This concern about spatial and institutional navigation and mediation is also the theme of Skinner and Watson (Chapter 11). They bring readers to the urban dimensions of informal work and argue for planning – national and urban – to look at the forms of investment and service improvements for such workers who are invariably sidelined. Skinner and Watson emphasise the need for a spatial sense in economic plans and policies to accommodate a wide range of work types visible in cities. Their chapter also emphasises the contradictions of public land use and working rights, themes of contradiction in public policy that are engaged in different ways by the authors across the chapters. Working rights and land ownership may clash constitutionally as well as procedurally in the dynamics of cities. They point to a wider unease with regulation as the primary instrument of development.

The contradictions may be compounded by a focus on interpreting urbanisation itself and its implications. In Chapter 7, Ivan Turok draws urbanisation firmly back into national economic growth and development patterns. His argument questions whether urbanisation can offer a common past historical thread for national development patterns and indeed whether within developing countries, such optimism about growth and development is warranted. By examining growth and urbanisation correlations and diseconomies of urbanisation, his push, as with other chapters, is towards improvements in public administration, governance capabilities, and new growth priorities. By questioning the interpretation of urbanisation in different ways, Turok questions the focus on cities as well.

Turok and Srinivas, for different sets of reasons, question the view that urbanisation inevitably can 'lift all boats' or that planning's priority should necessarily be cities. Srinivas (Chapter 10) further argues against a 'Southern' label in economic development by drawing attention to national transformation and specific plan priorities in the twentieth and twenty-first centuries, which have had diverse sub-national assumptions as well as effects. Her chapter emphasises structural and institutional diversity within and between groups of 'breakaway' nations. The historically diverse clubs of geopolitics helped determine whether governments were able to steer and manage domestic development gains. Many 'breakaway' technologies, industry sectors and sub-national regions, from agriculture to services, gained from public priorities and long investments in learning. Srinivas argues that 'global South' and 'Southern' labels create a false economic and theoretical solidarity where none may exist and the specifics of the economic plans are vital.

A similar question on national development patterns and urban gains is framed in different language by Assaad (Chapter 9). His Egypt-focused chapter dwells on the need to move forward policy design to incorporate informal work and especially youth and women workers. Urbanisation forms the backdrop for his labour market dynamics, but is not the sole focus of possible improvements. His chapter looks at the design of equitable labour policies in Egypt, which requires attending to both the measurement (unemployment rates) and policy instruments (job training, search, insertion subsidies). These must be shaped by actual gender-differentiated urban work patterns rather than a one-size-fits-all approach to labour markets.

There are revealing questions that remain: who is responsible for productivity, skills and learning in a traded world? Decisions about physical spaces and investments are often governed at higher levels of government, and public finance, jobs-focused, or environmental planners can often argue this both ways depending on the specific issues at risk. Similarly, what political economy – national, or more regional – would expand the fiscal base and push for greater ethics and accountability norms? Given multiple new national handshakes in multilateral and bilateral trade that are ongoing, how are we to view 'local' planning contexts and their autonomy? Is the crux the effective planning outcomes or democratic participation in plans per se? The chapters here have provided a new threshold and pressing questions for economic debate.

Part III: New drivers of change: ecology, infrastructure and technology

Planning in rapidly growing cities and regions is increasingly required to respond to concerns about resource limits and ecological threats while at the same time coping with developmental imperatives and growing demands from urban residents and investors. New urban technologies sometimes hold out the hope of greater efficiency and risk reduction, but embedding these in societies and governance structures, inevitably shaped by their place and history, makes this a complex task. While this section of the book is not exhaustive in terms of current drivers of change, the issues of environment, resources and technology are having a profound impact on space, economy and society across the globe. They reflect forces that are giving rise to growing inequalities, but at the same time, all the chapters demonstrate the possibility of intervention to ameliorate, if not to overcome, these challenges.

Eric Chu, Isabelle Anguelovski and Debra Roberts (Chapter 13) consider how to plan urban climate change adaptation measures in global South cities where exposure to impacts and lower capacities to respond are pronounced. In such contexts, municipalities have to find ways to balance climate adaptation needs with development priorities through institutions that are dealing with conflicting priorities and practices. In Chapter 14, Jia-Ching Chen explains how China has attempted to reconcile urbanisation policies that aim to stabilise economic growth and improve living standards, with environmental governance to curb pollution and optimise natural resources. The party-state has named this policy "the construction of an ecological civilisation" and considers it as a pillar of its ideology of socialist development. Chen explains how this policy can be seen as an emergent mode of territorialisation and as a regime for producing and governing a national spatial structure and nationally scaled environmental values. From a somewhat different perspective, and from Brazil, Mônica Haddad (Chapter 15) examines planning strategies used to minimise the impact of urbanisation on the natural environment. She shows how public sector practices shape the way in which urban planning tools are used, with political influence and corruption as well as a fragility of supporting policies and lack of public participation, hindering implementation.

Rapid urbanisation in Southern contexts also gives rise to a range of resource limitations, one of which is food. However, as Jane Battersby explains in Chapter 16, this is not a simple issue of food supply, as long-held food productionist policies would have it. Research in African cities shows very high levels of household food insecurity largely arising from poor physical access to affordable food outlets. As supermarkets have increasingly monopolised urban food distribution, informal food sellers – traditionally the main source of cheaper and fresh food – have been marginalised both economically and from public spaces accessible to low-income households. Planning has been used by governments as a mechanism for this marginalisation, yet planning also provides the potential to address food insecurity. A key issue for urban planners here is to develop and secure public spaces and infrastructure for the informal food retail market.

Technology is often promoted as the 'fix' to deal with resource limitations and threats of various kinds. 'Smart city' solutions are frequently promoted, usually by corporate interests, across many global South regions, but rarely are these ideas embedded in, and appropriate to, their places of adoption. Nancy Odendaal (Chapter 17) argues that socio-economic development cannot be solved by technology alone. The multifaceted nature of the digital divide remains evident where projects underestimate the importance of human and social systems in technology adoption and use. However, there are opportunities for planning (as a form of spatial governance) to encourage the use of technologies to empower communities and to make planning more participatory.

Technology systems of various kinds have long shaped city development, and perhaps none more profoundly than motorised transport. In the last chapter of this section (Chapter 18), Anjali Mahendra considers the challenges of urban transport planning in global South cities. Where both population numbers and incomes have risen rapidly, vehicle use has escalated dramatically, but with little state attention to public transport or mobility needs of the poor. For transport planning to be effective, she suggests, key governance issues and priorities must be addressed – particularly integrating transport with urban land and economic development and placing far greater emphasis on the mobility needs of the poor majority through public transport and non-motorised transport.

Part IV: Landscapes of citizenship

This section brings together two foundational axioms in thinking about cities, planning and the relationship that binds them. The first suggests cities as sites that hold and heighten both the promise and perils of citizenship. The second seeks to root planning within an ethos of social justice and equity that can tilt cities towards the former rather than the latter. Across the diversity of what they see as 'planning', these seven chapters are, in a sense, a snapshot of the current state of these relationships in a set of specific, Southern geographies. They do so, in part, through different ways of thinking about the 'public'. All valences of the idea of the public are at play here: the public as the polity and the people; as the universal; as the state; as public good and goods; as public space; as public institutions. They suggest to us that the public can act as a way to read the thread of citizenship and its attendant promises of rights, equality, and dignity within contemporary cities.

For Fernando Carrión and Alexandra Velasco (Chapter 23) as well as Mercy Luthango-Brown and Elena Reyes (Chapter 24), the public is fractured by violence. Both chapters grapple with violence and what it does to cities. Even as both agree that it is spatial – and thus invokes both planning and design in particular ways – they remain uncertain on its 'Southern' location. For Carrión and Velasco, it is clear that there is a geography of violence that comes out of a "social division of space and a particular logic of urbanism" (Chapter 23, this volume) yet while they speak from the Latin American city, they remain more concerned with the political economy that shapes urban violence than its particular 'Southernness'. Brown-Luthango and Reyes, similarly, wonder if the "accentuated inequality" of Cape Town shapes the nature of violence in a particular way there, but focus on a broader question that applies clearly to all cities and to an understanding of planning as a mode to shape the built form. They ask: what role does physical design play in the mitigation of existing violence?

As both authors move across scales from the inter-personal to the structural, however, they mark the conditions of precarity in the cities they write from as a form of violence in themselves, an argument that Abdou Maliq Simone echoes in a different register. If precarity and uncertainty are the defining conditions of the 'South', then how does one speak of a public or rights? For Simone, planning then becomes an assemblage of practices of survival and improvisation. "If you are one of the scores of millions of residents across much of the urban 'South'", he asks,

> who cannot depend upon one specific job to earn your keep, that lacks sufficient documentation to secure a place to live over the long term, or that can't afford to get sick or into any kind of trouble, what is it that you pay attention to in order to know something about what to do?
>
> *(Chapter 21, this volume)*

Bhan, Goswami and Revi (Chapter 20), as well as Mackintosh and Tibandebage (Chapter 25), then shift both planning and the public back to the institutions of the state. For Mackintosh and Tibandebage, planning is a process of institutional (re)design, drawing upon a conceptual vocabulary of "social norms, incentives (market and non-market), institutional capabilities, and feedback loops" (Chapter 25, this volume). Writing about planning for access to health care in Tanzania, they argue for a key role for public institutions especially in a Southern context they mark through both a geo-economic history of structural adjustment, and the persistence of deep inequalities in access and effective demand. Bhan *et al.* (Chapter 20), on the other hand, take on the imagination of the 'universal' within the public in Indian cities, thinking about how the provision of basic needs like water and secure tenure can be attained in a context where most urban residents struggle with some form of spatial illegality that results in both *de jure* and *de facto* exclusions. These chapters are then complemented by Bonner *et al.* (Chapter 22) who argue for publics across scale and jurisdiction, showing how shared conditions and a wide demographic presence of informal work across the global South has enabled and compelled the formation of transnational advocacy networks for rights and entitlements among home-based workers and street vendors, among others.

The strongest assertion for a distinctive planning paradigm from the South comes from Oren Yiftachel (Chapter 19). For his study of Israel/Palestine, the public is shaped in the shadow of an ethnocratic state. He writes of the use of *terra nullius*, the notion of a land 'empty' of rights, as a key mode of dispossession in terms of the 'Southeast'. He argues that though this mode is prevalent across a large geography in the word, it remains under-theorised and absent within most planning theory. This, he argues, is because most planning thought "emerges from liberal democracies in which citizenship, governmental liberalism, legal (de jure) equality, human rights and the rule of law are basic norms which operate within a hegemonic capitalist framework" (Chapter 19, this volume). Any concepts that run against the gain of such liberalism, he argues, remain marginal.

Part V: Planning pedagogies

This last section of the book asks: if changing contexts and demands on planning raise the need for new ways to train professionals, what forms and approaches to pedagogy could be considered? Cities and regions of the global South are generally not known quantities for which extensive and reliable data is available to inform planning (although this is of course variable). We have argued that planning work requires deep understanding of context (preferably co-produced with groups that extend well beyond those regarded as planners) to shape interventions that are innovative and at the same time highly sensitive to fragile livelihoods and environments. The chapters in this section all support the view that deep learning about the complexities of a place is an essential precondition for any kind of intervention; hence how 'we' (teachers, students, practitioners, communities) learn together is a key factor in thinking through planning action. Authors in this section examine a range of ways in which planners and other 'professionals' can learn in context and also how they can teach.

Colin McFarlane (Chapter 26) opens the debate on pedagogy with the provocation that while urban research has conventionally asked the question of how we might come to *know* the city, the question of learning itself has remained black-boxed. It has ignored the ways in which learning is caught up in the production of urban knowledge, policy, planning, ways of seeing and forms of action; and has ignored its potential to be transformatory. In this chapter McFarlane proposes a critical urbanism of learning, in which more collaborative forms of planning involve two forms of learning: both about the city and between different knowledges of the city. This requires a dialogic exchange in the context of often profoundly unequal power relations between planners, policymakers, researchers, residents, civil society groups, and others.

The following chapter by Alessandro Petti on Palestinian refugee camps (Chapter 27) offers a startling illustration of how ways of knowing can be different and transformative. After 60 years, these are neither fragile tented camps nor cities as we usually understand them. The chapter relates an initiative to establish a campus in the camp as an experimental education programme to transgress the borders between an 'island of knowledge' and an island of 'social marginalisation'. The camp location enabled the university, through architectural expression, to open its doors to other forms of knowledge and to experimental and communal learning able to combine critical reflection with action.

University–community engagements can open up new ways of knowing. They potentially offer students important ways of learning as they expose them to real-world complexities, develop skills and values that are difficult to teach in a classroom setting, and co-produce context-specific knowledge that benefits both students and communities. However, on occasions they can also fail to meet expectations. In Chapter 28, Tanja Winkler illustrates how a well-intentioned engagement between planning students and community leaders through a studio project in Cape Town was beset by complex community dynamics, inadequate institutional support and student anxieties about grades and academic success. She concludes that in order for such engagements to be transformative rather than simply instrumental requires project choice where there is established capacity to engage, political support and careful attention to student concerns. Achieving transformative engagements, she suggests, has to encompass values of democracy, reciprocity, power sharing and social justice through the co-production of knowledge.

Adriana Allen, Rita Lambert and Christopher Yap (Chapter 29) are also concerned about how urban and planning professionals learn and along with the other chapters in this section they too seek to open up the equivalent of McFarlane's 'black box' of the learning process. Their chapter asks how we learn a city and relate it to learnings of other cities, how this learning can produce actionable knowledge in context, how we link theory and practice and how this is different from global North academia. The chapter draws on the four-year Learning Lima project, described as a co-learning alliance between a London-based Master's programme and various institutions and collectives of the urban poor in Lima. It describes a pedagogical undertaking that is a fundamentally political process that opens new ways of conceiving, perceiving and living the city; that contrasts and interrogates preconceptions; and that ultimately connects urban theory and planning praxis.

The final chapter by Jigar Bhatt (Chapter 30) tackles the issue of learning from a different perspective. Starting with the assumption that public and programme-evaluation planners want to know if they are 'doing good and being right' he explores an evaluative tool that has come to dominate the way development projects for poverty alleviation are conceived and understood. This is the randomised control trial, or RCT, which he describes as the evaluative 'gold standard' in development. This approach, he argues, has drastically shrunk the space of creative knowledge production and the potential to learn from such development projects. Planners therefore need to reclaim the space that RCTs have monopolised by developing more relevant intervention and learning models to restore balance in thinking and practice. The chapter outlines the rise of RCTs and then takes issue with the internal-versus-external validity debate in the field. Bhatt instead proposes a model that borrows from the 'whole systems approach', which he says does not dispense with RCTs but offers a 'Southern' inspired approach to development projects' design and evaluation.

References

Bhan, G. (2016) *In the Public's Interest: Evictions, Citizenship and Inequality in Contemporary Delhi*, Athens: University of Georgia Press; New Delhi: Orient Blackswan.

Caldeira, T.P. (2016) "Peripheral urbanization: Autoconstruction, transversal logics, and politics in cities of the global South", *Environment and Planning D: Society and Space*. DOI: 0263775816658479.

Caldeira, T.P.R. and Holston, J. (2005) "State and urban space in Brazil: From modernist planning to democratic interventions". In A. Ong and S.J. Collier (eds) *Global Assemblages: Technology, Politics, and Ethics as Anthropological Problems*, Malden, MA: Blackwell, pp. 393–416.

Comaroff, J. and Comaroff, J.L. (2015) *Theory from the South: Or, How Euro-America is Evolving Toward Africa*, London: Routledge.

Connell, R. (2014) "Using Southern theory: Decolonizing social thought in theory, research and application", *Planning Theory* 13(2): 210–223.

Dados, N. and Connell, R. (2012) "The global South", *Contexts* [American Sociological Association] 11(1): 12–13.

Ferguson, J. (2013) "Declarations of dependence: Labour, personhood, and welfare in Southern Africa", *Journal of the Royal Anthropological Institute* 19(2): 223–242.

Graham, M. (2015) "The geography of academic knowledge", Geonet blog, University of Oxford. Available: http://geonet.oii.ox.ac.uk/blog/the-geography-of-academic-knowledge/ [accessed 27 February 2017]

Lund, F. and Srinivas, S. (2000) *Learning from Experience: A Gendered Approach to Social Protection for Workers in the Informal Economy*, Geneva: International Labour Organization.

Merrifield, A. (2014) *The New Urban Question*, London: Pluto Press.

Paasi, A. (2005) "Globalization, academic capitalism and the uneven geographies of international journal publishing spaces", *Environment and Planning A* 37(6): 769–789.

Papaioannou, T. and Srinivas, S. (2016) "Innovation as a political process of development: Are neo-Schumpetarians value neutral?" Paper presented at the Science Policy Research Unit (SPRU), University of Sussex, 50th Anniversary conference, September 2016.

Parnell, S. and Pieterse, E. (2010) "The 'right to the city': Institutional imperatives of a developmental state", *International Journal of Urban and Regional Research* 34(1): 146–162.

Simone, A.M. (2004) *For the City Yet to Come: Changing Life in Four African Cities*, Durham, NC: Duke University Press.

Srinivas, S. (2010) "Industrial welfare and the state: Nation and city reconsidered", *Theory and Society* 39(3–4): 451–470.

Srinivas, S. (2012) *Market Menagerie: Health and Development in Late Industrial States*, Palo Alto, CA: Stanford University Press.

Srinivas, S. (2014) "Demand and innovation: Paths to inclusive development". In S. Ramani (ed.) *Innovation in India: Combining Economic Growth with Inclusive Development*, Cambridge: Cambridge University Press, pp. 78–106.

Srinivas, S. (2015) "Healthy industries, unhealthy populations: Lessons from Indian problem-solving". In M. Mackintosh, G. Banda, W. Wamae and P. Tibandebage (eds) *Making Medicines in Africa: The Political Economy of Industrializing for Local Health*, Basingstoke: Palgrave Macmillan, pp. 183–199.

Srinivas, S. (2017, forthcoming) "Evolutionary demand, innovation, and development". In D. Nathan, S. Sarkar and M. Tewari (eds) *Upgrading and Innovation in GVCs in Asia*, Cambridge: Cambridge University Press.

Srinivas, S. and Sutz, J. (2008) "Developing countries and innovation: Searching for a new analytical approach", *Technology in Society* 30: 129–140.

Urban Poverty and Inequality Collective. (2015) *Making the Urban Welfare State: New Questions from the South*. Berkeley, CA.

Watson, V. (2006) "Deep difference: Diversity, planning and ethics", *Planning Theory* 5(1): 31–50.

Watson, V. (2014a) "African urban fantasies: Dreams or nightmares?", *Environment and Urbanisation* 26(1): 213–229.

Watson, V. (2014b) "The case for a Southern perspective in planning theory", *International Journal of E-Planning Research* 3(1): 23–37.

Part I
Planning and/as the state

1

Spatial rationalities and the possibilities for planning in the New Urban Agenda for Sustainable Development

Clive Barnett and Susan Parnell

Introduction

This chapter traces the emergence of global debates about urban policy and planning in multi-lateral governance forums that put enormous, possibly untenable, pressure on planners to deliver sustainable development. In the ushering-in of a more city-centric and pro-planning era that has taken place over the last few years we focus in particular on the process through which a dedicated 'Urban SDG' was adopted by the United Nations' Sustainable Development Goals (SDG) process in 2015 and the subsequent development of a 'New Urban Agenda' (NUA) for global development policy, adopted by UN-Habitat's bi-decennial conference, Habitat III, in Quito in Ecuador in October 2016.[1] The chapter argues that these initiatives are just the start of a process of implementing and monitoring priorities for housing provision, infrastructure investment and land development that have significant potential for reconfiguring the role and understanding of local government in general and the more specific role of spatial planning. Global level agendas, we argue, are key to understanding the contemporary rehabilitation of national and local territorial and strategic planning practice both politically and professionally, especially in rapidly urbanising contexts like Africa and Asia.

To appreciate the importance of the global agenda for the profession over the decades to come, we adopt an expansive understanding of 'planning' as a set of practices concerned with "the linking of knowledge to action" (Friedmann 2011: 208). In turn, like Forester, we understand 'planners' to refer very generally to "all those who need to learn about their environments – public or private, social or natural – in order to change them" (Forester 2006: 124). Understanding planning and planners in this sense, rather than the narrower certified professionalisation such as that associated with organisations like the Royal Town Planning Institute, helps us frame our account of how the emergence of global urban policy initiatives represents a distinctive reconfiguring of planning, that is at the same time more inclusive and also dangerously all embracing. Insofar as a wide variety of public and private actors are increasingly involved in the production, distribution and application of spatial knowledge about urban processes, the conventional professions and disciplines of (urban and regional or town) planning are now located as just one element in a complex field of knowing and learning about urban settlements 'in order to change them'. This is the first paradox of our time – that at exactly the same time as planning is receiving an expanded mandate, the definition of who is a planner has become less clear, making it difficult to either educate or hold to account the new cohort of spatial advocates and practitioners.

There is a second paradox. Having achieved a global commitment to a bold urban agenda, the current formulation of the profession may be unable to deliver on this expectation. To understand the unwieldy scope of the urban agenda and its implications for planning, the chapter starts by identifying how a dedicated Urban SDG was successfully included in the 2015 SDGs and briefly explores how this legacy was then unevenly absorbed in the NUA preparatory process. We then consider the complex processes of lobbying, consultation and negotiation through which urban issues, and by implication also the expectations of planners, have been institutionalised in global development agendas. Appreciating the inherently political processes through which global urban policy is constructed is important to understanding the potential and limits for creative engagement by planning professionals, local state actors, social movements and NGOs with the new opportunities opened up by the SDGs and the NUA in particular. Noting the somewhat chaotic conception of the urban in the NUA, we then discuss the lack of a distinctive meaning of 'the city' in the global policy discourse. We suggest that as planners are asked to lead local and national initiatives that engage the new global urban policy imperative, most likely through the mechanism of National Urban Policies, the diversity of definitions of 'the city' and 'urban' should be treated operationally as different *spatial rationalities*, crystallising distinctive practical understandings of how managing spatial processes can bring about change. In concluding, we identify key tensions that will characterise the unfolding of the post-2015 urban agenda at a number of scales.

Locating 'the city' in the SDGs

The city is a scale of action not hitherto acknowledged by the multilateral community. In September 2015, however, the 69th General Assembly of the United Nations formally approved the Sustainable Development Goals (UN 2015). Urban issues had acquired heightened visibility during the process of negotiating a new development agenda, overseen by the UN, to replace the Millennium Development Goals (MDGs) framework. Among the 17 agreed upon SDGs, Goal 11 declared a commitment to "Make cities and human settlements inclusive, safe, resilient, and sustainable". The inclusion of a dedicated Urban SDG was the culmination of an explicit, public campaign to have urban issues recognised as core to future development agendas, although this campaign drew on a longer history of urban thought in global development policy (Cohen 2016; Parnell 2016). The mobilisation of expertise to secure an Urban SDG in global development frameworks was followed soon after by the formulation of the so-called NUA, led by UN-Habitat, the UN's flagship human settlements agency. At issue in these and other multilateral agreements (especially the 2015 UNFCCC COP21 Paris Agreement on climate and the Sendai Framework on Disaster Risk Reduction) is how the UN system will frame the ways in which nation-states, city and regional governments, UN agencies, international donors and civil society actors problematise and address urban issues for the following 20 years and beyond.

Although there are now multiple UN policy documents in which the city emerges as a critical site of action, from the perspective of planners the most directly relevant of these is the NUA where there is an overt concern with human settlement improvement and territorial development. The negotiations in 2016 leading up to Habitat III were slow to get going, with delays in releasing drafts and finally outright conflict in the final scheduled preparatory meeting held in

Surabaya, Indonesia. The most important tensions to emerge did not centre, as was widely predicted, on the inclusion of wording around the 'right to the city'.[2] Rather, they revolved around the role of UN-Habitat and the NUA relative to that of the SDGs,[3] in and on the removal of the proposed Multi-Stakeholder Panel.[4] Despite these disputes, in the bulk of the text of the NUA, where the ambitious agenda of what better planning might achieve is set out (alongside stronger local government and sub-national fiscal and legal reform), there were remarkably few changes made. Accepting that the role of planning would need to be enhanced in realising the 2030 vision was less an endorsement of the profession, however, than it was an argument that flowed directly from acknowledging the city as a driver of sustainable development.

The campaign for an Urban SDG and the development of the NUA prior to the Habitat III process represent an unprecedented recognition of urban issues in global development policy. The inclusion of a stand-alone urban goal "to make cities safe, inclusive, resilient and sustainable" is genuinely path breaking. In important respects, the inclusion of Goal 11 in the SDGs framework reflects a significant, and broadly dispersed, conceptual shift towards thinking of global processes as necessarily working horizontally through places, localities and regions, rather than vertically impacting upon them as external forces (see Barnett and Parnell 2016). The campaign for the inclusion of a dedicated Urban SDG combined four distinct claims about contemporary urbanisation processes, which when combined made a case for the centrality of urban-based action (and by proxy more planning) in relation to the over-arching aims of the new sustainable development agenda. First, there was an empirically led claim that the key problems to be addressed by the sustainable development goals, not least poverty, are increasingly concentrated in urban areas. Second, there was a conceptual claim that the dynamism of cities as economic agglomerations of growth and social clusters of innovation presented an opportunity that must be harnessed to achieve the SDGs. Third, cities were identified as pathways of global environmental change. And fourth, there was a claim that cities and other localities represent the most effective political scale for coordination and decision-making to deliver the SDG agenda. It should also be emphasised that the campaign for an Urban SDG and the development of the NUA are not simply a matter of recognising contemporary urbanisation trends and demographic facts of living in an urbanised world. The assertion of an urban frame for addressing development challenges represents a significant shift within UN-level governance processes. It reflects not only recognition that urban settlements are crucial pathways to sustainable development, but also an acknowledgement of the crucial developmental role of sub-national government actors and multi-level governance (see SDSN 2013). In short, what is most significant about these overlapping initiatives is the recognition of local and regional territorial action as an important dimension of global changes. This is a dramatic departure from past development thinking that has tended to privilege national scale actors and strategic and fiscal planning over spatial thinking.

But to fully appreciate the significance of the inclusion of an urban lens in global development policy, it is necessary to recognise the more general significance of the SDGs as a whole. It is these normative changes that planners will be called upon to operationalise in specific places in ways that will make the transformative changes implied by the 2030 agenda. The SDGs represent a fundamental shift in the aspirations and practices of global development policy. There are five dimensions to the shift embodied by the SDGs. First, unlike the MDGs, which focussed primarily on alleviating poverty in the global South, the SDGs are truly global in their ambition, setting out single, minimum standards for all nations. Second, the SDGs are premised on the developmental interdependence of social, economic and environmental values. They give much greater weight than ever before to the ecological limits of human existence and the dangers of climate change. Achieving this integrated vision of sustainable development will require a fundamental transformation of most accepted practices of urban management. Third, the SDGs emphasise reducing *inequality* as well as poverty. Fourth, the SDG monitoring and reporting framework, enabled by innovations in geospatial science, complex statistical modelling and big data analysis, allows the integration of spatial and statistical analysis and the nesting of local, national and global indicators. This technological revolution in data analysis allows greater flexibility in indicator selection and reporting, and so promises to refashion the measurement of global development and to facilitate heightened importance of cities and localities in these monitoring processes. Finally, the global development agenda is now being debated alongside issues of institutional capacity building and the provision of finance, which again draw into focus the importance of reconfiguring urban scale institutions and infrastructures.

It is in this broader context of the reframing of development agendas that the heightened significance accorded to cities and urban issues, and with this to spatial planning, needs to be placed. They enable us to see that the assertion of an urban perspective in development policy extends far beyond simply implementing and monitoring Goal 11 of the SDGs. The approval of the Urban SDG and the roll-out of the discourse of the NUA represents just one aspect of the consolidation of a distinctively pro-urban, spatially nuanced approach to issues of poverty alleviation, sustainability and basic needs provision in global governance agendas. It runs alongside, for example, an increasing recognition of the importance of urban-scale responses to climate change issues, indicated by the inclusion of urban issues in the Intergovernmental Panel on Climate Change (IPCC) in 2014, and the heightened significance of urban issues in the UN Climate Change Conference, COP21, in Paris in 2015. In these and other global initiatives, cities and urban processes are now routinely framed as bearing a double responsibility. On the one hand, how cities and towns are planned and managed is presented as causes of myriad contemporary challenges (from global warming to obesity, and from financial instability to social exclusion). On the other hand, they are also presented as providing opportunities to act in response to those challenges (see Birch and Wacheter 2011).

The idea that urban areas are the locations of all sorts of problems is of course an old, established theme in social thought and public policy. In no small part, modern urban and regional planning has its origins in this idea. But urban issues have traditionally been thought of as symptoms, as the place-specific manifestations of more general processes. By contrast, in the twenty-first century, urban processes are now ascribed causal significance in generating global challenges including ecosystem degradation, climate change, peak oil, systematic inequality and persistent poverty, food insecurity and energy transitions (see Swilling 2011). These challenges are now understood to have their roots in specifically urbanised patterns of accumulation, consumption and interaction. But at the same time, cities and localities are now presented as having all sorts of opportunities and potentialities for reconfiguring those global processes. With the elevated status of the city the responsibilities of the planner grow exponentially.

The campaign around the Urban SDG and the elaboration of the NUA is one example of broader process of the *urbanisation of responsibility* (see Barnett 2012). Placing these debates and initiatives within this broader understanding is important as it allows us to see how ideas about the tasks and agents of planning are currently being transformed. The role of cities and urban processes is no longer confined to an image of spatial planning as a residual field for managing the externalities generated by more general processes. What is distinctively new about the contemporary ascendancy of place-based, city-centric policy visions at national and international levels is the proposition that urban-scale institutions, infrastructures, and communities of interest are now empowered to creatively shape multiple global challenges associated with the growth in the size and complexity of urban settlements, the increase in urban populations, and the generalisation of conditions of urban living (Gordon and Buck 2005). The conceptual shift that lies behind the campaign for Goal 11 of the SDGs, and is explicitly asserted in the NUA in particular, involves thinking of global processes working horizontally through places. And it is associated in turn with the idea that cities and regions are the sites for the experimental development of new models, practices and solutions to global problems that can be translated to other places.

We have identified a series of conceptual shifts lying behind the assertion of an urban imagination in global development policy and with this the amplified role of planning. While these shifts are informed by academic fields of inquiry, it is important to recognise that their significance for fields of practice such as planning relies on a wider complex of knowledge and thought coordinated by UN agencies through which global urban policy has emerged. Before considering the spatial rationalities underwriting the NUA's assertion of the importance of city-level action on global issues, it is therefore important to consider the institutional dynamics through which this sort of global agenda is developed.

Making global urban policy

We have suggested that the inclusion of Goal 11 in the SDGs and the development of the NUA can be seen to mark a decisive shift in the status of urban-related fields of practice in global development agendas. In order to appreciate the potential for creative engagement by planning professionals, local state actors, social movements and NGOs with the new opportunities afforded by this shift, it is important to acknowledge the inherently political processes through which global urban policy has been and will continue to be made. Our understanding of this politics is not restricted to political parties but extends to the wider institutional structures through which planners are mobilised.

As already indicated, the campaign for an Urban SDG involved the concerted mobilisation of networks of local government and urban and regional planning, such as ICLEI (Local Governments for Sustainability) and UCLG (United Cities and Local Governments), partnering with UN-level organisations such as the Cities Alliance. It has also involved academic networks, such as the Swedish funded Mistra Urban Futures programme; networks of climate change governance, such as the C40 group; and commercial actors, such as the Urban Land Institute, the international network representing real estate capital (Peirce 2014). In short, the approval of the Urban SDG is a product of what one might call an *urban thought collective* – a dispersed network of individuals and organisations exchanging ideas and knowledge about urban processes in an ongoing set of processes of learning, application and reflection (see Fleck 1979, p. 39). In short, a fluid alliance of interests and organisations has generated a coherent discourse through which to assert the need for a cross-sectoral policy focus on cities in future development policy agendas. While the basis for this mobilisation was never to advance the place of planning, it may have been the unintended outcome.

The emergence of a global agenda of urban development is the outcome of a complex field involving transnational advocacy networks and international NGOs, but also sectoral interests representing multinational companies from sectors such as transport, real estate, water, electricity and other utilities, and information technology (Gilbert 2011). The production of coherent-looking policy agendas of global governance involves an ongoing choreography of consultation, summitry, reporting, drafting and ratification (Marx *et al.* 2012). Global development policy-making involves a highly complex process of proposal making, agenda setting and position-taking. The institutionalised commitment to participation in UN processes means that there is a range of opportunities for articulating opinions and developing partnerships for a wide range of governmental actors, civil society and non-governmental actors and business interests.

The campaign for an Urban SDG, and the subsequent development of the NUA exemplify this aspect of global development policy-making. Both processes made use of open access web platforms as the entry-point for a wide range of stakeholders to make propositions and lobby about urban futures. However, this formal commitment to consultation makes the imperative of developing coherent agendas with intelligible narratives and actionable proposals all the more crucial. The process of editing and selecting from a huge and diverse range of opinions and expertise is an intensely *political* one, involving the arts of bargaining, compromise and deferral through which competing interests and conflicting positions are aligned. The development of a new urban agenda has therefore been channelled through and coordinated across multiple channels of engagement.

As a response to the problem of organisational complexity, a new layer of UN-coordinated knowledge brokerage was developed to ensure that the SDG process was transparent and consultative, involving the creation of the Sustainable Development Solutions Network (SDSN).[5] The SSDN campaign for an Urban SDG drew together participants from across the UN's Major Groups in support of the urban agenda within the UN system. The successful campaign that culminated in the approval of Goal 11 depended on overcoming an entrenched anti-urbanism in development policy thinking. However, beneath the temporary agreement across sectors and interests on the need for an urban focussed goal, very different interpretations of how cities are significant for achieving sustainable development will become more evident as the politics of implementing the SDGs unfolds. Locally, in the myriad of cities and towns across the world that will seek to align their development trajectories with the national urban policies that will in turn reference the global agenda, the task of navigating conflicting rationalities will invariably fall to the budget office and the urban planner.

'The city' and the emerging remit of planning

If increased responsibility for achieving global aspirations for sustainable development is to be placed ever more heavily on the shoulders of national and local planners, it is important to be able to set out the parameters of this responsibility. In light of the complex political processes through which Goal 11 came to be approved, and through which the NUA was developed in the lead up to Habitat III in 2016, it is perhaps hardly surprising that there is no single definition of the city at work in fields of global urban policy. However, there is a series of recurring themes in this field that suggest that the meaning of 'the city' is closely associated with practical understandings of how spatial practices can be reconfigured in pursuit of global objectives. We suggest that there are three broad understandings of 'the city' at work in global urban policy, each of which illustrates the ways in which place-based processes are presented in this field as sites within spatially extensive networks and flows.

First, there is an idea of the urban in terms of the *clustering* together of proximate activities. This idea is found in arguments about the agglomeration efficiencies that characterise urban economies, in arguments about the socio-cultural benefits that follow from the concentration of diverse populations in urban areas, and in arguments about the potentials for innovation that follow from city-region specialisation.

Second, alongside this sense of the city as a space of proximity, there is an understanding of the city as a *hub*, a node in wider networks, including urban–rural relations of migration or trade, as well as environmental relationships that extend beyond the scale of any single settlement. This emphasis in part reflects sophisticated scientific and social-scientific understandings of the multi-scalar dynamics of environmental processes, or the dynamics of the urbanisation of population. But it is also a trace of a 'diplomatic' imperative to find a way of asserting the importance of urban issues without seeming to marginalise rural-based issues and constituencies or those concerned with global environmental processes.

Finally, there is a strong claim concerning the role of the city as a *scale* for the holistic integration of various processes into a systematic approach to management, planning and regulation. This third aspect of the discourse of the NUA is crucial, for it shifts attention from a passive construction of urban spaces as bearers of problems and opportunity, towards an idea of the city as a scene for creative concerted action.

The different meanings ascribed to the city and urban processes in the SDGs, the NUA and other strands of contemporary development thinking are part of a much larger pattern of contemporary urban thought. In the proliferation of urban concern evident across different fields, it is noteworthy that the meaning of 'cities' and 'urban' has become highly variable (see de Jong *et al.* 2015; Parnell *et al.* 2016). The urban can refer to a very broad range of concerns or issues: to a site of sociability valued in terms of community relations and cohesion; to a technological infrastructure, problematised in terms of the circulation of material objects and information; to a privileged site of democratic practice, a scene of participation, experimentation and accountability; or to a field of innovation and potential economic growth.

There is a temptation to think that there must be, or should be, some common strand running across all these different understandings. But rather than trying to identify a theoretically singular definition, we suggest that each of the distinct understandings of cities, localities and the urban evident in contemporary public debates should be thought of as crystallising a distinctive spatial rationality. By spatial rationalities, we are referring to the assumptions about the causal power of the spatial forms, built environments and designed settings in which human action takes place that are operative in particular fields of practice. For example, planning practice and planning theory and other spatial disciplines such as architecture tend to ascribe significant causal power to spatial arrangements and forms in configuring certain preferred outcomes (Huxley 2006). It is often assumed that the design, classification and ordering of spaces and environments can comport individuals or whole populations to conduct themselves in particular ways. Traditionally, this assumption has been associated with attempts to control disorderly practices, to design out crime, to uplift people culturally, or to use environmental features to improve the health of populations. The assumption is in turn based on the idea that particular spatial configurations or environments generate forms of disorder, decay or disease.

Spatial rationalities are not restricted to urban and regional planning narrowly defined. The same types of causal postulates are operative across a range of fields of policy and practice in which spatial configurations are presented as a medium for acting on people's actions. In all sorts of fields, particular understandings of the causal power of spatial forms, spatial relations or environmental configurations inform institutionalised attempts to manipulate, design and manage spaces and environments for different ends (Barnett 2012).

The variable understandings of 'the city' and the urban in the SDG and NUA process are just one example of these sorts of spatial rationalities. To be more precise, we suggest that the elaboration of a global urban development agenda is indicative of a significant shift in the prevalent spatial rationalities defining cities and urban processes as potential objects of planning. In classical models of planning, it was assumed that reconfiguring and redesigning could help to bring about various forms of desired change. Increasingly, in contemporary models of decision-making, evident, for example, in the emergence of programmes of urban resilience and adaptation or of 'urban ecological security', the aim of planning practices is not to pursue a linear model of progressive social change, economic growth or welfare provision. It is, rather, to bolster the capacity of urban environments to withstand change, in the form of various shocks, whether global financial instability, terrorist attack, catastrophic physical disaster, or impending 'transitions' such as climate change or peak oil. But at the same time, the aim is also to enhance the potential of cities and other places to generate as yet unimagined future pathways of innovation and experimentation.

In order to grasp the significance of this shift in the prevalent spatial rationalities of urban management programmes in the twenty-first century, and the role of knowledge, learning and planning in those programmes, it is helpful to consider the career of a core concept of twentieth-century planning theory, the idea of 'wicked problems'. The idea of wicked problems refers to the notion that certain sorts of issues are inherently characterised by causal complexity, predictive uncertainty and political contestation, making any easy 'one-shot' solution to them by the application of scientific knowledge impossible. It was an idea originally developed in relation to the specific features of urban issues such as inner city poverty, housing provision or spatial concentrations of crime (Harrison 2000; Rittel and Webber 1973). And it was also an idea originally developed as a critique of the overly optimistic promises of technocratic social analysis to be able to solve complex issues through the application of data-systems (Goodspeed 2015).

The idea of wicked problems has been revived in the twenty-first century as a way of describing a set of issues associated with new forms of complexity science and resilience thinking, from climate change to public health to terrorism (see Head and Alford 2013; Zellner and Campbell 2015). It is notable that in recent invocations of the idea, the affirmation of professional modesty originally associated with the idea is often missing. New forms of data-led science assert that wicked problems can be addressed by breaking down and modelling complexity, in order to deliver clear solutions to decision-makers (Roberts 2000).

What is most notable about the revival of the idea of wicked problems as a relevant way of framing all sorts of twenty-first challenges is how this is associated with claims that cities and other urban settlements are the privileged sites for addressing those challenges. Tracing the 40-year career of the idea of wicked problems helps us see how cities are no longer thought of as merely locations impacted by global processes or as the locations of distinctively 'urban problems'. Rather, they are now presented as the key sites for the development, trialling and translation of all sorts of practices in order to address all sorts of problems that extend far beyond the traditional scope of urban policy or urban and regional planning. Whereas 'wicked problems' was once the name given to a set of specifically urban issues, cities are now understood as the sites where all sorts of global challenges might at least be resolved. In a sense, then, 'the city' now appears as the variable form that can serve as the arena in which claims of expertise can be squared with acknowledgements of contingency and uncertainty.

The ambivalence about expertise – about just how knowledge can and should be linked to action – that is central to the idea of wicked problems is reflected in the way in which the proliferation of concern with cities and urban issues in contemporary scientific, policy and public debate goes hand in hand with the observable trend to think of specific places as *experimental* sites for policy, technological and social interventions. There are in fact a wide range of notions of the experimental at work in contemporary urban thought: these include ideas of using towns and cities as test-beds for bounded experiments; as locations for piloting targeted policy interventions; as sources of best practice case studies; or as sites of comparative learning between places (see Evans *et al.* 2016; Halpern *et al.* 2013; Karvonen and Van Heur 2014; Patel *et al.* 2015). More abstractly, the experimental is sometimes invoked to bring to mind images of creativity, edginess and innovation – of doing things differently; sometimes to bring to mind more positivist ideas of establishing 'what works' or establishing 'proof of concept'. But in both senses, there are a series of layered understandings of how things learnt in one place might be applied or translated to others, for various purposes.

The crystallisation of a global urban development agenda is, then, one example that marks a decisive shift in the way in which 'the urban' is located with various fields of policy problematisation. No longer seen as the site of specifically defined 'urban problems' to be addressed by fields of urban policy or spatial planning, the city is now presented as a hub, driver and node through which all sorts of global challenges can be addressed in practicable ways. We have argued here that the forms of knowledge circulating in fields of experimental urbanism serve as so many epistemologies of action: they can be thought of as embodiments of specific causal rationalities that ascribe agency to spatial imaginaries of 'the city', the 'urban–rural continuum', 'networks' and other key concepts of management, regulation and planning. When seen as one index of the refashioning of the city as an experimental site for resolving wicked problems, the city emerges across the breadth of global urban development policy as a site for interventions to promote sustainability, a scene for the coordination of policies and a pathway for economic or social change.

Conclusion

We suggested at the outset that planning should be understood as a practice concerned with linking knowledge about environments of action to projects of change. In this sense, planning is a broadly distributed set of practices not to be restricted to fields conventionally defined as planning. Understood in this way, the emergence and consolidation of urban agendas in global governance initiatives can be seen to involve an enhanced role for planning practices, as long as these are thought of as spread across a whole series of public and private actors who are now involved in the production, distribution and application of spatial knowledge.

The argument we have developed here suggests that recognising the potentials opened up for new modes of planning practice, understood in the broad sense with which we began this chapter, requires us to appreciate the ways in which conceptual claims are articulated with practical pathways of action. Satterthwaite has argued, for example, that among a whole range of previous 'new urban agendas' in development policy since the 1970s, those that have been most effective in facilitating urban-scale transformations were those that presented "do-able local actions" aimed at urban issues and implementable by urban-scale institutions (Satterthwaite 2016: 4). It should be said that the examples that Satterthwaite has in mind – the Healthy Cities Movement started in the 1980s; sustainable development policies under the umbrella of Local Agenda 21 after 1992; participatory budgeting practices since the 1990s and disaster risk reduction programmes promoted by the UN's Making Cities Resilient programme – are all relatively 'low-cost' initiatives. They primarily involve reconfiguring the existing regulatory powers of local government apparatuses, empowering existing professional and expert actors in urban governance, and relying on their capacity to build networks, negotiate partnerships and encourage participation.

It is in this light that the significance for future planning practices of the SDGs and the NUA should be approached, not least because to do otherwise would tax the existing certified planning professionals beyond their numerical capacity and engage them in action far beyond their traditional mandates. The SDG process is one vector through which the emergence of an apparatus of global urban governmentality is already developing. Definitions of indicators of progress and the provision of systems for monitoring those indicators are quite central to the delivery of all of the SDGs, and cities have been ascribed a central role in providing the necessary technical and professional systems that this requires (see Simon *et al.* 2016). Furthermore, the NUA also depends on the capacities of local-scale actors to address issues of poverty, inequality, economic development and infrastructural investment. In these ways, then, the latest 'new urban agenda' inaugurated by the post-2015 SDG process implies cities and other places have much enhanced fiscal capacities than is often currently the case. The question of the financing of urban development and fiscal autonomy of local governments has been and will likely continue to be among the most contentious issues throughout the SDG and NUA process. How this issue is resolved will be quite crucial in determining the sorts of 'do-able actions' that are opened up as potential fields of engagement for planners, in the broadest sense.

The approval of Goal 11 of the SDG framework and the development of the NUA mark an important step in the longer history of UN engagement with issues of multi-scale government, multi-actor governance and multi-sectoral or integrated development by institutions of global governance (Parnell 2016). In the 1980s, cities were seen primarily as sites of localised problems. By the time of the 1996 Habitat II meeting, cities were seen as key strategic nodes of intervention. By the time of Habitat III in 2016, arguments were presented that placed cities at the heart of a whole global environmental system, a system that was now represented as a fundamentally urban system (Acuto and Parnell 2016). And it is important to emphasise that the further development and implementation of the UN-brokered urban agenda will be shaped by conflicts and debates that were, for a time, effectively dampened in the construction of the pro-urban consensus that drove the Urban SDG campaign. There are different perspectives for thinking about the importance of cities, including thinking of cities as one of many sites for pursuing sustainable development (alongside oceans, forests and farmlands); thinking of cities as the new centres and most important locations in which distinctive challenges of sustainable urban development must be supported; or thinking of cities as the key drivers of sustainable development in the urban Anthropocene. How the relations between these different viewpoints are negotiated in future will also help shape the pathways through which future planning practices evolve.

The key axis around which the relations between these viewpoints revolve is a dispute between two different interpretations of just what the SDG process and the NUA imply for models of urban politics. One view suggests that urbanisation needs to be approached positively and that every place must be much better run if sustainable development is to be achieved across the rapidly expanding urban population. A second, more radical view is that every urban citizen's lifestyle, every city service, every city, every city region and the system of cities itself needs to be run in an entirely different way to guarantee that the collective urban condition secures, rather than precludes, global sustainability. The distinction might be described as a difference between making cities and territories work *better* to achieve sustainability, and a vision of the radical *transformation* of settlements as the mechanism to meet global challenges such as the ending of poverty, the reduction of inequality, gender equity, deradicalisation, dematerialisation, enhancing ecosystem integrity, decarbonisation and others. This is an old division within planning practice and theory, of course, between a view of planning as an ameliorative practice and a view of planning as a transformative practice. And it is a division that is now writ large in global development agendas.

Notes

1 For further details of these processes that were ongoing as we prepared the chapter, see http://citiscope.org/habitatIII [accessed 20 February 2017].

2 See http://sd.iisd.org/policy-updates/the-right-to-the-city-and-the-new-urban-agenda/; www.uclg-cisdp.org/en/right-to-the-city/Habitat-III/new-urban-agenda;www.habitat3.org/the-new-urban-agenda/policy; http://citiscope.org/habitatIII/news/2016/06/new-urban-agenda-second-draft-released-drawing-praise-and-criticism [all accessed 20 February 2017].

3 See http://blog.felixdodds.net/2016/07/is-clos-killing-un-habitat.html [accessed 20 February 2017].
4 See http://citiscope.org/habitatIII/commentary/2016/07/habitat-iii-loses-proposed-multi-stakeholder-panel-now [accessed 20 February 2017].
5 See http://unsdsn.org/about-us/vision-and-organization/ [accessed 20 February 2017].

References

Acuto, M. and Parnell, S. (2016) "Leave no city behind", *Science* 352 (6288): 873.

Barnett, C. (2012) "Changing cities". In M. Butcher, N. Clark, J. Smith and R. Tyszczuk (eds) *Atlas: Geography, Architecture and Change in an Interdependent World*, London: Black Dog Publishing, pp. 72–79.

Barnett, C. and Parnell, S. (2016) "Ideas, implementation and indicators: Epistemologies of the post-2015 urban agenda", *Environment and Urbanization* 28(1): 87–98.

Birch, E.L. and Wacheter, S.M. (eds) (2011) *Global Urbanization*, Philadelphia, PA: University of Pennsylvania Press.

Cohen, M. (2016) "From Habitat II to Pachamama: A growing agenda and diminishing expectations for Habitat III", *Environment and Urbanization* 28(1): 3–48.

De Jong, M., Joss, S., Schraven, D., Zhan, C. and Weijnen, M. (2015) "Sustainable-smart-resilient-low carbon-eco-knowledge cities: Making sense of a multitude of concepts promoting sustainable urbanization", *Journal of Cleaner Production* 109: 25–38.

Evans, J.P., Karvonen, A. and Raven, R. (eds) (2016) *The Experimental City*, London: Routledge.

Fleck, L. (1979) *Genesis and Development of a Scientific Fact*, Chicago, IL: University of Chicago Press.

Forester, J. (2006) "Policy analysis as critical listening". In M. Moran, M. Rein and R. Goodin (eds) *The Oxford Handbook of Public Policy*, Oxford: Oxford University Press, pp. 124–151.

Friedmann, J. (2011) *Insurgencies: Essays in Planning Theory*, London: Routledge.

Gilbert, A. (2011) "Who sets the global urban agenda? The role of sectoral lobbies, local politicians and global institutions", *The Global Urbanist*, 20 December. Available: http://globalurbanist.com/2011/12/20/who-sets-global-urban-agenda [accessed 30 June 2016].

Goodspeed, R. (2015) "Smart cities: Moving beyond urban cybernetics to tackle wicked problems", *Cambridge Journal of Regions, Economy and Society* 8(1): 79–92.

Gordon, I. and Buck, N. (2005) "Introduction: Cities in the new conventional wisdom". In N. Buck, I. Gordon, A. Harding and I. Turok (eds) *Changing Cities: Rethinking Urban Competitiveness, Cohesion and Governance*, Basingstoke: Palgrave Macmillan, pp. 1–21.

Halpern, O., LeCavalier, J., Calvillo, N. and Pietsch, W. (2013) "Test-bed urbanism", *Public Culture* 25(2): 273–306.

Harrison, T. (2000) "Urban policy: Addressing wicked problems". In H. Davies, S. Nutley and P. Smith (eds) *What Works? Evidence-based Policy and Practice in Public Services*, Bristol: Policy Press, pp. 207–228.

Head, B.W. and Alford, J. (2013) "Wicked problems: Implications for public policy and management", *Administration and Society* 46(6): 711–739.

Huxley, M. (2006) "Spatial rationalities: Order, environment, evolution and government", *Social and Cultural Geography* 7(5): 771–787.

Karvonen, A. and Van Heur, B. (2014) "Urban laboratories: Experiments in reworking cities", *International Journal of Urban and Regional Research* 38(2): 379–392.

Marx, C., Halcli, A. and Barnett, C. (2012) "Locating the global governance of HIV and AIDS: Exploring the geographies of transnational advocacy networks", *Health and Place* 18(3): 490–495.

Parnell, S. (2016) "Defining a global urban development agenda", *World Development* 78: 529–540.

Parnell, S., Crankshaw, O. and Acuto, M. (2016) "2030 policy endorsement of a sustainable future: Implications for urban research". Available: www.urbantransformations.ox.ac.uk/news-debate/key-debates/#sthash.MDCjGfaa.dpuf [accessed 24 August 2016].

Patel, Z., Greyling, S. Parnell, S. and Pirie, G. (2015) "Co-producing urban knowledge: Experimenting with alternatives to 'best practice' for Cape Town, South Africa", *International Development Planning Review* 37(2): 187–203.

Peirce, N. (2014) "Behind the push for an 'urban SDG'", *Citiscope*, 14 August. Available: http://citiscope.org/story/2014/behind-push-urban-sdg [accessed 30 June 2016].

Rittel, H.W.J. and Webber, M.J. (1973) "Dilemmas in a general theory of planning", *Policy Sciences* 4: 155–169.

Roberts, N. (2000) "Wicked problems and network approaches to resolution", *International Public Management Review* 1(1): 1–19.

Satterthwaite, D. (2016) "A new urban agenda?", *Environment and Urbanization* 28(1): 13–34.

SDSN Thematic Group on Sustainable Cities (2013) "Why the world needs an urban development goal". Available: https://sustainabledevelopment.un.org/content/documents/2569130918-SDSN-Why-the-World-Needs-an-Urban-SDG.pdf [accessed 3 August 2015].

Simon, D., Arfvidsson, H., Anand, G., Bazaz, A., Fenna, G., Foster, K., Jain, G., Hansson, S., Evans, L.M., Moodley, N., Nyambuga, C., Oloko, M., Ombara, D.C., Patel, Z., Perry, B., Primo, N., Revi, A., Van Niekerk, B., Wharton, A. and Wright, C. (2016) "Developing and testing the Urban Sustainable Development Goal's targets and indicators: A five-city study", *Environment and Urbanization* 28(1): 49–63.

Swilling, M. (2011) "Reconceptualising urbanism, ecology and networked infrastructures", *Social Dynamics* 37(1): 78–95.

United Nations (2015) "Transforming our world: The 2030 Agenda for Sustainable Development. Finalised text for adoption". Available: https://sustainabledevelopment.un.org/content/documents/7891TRANSFORMING%20OUR%20WORLD.pdf [accessed 3 August 2015].

Zellner, M. and Campbell, S.D. (2015) "Planning for deep-rooted problems: What can we learn from aligning complex systems and wicked problems?", *Planning Theory and Practice* 16(4): 457–478.

2

Growth and inclusion in the mega-cities of India, South Africa and Brazil

Patrick Heller

Introduction

Cities encapsulate the central social contradictions of global modern capitalism: increased returns to global connectedness and the hyper-commodification of land have become inextricably linked to new forms of social marginalisation, most notably the increasing informality of labour and life. The Southern mega-city of the twenty-first century brings the classic development challenge of reconciling growth and inclusion into dramatic relief. It also highlights the centrality of city governance and effective planning in coordinating and managing growth and inclusion.

A growing literature argues that in the era of neo-liberal globalisation, the local state plays an important role in promoting marketisation. In particular, by privileging growth sectors through processes of land-use planning and targeting of services and infrastructure, state interventions effectively ration the city's resources and condemn large majorities to informality. This chapter challenges this structuralist account of governance on two counts. First, I bring politics and institutions back in as independent explanatory concepts by showing that even when cities do indeed ration city resources and services in favour of growth and at the expense of inclusion, they do so in ways that cannot simply be read off from global pressures. Indeed, I contrast the cases of India and South Africa to show that there is tremendous variation in pro-growth regimes, both in terms of how they are politically constituted and the effects that they have. I then turn to a third case – Brazil – to show that even under very similar structural conditions, a particular configuration of state authority and civil society mobilisation can strengthen inclusive planning and move the pendulum from the rationed city to a more inclusive city.

The empowered and embedded city

Any discussion of governance capacity has to begin with the basic question of how much 'state' the city actually has. The developmental state literature has emphasised the Weberian attributes of a modern bureaucracy, or in Michael Mann terms, the infrastructural power of the state. But Mann also insisted that there is a second dimension of state capacity that has to do with the capacity of the state to issue authoritative commands. He moreover makes a key distinction between *distributive power* – the power of one actor to get another actor to do something they otherwise would not do – and *collective power*, "whereby persons in cooperation can enhance their joint power over third parties or over nature" (Mann 1988: 6). Collective power in this sense implies active planning and refers to a state's ability to resolve the collective action problems of social and private actors as well as managing the technical challenges of reconciling social, market and environmental needs. Collective power, or more specifically the power to coordinate, is the key to the success of inclusive planning. As conceived here, planning matters in two distinct senses: (a) the technical and institutional capacity to work across levels of government (national, provincial, regional and city) and governmental agencies to coordinate state actors and mobilise the necessary resources over time; and (b) the ability to work with societal partners to co-produce outcomes.

These institutional dimensions of power are politically constituted, that is, the product of historical processes of contention and political realignment that instantiate specific distributions of power. In its planning functions, the local state's planning capacity inheres not simply in its internal coherence and organisational structures, but also in the manner in which it is articulated with higher levels of the state (regional, provincial, centre) and with organised social actors. No local state can act effectively and in particular tackle problems of coordination without a certain degree of functional autonomy from higher levels of state power. This is true for two basic reasons: first, if local government is simply an extension of higher levels of government it is difficult to secure any degree of internal, inter-agency coordination at the local level. Second, if local government does not constitute a political space on its own terms, there is little room or incentive for locally organised actors to contest and co-ordinate at the local level. No local government, no local politics.

Finally, local states can have capacity and political autonomy, but in the absence of ties to subordinate actors, local states have neither the incentives nor the information and partners through which to promote inclusionary policies. First developed to explain the East Asian developmental state, Evans defines embeddedness as "a concrete set of social ties that binds the state to society and provides institutional channels for the continual negotiation and renegotiation of goals and policies" (1995: 12). Evans identified the East Asian state's close ties to emerging industrial elites as the key to its success in orchestrating economic transformation. But in re-rethinking the role of the state in promoting social inclusion the concept of embeddedness has to be broadened. As Evans and I have recently argued (Evans and Heller 2015), the extent to which the state is embedded in civil society is especially critical to underwriting the coordinated power that is essential to socially inclusive forms of planning, and in particular to delivering capability enhancing services such as basic education, health, housing and core infrastructure (water, sanitation, public transport).

An ideal type of democratic inclusive planning can now be summarised. In order to be effective, the local state must enjoy bureaucratic, fiscal and coordination capacity. Bureaucratic and fiscal capacity gives the local state distributive power. But development, both in terms of growth and inclusion, also calls for collective power. This coordination capacity depends critically on how and with whom the state is embedded. In the democratic ideal it is embedded in a broad-based civil society through both the representational mechanisms of electoral politics as well as more continuous, instituted and direct forms of accountability and co-production with its citizens.

In the rest of this chapter I put these concepts to work by analysing the efficacy and effects of urban governance in the *modal* mega-city in Brazil, India and South Africa. Within each country there is of course enormous internal variation, but on average, when mega-cities are compared across these countries, clear differences emerge and point to analytically useful typologies of the growth cabal (India), the growth machine (South Africa) and the social city (Brazil).

India: the growth cabal

The Indian city presents something of a paradox. On the one hand, Bhan (2013) has demonstrated just how planning – in the form of successive master plans in the case of Delhi – has left the majority of Delhi's residents with little option but to settle in informal slums and 'unauthorised' colonies where they receive limited services. On the other hand, if planning in India has served in effect to ration the city's resources, it has not underwritten a growth machine in Logan and Molotch's (1988) sense of the term. The Indian city is growing, but that growth is driven more by rents and speculation than by enhancing productive capacity and the sustainability of even this limited form of growth is threatened by massive failures of coordination and infrastructure delivery (Heller *et al.* 2016).

At the most basic level, planning in Indian cities has failed to deliver basic public goods. From sewage to transport and public health, the level and quality of public goods and social services – which was low to begin with – has fallen behind the pace of social and economic transformation. Overall levels of infrastructural investment in Indian cities have been low by any comparative measure (GOI 2011). As countless government reports have pointed out and as reformers have long argued, the problem here is certainly one of limited local governance capacity, but these problems in turn reflect the underlying institutional and political equation.

In contrast to the Brazilian and South African constitutions, the Indian constitution never specifically recognised local government in India as a third sphere of government. The Centre and States have never transferred significant authority or resources to cities.[1] As a result, the Indian city is not only almost entirely dependent on resource transfers from above but it also does not constitute a governed space, at least not in the formal sense of the term. The problem is even more acute in India's mega-cities. Because State-level politicians and bureaucrats have a strong incentive to retain control over urban wealth and votes, efforts at decentralisation have been systematically thwarted and urban 'governance' reduced to a patchwork of overlapping jurisdictions, fractured lines of command, uneven and highly contested forms of authorisation and porous legality.

Most notably, land development is a State subject and as a result almost all critical planning functions are controlled by State-level departments or actors. The resulting hodgepodge of line bureaucracies and executive bodies that answer to the State and Centre and municipal departments that have their own lines of command, has undermined both the institutional and political capacity for effective coordination. Planning as a result has been more geared to rationing than to inclusion. This is most evident in the pattern of layered exclusions that dominates the urban spatial form. Indian mega-cities are a mosaic of highly differentiated land tenure regimes ranging from fully legal and planned settlements to illegal slums, with many gradations in between.

In a context of such liminal governance, urban citizenship is highly compromised. Though legally constituted as citizens, urban denizens are "citizens without cities" (Appadurai 2002). On the one hand, citizens have at best very weak forms of representation. Representatives elected at the municipal level have very limited authority and mayors are ceremonial. Most of the day-to-day governance power lies in the hands of bureaucrats appointed by the State government (or the Centre in the case of Delhi) and all the significant legislative and fiscal power resides in State and Central level politicians. On the other hand, engagements with the local state are dominated by the dependency and clientelism that flows from insecurity of tenure and differentiated citizenship. There are few formally constituted channels or spaces for civil society to engage with municipal government.

If the modal mega-city in India is characterised as high growth and high exclusion, the outcome has as much to do with the nature of the urban regime in India as with the structural effects of neo-liberal globalisation. The predominance of informal networks of power, the fragmentation of governance capacity and the porous nature of the bureaucracy, coupled with the absence of both effective electoral and participatory counterpower, have underwritten what I call a 'growth cabal', which has two dimensions. The first is simply the increased power and influence of rent-seeking elites, especially those organised around land development, securing government contracts and assorted brokerage functions. The poor have a voice, but representative authorisation is fully subordinated to patronage politics and there is little institutional space for participatory engagement. This political-governance logic all but rules out the possibility of developing the broader partnerships and the forms of coordination that could make for a more inclusive city. But what is equally remarkable, and in a comparative sense rather unusual, is the extent to which Indian cities are not growth machines. Growth of course is taking place, but instead of resting on the state–business coalitions of the growth machine, which while exclusionary, nonetheless exert a form of collective power (Logan and Molotch 1988), it is fuelled by an assemblage of highly discretionary, negotiated transactions (as opposed to pacts) between state actors and elites around specific rent-seeking opportunities.

In the absence of coordinated state action, urban land in India is being valorised ("unlocking land values" in the language of state policies) but the process is far more discretionary, brokered and extra-institutional than the idea of a machine implies. The role of state land development authorities has in fact been described as more akin to the predatory state than the developmental state (Centre for Policy Research 2013). The distinction here is not moreover just between formal and informal, legal and illegal. It is also a distinction of political economy. Coordinated growth means that the process of developing land is managed, that is that services and infrastructure keep pace with growth, that planning effectively integrates land management, basic services and transportation networks, and that overall land development is aligned with long-term sustainable growth. In India, land appropriation and development as such is achieved against or despite – rather than in concert with – all efforts at planning. What is referred to popularly as the 'land-grab raj' is more growth cabal than growth machine.

South Africa: the growth machine and contained citizenship

South African cities are marked by two implacable legacies. The first is the spatial form of the apartheid city. The apartheid city was designed to segregate and exclude, and it is doubtful that a form of durable inequality has ever been so carefully and systematically engineered. But the apartheid city necessarily required enormous state capacities, not only of repression, but also for building carefully calibrated and impeccably differentiated systems of race-based housing and service delivery. As exclusionary as the apartheid city was, it was also highly coordinated, delivering cheap and docile labour to a capitalist sector and a generous welfare state to the ruling white population.

At the metropolitan level, South Africa is quite unique in the global South having inherited municipal structures that, in comparative terms, enjoy significant governance capacities and fiscal autonomy, especially in the three mega-cities of Johannesburg, Cape Town and Durban. It is even possible to talk of a local developmental state (van Donk *et al.* 2008). The democratic character of that state is another matter. If the South African city has been effective in promoting growth, a lack of embeddedness has limited its capacity to promote social inclusion and this despite a political commitment to 'deracialising' the apartheid city. The African National Congress (ANC) is a programmatic Left party, and could in principle have been a vehicle for an inclusive social democratic project. But because of its electoral dominance and its relative organisational insulation, the ANC has distanced itself from civil society and from the ideals of participatory democracy that that anti-apartheid movement embraced. Thus, at the level of the city, high capacity and coordination have sustained a growth machine based on a class coalition organised by the ANC, but in the absence of embeddedness in civil society, the inclusion project of the post-apartheid period has faltered.

South Africa's metropolises have the core capacities for inclusive planning. South Africa's largest cities collected 56 per cent of their own resources in 2008 (South African Cities Network 2011: 146). That the biggest chunk of these revenues come from property taxes (37 per cent) that still come overwhelmingly from the white community not only speaks to the local state's infrastructural power but also to its redistributive power. Cities have mobilised these resources to roll out services to the black townships, virtually closing the 'service delivery gap' between white and black areas (Kracker-Selzer 2012). The state has also subsidised and managed the delivery of close to 2.8 million houses for 11 million people between 1994 and 2009, a figure that is four times the volume of private sector housing (South African Cities Network 2011: 50).

But despite concerted efforts to 'de-racialise' the city through spatial planning and investment priorities, South African cities remain as segregated as ever. If any change has taken place it is that a slight decline in racial segregation (which under apartheid was almost absolute) has been countered by an increase in class-based segregation (Kracker-Selzer and Heller 2010; Schensul and Heller 2011). Spatial analysis of occupational groups in Johannesburg shows that the upper middle class of professionals are more concentrated than ever, and have quite literally cloistered themselves in the northern suburbs of the city even as Soweto has been virtually depleted of its own sizeable black professional middle class (Kracker-Selzer and Heller 2010). Private developers have dominated the spatial reconfiguration of the South African city by pushing large-scale, high-end greenfield developments of residential areas and shopping malls (Beavon 2004). Planners complain bitterly that city budgets have been eaten up by the need to provide bulk infrastructure to new suburban developments. This reflects market forces but also planning failures. The government's housing policy for the urban poor has reproduced the spatial form of the apartheid city (Harrison *et al.* 2008: 12) as developers, rather than housing officials, have dictated the location of new projects in peripheral areas of the city. Moreover, a failure to embrace policies for in situ development of informal settlements has seen the size of the latter grow (South African Cities Network 2011: 50).

The shortcomings of inclusive planning in South African cities reflect the limits imposed by market forces. But it also reflects the ANC's determination to exert political control and impose a technocratic vision of development on cities. In the name of efficiency and more rapid delivery, the ANC has managerialised decision-making processes and reduced the quality and scope of participatory processes created under the Reconstruction and Development Programme (RDP). The introduction of single-tiered metropolitan government into a Unicity structure that dismantled traditional white-dominated local authorities did significantly enhance the coordination capacity of city government. But it was also accompanied by the dismantling or hollowing out of community development forums and the insulation of the Integrated Development Planning (IDP) process from participatory inputs, marking the consolidation of a bureaucratic and corporatist vision of urban governance (Beall *et al.* 2002). The privatisation or out-sourcing of many government functions and increased reliance on consultants has virtually crowded out community structures.

This narrowing of the democratic space has moreover clearly led to a crisis of legitimacy. Since 2004 there has been a sustained explosion of urban protests. Some have taken the form of organised action to opposed service delivery cut-offs for non-payment and measures to privatise service delivery. But many have taken a much more inchoate and often violent form. A South African Police Service data source identified an average of 9,348 'gatherings' a year between 2004 and 2005 and between 2011 and 2012, including over 1,000 instances of 'unrest' in 2011–2012 (Von Holdt 2013). The press has dubbed these protests 'service delivery protests' but clearly they are much more political. Service delivery in urban areas has actually expanded rapidly, but the sense of social exclusion has grown, all the more so because of the strong rights-based political culture that drove the anti-apartheid movement. It is significant that many of the protests have targeted ANC ward-level councillors and been triggered by reports of local government corruption. Pithouse (2008) also rejects an economistic interpretation and argues instead that the protests are about "citizenship" understood as "the material benefits of full social inclusion . . . as well as the right to be taken seriously when thinking and speaking through community organization".

The South African city as such neatly fits the growth machine typology. The mega-cities of South Africa have experienced dramatic spatial transformation since the mid-1980s following a pattern that resembles the sprawl of US cities. In Johannesburg and Durban, inner cities have been more or less abandoned by white residents fleeing to gated suburbs and downtown Cape Town has been radically gentrified. As in India, land development has become a dominant source of rents for the private sector. A wide range of elite interests including developers, the ANC, new emergent business interests promoted by Black Economic Empowerment (BEE) state policies, an expanding service and information sector, a sophisticated banking sector inter-linked with mining interests and middle-class factions building enclaves of privilege have coalesced to provide a powerful pro-growth coalition.

If South African cities earn the label of 'machine', which escapes Indian cities, it is essentially for three reasons. First, the rentier coalition in South Africa is stable and carefully managed by a dominant party, the ANC. Varied interests within the coalition are well coordinated, and developer interests in particular have not come at the expense of business interests, as is the case in India. Second, land development in South Africa is organised, planned and supported by public policy and, compared to the Indian case, much more sustainable. Concretely, even though the development of the city has if anything exacerbated spatial inequalities, developments are legal, to code and integrated into the service delivery and transportation networks. Even though the urban poor are still spatially excluded, they do receive basic services. Third, if informal networks and deal-making are inevitably part of the machine, the pursuit of land rents is a relatively orderly affair, conducted largely within the purview of the law.

Brazil: decentralisation and the social city

Brazilian cities have until recently followed a path that has striking similarities to the patterns of urban exclusion that mark the South African and Indian city. Under the military-authoritarian regimes, the major cities were subject to highly bureaucratic and exclusionary forms of planning. Brazilian cities were moreover dominated by oligarchical parties that built vast and highly effective patronage machines. The result was highly rationed cities marked by levels of racial and class segregation that fully reflected Brazil's extreme levels of income inequality and its history of 'graduated' citizenship.

But beginning in the 1980s, broad-based urban social movements that were articulated with the wider democratic movement triggered a process of political transformation that reshaped the city in three fundamental respects. First, these movements directly challenged the clientelistic practices of political parties and cultivated a rights-based culture of "insurgent citizenship" (Holston 2008). Second, a diverse and autonomous set of civil society actors, ranging from activists planning NGOs to social movements in the health, housing and transport sectors experimented with new participatory structures, demanding a direct role in policy-making. Third, these movements played a direct role in framing the 1988 constitution, which fundamentally reconfigured the institutions of Brazilian democracy and the nature of state–society relations.

The new constitution gave municipalities a much more autonomous role in shaping local development. As del Rio notes, the constitution "relied on the city as a locus for the redistribution of wealth and the re-democratization of society" (2009: 25). Most concretely, it conferred upon municipalities "the power for the enactment of laws governing the use and development of urban space, in order to guarantee the 'full development of the city's social function' and the 'welfare of its inhabitants'" (Fernandes 2002: 120). A wide range of federal programmes that started under Cardoso and were extended under the Workers' Party (PT – Partido dos Trabalhardores) governments of Lula took the form of direct transfers to municipalities. As Arretche (2012: 16) points out, these were carefully calibrated to provide cities with resources while imposing standards of delivery. Formulas for fiscal allocations were not only insulated from political deal-making, but also reduced local revenue inequalities by favouring revenue-poor cities. 'Municipalisation' marked a dramatic reconfiguration of power between the centre and cities and was made possible, both politically and in practice, by a mobilised civil society that saw decentralisation as critical to building local participatory democracy and two fairly programmatic parties (Cardoso's PSDB and Lula's PT) who recognised that reforming the Brazilian state called for bypassing political elites entrenched at the State level.

The new constitution also embedded the city, through a web of legally mandated participatory structures, in civil society. This has produced tremendous variation in local state capacity and developmental outcomes. There is now an extensive body of research that shows that the design and effectiveness of a range of participatory institutions varies significantly across Brazilian cities (Baiocchi *et al.* 2011; Wampler 2015). But overall, and particularly in comparison to India and South Africa, the increased autonomy of cities and the greater opportunities for civil society engagements have supported the expansion of both federal and local policies of social inclusion. The resulting modal urban regime is best described as the social city, that is, a city in which there are concerted efforts, in large part aligned with national government initiatives, to address legacies of unequal access to the city, including basic services, infrastructure development and formalisation of the informal.

It is impossible to tell the story of the expansion of the Brazilian welfare state over the last two decades without highlighting the pivotal role played by cities. Because the transfers were initiated by two fairly programmatic parties (Cardoso's Social Democratic Party – PSDB – and Lula's PT), and because the strategy of building the welfare state explicitly bypassed political elites entrenched at the State level (Fenwick 2009), the influence of traditional clientelistic parties was significantly curtailed. The synergistic nature of this Centre–City alliance is nicely illustrated by the anti-poverty Bolsa Familia programme. On the one hand the federal government has been the driver, designing the programme and providing the resources. On the other hand, programme conditionalities have required active participation by municipalities and an expansion of their educational and health activities. The programme's extraordinary targeting success can be attributed to the close collaboration between the national implementing agency and city officials and the fact that the programme itself evolved through a scaling up of local experiments (Fenwick 2009).

Embedding the city

The demand for participatory democracy that emerged from the democracy movement was translated into novel institutional designs, many enshrined in the new constitution. The most significant participatory reforms came in the form of the various sectoral councils (health, transport, education, environment) that were mandated by the constitution. At the national level as well as in all municipalities the councils include representatives from sectoral interests, government and civil society. The councils are essentially neo-corporatist deliberative arenas with significant counter power, most notably the right to veto the allocation of federal monies to municipal budgets. Over 600 Brazilian cities have also now adopted some form of participatory budgeting (Wampler 2015: 262). But the most direct effort to curtail market forces comes in the form of the 2001 *Estatuto da Cidade* (City Statute), which not only "incorporates the language and concepts developed by the urban social movements and various local administrations since the 1970s" but requires that all urban policies be subject to popular participation and "introduces a series of innovative legal instruments that allow local administrations to enforce the 'social function'" (Caldeira and Holston 2005: 405–406). From a planning perspective, the Statute marks a significant consolidation of what were dispersed instruments of urban land use control, plan implementation and development control, land-tenure regularisation and special zoning for favelas (del Rio 2009: 31). While implementation remains problematic and the planning process is often dominated by developers, it has afforded civil society organisations new points of leverage over the planning process.

The transformative effects of embedding the local state in civil society have occurred along two axes. First, the institutional surface area of the state has expanded, increasing the points of access and coordination between civil society and the local state. The participatory budgeting process is a clear example, with citizens directly formulating the state's budgetary priorities. Participatory budgeting not only significantly democratised the budgetary process, but also markedly increased the access of civil society organisations (CSOs) to broader decision-making processes (Baiocchi *et al.* 2011). The record of sectoral councils has been more mixed, but across a range of policy arenas the councils have given movements critical points of leverage over local government. Second, these new institutional interfaces coupled with broader changes in political practices have also transformed the nature of engagement with the state.

It is possible to highlight some more direct evidence of the link between embedding and social outcomes, including evidence of co-production. A first example comes from the dramatic expansion of universal primary health care services. As Gibson (2014) has shown, this required significant resources and institution-building efforts from the federal government, but the key reform has been building and staffing community clinics in the neglected areas of cities. Gibson's data shows that there is significant variation across cities on this measure but that the outcome is positively correlated with the strength of the local Sanatarista (health) movement and the degree to which it has penetrated the local state. Similarly, Pires (2011) has collected extensive data on the formation of sectoral councils and the size of the local social state. He finds a very strong correlation between the degree of institutionalised participation (measured as the number of active councils) and increases in social and health personnel.

Conclusion

For global cities of the South the default mode might be said to be the growth machine. Given the pressures of heightened competition, increased wage arbitrage, capital mobility and globalised land markets, the urban form is more than ever subject to intense market forces. Much of the global cities literature has simply assumed that under these conditions urban governance and planning are mechanically hitched to growth imperatives, accelerating commodification and exclusion (Brenner and Theodore 2002). But this facile causal linkage completely discounts politics, and in particular underestimates the countervailing effects of democratic contention and the possibilities of more inclusive forms of planning. It cannot moreover explain the range of governance regimes I have identified in Brazil, India and South Africa.

In India, elite dominance of cities has certainly paved the way for urban policies that have unlocked tremendous land rents. But the resulting regime falls far short of a growth machine because it is neither coordinated nor capable of sustaining productive growth. The city lacks state capacity and political autonomy. As a result, it is embedded in society mostly through the particularisms of elite capture and popular clientelism. In the absence of coordination capacity, politics is dominated by a cabal of bureaucrats, politicians and developers, who extract enormous rents from the city. This is a cabal and not a machine because it is not built on a stable dominant coalition that harnesses the coordinating powers of governance but quite to the contrary thrives on destabilising governance and exploiting the resulting institutional failures to engage in a frenzied process of accumulation by dispossession.

In South Africa, a high capacity and relatively sovereign local state has provided a fairly stable platform for a growth machine. As the traditional sources of mining and industrial profits have been squeezed, capital has found new outlets in land development. The ANC has been the cement of this coalition, not only by promoting policies that have aligned the interests of white capital and a rising black middle class, but also by exerting tight control over the increasingly restive urban poor through a mix of populist policies, targeted patronage, nationalist discourse and selective repression. Because ANC political power trumps local sovereignty, possibilities for a more embedded and inclusionary urban regime that were promising at the time of transition have been dashed by a dominant party that has established itself as gatekeeper between society and an increasingly technocratic and insulated state. If the dominant urban regime is built on a solid and relatively efficient institutional platform and enjoys a significant degree of planning capacity, in the absence of broad-based embeddedness the local state can neither effectively coordinate across conflicting interests nor co-produce with civil society actors. In the absence of genuine avenues of participation, material grievances and growing inequality are leading to increasingly contentious forms of popular engagement with the local state.

Brazilian cities were long held up as paragons of fragmentation and social exclusion. Inequalities persist, but there is no denying the dramatic institutional and political transformations that the city has undergone. Broad-based urban movements that scaled up by penetrating and revamping the state and scaled out by building citizenship through participatory structures have transformed the nature of state–society engagements. As important as the federal government has been, it is an increasingly sovereign and embedded city that has been the critical locus of transformation. The success with which the "insurgent citizens" (Holston 2008) of the urban periphery have successfully demanded and secured a right to the city draws attention to the possibilities of participatory politics and more broadly is a reminder of how an active democracy can serve as a countervailing force to market pressures.

Capitalism is marked by social-spatial contradictions, and nowhere are these more pronounced than in the mega-cities of the twenty-first-century global South. The challenges of sustaining growth and promoting inclusion under conditions of heightened global competitiveness and accelerating commodification require an active capacity to coordinate not only *functionally* across sectors but also *politically* across and with interest groups. In this sense, the effectiveness of planning relies both on core institutional capacities, including a degree of decision-making that is properly scaled, as well as specific patterns of engagement with social actors. The variation in developmental trajectories of cities in India, South Africa and Brazil highlights the political nature of the institutional conditions for effective planning. In India, a highly centralised regime, supported by vested political interests, has resulted in a fundamental misalignment of democratic authority and institutional power that has all but eviscerated any possibility for effective planning at the level of cities. In the absence of effective planning, unco-ordinated and unfettered growth is exacerbating socio-spatial exclusions. In South Africa, cities have inherited a much higher level of resources and planning capacity and this has enabled cities to provide the basic infrastructural and regulatory conditions for growth. But in the absence of more institutionalised forms of engagement with the urban poor, these core capacities for planning and governance have not been extended to secure socially inclusive policies. In Brazil, the political developments of the last three decades have significantly leveraged the possibilities and effectiveness of planning. In general, democratisation of state–society relations has given the state fundamental new powers of intervention. Cities have acquired new resources and greater autonomy for local decision-making. But just as critically, the increased surface area of the local state and a range of rights-based forms of engagement have deepened and broadened the local state's embeddedness, enhancing its social function.

Note

1 The 74th constitutional amendment passed in 1993 has successfully enforced municipal-level elections, but other provisions to devolve authority and developmental responsibilities have been blocked at the state level.

References

Appadurai, A. (2002) "Deep democracy: Urban governmentality and the horizon of politics", *Public Culture* 14(1): 21–47.
Arretche, M. (2012) "State effectiveness in contemporary Brazil", *LASA Forum* XLIII(4): 15–17.
Baiocchi, G., Heller, P. and Silva, M.K. (2011) *Bootstrapping Democracy: Transforming Local Governance and Civil Society in Brazil*, Stanford, CA: Stanford University Press.
Beall, J., Crankshaw, O. and Parnell, S. (2002) *Uniting a Divided City: Governance and Social Exclusion in Johannesburg*, London: Earthscan.
Beavon, K. (2004) *Johannesburg: The Making and Shaping of the City*, Pretoria: University of South Africa Press.
Bhan, G. (2013) "Planned illegalities: Housing and the 'failure' of planning in Delhi: 1947–2010", *Economic and Political Weekly* XLVIII(24): 58–70.
Brenner, N. and Theodore, N. (2002) "Cities and the geographies of 'actually existing neoliberalism'", *Antipode* 34(3): 349–379.
Caldeira, T. and Holston, J. (2005) "State and urban space in Brazil: From modernist planning to democratic intervention". In A. Ong and S.J. Collier (eds) *Global Anthropology: Technology, Governmentality, Ethics*, London: Blackwell, pp. 393–416.
Centre for Policy Research (2013) *How to Govern India's Mega Cities: Towards Needed Transformation*, New Delhi: Centre for Policy Research.
del Rio, V. (2009) "Reclaiming city image and street livability: Projeto Rio Cidade, Rio de Janeiro". In V. del Rio and W.J. Siembieda (eds) *Contemporary Urbanism in Brazil: Beyond Brasília*, Gainesville, FL: University Press of Florida, pp. 224–245.
Evans, P.B. (1995) *Embedded Autonomy: States and Industrial Transformation*, Princeton, NJ: Princeton University Press.
Evans, P. and Heller, P. (2015) "Human development, state transformation and the politics of the developmental state". In S. Leibfried, F. Nullmeier, E. Huber, M. Lange, J. Levy and J. Stephens (eds) *The Oxford Handbook of Transformations of the State*, Oxford: Oxford University Press, pp. 671–701.
Fenwick, T.B. (2009) "Avoiding governors: The success of Bolsa Família", *Latin American Research Review* 44(1): 102–131.
Fernandes, E. (2002) "Providing security of land tenure for the urban poor: The Brazilian experience". In A. Durand-Lasserve and L. Royston (eds) *Holding Their Ground: Secure Land Tenure for the Urban Poor in Developing Countries*, London: Earthscan, pp. 101–126.

Gibson, C. (2014) "Developing health: Pragmatist civil societies and social development in urban Brazil", unpublished book manuscript, Simon Frasor University.

Government of India (GOI) (2011) "High powered expert committee report on Indian Urban infrastructure and services", New Delhi. Available: http://icrier.org/pdf/FinalReport-hpec.pdf [accessed 20 November 2015].

Harrison, P., Todes, A. and Watson, V. (2008) *Planning and Transformation: Learning from the Post-Apartheid Experience*, London: Routledge.

Holston, J. (2008) *Insurgent Citizenship: Disjunctions of Democracy and Modernity in Brazil*, Princeton, NJ: Princeton University Press.

Heller, P., Mukhopadhyay, P. and Walton, M. (2016) "Cabal city: Regime theory and Indian urbanization". Watson Institute for International and Public Affairs Research Paper No. 2016-32. Available: http://papers.ssrn.com/sol3/papers.cfm?abstract_id=2803570 [accessed 8 February 2017].

Kracker-Selzer, A. (2012) "Enhancing capabilities or delivering inequality: Contestation and service delivery in urban South Africa", PhD Dissertation, Department of Sociology, Brown University.

Kracker-Selzer, A. and Heller, P. (2010) "The spatial dynamics of middle-class formation in postapartheid South Africa: Enclavization and fragmentation in Johannesburg", *Political Power and Social Theory* 21: 171–208.

Logan, J.R. and Molotch, H.L. (1988) *Urban Fortunes: The Political Economy of Place*, Berkeley and Los Angeles, CA: University of California Press.

Mann, M. (1988) *States, War, and Capitalism: Studies in Political Sociology*, Oxford and New York: Blackwell.

Pires, R. (2011) *Efetividade Das Instituições Participativas No Brasil: Estratégias de Avaliação*. Brasilia: IPEA.

Pithouse (2008) "The University of Abahlali baseMjondolo". Available: www.libcom.org/library/the-university-abahlali-basemjondolo [accessed 11 February 2016].

Schensul, D. and Heller, P. (2011) "Legacies, change and transformation in the post-apartheid city: Towards an urban sociological cartography", *International Journal of Urban and Regional Research* 35(1): 78–109.

South African Cities Network (2011) "State of the cities report". Available: www.sacities.net/knowledge-centre/research/publications/25-what-we-do/socr/607-socr-2011-report [accessed 20 November 2015].

Wampler, B. (2015) *Activiating Democracy in Brazil: Popular Participation, Social Justice, and Interlocking Institutions*, South Bend, IN: University of Notre Dame Press.

van Donk, M., Swilling, M., Pieterse, E. and Parnell, S. (2008) *Consolidating Developmental Local Government: Lessons from the South African Experience*, Cape Town: University of Cape Town Press.

von Holdt, K. (2013) "South Africa: The transition to violent democracy", *Review of African Political Economy* 40(138): 589–604.

3

Urban planning at a crossroads

A critical assessment of Brazil's City Statute, 15 years later

Edesio Fernandes

Introduction: the City Statute and urban planning in Brazil

It is increasingly acknowledged that the combination of socio-spatial segregation and informality that has profoundly marked urban development globally has, to a significant extent, resulted from the exclusionary nature of prevailing urban legal systems. Policy makers, urban managers and social movements committed to the urban reform agenda ask a fundamental question: what does it take to turn national and local urban legal systems into effective factors of socio-spatial inclusion instead? A growing socio-political movement has vigorously argued that the promotion of legal reform is necessary to support urban reform. As a result, new urban laws governing land rights and management, territorial organisation, planning and housing have been recently enacted in several countries and cities, and serious investment has been made by institutions such as UN-Habitat and the World Bank towards the formulation and approval of inclusive urban legal systems. In this context, Brazil's celebrated[1] national urban policy law – the 2001 City Statute – has been widely regarded as a groundbreaking regulatory framework that is conducive to providing legal support to urban reform. Lauded internationally, the ambitious City Statute has been proposed as a paradigm to be considered internationally.

This law places special emphasis on urban planning. Long discredited following decades of nonexistence, irrelevance and inadequacy, urban planning was revived in the 1980s and 1990s, especially at the municipal level, as part of the intertwined processes of political democratisation and institutional decentralisation, and within the context of the emerging urban reform socio-political process. The 1988 Federal Constitution gave urban planning an enormous boost as it declared that property rights are only to be recognised when land and property fulfil those social functions determined by municipal master plans and other urban and environmental laws. This heralded a new chapter for planning, and for planners who were committed to changing the nature of the urban planning process. Traditional 'Urban Planning' became 'Inclusive Planning', and old 'master plans' became 'participatory master plans'. The City Statute entrenched these shifts and required municipalities to formulate new municipal master plans (MMPs) according to the new planning and management principles.

But, what exactly can be expected of these new urban laws? What is required for them to be fully enforced, and socially effective? What are the nature, possibilities and constraints of progressive urban laws vis-à-vis the broader socio-political process? What has effectively happened to this new wave of urban planning? This chapter explores these questions through a discussion of the City Statute. Almost 15 years have passed since its approval, and a comprehensive assessment of the urban land governance framework it proposed – and especially of the municipal initiatives that sought to implement the Statute's vision – is urgently necessary. Such a critical assessment of the conditions of the Statute's enforcement should provide important elements for the more general discussion on the expectations of newly approved urban laws, and promote a critique of the roles of all involved stakeholders, so as to correct mistakes, change courses and advance the urban reform agenda. Above all, this assessment is necessary to determine *if* and *how* the new generation of MMPs has effectively translated the general principles of the City Statute into rules and actions, as well as discussing what the main legal and social obstacles to the full implementation of the national law have been. It is also necessary to discuss *if* and *how* Brazilian society has made effective use of the many legal possibilities for the recognition of the range of social rights created by the new legal-urban order. In particular, this assessment has to take into account the broader context of the deep political-institutional crisis Brazil has experienced over the last three years – which has already led to the President's impeachment.

A new urban land-governance framework

The enactment of the 2001 law was the result of a nationwide process of social mobilisation. The City Statute regulated the original chapter on urban policy introduced by the 1988 Federal Constitution, which had itself been preceded by an unparalleled socio-political mobilisation, especially through formulation of the Popular Amendment on Urban Reform. I have discussed both the constitutional chapter and the City Statute in detail elsewhere (see Fernandes 1995, 2007, 2011; Fernandes and Rolnik 1998); for the purposes of this chapter, it should be stressed that the main dimensions of the City Statute are as follows:

- It firmly replaced the traditional legal definition of unqualified individual property rights with the notion of the social function of property so as to support the democratisation of the access to urban land and housing.
- It defined the main principles of land, urban and housing policy to be observed in the country.
- It created several processes, mechanisms, instruments and resources aiming to render urban management viable, with emphasis placed on the capture for the community of some of the surplus value generated by state action that has been traditionally fully appropriated by land and property owners.
- It proposed a largely decentralised and democratised urban governance system, in which intergovernmental articulation as well as state partnerships with the private, community and voluntary sectors are articulated with several forms of popular participation in the decision- and law-making process.
- It recognised the collective rights of residents in consolidated informal settlements to legal security of land tenure as well as to the sustainable regularisation of their settlements.

Together, these intertwined dimensions of the City Statute constituted a new urban land governance framework in Brazil.

Federal Law 10.257/2001 belongs within the context of a broader legal-urban reform process that has been ongoing for some 30 years. A new legal-urban order has been consolidated – sophisticated, articulated and comprehensive – including the constitutional recognition of

Urban Law as a field of Public Law with its own paradigmatic principles, namely, the socio-environmental functions of property and of the city and the democratic management of the city. The collective right to sustainable cities was explicitly recognised, and there is a clear commitment in the legal system to the urban reform agenda. These structural legal changes have been expanded at all governmental levels – federated states and especially municipalities. This new legal-urban order has been supported by the creation of a new institutional order at federal level, with the creation in 2003 of the Ministry of Cities; National Conferences of Cities have been promoted every two years since then; the National Council of Cities meets regularly; and Caixa Economica Federal – the world's largest public bank – has promoted several federal plans and projects, especially Plan to Accelerate Growth (PAC) and My House, My Life National Housing Programme (MCMV). All these combined amount to the largest social programme in the history of Latin America. Both the legal and the institutional orders are fundamentally social conquests, having largely resulted from a historical process of socio-political mobilisation involving thousands of stakeholders – associations, NGOs, churches, unions, political parties, universities and sectors of land and property capital – which since the late 1970s have claimed for the (rather late) constitutional recognition of land, urban and housing questions, as well as for the decentralisation and democratisation of, and popular participation in, law and decision-making processes.

Given the highly decentralised nature of the federative system, the materialisation of this legal framework was largely placed in the hands of the municipal administrations through the formulation of MMPs. Prior to the enactment of the new law, the majority of municipalities did not have an adequate regulatory framework in place to govern the processes of land use, development, preservation, construction, regularisation, etc. Most of them did not have basic information, maps, photos and other relevant materials either. Of the 1,700 municipalities that had a legal obligation to approve MMPs so as to apply the City Statute, some 1,450 have already done so, which is remarkable in itself.

Despite this, urban-environmental problems have worsened in the main cities, street protests and land and housing conflicts have increased in urban areas, and a growing number of people have questioned the validity of urban planning as a means of promoting socio-spatial inclusion, as well as the efficacy of the City Statute as a means of giving meaning to the constitutional notion of the social function of property.

Since the enactment of the City Statute, cities have undergone significant changes. The rates of urban growth have decreased but are still relatively high, especially in middle-sized and small cities, thus leading to the formation of new metropolitan regions. Economic development and the emergence of a so-called 'new middle class'/'precarious working class' have aggravated further long-standing urban problems of transportation, mobility, environmental impact and urban violence. Infrastructure and energy provision problems have increased, and the fiscal crisis of the public administrations is widespread, especially at the municipal level. Above all, the land and housing crisis has escalated. The housing deficit remains enormous (between 6 and 7 million units), and, despite the impressive number of units already built/contracted, MCMV has not reached the poorest families and has been criticised for reinforcing long-standing processes of socio-spatial segregation. While the levels of land, property and rental appreciation – and speculation – have broken historical records, there is an enormous stock of vacant serviced land, abandoned/under-utilised properties (calculated as 5.5 million units), as well as of public land and property without a social function. Informal development rates are still high, with the densification/verticalisation of old settlements and the formation of new settlements usually in peripheral areas – while it has also been taking new shapes – backyarders, informal rental transactions, etc. The proliferation of gated communities in peripheral areas/other metropolitan municipalities means that for the first time, rich and poor are competing for the same space. Urban development in the new economic frontiers – especially in the Amazon – has largely taken place through informal processes, and there are a growing number of land disputes and socio-environmental conflicts.

Moreover, over the last two decades or so, significant public resources – land, fiscal incentives, various credit, tax exemptions, building and development rights – have been given to land developers/urban promoters/builders, usually within the context of urban renewal/revitalisation programmes, rehabilitation of downtown areas/historic centres, large scale projects and modernisation of harbours/ports/infrastructure. The number of recent forced evictions – the World Cup was estimated to have evicted 250,000 people alone – is staggering, not only in Rio de Janeiro and São Paulo, but even in municipalities such as Belo Horizonte and Porto Alegre that were long committed to the urban reform process. This process, which was so vivid in the 1980s and 1990s, seems to have lost momentum, and stakeholders have questioned who has benefitted from the transfer of public resources. They have increasingly denounced the growing process of property speculation; the elitist utilisation of the enormous amount of financial resources newly generated especially through the sale of building and development rights in public auctions; the way the so-called 'unlocking of land values' by large projects and events has reinforced socio-spatial segregation; the recurrent abuse of the legal arguments of 'public interest' and 'urgency'; and the enormous socio-environmental impact of federal programmes and others.

Growing land conflicts, rental prices, urban informality, numbers of evictions and removals, worsening of transportation, mobility and sanitation problems, but especially a growing process of commodification characterise Brazilian cities, which are currently both venue and object of post-industrial capitalist production. This new stage of urban development and financialisation of cities has required the strengthening of the individualist and patrimonial legal culture that had long prevailed prior to the enactment of the City Statute: property viewed merely as a commodity; consideration of exchange values but not of use values; and the right to use/enjoy or dispose of property often meaning the right not to use/enjoy/dispose of – in other words, to freely speculate.

What has happened to the urban reform process? How to explain the growing gaps between the progressive new legal order and the exclusionary urban and institutional realities? The legal-urban order is still largely unknown to jurists and society, when not objects of legal as well as socio-political disputes. The legal and social efficacy of the implementation of the City Statute remains a challenge. There is also a gap between the institutional order and the urban and social realities. The Ministry of Cities has often been preempted or bypassed by the federal budget or by other ministries, and the National Council of Cities has often been bypassed by Ministry of Cities or other Ministries, having had difficulties to renew the levels of social mobilisation. When there is not a lack of projects, duplicity, inefficiency, waste, lack of continuity and corruption have marked the fragmented urban management at all governmental levels.

It is in this context that there is a growing skepticism among planners, managers, academics and society regarding the City Statute. The federal law has been demonised by some, who have blamed it for recent processes of socio-spatial segregation, the fact that the new urban management tools have been appropriated by conservative sectors, as well as for the fact that new forms of old processes of "socialisation of costs and privatisation of benefits" have emerged with the re-concentration of public services and equipment. Has the City Statute failed, as the skeptics believe? Rather than contributing to the promotion of socio-spatial inclusion, has it perversely contributed to the current escalating process of commodification of cities, and to the further peripheralisation of the urban poor, as some have argued?

Is the critique legitimate? An assessment of MMPs

The new legal-urban order consolidated by the City Statute placed law and planning at the heart of the socio-political process, especially at the municipal level, and it is the very quality of this process that will determine the meanings and reach of the notion of the social function of property. It is unquestionable that, for all its sophistication and successive developments, the legal-urban order still has significant limits: there are several bottlenecks in the judicial system, including the length and costs of judicial procedures; the difficulties with the registration system remain challenging; MMPs have not been articulated with an adequate urban management system; municipalism is exaggerated and often artificial, and there is not a properly defined metropolitan/regional dimension; and the different realities of middle-sized and small municipalities, and especially the different realities of North and North-East, have not yet been properly contemplated by the legal order. A crucial aspect is the reduction of the notion of spatial organisation to that of 'municipal' organisation, 'local' government being reduced to 'municipal' government – when it should be at least 'metropolitan'. Nonetheless, the progress of the legal-urban order is undeniable.

It is in this context that one should ask: is the federal law the real problem? Or, has there been an adequate understanding of the new legal-urban order by lawyers, urban planners, public managers and society? Have the new legal and politico-institutional spaces been occupied? Have the new legal principles been translated into urban policies? Have the new legal rights been claimed by the population? Have the new legal principles been defended by the judicial courts?

There are some important surveys, case studies and comparative studies already available, especially Cymbalista and Santoro (2009), Santos Jr. and Montandon (2011) and Schult *et al.* (2010). There are also several published case studies, and a "Bank of Experiences" has been created by the Ministry of Cities.[2] These existing studies have clearly shown that there has been progress on many fronts: the general discourse of urban reform has been adopted by most MMPs; specific sectors – environment, cultural heritage – have been addressed; Special Zones of Social Interest (ZEIS) have been created in areas occupied by existing informal settlements; and, whatever the variations, the participatory nature of the discussion of MMPs was remarkable. Perhaps the main achievement has been the production and recording of data about cities.

However, there are several problems of legal efficacy undermining the new MMPs: excessive formalism and bureaucracy of municipal laws; requirement of further regulation by several subsequent laws for full enforcement; punctual changes have been promoted without participation; and both the obscure legal language and the imprecise technical legal writing (urban laws are rarely written by legal professionals) have widened the scope for legal and socio-political disputes. There are also serious problems of social efficacy undermining the new MMPs: most plans remain 'traditional', merely technical and regulatory, often failing to territorialise the proposals and intentions, or to intervene in the land structure and the land and property markets. The emphasis on the new tools has been placed without a clearly defined project for the city. The majority of MMPs have failed to recapture any surplus value resulting from state and collective action, and when this has happened, there has been no or limited social redistribution of the newly generated financial resources. Moreover, most MMPs have placed little to no emphasis on social housing in central areas, having failed to earmark central, serviced, vacant land for social housing. Generally speaking, there are no specific criteria for the expansion of urban zones; public land and property have not been given a social function; and there has been no clearly articulated socio-environmental approach. Large projects have often bypassed MMPs, and presumed collective eviction. Above all, land, urban, housing, environmental, fiscal and budgetary policies have not been integrated, and the regularisation of informal settlements is still largely viewed as an isolated policy, with most MMPs imposing enormous technical difficulties to the legalisation of informal settlements. Bureaucratic management and technical complexity have also meant that there has been a widespread lack of administrative capacity to act at municipal level. Many MMPs are copies of models promoted by an 'industry' of consultants. Obscure planning language has been as problematic as obscure legal language.

But, it is the country's broader politico-institutional context that requires further understanding.

The City Statute in context

Since the 1980s, significant progress has been made towards the creation of a legal framework to govern urban development nationally. However, both the promising progress at the municipal level, especially in the 1990s, and the reach and implications of the more recent federal laws and relevant federal programmes have, over the last 15 years, been undermined by the tensions inherent in the country's politico-institutional system and renewed socio-political disputes within civil society, as well as jeopardised further by the ongoing political crisis that has led to the President's impeachment. From the perspective of the urban reform agenda, this last period has been particularly difficult as there has been a gradual, notable backlash at all governmental levels insofar as the articulated processes of urban development, policy, planning and management are concerned. The enormous public investment in cities made by the government especially since 2003 has been jeopardised, if not partly wasted, by the lack of a clearly defined and integrated conceptual framework and a corresponding institutional context governing the overall treatment of the 'urban question'. Far from redressing long-standing urban, social and environmental problems, the nature of governmental action at all levels has worsened the pattern of urbanisation, that is: a perverse pattern of combined socio-spatial segregation, environmental degradation, economic inefficiency, fiscal crisis, administrative irrationality, social insecurity, as well as rampant land and housing informality. The current process of 'urban spoliation' can no longer be blamed on the lack of laws, planning, or financial resources.

The last 15 years were strongly characterised by intertwined processes of politico-institutional tensions and socio-political conflicts, as well as being increasingly marked by a manifest conceptual socio-political dispute regarding the definition of *what cities are, for whom they are managed, who makes decisions and how, and who pays, and how*, for the financing of urban development. Federal policies on and in cities were, and remain, sectoral, isolated and fragmented, thus reflecting the same institutional fragmentation that exists among several federal ministries and even within the Ministry of Cities itself. In conceptual terms, the federal government has failed to understand what cities really are in the contemporary world, especially given the current stage of post-industrial capitalism in the country. Nor has the federal government understood the nature and implications for the country of the ongoing process of urban development, especially within the context of rapidly globalising land, property and rental markets. While most political parties, politicians and managers do not have or follow a clearly defined urban policy, at the federal level especially investment in cities has been viewed merely as a means of 'creating infrastructure for economic development' and/or 'formulating social policy'. The enormous investment of public resources, especially through the PAC and MCMV, has taken place without the previous definition of an integrated land, territorial, and urban national policy. The equally massive Bolsa Familia social poverty eradication/income redistribution programme has also been largely conceived without a solid understanding of its impact on urban areas and on the overall pattern of urban development and management. Moreover, several economic policies – such as incentives to the national automobile manufacturing industry – were implemented with little understanding of their impact on cities, while other economic development policies – such as the construction of a system of dams – had little concern for their environmental impact. The institutional and legal action of the federated states has also been very limited.

The fact that decreasing poverty rates co-exist with growing informal development rates has clearly shown that poverty can no longer be viewed as the sole reason for informal development, or even the main one. Other factors to be considered have largely to do with the nature of the territorial organisation order and its relation to the land structure: there lies the reason for the state's structural inability to provide accessible, adequate, sufficient, well-located and affordable access to serviced urban land and housing. The current public services crisis has demonstrated that the building of walls, imaginary or concrete, is insufficient to protect the more privileged socio-economic groups. If in the past the lack of basic sanitation only affected the urban poor, today's public health crisis has no borders, especially given the combined impact of the ongoing Zika, Dengue and Chikungunya pandemics. Even if the urban poor remain more directly affected, the impact has been widely felt. The same applies to the growing failure of other public services and infrastructure systems, such as the widespread energy crisis, especially given the saturation of the electricity and water provision models. The serious problems now regularly experienced by the urban population concentrated in São Paulo, Minas Gerais and Rio de Janeiro, among others, tend to be aggravated further by severe droughts and other consequences of extreme weather patterns. And yet, urban policies, plans, projects and laws at all governmental levels have failed to seriously take into account these new environmental scenarios, and economic development is still promoted in an unqualified manner.[3]

All in all, the inadequacy of the politico-institutional system and resulting governance processes is manifest, from the lack of a metropolitan sphere, intergovernmental articulation and a national territorial policy/system of cities, to conflicting 'green' and 'brown' agendas and inefficient environmental policies regarding coastal protection, river basins, vegetation, global warming, gas emissions and the Amazon.[4] At the root of these problems is the lack of a comprehensive and articulated land policy. The 'urban question' has been up for grabs, when it has not been auctioned off by the government.

Had the federal government understood that, perhaps the current political crisis would not have taken place. Amid all the uncertainties, two things are certain: the street demonstrations have an urban nature and – consciously or not, directly or indirectly – they are ultimately about the nature of the social process of production of urban space. Even when they are conveyed in specific or narrow terms, their claims ultimately address and condemn the general urban development pattern: socio-spatial segregation that affects the 84 per cent urban population (Worldometers 2016); the increasing peripheralisation of the urban poor; the concentration of public services, equipment, facilities and opportunities; growing taxation and limited access to public services. Cities are the socio-spatial expression of an exclusionary and perverse socio-political pact.

Has this exclusionary urban development pattern resulted from the lack of urban planning, as many have argued? The answer is no. Brazil's urbanisation has largely been a state-led process. What is at stake is the kind and nature of traditional urban planning (theory, education and, above all, practice), which has long been viewed merely as a 'technique of territorial organisation', as well as being 'neutral' and 'objective' socially and politically. The fragmented approach has dissociated urban policy from land policy and housing policy, as well as from transportation, environmental, fiscal and budgetary policies. Urban planners and managers have no understanding of the exclusionary and speculative dynamics of the property markets they create, seeing themselves as poor hostages of such aggressive markets. As a result, this elitist planning tradition has led to informality. Moreover, there is no attempt to share with the community some of the significant surplus value resulting from state action (through public works, services and urban laws) and, when there are attempts, as in the case of São Paulo, they reinforce socio-spatial segregation as most of the newly gained resources are invested in the same areas in which there is already a higher concentration of services and equipment. Mistaking effects for causes, well-intended governmental actions have had bad effects: as described, the social programme implemented by the federal government has already built over 2 million houses in precarious peripheries, and costly, but isolated, regularisation programmes have led to higher land, property and rental prices and thus to evictions.

Urban planning has also long been dissociated from urban management. Lengthy bureaucratic procedures, the lack of intergovernmental articulation as well as of transparency and accountability together have significantly contributed to the current urban crisis, particularly at the local level, given the widespread lack of municipal capacity to implement more complex proposals. Although nominally recognised and even required by the legal-urban order in force, popular participation has not taken place in all stages of decision-making and it has often been manipulated, reinforcing the long-standing tradition of political patronage.

The period was also marked by a national context of intense, excessive and rather artificial politico-institutional decentralisation, as well as by tense, volatile and largely manipulated intergovernmental articulations. While their legal obligations have increased over the years, municipalities still fundamentally lack the capacity to act and thus to formulate, implement and monitor urban policies. The lack of sufficient resources explains only part of the problem, however, as there has also been widespread poor utilisation of existing fiscal and financial resources by municipal administrations. Municipalities have often opted for maintaining their dependence on federal and federated-state financial transfers for political reasons, especially regarding their hesitation to make full use of their legal power to implement land and property taxation and thus antagonise powerful local groups. By doing so, they have reinforced historical patterns of political patronage, as well as renovating long-existing dynamics of political clientelism.

In this confused politico-institutional context, only in recent years has there been a more critical discussion on the inadequacies of the prevailing 'federative pact' and the shortcomings of the formal municipalism in place, thus questioning the notion that the role of federal government is merely to provide financial support to municipalities. According to this still dominant view, even the sizeable federal resources from both PAC and MCMV have to be spent by municipalities, and by municipalities alone. In many cases, significant financial resources have been unused or have even returned to the federal government given the lack of municipal capacity to formulate projects, open tendering processes and monitor the implementation of public contracts. In cases where NGOs and the private sector have been included in the process of urban management, through public–private partnerships (PPPs) and other schemes, mismanagement and corruption have also been common. There has been a growing call for the formulation and implementation of truly national policies to govern the phenomenon of urban development and all its implications and consequences.

To complicate a chronically difficult situation even further, for all its long-standing shortcomings, contradictions and constraints, the federal government actions in cities have been severely affected by the ongoing, and worsening, political, institutional, economic and fiscal crisis that virtually paralysed the country from January 2015, when President Dilma Rousseff took office to undertake her second mandate, until her impeachment in August 2016. This is a lamentable situation that has already had serious implications for the existing set of sectoral federal policies: the prospects for the future of the country's patchy national urban policy are now even worse, as ultimately what is at stake at the core of the impeachment process are two different national socio-political projects with the one opposing the Rousseff government being more unambiguously in favour of neoliberal policies, privatisation schemes, flexibilisation of labour relations, unqualified property rights and the deregulation of economic activities and urban development processes.

In 2006, the Ministry of Cities had already been sacrificed in the name of 'governability', handed over to a conservative party, and gradually turned into little more than a 'clientelistic business desk'. While the new President has promptly signalled his support for further orthodox fiscal and economic austerity measures, several decisions taken by the elected government over the last 12 years, especially those of a social nature, have been revoked without any discussion or consultation. This backlash has meant that significant cuts have been announced to the main federal programmes, including Bolsa Familia, PAC and MCMV. The newly appointed Minister of Cities seems to be strongly committed to a widespread programme of privatisation, PPPs and support to land developers and property promoters. His very first measure has been to abolish the dimension of MCMV that encouraged collective self-construction by housing cooperatives and residents' associations. Such changes have been met with increasing hostility on the part of a growing number of people and organisations, and socio-political protests have occurred. As a result, some decisions have been fully or partly reversed or suspended, worsening the general political uncertainty. It is impossible to say with any degree of clarity what the future might hold, although new socio-political pacts are forming and the level of social mobilisation has gradually decreased. It can be expected that Brazilian cities will have an even harder time in the near future. While daily revelations of the extent of corruption practices have gripped the nation's attention, the political crisis has worsened the economic and fiscal crises.

Conclusion

The last 15 years have been marked by ups and downs, advances and backlashes, euphoria and depression, but the current mood is one of uncertainty. While there has been greater public intervention in urban areas at both the federal and the municipal levels, especially through a number of laws, programmes and plans, the lack of a consistent, articulated urban policy framework has undermined much of the efforts, often reinforcing patterns of waste, inefficiency, exclusion and segregation, rather than promoting productivity, sustainability and inclusion. Long-existing structural problems, obstacles and bottlenecks have not been removed, thus determining the limited, elitist nature of urban governance, and long-standing tensions and disputes have gradually come to the fore. Given the evolving, spiraling political and economic crises, much of what had been achieved – especially through poverty reduction policies – has turned out to be fragile.

The post-City Statute confirmation of old socio-spatial segregation processes at all governmental levels, despite the possibility of significantly changing the course of things through the formulation of profoundly different and inclusive MMPs, seems to demonstrate that, with the support of legal professionals, urban planners and public managers remain, and have seemingly increasingly become, hostages to exclusionary land and property markets that they have created and fomented in the first place, as well as to segregating public policies that they have implemented. To break with this perverse logic, a concentrated effort needs to be promoted to provide more information – as well as better formation – to planners and legal professionals, judges, prosecutors and registry officers, as well as society as a whole, on the nature and possibilities of the new legal-urban order that the City Statute symbolises. If judicial courts need to follow Urban Law principles when interpreting property-related conflicts, rather than embracing obsolete unqualified private law ideas, civil society also needs to claim more vigorously for the recognition of social and collective rights.

Brazil's legal-urban order has significantly changed, but have the jurists understood that? Has the nature of urban planning been changed accordingly? Have urban managers assimilated the new principles? Has civil society awoken to the new legal realities? The answer is no. To play the game according to the new rules is fundamental for the collective construction of sustainable and fairer cities for the present and future generations. The future of the City Statute requires a thorough renewal of the socio-political mobilisation process around land, urban, housing and environmental matters so as to advance urban reform nationally. It is a task of all to defend the City Statute from the proposed, essentially negative, changes being discussed at the National Congress, overcome the existing obstacles and improve the legal order further, but above all, to fight for its full implementation. However, the street demonstrations have told an important cautionary tale: legal reform is not sufficient. For all the undeniable progress towards confronting poverty and inequality, there is still an enormous amount of work to be done to redress several forms of historical injustices, to provide better public services and to promote effective inclusive socio-economic and urban policies. This requires an articulated set of public policies, ranging from the creation of a truly redistributive tax system – in a country dominated by regressive and indirect taxation – to more incisive land policies.

If 'bad laws' can make very difficult both the recognition of collective and social rights and the formulation of inclusive public policies, 'good laws' per se do not change urban and social realities even though they express principles of socio-spatial inclusion and socio-environmental justice, or even, as is the rare case of the City Statute, when the legal recognition of progressive principles and rights is supported by the introduction of the processes, mechanisms, tools and resources necessary for their materialisation. If decades of socio-political disputes were necessary for the reform of the legal-urban order and for the enactment of the City Statute, a new historical stage has been opened ever since, namely that of the socio-political disputes at all governmental levels, within and outside the state apparatus, for its full implementation. The fact is that Brazil, and Brazilians, have not yet done justice to the City Statute.

Once the current political crisis is over, the promotion of urban reform will take time and will require continuity and systematic responses at all governmental levels in order to address existing problems, as well as other fundamental factors such as capacity building, approval of articulated policies according to a clearly defined urban agenda and the allocation of the necessary resources. There is still a long way to go and serious obstacles to overcome. The rules of the game have already been significantly altered; what remains to be seen is whether or not the newly created legal and political spaces will be used in such a way as to advance the urban reform agenda in the country. The right to the city is still to be conquered. There are many important lessons there for scholars, policy makers, managers and activists elsewhere.

Notes

1 The law was approved following 12 years of intense discussion and fierce disputes within and outside the National Congress. Now acclaimed internationally, in 2006 Brazil won UN-HABITAT's Scroll of Honour for having approved the law.

2 For more information, see www.cidades.gov.br/index.php/planejamento-urbano/392-banco-de-experiencias [accessed 2 November 2016].

3 For example, public works such as the Belo Monte dam and many PAC-sponsored projects do not have a clear environmental impact dimension, and the deforestation of the Atlantic rainforest, as well as of the Amazon, continues at alarming rates.

4 The largest socio-environmental disaster in Brazil's history has developed for the last nine months – initially in Mariana, Minas Gerais, as a result of the waste water released by the bursting of a dam, which subsequently affected 40 cities of two federated states, a large stretch of the Atlantic Ocean, the valley of the Rio Doce, several communities and economic activities and has yet to receive proper governmental attention.

References

Cymbalista, R. and Santoro, P.F. (2009) *Planos Diretores: Processos e Aprendizados*, São Paulo: Instituto Polis.
Fernandes, E. (1995) *Law and Urban Change in Brazil*, Aldershot: Avebury.
Fernandes, E. (2007) "Constructing the 'right to the city' in Brazil", *Social and Legal Studies* 16(2): 201–219.
Fernandes, E. (2011) "Implementing the urban reform agenda in Brazil: Possibilities, challenges, and lessons", *Urban Forum* 22(3): 299–314.
Fernandes, E. and Rolnik, R. (1998) "Law and urban change in Brazil". In E. Fernandes and A. Varley (eds) *Illegal Cities: Law and Urban Change in Developing Countries*, London and New York: Zed Books, pp. 140–156.
Santos O.A. Jr and Montandon, D.T. (eds) (2011) *Os Planos Diretores Municipais Pos-Estatuto da Cidade: balanco critico e perspectivas*, Rio de Janeiro: Observatorio das Metropoles/Letra Capital.
Schult, S.I., Momm, S.C. and Souza, L.A. (eds) (2010) *Experiencias em planejamento e gestao urbana: Planos Diretores Participativos e Regularizacao Fundiaria*, Blumenau: Edifurb.
Worldometers (2016) Brazil Populative (live). Available: www.worldometers.info/world-population/brazil-population/ [accessed 2 November 2016].

4

African urbanisation and democratisation

Public policy, planning and public administration dilemmas

Dele Olowu

Introduction

Although urbanisation is occurring rapidly, the democratisation and urbanisation processes in Africa remain disconnected and, in most cases, actually work contrary to each other. This chapter argues for a different approach by politicians and planners.

African countries have urbanised rapidly since their political independence, mostly in the 1960s. For much of this period, the dominant political and management philosophy articulated by most political leaders and by development partners and advisers from outside the continent was a highly centralised development management paradigm, referred to as monocracy (Ostrom 1990; Wunsch and Olowu 1990). As much of the rest of the world swung in favour of democratisation since the late 1980s, African countries, especially those south of the Sahara, also moved in tandem, with the southern and northernmost tips joining later. Democratisation and its twin counterpart, decentralisation, have worked in opposing directions and have thus failed to build the critical fiscal and political incentives to make African cities political and economic drivers of change and dynamism throughout the continent.

The chapter first defines the key concepts, then provides evidence that democratisation and urbanisation have worked in contradictory directions. Finally, it proposes suggestions of how this could be different if political, administrative and policy advisers, and planners at regional, national and local levels, would seek to connect African urbanisation and democratisation differently, using the prism of decentralisation. The chapter provides three cases that illustrate the problems and possibilities on this continent.

Urbanisation and democratisation in Africa

In spite of arguments that urbanisation estimates should be revised downward (Potts 2012), African levels and rates of urbanisation are second only to those in Asia. In comparison with other world regions, Africa has the lowest urbanisation levels but its rate of urbanisation remains higher than most other parts of the world. The proportion of those who live in urban centres had increased to 40 per cent by 2009 and the United Nations Department of Economic and Social Affairs (2014) projects that this will be 56 per cent by the year 2050. This would change the profile of the sub-continent from a rural to a mostly urban population.

In contrast to the historical experiences of other parts of the world in which urbanisation is associated with increasing levels of employment, income, manufacturing, economic and social infrastructure, over 60 per cent of Africa's urban population live in slum conditions. However, the proportion of those in slums has fallen from a high of 71.9 per cent in 1990 to the present rates of 61 per cent (Stren 2014). Further, new types of urban-based activities are being generated by the predominantly informal economy that demonstrates entrepreneurship and resourcefulness of especially the youthful population that dominate in Africa's cities. The small- and medium-scale operations have become major employers of labour and have also managed to combine informal connections between governance, politics, religion, family life and business to create and sustain networks for problem-solving in the face of inadequate institutional and financial structures in many African cities (Simone 1999). They also used these networks to produce and sell their products (Turok 2014). For instance, by 2010, 90 per cent of the real estate valued at USD 48.2 billion in Lagos metropolis was in the informal sector (Lagos State 2012: 21).

There are also community-based associations and other groups, especially faith-based organisations, that are making crucial contributions to social production in the cities: some of them run schools, electricity, water and even have their own self-governing communities. They excel and succeed where the public and even private providers have failed (Ter Haar 2011).

The involvement of both the formal and informal sectors in the production of critical infrastructure and services would make urbanisation a more efficient, sustainable and inclusive process, building on the experiences of countries within and outside the region. Sadly, the discourse from the public managers has not supported this. The tendency in policy declarations, writings and approaches to tackling these multiple challenges is to come up with a litany of diverse proposals, many of which are more theoretical than realistic. For instance, the regional development bank (African Development Bank 2012), in addressing the African urban challenges, urged governments in the region to do the following:

- Upgrade informal settlements through the provision of integrated infrastructure services that target the marginalised groups including the poor, youths, women, elderly people; and to define and implement clear urban development strategies.
- Mobilise urban financing from local and foreign investors, efficiently allocate such resources between central and local governments' urban projects and also strengthen municipalities.
- Improve the human capital through equal access to education and health care services for all categories of citizens in order to meet labour market needs.
- Diversify economic activities through the creation of new economic hubs oriented towards sustainable and value-added production and exploitation. These reforms should be more inclusive, embracing all citizens irrespective of their age, gender, ethnicity and socio-economic conditions.

If the governments could do all of the above, it seems evident that the cities would not be in such a poor state.

In view of this disconnect at practical and discourse levels on African urbanisation, I argue that if properly conceived and implemented, two key reforms that are already popular on the continent could make a significant difference to how Africa's cities and urbanisation are managed. These two reforms have been at the forefront of African development discourse and practice especially since the 1980s. These are democratisation and decentralisation. Unfortunately, there has been a total disconnect in the implementation of these policies and urbanisation on the continent. This must change if future outcomes are to be different.

Democratisation and the management of African cities: the record of African decentralisation

Democratisation and African cities

In the late 1970s and most of the 1980s, many African countries experienced different forms of autocracy – traditional, civilian and military – after a few years of post-independence civilian democracy. However, with the end of the Cold War and the fall of the Berlin Wall, a fresh round of democratic experiments took place on the continent. Civilian and military authoritarian governments gave way to civil democracies and many countries experienced successful replacement of incumbents at the ballot box at national and sub-national levels. For instance, by 1989, only 3 countries (6 per cent) in sub-Saharan Africa were classified as free (judging by electoral and civil liberties) and 33 countries (or 70 per cent) as not free. By 2000 the comparative data had jumped to 9 countries (19 per cent) free, 24 (or 50 per cent) partly free and 15 countries (or 31 per cent) not free (Freedom House 2015).

There are many reasons why cities should have been important centres of democratisation. First, they have a well-motivated population that understand and are readily mobilised for the political process, even by national politicians. Second, there are substantial economic resources to pursue political agendas and the middle classes are concentrated here (Resnick 2014). Unfortunately, however, the medium by which the democratic process should have been translated to the local and urban centres is through the reform processes associated with decentralisation, which is discussed next. This process has unfortunately not worked well to support the democratisation of the cities for a variety of reasons.

The interface between democratic decentralisation and the management of African cities

Decentralisation is an institutional reform that consists usually of policies designed to transfer responsibilities, resources, accountability and authority from the central government to a sub-national authority. The programme takes one of three main forms as shown in Table 4.1.

Theoretically, the degree of control or autonomy that sub-national governing institutions have over the four key elements transferred to them (authority, responsibility, accountability and finances) determines the form of decentralisation: deconcentration, delegation or devolution. Decentralisation can be introduced comprehensively or sectorally, in a staggered manner or with an across the board 'big bang' in a country with all actions expected to take place simultaneously. There are advantages and disadvantages to each of these policy choices.

The three main components of decentralisation are: administrative, fiscal and political decentralisation (World Bank 2004).

Most countries in Africa have had decentralisation programmes going on for two or three decades or more. The reality, however, is that whereas responsibility is readily decentralised, there is greater reluctance in decentralising financial resources or accountability, and in some cases, even the capacity in terms of the appropriate institutions and personnel. The result is the patchy impact of decentralisation and the resort in some cases to re-centralisation (Dickovick and Wunsch 2014; Olowu and Wunsch 2004). In particular, the cities that should have been

Table 4.1 A typology of decentralisation

Transfer of:	*Deconcentration – administrative decentralisation or field administration*	*Delegation or agencification*	*Devolution or democratic decentralisation*
Decision-making authority	Low	Narrow but broad in specified sector	High
Responsibilities	High	Specified and focused	Broad but specified by law
Finance and human resources management	Centralised	Broad and separated	Mixture of central and local sources/ management
Accountability arrangement	Upward (control)	Upward and to clients (control and voice)	Upward, downward and lateral (control/voice)

Source: Olowu (2015).

the recipients of wider powers of decentralisation of resources and responsibilities as well as capacity and accountability, as happened during the colonial and immediate post-colonial eras, have not benefitted from many of these decentralisation reforms for various reasons, which will be discussed next.

First, some of the countries have transferred substantial expenditure and human resources from the national governments to the locality, but income or revenue resources and accountability in most instances remain with the national government. In Ghana, for instance, there has been a local government law in place since 1989 but local governments are still largely run by centrally appointed district officers who actually preside over the councils, even though there are elected mayors and councillors (Ayee 2004). In other places like Uganda and Rwanda, political competition at the local level is restricted to non-partisan forms even though the national governments had conceded multi-party politics at the national levels. In Uganda a process of re-centralisation has commenced as the national government feels threatened by much stronger opposition coming from the cities, an issue that has also played itself out in other countries, as discussed below (Dickovick and Wunsch 2014).

Second, most countries in the region have tended to commit to a policy of the most advanced form of decentralisation – devolution (referred to as democratic decentralisation) – but continue to practice deconcentration (field administration – see Table 4.1). In an age when it had become politically correct to have a democratic decentralisation reform both as a result of domestic pressures from the citizens and especially the civil society campaigners and the business community, as well as external pressure from donors who have raised decentralisation to the status of a conditionality for providing external development assistance, decentralisation becomes a convenient smokescreen. Richard Stren (2014: 23) showed that in spite of the structural fiscal reform as it impacted on urban governance in the Anglophone countries (37 out of 45 on the World Bank list),

> ultimate control over the senior officials as well as the financing of the local body were left to the minister of local government at the national, or provincial/State, level. The francophone equivalent was a communal structure similar to that found in France, with an executive mayor responsible to the national Minister of the Interior, elected representatives, a small number of departments, and a relatively limited range of service responsibilities. Until the 1990s, most major services in a typical francophone city were carried out either by agencies of the central government or by semi-autonomous local agencies supported by large multinational corporations.

Stren (2014) noted that these reforms further increased central dominance either through the creation of agencies that manage municipal services or commissions that take over the responsibilities of elected local bodies. One notable positive outcome is the increase in locally elected mayors who though lacking domestic power at home meet at international fora (e.g. Africities, Habitat), providing them opportunities to learn from each other's experiences.

A third tendency is for decentralisation reforms and policies to completely ignore urban and urban management issues or transfer them to agencies of the national government. The prevailing approach is one that views urbanisation as an aberration to a predominant local economy that is largely rural, rather than one that acknowledges the growing and strategic importance of urban centres. Only a few countries – like Mozambique, due primarily to her own historical evolution – sought for a dual approach that separates rural from urban areas (Kathyola and Job 2011).

Fourth, the dominant approach to fiscal decentralisation, encouraged by many development partners, is to transfer financial resources from the centre to the localities for enhancing the capacity of local government through some kind of revenue sharing or grants. This is not bad in itself but since it also means the total neglect of the development of local or municipal government's own source revenues, it has made recentralisation and control easy for national governments using the fiscal tools available to them. The cities have lost out more because they have a high potential, which remained untapped for diverse reasons. Some of them had robust own-source revenues even in the colonial and in the immediate post-colonial years based on their highly developed property tax systems (Olowu 2004): cities like Lagos, Nairobi and Dakar floated bonds in Paris and London.

Finally, the above problem compounded what is perhaps the most pernicious manner in which the present approach to decentralisation has impacted negatively on urban governance, planning and management. This is with respect to the relationship between central governments and the political opposition, based mainly in the cities, where democratic elections have been allowed as a result of democratic decentralisation. Democratic decentralisation has resulted in a situation in which the cities have become the most critical centres of opposition to central governments in some countries. Decentralised powers of popularly elected municipal councils have constituted a critical power base for political contestation against the central government and indeed in some cases in unseating these central governments and their being replaced by the opposition. This has played out in Dakar with the opposition also unseating the incumbent government in Senegal, Lusaka in Zambia and very recently in Nigeria with the opposition party in power at the (Lagos) state level unseating the national government at the March 2015 elections.

To avert this competition for popularity and possible loss of face, the central government resorts to what Resnick (2014) has referred to as "subversive strategies". To attenuate this, some countries have, under pressure from civic groups, constitutionalised the powers of local government, a major departure from the global North countries' practice, which regards local governments as statutory creations of the central government (Olowu 2012). But central governments can also have a battery of other political, administrative and fiscal possibilities to subvert or discredit the political opposition (see Table 4.2).

This vertically divided authority makes rational solutions to urban and municipal challenges difficult to tackle and underscores why many years of democratic decentralisation effort have had few outcomes. The political nature of the urban terrain is often ignored by most technical analysts. Together with the colonial heritage of urban segregation, the allocation of resources and infrastructures explains why the basic elements for improving urban governance are absent in many African countries and cities (Mabogunje 1990; Rakodi 1997). These basic elements include a lack of basic urban land cadastre and revenues that can be regarded

Table 4.2 Strategies of subversion

Domain	*Central level actions*
Political	Resist relinquishing control to subnational officials from opposition through increase in appointed members of the councils
Administrative	Divest certain administrative responsibilities unless they offer high levels of visibility to key voters
Fiscal	Limit ability of subnational officials to finance their administrative responsibilities by reductions and restrictions in intergovernmental transfers and/or limitations on taxation or delays to improving donor funding

Source: Resnick (2014: 58).

as exclusive for the city. In the Western world and some of the transitional and modernising countries, the property tax is one such resource that has been overexploited to the point of becoming regressive. By contrast, in many African cities there are few such taxes on the rich at all, or if they are taxed some of these are appropriated by the national or state governments, not the local or municipal government (see below).

Without autonomy, protected responsibilities, accountability or resources it becomes clear why the local population is helpless in tackling the challenges that surround them, even in an age when these countries claim at international fora to have committed to democratic decentralisation ideals. Most assessments confirm that although structures are created and elections take place in local communities, in the most important areas that matter, fiscal and political accountability, most decisions concerning cities are still made by central governments in most African countries (Smoke 2015; Stren 2014).

The next section discusses some specific actions that might help to reverse the present trends.

Towards a different approach to African municipal management

Decentralisation is a fundamentally political activity with entrenched interests promoting and constraining reform. Given this political economy context, as already discussed above, it becomes clear why blanket devolution of fiscal and political powers to all communities has not worked and cannot work. African governments have perfected the art of stonewalling genuine devolution, especially to cities (Awortwi and Helmsing 2014). I suggest three principles that constitute a different approach to decentralisation on the continent that might be more effective in tackling the urban fiscal, infrastructural and institutional or accountability challenges.

The first is *asymmetric decentralisation*. This involves an appreciation that democratisation and decentralisation are cultural concepts that take time and critical resources to become inculcated in the local population and their leaders. Hence, these concepts should be implemented differentially in urban as against rural areas. Planners and policy makers – that is, political and senior administrators as well as economic and physical planners – operating at differing levels in the continent would need to implement democractic decentralisation differently between urban and rural communities so that the urban centres can play the historical roles of cities as centres of innovation, production and change. This means a commitment to *asymmetric decentralisation*. This implies that for most rural areas, the predominant governance arrangement would be field administration or deconcentration, which over time can pave the way for devolution. But there should be immediate introduction of devolution in the cities, with cities graded based on their fiscal and infrastructure indicators.

The second principle is the *development of and reliance on property taxes in municipalities*. The major revenue source for ensuring that municipalities generate their own source revenues, for which they would be accountable to the electorate, is the property tax. The most important arguments in its favour can be articulated as follows:

- There are dangers of depending on grants either from national government or donors. Though important in making substantial resources available to local levels of governance, it leaves municipal governments at the total mercy of the national government. Moreover, it also implies that the financing of municipal expenditures would be subject to the wide swings of the global economy, especially for countries that are dependent on natural resource extraction, which includes many African countries. By contrast, most of the responsibilities for which municipal governments are responsible are basic and require some stability.
- While grants and other fiscal transfers enhance upward accountability to the national government, downward accountability to the electorate is often constrained even though there are a range of mechanisms that some governments in the developing world have used to improve downward accountability, e.g. publishing records of monthly grant transfers (Nigeria), participatory budgets (Brazil), score cards (India), etc.
- Property tax helps in the development of municipal bank infrastructure and the mobilisation of municipal credit in the commercial market and reduces the high economic and social inequalities as the rich are taxed to support infrastructures used by all city residents. This can also be further complemented by some informal forms of taxation through donations and civic actions by community and faith-based groups used to construct and maintain infrastructures – schools, health centres, etc.

In a major review sponsored by the World Bank, which has been the main lender to municipal governments in developing countries, a review of data from 49 cities in 22 countries showed that where local government taxes play an important role in financing urban public services, property taxes are an important revenue source (Bahl and Linn 1992: 81). This and other research established that contrary to the argument in many industrialised countries that the tax is regressive, it is progressive in most developing countries because of the high inequalities and the poor development of the tax instruments.

My own research in nine cities of four developing Commonwealth countries (Harare and Kariba in Zimbabwe; Cape Town and Durban in the Republic of South Africa; Delhi, Bombay and Hyderabad in India; and Lagos and Kano in Nigeria in 1993 and 1994) confirmed most of the above points in favour of the tax (Olowu 2004). Moreover, the cities with the property tax as a proportion of local or internally generated revenue (IGR) also had higher overall revenue per capita, an important issue in Africa where the overall per capita revenues were particularly low. In further work (Olowu 2004), attention was focused on 10 cities in one country and found the following.

First, although the tax had a long history since colonial times in Nigeria, it was only well developed in one city, Lagos, the capital city (until 1992). Second, all Nigerian local governments, urban as well as rural, had become hopelessly dependent on grants from the federal government from 1976 when a nation-wide local government reform gave local governments at first 3 per cent, rising over time to 20 per cent of the national revenue, which was dependent mostly on oil and gas. Some 30 per cent of VAT taxes were added in 1994. Whereas local governments in 1972 raised 94 per cent of their revenues, this reduced to 6 per cent by 1999. Most of the local governments that did not raise property taxes had very poor and dated records of properties in their domain. But the most astonishing finding from this part of the research was the fact that contrary to the prevailing scholarly and public opinion, the councillors who were interviewed in these municipalities did not see property tax as an unmanageable political risk and 80 per cent of rate payers interviewed – in cities that used and did not use the tax – signified their willingness to pay the tax if there was demonstrable benefit associated with the tax in the form of better services. Finally, the introduction of this tax helps to stimulate the development of the land cadastre, which is missing in many developing countries, and this helps to bring land into the market.

Broader institutional base for city management

A final element of the strategy for revitalising African cities is the engagement of the three key stakeholder institutions in urban governance: state, business and non-governmental or non-profits. They play especially large roles in expanding the democratic space in many ways and have even greater potential. Decentralisation must transcend the traditional institutional confines of state and business, as the non-profits have huge political and financial clout. This is also because the most intransigent opposition to the property tax is the belief that people in the civil society have in governments and their officials based on their hard experience. It is the fear that property taxes would be collected and would not translate into desired infrastructure. The involvement of key stakeholders in city management helps to attenuate this problem and is critical to the process of institutional and financial transformation. The next section of the chapter presents three case studies.

Three case studies

1 Exceptionalism in property taxation, Nigeria

Lagos is sub-Saharan Africa's largest conurbation and city-state. It is Africa's second largest mega-city and has been an island of success in terms of its ability to mobilise substantial sums from its own domestic sources in contrast with the other 35 states in Nigeria's federal state. Nigeria introduced an ambitious local government reform in 1976 under a military dispensation as a part of the process of democratisation. Services taken over by central governments were returned to local governments and provision was made for the election of local councillors and mayors. This reform transferred substantial financial resource allocations from the central government's oil-rich coffers through the 36 state governments to the nation's 774 local governments, boosting local government revenues nationally to 6 per cent of the country's GDP and 12 per cent of total public expenditure. Local government had 30 per cent of public sector employees. These represented a huge amount even by international, and ever more so by African, standards (Olowu and Wunsch 2004). However, the reform undermined local revenue generation and the consequent enforcement of local citizen accountability. Only Lagos state and its mostly urban governments differed and the main reason was that it had managed to continue the practice of raising substantial sums from property and other local taxes. This benefitted both the local and state governments.

One significant consequence of this relative independence and leverage of local revenues is that this state has managed to elect 'action governors' (chief executives at the state level who were high achievers) since 1979 to the present time who have provided the city state with quality infrastructure, have undertaken planning of slums and chaotic transportation and have brought about substantial changes to the city's landscape through an active collaboration with private, non-privates and even the international community. Lagos State Government was also able to hire high-quality civil servants especially compared to other states and the federal government (Lagos State Government 2012; Olowu 1990; Oyelaran-Oyeyinka 2006).

As a result of the fact that the state is 75 per cent dependent on own source revenues, it was able to secure favourable capital loans from the World Bank in 2006, to support the Lagos Metropolitan Development and Governance Project: USD 205.69 million, which has been used to develop city transportation, entrepreneurship for people in the large informal sector and enhance accountability structures. Important components of the project included community participation, workshops and city-wide consultative forums, communications with stakeholders "to promote good urban governance, accountability and transparency practices in LASG [Lagos State Government] operations, conflict resolution and resettlement, strengthening Citizens' Mediation Centres and an Office of the Public Defender for poor communities" (World Bank, 2006: 49–50, cited in Stren 2014: 19–37). The most important aspect of this exceptional state is how it has used performance in terms of infrastructure and institutional development to challenge the national government for civic popularity becoming in the process a bastion of opposition and the ouster of that ruling political party at the March 2015 national elections. The governor of that state has become one of the ministers at the federal level.

2 Impact of value added taxes for Ethiopian decentralisation

Reform of value added taxes (VAT) has been one of the standard reforms especially aimed at boosting the revenues of many poor or developing countries (Fjelstad and Moore 2009). Ethiopia introduced a nation-wide VAT for the funding of decentralised organs of government in 2002–2003 as did Nigeria in 1994 with local governments receiving 30 per cent of all collections. What is significant in the Ethiopia case is that 2000–2011 panel data showed that the intended recipients of the reform (importing firms) performed better in terms of employment and sales *after* the introduction of VAT in decentralised cities with greater autonomy than in non-decentralised cities. In other words, granting city-level powers away from the regional governments actually determined how firms responded to tax incentives.

Increasing the administrative powers of the mayor played an important role in making Ethiopian cities more competitive. This meant that financial devolution enabled mayors to transmit the effect of national-level reforms more efficiently. In addition, decentralisation led to convergence between poor-performing and high-performing cities. This promoted spatial equity across the country (Chaurey and Mukim 2015).

3 Decentralisation of taxation powers in Rwanda

Rwanda is Africa's least urbanised country but is urbanising very rapidly, from 4.6 per cent in 1978 to 16.5 per cent in 2012. The Rwanda Vision 2020 anticipates 35 per cent urbanisation in 2020 (Ministry of Infrastructure 2015). One half of the urban population resides in the capital city of Kigali for a variety of reasons (Goodfellow 2014). Phenomenal opportunities for raising land taxes abound and the country has had a good record of increased revenue performance, especially since establishing the Rwanda Revenue Authority in 1998. As part of its ambitious decentralisation and city transformation reforms, three taxes previously collected by the national government were decentralised. These were the trading licence, rental income and property tax. The trading licence has done very well, with 89–95 per cent of businesses paying for their licences regularly. On the other hand, the two other taxes that are directly related to property and that were held mostly by the rich and powerful have done very poorly. A sample of only 66 out of 30,000 properties in one of the city's three main quarters registered for property taxes.

A recent study (Goodfellow 2014) provided some explanations for why the property taxation in Kigali is responsible for less than 3 per cent of local revenue whereas in the neighbouring country of Uganda, Kampala raises 20 per cent of local revenue from the same tax source. Among the reasons given for this were, first, that only a few property owners reported for registration. The law required property owners to self-register. They were also asked to self-evaluate themselves given that this is a country where there were few valuers. The second broad reason was the sustained resistance by the property owners, who happen to be those in power, to the improvement of this tax. A 2006 new law on property taxes was brushed under the carpet. The study concluded that the lack of effective property and rental taxation represents a double missed opportunity for progressive taxation. It enhanced the creation of three types of enclaves: high-income areas, poor slums and empty space; and it resulted in a lack of the resources needed for affordable housing. A tax source with potential to contribute 5–6 per cent of the country's GDP raises only 0.0009 per cent of the GDP (Goodfellow 2014).

All of the above underscore the importance of planners at all levels refocusing on African cities. Unfortunately, a recent review showed that between 2012 and 2015 the existence of urban strategies has received the least attention among national and local authorities (Cities Alliance 2015).

Conclusion

A new approach towards African urbanisation in an age of democratisation and decentralisation is required. It must be an approach that is shared widely by the main stakeholders responsible for development planning in this continent at regional, national, metropolitan and community levels. The approach would seek to address the issue of urban planning and infrastructure financing through a concerted focus on how to deal with the vexed issue of the high inequalities in Africa's cities, using the instrumentalities of property taxes. The success of revenue agencies at the central level and in a few countries at the subnational level, point to the possibility of using this mechanism to address most of the conventional problems of the tax. Without this, current approaches to planning based on outmoded models would only continue to constrain the potential forces for the possible development of more equitable and truly modern African cities. It is heartening to note that at least a few islands of excellence already exist that can point the way forward to a future in which cities are dynamic centres of change throughout the continent.

References

African Development Bank (2012) *Urbanization in Africa*, Tunis: Tunisia.

Awortwi, N. and Helmsing, B. (2014) "In the name of bringing services closer to the people? Explaining the creation of new local government districts in Uganda", *International Review of Administrative Sciences* 80(4): 766–788.

Ayee, J. (2004) "Ghana: A top-down initiative". In D. Olowu and J. Wunsch (eds) *Local Governance in Africa*, Boulder, CO: Lynne Rienner, pp. 125–154.

Bahl, R.W. and Linn, J.F. (1992) *Urban Finance in Developing Countries*, Oxford: Oxford University Press.

Chaurey, R. and Mukim, M. (2015) "Decentralization in Ethiopia: Who benefits?", World Bank Working Paper No.101709. Available: https://openknowledge.worldbank.org/handle/10986/23574 License: CC BY 3.0 IGO [accessed 14 July 2016].

Cities Alliance (2015) "Assessing the institutional environment of local governments in Africa", Tangier, Morocco. Available: www.citiesalliance.org/sites/citiesalliance.org/files/CA_Docs/LGAfrica.pdf [accessed 14 July 2016]

Dickovick, T. and Wunsch, J. (eds) (2014) *Decentralization in Africa: A Comparative Perspective*, Boulder, CO: Lynne Rienner.

Fjelstad, O. and M. Moore (2009) "Revenue authorities and public authority in sub-Saharan Africa", *Journal of Modern African Studies* 47(1): 1–18.
Freedom House (2015) *Freedom in the World*. Available: https://freedomhouse.org/report/freedom-world/freedom-world-2015 [accessed 22 February 2017].
Goodfellow, T. (2014) "Rwanda's political settlement and the urban transition and taxation in Kigali", *Journal of Eastern African Studies* 8(2): 311–329.
Kathyola, J. and O. Job (2011) *Decentralisation in Commonwealth Africa: Experiences from Botswana, Cameroon, Ghana, Mozambique and Tanzania*. E-book available: https://books.thecommonwealth.org/janet-kathyola [accessed 22 February 2017].
Lagos State Government (2012) *The Lagos Policy Review* 1(1), September.
Mabogunje, A. (1990) "Urban planning and the post-colonial state in Africa: A research overview", *African Studies Review* 33(2): 121–203.
Ministry of Infrastructure (2015) "National urbanisation policy", Republic of Rwanda. Available: www.mininfra.gov.rw/fileadmin/user_upload/Rwanda_National_Urbanization_Policy_2015.pdf [accessed 22 February 2017].
Olowu, D. (1990) *Lagos State: Governance, Society and Economy*, Lagos: Malthouse Press.
Olowu, D. (2004) "Property taxation and democratic decentralization in developing countries", Working Paper Series No. 401, October, Institute of Social Studies, The Hague, Netherlands. Available: www2.ids.ac.uk/gdr/cfs/pdfs/Olowu2.pdf [accessed 1 February 2016].
Olowu, D. (2012) "The constitutionalization of local government in developing countries – analysis of African experiences in global perspective", *Beijing Law Review* 3(2): 43–50.
Olowu, D. (2015) "Assessing decentralization and local government for development", UNRISD Conference Paper, January, Geneva.
Olowu, D. and Wunsch, J. (eds) (2004) *Local Governance in Africa: The Challenge of Decentralization*, Boulder, CO: Lynne Rienner.
Ostrom, E. (1990) *Governing the Commons: The Evolution of Institutions for Collective Action*, Cambridge: Cambridge University Press.
Oyelaran-Oyeyinka, B. (2006) *Governance and Bureaucracy: Leadership in Nigeria's Public Service: The Case of the Lagos State Civil Service (1967–2005)*, Maastricht: Datawyse.
Potts, D. (2012) *Whatever Happened to Africa's Rapid Urbanization?* London: Africa Research Institute.
Rakodi, C. (1997) *The Urban Challenge in Africa*, Tokyo: United Nations University Press.
Resnick, D. (2014) "Urban governance and service delivery: The role of politics and policies", *Development Policy Review* 32(1): 3–17.
Simone, A. (1999) "Thinking about African urban management in an age of globalization", *African Sociological Review* 3(2): 69–98.
Smoke, P. (2015) "Rethinking decentralization, assessing challenges to a popular public sector reform", *Public Administration and Development* 35: 37–112.
Stren, R. (2014) "Urban service delivery in Africa and the role of international assistance", *Development Policy Review* 52: 19–37.
Ter Haar, G. (ed.) (2011) *Religion and Development: Ways of Transforming the World*, London: Hurst & Company.
Turok, I. (2014) "Linking urbanization and development in Africa's economic revival". In S. Parnell and E. Piertersie (eds) *Africa's Urban Revolution*, London: Zed Books, pp. 60–81.
United Nations Department of Economic and Social Affairs, Population Division (UN DESA) (2014) *World Urbanisation Prospects, The 2014 Revision Highlights*, New York: UN DESA.
World Bank (2004) *World Development Report*, Oxford: Oxford University Press.
World Bank (2006) "Project appraisal document on a proposed credit in the amount of SDR 13 8.10 million (US$200.00 million equivalent) to the Federal Republic of Nigeria for the Lagos Metropolitan Development and Governance project". Available: www-wds.worldbank.org/external/default/WDSContentServer/WDSP/IB/2006/06/19/000160016_20060619104001/Rendered/PDF/36433.pdf [accessed 14 July 2016].
Wunsch, J. and Olowu, D. (eds) (1990) *The Failure of the Centralized State: Institutions and Self-Governance in Africa*, Boulder, CO: Westview Press.

5

Data on rapidly growing cities

Lessons from planning and public policies for housing precarity in Brazil

Eduardo Marques

Introduction

This chapter discusses data and information for policies and planning on housing precarity and precarious settlements, considering the experience of Brazilian metropolises and their public policies. The majority of international debates about data on housing and cities focus on the availability and accuracy of information, considering the main technical elements associated with urbanism, infrastructure, housing stocks and production. Any data strategy, however, departs from representations about the cities it intends to picture, as well as from mainstream visions of what a city, a settlement and a house should be. In fact, with the aim of making the urban readable and possible to be understood and acted upon, information strategies and data production construct the subject of their intervention (Scott 1999).

For data on housing precarity, this represents a great problem, since precarious regions tend to be very distant from the normative ideas implicitly embedded in technical standards and models. These housing situations are generically called informal. The category 'informal', however, hides at least three different situations, considering their distances from the recognised patterns in three different (but interconnected) dimensions: their degrees of conflict with laws (property laws, particularly, what makes them illegal), their distances from administrative codes (parcelling regulations, for example, which may make them irregular) and their deviance from environmental, sanitation or housing standards (which makes them precarious). The problem is even greater for studies and interventions in large cities of the global South, which have vast regions of their urban tissues occupied by forms of urbanisation, land tenure and housing features that do not fit the patterns usually considered by mainstream standards. Although this chapter is mainly concerned with data for policies and planning, the issue is obviously also political, since the knowledge about city spaces – their 'representation' and 'cognition' – is associated with forms of (their) 'recognition'.

To discuss these issues, considering the conditions prevalent in Brazilian cities and policies is the goal of this chapter. Some features of this country and its cities are common to other Latin American and Southern countries and metropolises, while others are not. The chapter does not intend to represent, in this case, the variability present in cities of the global South. On the contrary, we understand that the phenomenon is characterised by intense and multifaceted heterogeneity. However, the discussion of the Brazilian case helps us to think about precarity in different situations and cities, taking into account both its similarities and its specificities.

Three sections develop the argument, in addition to this introduction and the conclusion. The first section starts by discussing some important elements associated with the definitions of informal urban areas, and their consequences for data and information for planning and policies associated with precarity. The second starts by presenting the variability of the phenomena and the types of urban and housing precarity in Brazil and other Latin American countries (considering the above-mentioned degrees of illegality, irregularity and precarity), and ends by summarising the policies developed in the recent period. The third section discusses the availability of data and the main challenges to providing detailed and accurate information for these areas, as well as some tentative solutions.

Data and the heterogeneity of precarity

Urban areas host many different types of housing and spaces. Some of them are defined broadly inside the loose and imprecise category of informality. In fact, the multiple and changing urbanisation patterns of informal areas include several different situations that must be analysed in their own terms and considering their diversity. The category informality, in fact, hides more than it illuminates, since it is defined by absences in relation to a pattern (the formal), being treated traditionally as a pathology (Valladares 2006). This is a product of political choices, since if the idea is to replace all informal urban tissue for standard occupation, there is no need to understand its heterogeneity. Consequently, the large majority of existing housing models fail to account for the many different housing situations that are developed vernacularly by the poor population worldwide. For the majority of its inhabitants, differently, informality is more a style of urbanisation than a concept that could allow us to precisely specify concrete situations (AlSayyad 2004). The situation becomes even more confusing with the international use of the terms 'slum' or 'shanty towns', by merging into single categories very distinct situations of urban insertion, residential segregation, access to infrastructure, land tenure, building conditions and urbanism. This problem affects authors located at both the right and the left sides of debates, since many of them homogenise the phenomena, regardless of the solutions they advocate.

Therefore, housing precarity is highly heterogeneous in many social and urban aspects worldwide, although traditional housing policies tend to homogenise it. The problem is caused by defining informality negatively, departing from the differences of a considered urban situation to some external and predefined standard of formality. Consequently, all the situations that differ from these patterns are included into one single category, as in the international use of the term 'slum'. The consideration of the complexity of the phenomenon depends on the understanding of the historical construction of its variability in each country and city in terms of at least three key features: urban location (and segregation), housing and urbanistic features, and land property and tenure. Grounded comparisons of precarious housing situations will certainly show commonalities between situations, but a departure from the specificities, and not from a general and abstract pattern, may lead to interpretations (and policies) that consider at the same time specific and common features. To highlight those specificities, the local names for different kinds of precarity in Brazil will be used throughout this chapter.

The definition of precarity is key not only for academic debates, but also for planning and public policy provision. This is key because the way data is conceived and organised influences what can (and cannot) be seen in the picture drawn by models, plans and policies, creating a strong connection between data and policy eligibility. In most extreme cases, irregular areas may even be removed deliberately from information systems. Policy bias, however, usually happens in a subtler way, since poor people and precarious areas of several sorts tend to be less visible to mainstream data strategies, creating socially regressive consequences for data collection and management (Torres 2002).

Variability of housing precarity and policies in Brazil

In the case of Brazilian cities (and the situation tends to be similar in several other Latin American countries, at least), housing precarity involves at least three different precarious solutions developed historically by the population who had no access to houses delivered by markets or by the State: precarious tenements (*cortiços*), *favelas* and irregular settlements.

The first ones – cortiços – received many regional names, but in all cases involved rented single rooms in collective buildings, where families shared kitchens, bathrooms and laundry areas. Land and buildings belonged to some private or (more rarely) public owner, who was usually absent and charged a rent for the occupation. This solution was characteristic of central areas in the last decades of the nineteenth and the first decades of the twentieth century. In the Latin American case, they occupied decayed colonial houses, which did not differ much from the forms of housing precarity prevalent in European cities in the nineteenth century, and resembled physically the subdivided *siheyuan* (traditional Chinese courtyard house) in the *hutongs* of contemporary China. They tend to be located in or near historical centres, but also recently in sub-centres of large metropolises. Easy access to services and jobs is a main element in the choice of this kind of solution by the population. Density tends to be very high, sanitary conditions are really poor and privacy is almost nonexistent, leading to constant conflicts between neighbours. Nowadays cortiços tend to present the worst situations in terms of sanitation and sociability, on average (Kowarick 2009).

Through the twentieth century, migration from rural areas boosted by both declining conditions in the countryside and the industrialisation processes in larger cities led to intense population growth rates in the largest cities. On account of a lack of land regulation or of housing public policies that could have provided affordable land, the majority of the poor and recently migrated population created a new precarious housing solution through the occupation of vacant land and the self-construction of shacks, later consolidated as brick houses. Although the name for this solution varies across countries (*barrios* in Venezuela, *villas miserias* in Argentina, *tomas* in Chile, for example) and even within Brazil, the best known term for these areas internationally is favelas. These are occupied areas (collectively or individually) that usually lack infrastructure and have irregular patterns of physical occupation. This is not, however, what defines their features; what specifies them is the lack of secure land tenure. Although the visual images of favelas usually show steeply sloped areas, they may be located in completely different places, depending on the city. Their location expresses, in fact, the areas that could not be developed by the private sector nor by the state at the time of the 'favelisation' process. While in several cities these correspond to sloped areas, in others they may be located in floodable areas, such as the river valleys or contaminated regions of industrial areas, among others. Their location may be segregated at the scale of the metropolis, but may also be close to rich areas, or both, as in the case of Rio de Janeiro (Perlman 2010).

A third precarious solution is sometimes confused with favelas, although the differences tend to be important for the specification of policies towards precarity. These are the irregular settlements. They were produced by private developers who, regardless of having sold plots destined for poor families, never had the settlement approved, usually by the local government. This may have happened for several reasons; for example, because of the existence of some irregularity on land property, or because the infrastructures were not constructed, or due to problems in the project submitted to the local planning authorities. In any case, the private entrepreneurs did not complete the legalisation of the settlement and the parcelling of the original plot, thus hampering the legal division of the plots and the issuing of separate land titles for each family. On those lands, the inhabitants have self-constructed their houses, and sometimes built simple infrastructure systems. Consequently, although the inhabitants have some proof of land claim (and land tenure, regardless of the presence of precarity), the plots can be commercialised only as informal markets with devalued prices. These settlements were disseminated in Brazilian cities from the 1950s onwards and, depending on the city, they are still the most prevalent precarious solution, such as in São Paulo (Ministério das Cidades/Centro de Estudos da Metrópole 2008). While favelas were typically constructed at the interstices of the urban tissue, both at central locations and in the peripheries, irregular settlements are typical of the peripheral expansion that marked the decades of highest urban growth rates in the 1960s, 1970s and 1980s. Therefore, irregular settlements are also usually segregated. As in favelas, housing and environmental conditions vary substantially, making more sense to talk about both favelas and irregular settlements in the plural (Valladares and Preteceille 2000), especially in recent decades.

It is also important to add that today many of these precarious solutions are mixed, with cortiços inside favelas that are inside irregular settlements, cortiços and favelas inside housing projects, housing projects that are in fact irregular, since the state did not complete their regularisation, etc. In any case, these three precarious solutions represent different situations that create consequences for sociability and for housing and environmental quality (Kowarick 2009), as well as for the public policies that can be enforced to fight precarity.

These forms of urban and housing precarity were already present for a long time in the twentieth century in Brazil, but the prevalence of each precarious solution varied. Since the first timid public housing policies in the 1930s, however, almost all the efforts were concentrated in the production of new housing units in peripheral projects for financed sale. Furthermore, these policies were marked by intense selectivity of beneficiaries (considering their income and occupations) and were always in insufficient scale to meet the demand. The great majority of peripheral areas in the largest cities of the country (and of Latin America) grew exponentially in the following decades with no planning and almost no infrastructure provision. In Brazil, mass housing production started only in the 1960s after the military coup d'état, but with very low quality, subjected to intense urban segregation and still relying on new housing units for financed sale.

Evictions of both cortiços and favelas were very common in the 1960s and the 1970s in large metropolises, especially in areas targeted by the development industry. In some cases, local residents resisted evictions politically. The majority of the social movements that marked the redemocratisation process, however, were associated with claims for recognition, tenure and infrastructure improvement. This is key for the understanding of the Brazilian case, since if the predominant policy was removal, the most probable strategy from residents would be invisibility. In Brazil, differently, the consolidation and the upgrading of precarious areas became increasingly frequent, turning visibility into the most common political strategy. This even led to the creation of civil society organisations dedicated to the consolidation and dissemination of information about precarious areas, such as the Observatório de favelas[1] and the Central Única das Favelas,[2] both about favelas and located in Rio de Janeiro, and the Centro Gaspar Garcia focused on cortiços in São Paulo.[3] Although initiatives of this kind produce localised studies and contribute to 'put favelas on the map' politically, they usually do not produce systematic information that can be used on a large scale for policy design and planning.

It was only at the end of the 1980s, with the demise of the federal housing system constructed by the military that alternative policies started to be developed. This happened initially in local governments and with intense participation of technical communities and housing movements, but reached the federal level at the end of the 1990s and beginning of the 2000s. This period created a broad set of policies, each of them targeted to different precarious situations. Among the new policies must be mentioned at least slum upgrading and regularisation, irregular settlement regularisation (both physical and juridical), socially redistributive zoning, self-construction processes controlled by civil organisations and social rent, among other programmes and initiatives. Participation in policy formulation and implementation was also introduced at Housing Councils (at the local level) and Cities' Conferences (at the federal level). Although state technical capacities continue to be insufficient, they were strongly enhanced at both the federal and the local level. Producing accurate data for precarious areas arose in this period, since evictions need no detailed information. It is in this direction that the Ministry of Cities, the National Bureau of Statistics (IBGE) and research institutions made important efforts to enhance the data available nationally on precarity, as will be discussed in the next section.

In what concerns precarity, the development of alternative policies obviously depends on the political preferences of different governments towards redistribution, but the availability of policy solutions, as well as the expertise necessary to develop them, has changed substantially in the last 30 years. In this sense, evictions continue to happen, especially in areas targeted by the development industry and/or included in large projects. The preparation of the Olympic Games in Rio de Janeiro, for example, evicted around 17,000 persons (CPCORJ 2014). However, since the return to democracy, slum upgrading, and not eviction, is the predominant policy for precarious areas in Brazil. These were produced initially with local resources, but have received a strong push from the federal government since the end of the 2000s.

In fact, the federal government incorporated many of these local alternative policies into its own policies in the 2000s, although still with low institutionalisation and poor insulation from pressures of the political system (Rolnik 2011). At the same time, the old policy agenda of new housing units for financed sale was resumed at the end of the 2000s (as part of anticyclical economic measures against the 2008 crisis). This policy is based on intense participation of the private sector and incorporated strong federal subsidies for very poor families (for the first time in the country). Therefore, after almost two decades of absence, the federal government returned to the scene with a strong programme that produced around 4 million housing units, although with problems of quality and project location in terms of segregation (Marques and Rodrigues 2016).

Summarising, it is fair to say that the policy scenario of the 2010s includes selective evictions, as well as the old policy agenda of housing construction for sale in peripheries (although now with subsidies), but also a new set of policies based on customised solutions for precarity, which may be mobilised depending on the government's policy preferences and on pressures from social movements. The development of these policies has created a demand for information on precarity that should be precise and accurate, and at the same time open enough to allow the representation of the diversity of situations present in cities.

Data for policy formulation, for planning and for implementation

As we have seen, the definition of precarity is key for planning and policies, since it suggests types of solutions and defines eligibilities. Considering the elements discussed in the previous section, a central element is the degree to which the definitions of the phenomena may recognise the heterogeneity of precarity and allow planners and decision makers to unpack the kinds of precarity present in each historical and geographical situation. Once again, the following discussion draws on the Brazilian case.

The difficulties for policies for precarious areas are not limited to their definition, however. Data collection itself has profound consequences for policy, since interventions and programmes are designed and implemented considering the characteristics, distribution and location of the problems they intend to face. Therefore, information may bias the solutions or even render some social groups and urban locations invisible. The problem is especially serious because poor people tend to be less visible to information systems and are more difficult to reach, since they have more restricted access to sources of information, tend to have poorer reading skills, as well as face more difficulties in understanding policy procedures and even correctly filling in policies' questionnaires. They also live in places that are sometimes not on official maps and databases due to problems with addresses. Therefore, there is a clear regressive logic in the relationships between information systems and public policies (especially social and urban policies), since poorer people are the ones in most need, but they are also the most difficult to find, reach and target (Torres 2002).

However, when we mention data for policies in precarious areas we are talking about several different kinds of information: surveys, documents and studies. There are at least two broad kinds of data about precarious areas that are necessary for policies: first, data and information for projects and concrete interventions such as slum upgrading or settlement regularisation; and second, data and information for planning and policy design. The first kind of data necessarily involves a much more detailed and updated description of the physical and social conditions present in specific favelas and irregular settlements, such as updated topographical and urbanistic surveys for each settlement. Satellite images may help the development of studies, but these depend mainly on detailed and on-the-ground surveys and fieldwork. This must include updated cartographies of streets, alleys, squares, the projections of existing constructions and the description of each house in terms of construction material, quality and maintenance. This information will be used for the development of blueprints and projects of urbanism, infrastructure (water, sewage, drainage and waste collection), localised solutions of geological and flooding risks (when present) and housing construction for relocation, if necessary, among others. Considering the degree of dynamics that characterises these settlements, the surveys and studies must be done just before the projects, and the latter, just before construction. I will not discuss this type of data targeted to projects and physical interventions in this chapter.

The second kind of information aims to provide a broad picture of the problems present in a specific situation, and are necessary, for example, for deciding which programmes should be created nationally and what features they should have. Each of them presents specific challenges that vary substantially from country to country, considering not only the existing kinds of urban precarities, but also the government structure, the responsibilities over urban policies and the information systems and data existent in each case. As in the previous section, the argument will be developed considering the Brazilian case.

Information for planning in Brazil may come from administrative data, from specific surveys or from periodic population surveys such as the Census. Administrative data comes from governmental programmes. The challenge in the case of precarious settlements is to find information that is standardised both nationally and for an entire region, as well as being detailed enough to be used for planning and policy design. Therefore, data standardisation is essential, but should be done in a way that allows the representation of regional and urban diversity. The problem comes from the fact that, at least in large federalist countries, spatially detailed data is usually produced by local levels of government, which tend to be many and tend to create different procedures and definitions of the phenomena. In Brazil, for example, there were 5,570 municipalities in 2010, each producing their own urban and housing policies independently (in fact, the large majority of them lacked any housing policy). Data on land tenure that may specify the kind of precarity makes the problem even greater, since they are usually available at the local level, but not at the executive branch of the government (in Brazil they are registered at local notary offices, with no centralised databases for information).

Another problem arises with data collected for different policies but that do not contain spatially detailed information, because space is not a key element for those other policies or due to policy fragmentation into different agencies and government levels. In this case, the solution depends on the possibility of mixing or superposing different sources of information. In the Brazilian case, the obvious example (although not the only one) of a wonderful source of data that unfortunately cannot be used for precarity is the Simple Registry (Cadastro Único) of the Bolsa Família programme. The Bolsa Família is a massive conditional cash transference programme (in the 2010s it was the largest in the world), which constructed an impressive registry of poor families all over the country (Campello and Neri 2014). A federal ministry manages the registry and it is operationalised by a federal bank, but local governments' social assistance bureaucracies in direct contact update it with the beneficiaries. In 2015 it included 23 million families in the registry and around 14 million in the programme (roughly a quarter of the country's population) and it is the reference not only for social assistance but also for other policies such as health and education. The problem is that this registry is not spatialised, since social assistance policies do not need this kind of detail. Additionally, this is also due to policy fragmentation – urban policies are local and disconnected from the health, education and social assistance federal systems. The use of information from the Cadastro for urban precarity, therefore, depends on its connection with spatial databases from local governments that may delimitate urban precarity. This has not yet been accomplished systematically, but represents a promising frontier.

The large majority of the spatially detailed information is produced by local governments with very low standardisation and is not available centrally at any level or agency. There are no centralised sources of street maps either. On the other hand, the best standardised information available for the whole country is produced by the National Bureau of Statistics (IBGE), as part of their strategy to conduct the census survey. In 2010, the country was divided into 316,574 census tracts, resulting in an average size of 630 inhabitants per tract in the whole country or 250 households (around 1,000 inhabitants) per tract in urban areas.

Through cartographic techniques inside geographic information systems and with the use of statistical analysis, the census tracts databases can be made compatible with existing street maps, although in some cases a lot of work may be required to fix the cartographies. Once this is completed, however, the socio-economic and demographic information of the census is available at a much disaggregated and quite homogeneous scale.

In terms of urban precarity, the census presents information that allows alternative strategies to find, quantify and characterise favelas and irregular settlements, although not cortiços, since these tend to be too small to be captured by cartographic strategies. The Brazilian Bureau of Statistics classifies the census tracts in several categories, to organise data collection. One of them is called 'subnormal' and corresponds, in theory, to precarious urban areas – there were 7,871 subnormal tracts in a universe of 215,811 tracts in 2000. The classification of these areas, however, is part of the preparation of the Census fieldwork, because the surveyors who work on these areas receive a different payment and technical training. Therefore, this gives us a standardised, national delimitation of precarious areas with reliable alphanumeric data attached, but it is always outdated since the classification precedes the application of the survey, being based on the previous Census or on other data available locally. Additionally, this is a classification of entire tracts, and small precarious areas in heterogeneous urban tissues tend be characterised as non-subnormal ('não especial', in the official terminology). This is the reason why indirect cartographic strategies with census data cannot provide information for cortiços, which are usually smaller than favelas and irregular settlements, becoming 'diluted' inside the tracts. In summary, this information is reliable, standardised nationally and detailed in spatial terms, but it tends to underestimate recent and small precarious areas.

In order to solve these problems but to use this important source of information, the Center for Metropolitan Studies and the Ministry of Cities have developed a methodology of estimation of precarious settlements (Ministério das Cidades/Centro de Estudos da Metrópole 2008). It was based on the assumption that the Census underestimated the phenomena, but delimited correctly the kinds of precarity present in Brazilian cities (except for cortiços). The use of this method in 2000 allowed the estimation of the population in precarity for 561 municipalities (which concentrated 98 per cent of the subnormal tracts) as being 6,050,258 inhabitants in 1,546,250 households, in addition to the 6,365,573 inhabitants and 1,618,836 households, respectively, already included in the subnormal tracts. Thus, the procedure doubled the estimated population in precarity from 7.5 per cent of the urban population considering just the subnormal to 14.3 per cent already including the reclassified tracts. Based on the digitalisation of the tracts cartographies, the study also drew estimative boundaries of precarious areas for 364 municipalities. More recently, the same method has been applied to the four metropolitan regions of the state of São Paulo to estimate the size, location and growth of precarity using the 2010 Census (EMPLASA/CEM 2013). This study showed that while precarity was relatively reduced in the metropolitan region of São Paulo between 2000 and 2010 (from 15 to 14.5 per cent of the population), it grew in the other three metropolitan regions, resulting in a relative increase from 13.5 to 14.3 per cent of the population for the four metropolises.

As mentioned before, this strategy does not solve the problem of providing information for cortiços. Due to their size and location within the urban tissue, these depend on direct surveys. Sampling techniques may be used in this case, but a previous understanding of the variability and general location of the phenomena is essential to avoid biases. A close collaboration of the local agencies with housing movements may provide important information on cortiços' locations, providing a basis for surveys based on sampling techniques. At the national level and for general planning, however, the existing information in the Brazilian case is not very reliable. The Bureau of Statistics estimates 29,582 cortiços in the whole country in 2010, but a study of the municipality of São Paulo estimated only 996 in its central area in 2010, suggesting that the national figures might be strongly underestimated. In the case of this precarious solution, the development of detailed studies seems to be crucial also for planning and policy design.

Conclusion

We have seen in this chapter that the issue of data for policy and planning in precarious urban areas is not just a technical problem involving methods and techniques of data collection and storage, but involves conceptual elements related to the definition of the phenomena that may account for its heterogeneity in each case. The issue involves considerable challenges, since the concepts used to define the phenomena establish policy eligibilities and effectiveness. Furthermore, the information systems tend to underrepresent exactly the social groups who are in deeper need of social and urban policies.

The presentation of the Brazilian case is intended to exemplify the many facets of the problem in the concrete situations of complex urban networks. Historically, poor populations in Brazilian cities have developed three main precarious housing solutions to face the challenge of inhabiting the city while earning very low salaries, having no access to formal housing markets and very limited and selective public policies. The policies created to address these problems have changed considerably in the last decades. Traditionally, the only existing policies have aimed at constructing new units for financed sale in urban peripheries, but the recent decades witnessed the creation of several policies which targeted specifically, and were customised to solve, each form of precarity such as slum upgrading and irregular settlements regularisation, although still in insufficient volume, poorly institutionalised and weakly insulated from political pressures.

These new policies asked for more precise data that could represent the variety of precarious situations, leading to new data and information strategies. These new data are used for concrete interventions but also for planning and policy design. This chapter did not discuss the first group in detail, but the second combines the use of administrative data from policies with official surveys processed by quantitative techniques inside geographic information systems. Through these techniques, data from different sources (and produced by different methods) may be crossed and mixed using their superposition and association in space, leading even to the production of new information. The combination of these methods has been successful in helping to fill the gap of information for policy, although sometimes with indirect and estimative strategies.

Notes

1 observatoriodefavelas.org.br [accessed 22 February 2017].
2 cufa.org.br [accessed 22 February 2017].
3 www.gaspargarcia.org.br [accessed 22 February 2017].

References

AlSayyad, N. (2004) "Urban informality as a 'new' way of life". In A. Roy and N. AlSayyad (eds) *Urban Informality: Transnational Perspectives from the Middle East, Latin America and South Asia*, New York: Lexington Books, pp. 7–30.

Campello, T. and Neri, M. (2014) *Bolsa Família Program: A Decade of Social Inclusion in Brazil: Executive Summary*, Brasília: Ipea. Available: www.ipea.gov.br/portal/images/stories/PDFs/140321_pbf_sumex_ingles.pdf [accessed 7 September 2015].

CPCORJ (2014) *Megaeventos e Violações dos Direitos Humanos no Rio de Janeiro – Dossiê do Comitê Popular da Copa e Olimpíadas do Rio de Janeiro*. Available: https://comitepopulario.files.wordpress.com/2014/06/dossiecomiterio2014_web.pdf [accessed 7 September 2015].

EMPLASA/Centro de Estudos da Metrópole. (2013) *Diagnosis of Precarity in the Cities of São Paulo Macrometropolis*, São Paulo: CEM. Available: www.fflch.usp.br/centrodametropole/en/1191 [accessed 7 September 2015].

Kowarick, L. (2009) *Viver em Risco: Sobre a Vulnerabilidade Socioeconômica e Civil*, São Paulo: Editora 34.

Marques, E. and Rodrigues, L. (2016) "Public housing production". In E. Marques (ed.) *São Paulo in the Twenty-First Century Spaces, Heterogeneities, Inequalities*, New York: Routledge, pp. 196–213.

Ministério das Cidades/Centro de Estudos da Metrópole (2008) *Assentamentos Precários no Brasil Urbano*, Brasília: Ministério das Cidades. Available: www.fflch.usp.br/centrodametropole/en/580 [accessed 7 September 2015].

Perlman, J. (2010) *Favela: Four decades of Living on the Edge in Rio de Janeiro*, Oxford: Oxford University Press.

Rolnik, R. (2011) "Democracy on the edge: Limits and possibilities in the implementation of an urban reform agenda in Brazil", *International Journal of Urban and Regional Research* 35(2): 239–255.

Scott, J. (1999) *Seeing like a State: How Certain Schemes to Improve the Human Condition Have Failed*, Yale, CT: Yale University Press.

Torres, H. (2002) "Social policies for the urban poor: the role of population information", UNFPA Country Support Team for Latin America and the Caribbean, Working Paper Series CST/ LAC No. 24. Available: www.fflch.usp.br/centrodametropole/antigo/v1/pdf/torres_unfpa.pdf [accessed 7 September 2015].

Valladares, L. (2006) *La Favela d'un Siècle à L'autre*, Paris: Editions de la MSH.

Valladares, L. and Preteceille, E. (2000) "A desigualdade entre os pobres – favela, favelas". In R. Henriques (ed.) *Desigualdade e pobreza no Brasil*, Ipea: Rio de Janeiro. Available: http://empreende.org.br/pdf/Programas%20e%20Pol%C3%ADticas%20Sociais/Desigualdade%20entre%20os%20pobres%20-%20favelas.pdf [accessed 7 September 2015].

6

A 'peripheries' view of planning failures in Kolkata and Hyderabad in India

Sudeshna Mitra

Commentaries regarding India's rapid and multi-dimensional urban transition often allude to 'unplanned' cities. Terms such as planning 'failure' and 'exception' are also common critiques. Such diagnoses have intuitive appeal, since urbanisation in India is dominated by disaggregated and disparate acts of political negotiation and auto-construction, which exceed and elude the terrain of governability made possible by formal planning tools such as master plans. However, in this chapter, diagnoses of planning 'failure' and planning 'exceptions' are used as provocations to examine the assumptions of power embedded in planning as a professional practice, particularly in expectations of normative outcomes, such as spatially ordered spaces and social and spatial equity. Critical research decrying the shortcomings of planning practice also often lapse into the same frame of expectations, even as they highlight the need to pay attention to the complexities of political society and incomplete nature of claim-making that marks state–society relationships in post-colonial India. This chapter asks for the suspension of expectations of particular outcomes as frames to analyse planning practice and suggests instead an engagement with contemporary exigencies and processes shaping the practice. It considers planning practice as is, with a focus on processes and instruments being deployed by planning bodies, in the face of multiple instrumentalities of state, society and market. The cases in this chapter highlight ways in which planning practice is evolving, particularly ways in which there is a break from static imaginations of territorial state power and absolute spatial control, embedded in planning instruments such as master plans. Further, the chapter highlights that the continuities and discontinuities visible in contemporary planning practice are visible historically. These highlight that planning has always been a politicised terrain, despite the static and imagined binaries of planned/unplanned urban spaces claimed both within the professional practice's own discourse, as well as research that foregrounds the lens of 'failure', in critiquing the role and impact of planning practice.

Overall, the chapter's argument is that planning, like the post-colonial state, needs to be understood beyond the lexicon of developmental 'failure' and in light of new scales, actors, negotiations and imperatives, manifesting in uneven power geographies, particularly around questions of land, investments and development partnerships. Planning mandates and planning failures need to be examined beyond desired planning outcomes, and as discrete questions of governance, law, instruments and processes.

This chapter uses the notion of 'periphery' in two ways: first, as a positionality to reassess planning practices within a relational milieu of power; and second, there is a focus on the geographic peripheries of cities, or peri-urban areas, where presumptions regarding territorial powers of the state are disrupted, by jurisdictions of multiple government bodies, as well as interests of land owners and investors.

The chapter focuses on the peri-urban transformations of two of India's largest cities, Kolkata and Hyderabad, after liberalisation of the national economy in 1991. In both cities, the peri-urban areas that are discussed in this chapter were located just outside the territorial boundaries of the cities' planning and development authorities, and were eventually brought under urban jurisdiction either through new authorities or through extension of jurisdictional boundaries of existing authorities. In both cities, the peri-urban areas transformed rapidly, with state-level governments actively remaking relationships with investors, landowners, businesses and residents, and instituting changes in planning and governance systems. The state governments' city-scale place-making objectives, consolidated after liberalisation, were explicit – to create externally legible and attractive investment destinations. In using the term 'place-making', I refer to state governments' objectives to create internationally legible destinations to attract and anchor external investment flows in particular cities, often in direct competition with projects and policies of other state governments. I am also highlighting how these objectives were operationalised by efforts of city-level government bodies and private sector players, particularly through the aegis of planning instruments and/or planning bodies. The two cases provide a window into the actions of planners and contemporary exigencies of planning practice, and are useful for reconsidering the historical assumptions regarding the territorial powers of the state, manifested through planning agencies. It also provides a frame to understand planning in peri-urban areas of India's mega-cities, often decried as particularly aggravated sites of planning failure, beyond a place-based autopsy of overlapping crises and as a relational practice within a milieu of diverse incentives and objectives, across the public–private spectrum.

The following sections of the chapter discuss Kolkata and Hyderabad's peri-urban experiences and examine the extent to which city-level planning authorities[1] and planners in these authorities were able to direct, exercise control and participate in place-making. Various forms of planning instituted to respond to state-level development visions, including spatial and infrastructural interventions, creation of new planning bodies and financial guarantees and negotiations with landowners and investors are discussed, along with their susceptibility to changing governance and market conditions. The experiences in both cities reveal emergent competencies of planners to engage with particular market conditions, albeit with outcomes that severely compromised on equity and sustainability. A brief contextualisation of the cities' planning experiences within the historical legacies that have shaped planning practice in India, highlights that rather than being artefacts of a neoliberal turn in India's development trajectory, the cities' experiences reveal continuities in underlying presumptions regarding the state, market, land, state–citizen compacts, and urbanisation's constitutive geography of power. The concluding section revisits the chapter's core argument that planning is shaped by urban power geographies and there is a need for planning practice to be analysed through frames that go beyond spatial and normative outcomes and binaries such as success/failure and rule/exception.

Place-making in the peripheries of Kolkata and Hyderabad

The peri-urban transformations in Kolkata and Hyderabad have yielded landscapes comprised of bounded islands of development with iconic state-led and developer-led projects. These pockets, often reflecting hyper-modern built form and investment aspirations, transcend locational specifics and constitute the advertised, imagined 'place' of billboards and investor meets and are part of larger narratives regarding the making of 'globally' competitive cities. New infrastructure corridors have made particular peri-urban areas legible to investors. Meanwhile, large swathes of land, situated between iconic developments and infrastructure projects are being consolidated, subdivided, transacted and developed by various scales of private actors, many local, who comply with extant regulatory regimes to differing degrees. The changes reveal a continuing cycle of catch-up to the peri-urban area's new image and imputed value in the eyes of local and external capital flows. Land markets are marked by price volatility and speculation, often linked to government decisions. These transitions have impacted urban–peri-urban linkages of food, water, energy and waste and the spatial, economic and social fragmentations in both cities have eventually reflected in the political constituencies of state-level politics also. The political parties responsible for the cities' initial transformations – the Communist Party of India-Marxist (CPI(M)) in West Bengal and the Telugu Desam Party (TDP) in Andhra Pradesh (united) – lost their respective state elections in the period following their most intensive efforts to create new 'global' images for Kolkata and Hyderabad, with pushbacks from the rural electorate, which felt marginalised by the state governments' attempts to bring in external investors to particular locations.

Yet, it would be disingenuous and reductive to characterise the peri-urban areas of Kolkata and Hyderabad as planning 'failures'. The idea of planning 'failure' is closely tied to normative expectations regarding planning *outcomes*, rather than analyses of the constitutive geography of planning *processes*. Expecting particular planning outcomes reveals assumptions regarding planning power (Forester 1988). Planning practice as an instrument to actualise particular visions of the modern state at the city scale continues to remain mostly unquestioned. Within planning practice there is also enduring belief in the planning ideals of 'organised' and planned urban spaces, over and beyond other considerations of use and value of urban land. Current planning practices and processes emerge from India's post-independence trajectory of applying 'travelling' planning ideas from mostly the US and the UK (Healey 2012), via the Ford Foundation (Banerjee 2009) among others. These practices and processes aim to create new urban and industrial destinations, including new capital cities such as Chandigarh, industrial townships such as Jamshedpur and the institution of Master Plans as the key instrument of spatial control for urban growth and land markets, with little consideration for existing formal-informal urban forms and activities (Holston 2008; Miraftab 2009; Vidyarthi 2010). Through this history and onto contemporary imaginations of professional planning practice, the power of the state to territorialise at the scale of the city-region remains assumed rather than constructed. However, the politics that emerge out of the complexity of urbanisation's constitutive geography of power, the voice and agency of public, private and community networks, and the rescaling of the city as a destination across the regional, national and international scales are now difficult to ignore in everyday practice encounters. This provides the imperative for conceptually unpacking assumptions associated with planning practice and the metrics used both in academic and policy research to assess its relative 'success'. These questions are relevant not only in India, but across countries, where planning authorities are struggling to come to terms with increasing land values and investment interests. The following sections detail planning as it was instituted in the peri-urban areas of Kolkata and Hyderabad. The first section highlights the post-liberalisation development motivations of the state governments consolidated at the city scale, which set the planning agenda in the two cities. Subsequent sections highlight the forms and instrumentalities that planning initiatives took on to materialise these visions, including spatial and infrastructural interventions, creation of new planning bodies and financial guarantees and negotiations with land-owners and investors.

Post-liberalisation state-government development visions

The state-governments' place-making efforts to attract external investors and gain greater financial autonomy became locationally focused on state capitals such as Kolkata and Hyderabad; particularly their peri-urban areas. These locational specifics reflect choices made by regional political parties who were in power at the state level, in response to decentralisation and rescaling of fiscal responsibilities that occurred after liberalisation. They were also prompted by virtue of being operationally simpler alternatives to the complexities associated with instituting deep, comprehensive and inclusive state-level market reforms.

The national economy was liberalised in 1991, in response to an International Monetary Fund (IMF) loan to remedy a balance of payments crisis. It was concurrent with the rise of economic and political regionalism (Sinha 2004). The Congress (I) party, which had dominated elections at both the national level and across states, eventually lost ground to regional parties in state elections. No single political party was able to secure a clear majority in the national elections between 1989 and 1998, and several coalitions with regional parties, such as TDP in Andhra Pradesh (united) and the CPI(M) in West Bengal, were used to form the national government. This gave regional parties voice and agency beyond their regional constituencies. Meanwhile, liberalisation created a terrain of new imperatives – financial autonomy, international consultancies and business lobbies – and regional parties had to learn to balance these against regional constituencies and mandates. Inter-regional competition for investments was coloured by state-level social and political constituency politics, including those of linguistic, social and regional identity, as well as the power that regional parties gained within a fractured national political milieu.

To navigate this terrain, both the TDP in Andhra Pradesh and the CPI(M) in West Bengal hired private consulting firms, including McKinsey, to devise post-liberalisation strategies. In Andhra Pradesh, McKinsey prepared the Vision 2020 (Ramachandraiah 2003), and in West Bengal, prepared a policy report to promote information technology (IT) and other knowledge economy sectors (WEBEL n.d.). These plans were meant to delineate a state-level economic vision, but in practice, they were abstract and without links to operational plans. McKinsey proposed a radical shift from agriculture to the services sector in both states. Both the TDP and the CPI(M) found this difficult to operationalise, given the importance of their rural constituencies – the TDP emerged through the support of the land-owning agricultural entrepreneurial class of the '*khammas*' (Kohli 1988) and the CPI(M) consolidated its position through a long process of agricultural land reforms (Bandyopadhyay 2003; Harris 1993). However, it was possible to operationalise plans for state capitals such as Kolkata and Hyderabad, in isolation from the rest of the state, by focusing on zones of improved connectivity, easy land availability, medical and educational facilities, and a more accessible and consumer-friendly governmental interface (Chacko 2007; Chakravorty 2000; Dabla 2004; Dupont and Sridharan 2007; Kennedy 2007; Ramachandraiah 2003). Spatialising these initiatives in the cities' peri-urban areas offered access to investors from high growth, 'global' sectors, such as IT, Information Technology Enabled Services (ITeS), Biotechnology and other knowledge economy sectors (NASSCOM McKinsey Report of 2005), looking for large land parcels to recreate United States suburban landscapes, without compromising on access to labour, services and political presence that India's mega-cities offered. In choosing the peri-urban areas of state capitals as priority and default locations, the TDP and CPI(M) exemplify the route that many state governments adopted to navigate a complicated political and economic terrain post liberalisation, within limited time frames.

Planning through territorialisation: the role of new parastatals

Roy (2003: xx) highlights: "rapid peri-urbanisation . . . unfolding at the edges of the world's largest cities is an informalised process, often in violation of master plans and state norms but often informally sanctioned by the state". The statement highlights both the 'informal' nature of the actions of developers and landowners, and the 'informal' nature of state presence and sanctions in these areas. The peri-urban transformations of Kolkata and Hyderabad however belie both invocations of the diffuse modality of 'informality' in this description. First, in both cities the conscious intent of the state governments to materialise particular development visions through particular interventions and in particular locations was clear. Second, the consolidations, developments and violations by owners, developers and investors that emerged in these areas reveal responses to land market structures and incentives that the instituted planning interventions put in place, thus highlighting particular logics, rather than diffuse patterns of 'informality'. Operationalising the development of Kolkata and Hyderabad's peri-urban areas involved state governments acting at the city scale, prioritising certain development visions and using planning instruments such as Eminent Domain and Master Plans to operationalise their visions, in ways that had been impossible with McKinsey's broad-brush plans, which nonetheless recommended deep structural transitions. In reverting to particular planning powers and instruments these cities' experiences reveal more continuities than deviations from historical trajectories of urban developmentalism, as well as planning practice in the country.

However, there were also discontinuities within the cities' planning trajectories. In both cities, new parastatal planning agencies were created, over and beyond existing institutional structures. New agencies were used to extend jurisdictions of existing authorities and expand into areas beyond the stipulations of previous Master Plans. There were also new politicised governance coalitions across the public–private spectrum. These could be termed neoliberal exception, but exception as a lens is a form of 'othering' a substantive pattern of how planning power is actualised in India's cities and discourages analyses of processes re-structuring state–market–society relations.

In Kolkata, the Housing and Infrastructure Development Corporation (HIDCO), under the state Housing department (not the Urban Development department, as is usual), and Bhangar Rajarhat Area Development Authority (BRADA) were created to develop the city's eastern periphery, where development was restricted by the basic development plan of 1966, because of its wetland characteristics. Yet the east offered proximity to the existing IT hub in Salt Lake (renamed Bidhannagar) and the creation of HIDCO and BRADA allowed circumvention of the earlier Master Plan. However, there were conflicts within the bureaucracies of Kolkata Metropolitan Development Authority (KMDA), Kolkata's existing planning and development authority, and between KMDA and HIDCO/BRADA in operationalising these new visions. Now, these dynamics are less visible. The Principal Secretary of the Urban Development department is the chairman of HIDCO, and BRADA was dismantled after the Trinamool Congress won the state elections, after 34 years of CPI(M) rule. BRADA's dismantling came with the acknowledgement of indiscriminate land speculation and violence associated with land aggregation in the new areas.

In Hyderabad also, new authorities were established for strategically important peri-urban areas. The Cyberabad Development Authority (CDA) was set up for a new IT sub-city to the north-west of the city, beyond Hyderabad Urban Development Authority's (HUDA) jurisdiction. In practice, CDA's planning work was undertaken by HUDA officials. However, a new authority allowed Cyberabad to have its own Master Plan. Hyderabad also used public–private partnership-based governance models. The GMR Hyderabad International Airport Limited (GHIAL) was a consortium between the state government and private partners, including the GMR group, to develop the Shamshabad international airport, considered to be a critical infrastructure to reposition Hyderabad in the national and international investment context. The project was bundled with 5,000 acres of land and declared a Special Economic Zone (SEZ) with autonomous governance. This facilitated financial structuring through the leveraging of land. Moreover, 40 per cent of the area designated as the SEZ was under the catchment zone of the Himayatsagar lake, one of the main sources of water for the city. The SEZ designation played an important role in navigating the politics that emerged, when a public interest litigation (PIL) was filed against the airport. The PIL was rejected (Ramachandraiah and Prasad 2004). Later the Hyderabad Airport Development Authority (HADA) was established for the airport's hinterland. The GMR group was part of the new authority. In Hyderabad also, these dynamics are now less discernible, as these authorities have been subsumed under the new Hyderabad Metropolitan Development Authority (HMDA).

Territorialising through new development corridors

Place-making in the peri-urban areas was also facilitated through new high-speed, limited access road connections between new investment destinations and the cities' international airports. These transport corridors were not meant to mitigate existing infrastructure deficits, as is often the case in India's planning initiatives, but meant to anticipate and direct future growth. These corridors provided spatial form, without using a bounded notion of a development area. In Hyderabad, an eight-lane, high-speed, limited access Outer Ring Road (ORR) connected the airport to Cyberabad and Gatchibowli in the west, Genome Valley in the north, and multiple SEZs. The new development corridor, encircling the city, opened up the zone between the boundaries of the existing city and the ORR as a land market and secured a high degree of private development interest. It legitimised land consolidations on the outer boundaries of the ORR as being part of the Hyderabad growth story, which were otherwise perceived as being too far. In Kolkata too, a new six-lane, high-speed road connection was created between New Town Rajarhat and the city's international airport, which bolstered New Town Rajarhat as an independent land market. In both cities, the new roads made it possible to enter and exit the city and conduct business in the peri-urban locations, without substantive engagement with the existing cities. This speaks to the intent to create a new ahistorical image for the cities. Such parallel spaces and economies allowed the cities, and by extension the respective state governments, to develop connections with external investors, without deeper and broader market reforms, post-liberalisation.

Building relationships with land owners and private sector partners

An enduring critique of planning practice has been its uncomfortable conceptual and operational relationship with urban land markets, tenure arrangements and private and community rights to own, transact and develop property. From an economic perspective, planning regulations are interpreted as 'constraints', skewing land markets and creating 'inefficient' zones of high prices. From a more operational perspective, planning instruments such as Master Plans almost never acknowledge property boundaries, spatiality of tenurial practices, and/or private and community interests in owning, transacting, using and leveraging property at various scales. In many Indian cities, the Master Plan's proposed land use map is created in isolation from the development dynamics of the city, with little or no relationality to individuals, firms, developers, institutions, business and other interest groups associated with producing urban space.

In contrast, planning authorities in both Kolkata and Hyderabad's peri-urban areas tried to navigate the terrain of property owners, land markets, corporates and developers, often through dynamic 'transversal' logics that Caldeira (2016) imputes as characterising the behaviour of peripheral residents, rather than the actions of the state itself. Planning authorities preferentially targeted a select elite group, who dove-tailed with the state's vision of post-liberalisation development. In both cities new generation public–private coalitions were built through process-stage engagements with actors seen as significant. The coalitions highlight changes and learning in planning processes, beyond assumptions regarding the monopoly power of planners to spatialise collective urban futures in regulatory detail.

Initially, HIDCO tried to control the peri-urban land market through outright acquisition in New Town. As per HIDCO, the compensation amounts of INR 36,000 per acre (USD720) were the highest in the country at the time. Nonetheless, there were mobilisations and protests as villagers claimed that compensations neither reflected the displacement value of fishing and five rotations of crops that the land supported, nor did it reach all the people who were part of the dense social network of livelihoods and tenure-based rights in the area. The failure of the New Town pricing experiment prompted a different modality in the Bhangar-Rajarhat area, where the state retreated from direct acquisition and private actors were allowed to aggregate land under 'laissez-faire' principles, with implicit support from BRADA. This was associated with a high degree of violence. Eventually, BRADA was dismantled. In Hyderabad, by contrast, the politically significant land owning class of '*khammas*' were active in Hyderabad's peri-urban areas, before liberalisation. Original landowners, engaged in low-return agriculture, were often easily displaced via financial compensations. In Hyderabad's peri-urban land markets, there were limited deployments of Eminent Domain and while facilitating access to land was a cornerstone of negotiations between the state government and preferred investors, state role in land aggregation was less obvious.

Beyond access to land, the public–private coalitions highlight the cities' particular histories with local capital. In Kolkata, the imperative to build coalitions with developers emerged from efforts to allay the city's negative business image. Powerful union politics, a communist state government, deindustrialisation and state-level economic stagnation made it difficult to reach out to external investors. The CPI(M) government started remaking their own and Kolkata's image by floating 13 public–private companies (Bengal Ambuja, Bengal Shrachi, etc.) in partnership with local capital, represented by developers from the *Marwari* community, to build affordable housing (Sengupta 2006). These efforts were expanded through investor meets and road shows to reach domestic and international investors. Developers who partnered with the government became unofficial advisors and facilitators of developments in Rajarhat and New Town. This bias towards developers is still visible in Kolkata. In contrast, Chandrababu Naidu of TDP leveraged the ambitions of the '*khammas*' who were prominent not only in Hyderabad's land and construction industry, but also constituted the diaspora connected to the IT industry in the US. In building his own legitimacy and Hyderabad's new image, Naidu aggressively wooed corporates, such as Microsoft, from the beginning, and allowed this to trigger medium and small-scale domestic, often local investment interest (Chacko 2007; Ramachandraiah 2003). Naidu convinced Microsoft to establish its first international offshoring centre in Gatchibowli. He built high-recall infrastructure, such as the Shamshabad international airport, explicitly in competition with Bangalore, which he identified as Hyderabad's closest locational rival for international investors. These actions of the Naidu government at the city scale shaped national discourses of good governance and normalised particular practices of land aggregation, compensation, plan-making, instrumentality of planning agencies, and public–private negotiations within planning (Basu 2007).

State guarantees in land markets

Braudel (summarised in Wallerstein 1991) highlights that the state typically acts as a regulator, particularly of price, to prevent monopolies. However, the state may also act as a guarantor of price, and exceptional profits are possible only when the state rescinds its role as regulator and becomes a guarantor. In both cities, the place-making efforts in the peri-urban areas were associated with state guarantees, such as continued support for land development in far-flung peri-urban areas and commitment to provide governance and infrastructure support to new investors. The significance of these explicit and implicit state guarantees are revealed in the volatile land prices that followed perceived shifts in the state government's intent, in both cities. In Hyderabad, the alignment of the ORR went through multiple changes. In 2006, the alignment was changed 115 times (*The Hindu* 2006) and land prices fluctuated with each change. In Kolkata New Town, HIDCO tried to become the monopoly supplier of developed land and capture land value increases, but the price markups were argued to be one of the highest in Asia. The implied state guarantees of well-planned spaces and infrastructure were not sufficient for the asking price and there were no takers for HIDCO land. In contrast, private actors were allowed to aggregate land in the Bhangar-Rajarhat area, with the state government guaranteeing conditions for this 'laissez-faire' private role in land markets. Land investors responded well to this form of state guarantee and prices soared. When BRADA was dismantled, the land market collapsed and prices dropped, almost overnight. In both cities, state guarantees directed and shaped investor interest in peri-urban land markets.

Developing relationships with investment logics

The state governments' partnerships with different categories of private sector players were associated with the leveraging of land in multiple ways. Some partnerships facilitated private access to aggregated land – a general 'sweetener', since urban regulations, complex tenurial relationships, incomplete land records and urban land ceiling legislations are often entry barriers for private actors, especially investors without local networks. In some cases, land was the government's equity share and effectively a subsidy, since it diminished the private partners' imperative to recover land costs, in order to break even. Both modalities shaped the first generation public–private partnership projects in Kolkata. Land was also collateral for institutional financing, for example, in Hyderabad's international airport. With the government facilitating land access, private partners were able to attract private equity partners interested in projects with assured access to land to reduce their risk exposure. Brokers and developers in both cities revealed that certain investors used land as a 'hedge' in portfolio strategies. Land, like gold, retains value well over time, with or without development. If inflation rates remain lower than the rate of appreciation of land value, land assets are able to 'hedge' other risky portfolio investments. Hedging is possible only with implicit and explicit state guarantees.

These modalities reveal certain sophistications in conceiving land beyond the static imaginations that mark planning legacies. However, by all accounts, they reveal the capture of this sophistication by certain preferred partners, in exclusion of other forms of use and value of land associated with existing and future users of land in the areas.

Continuities with planning legacies

Kolkata and Hyderabad's experiences focus on a particular moment of India's planning history, post-liberalisation. While it may be convenient to explain these experiences as artefacts of a neoliberal turn in state–society–market relations, a brief look at India's modern town planning history, highlights many continuities. Modern town planning in India through the nineteenth and twentieth centuries, reveals continuing presumptions regarding the power of the state to territorialise at the city scale through planning instruments, often going over and beyond existing political configurations of state–society–market relations. Moreover, planning has remained a key terrain of state practice to materialise and mobilise development visions at the city scale. There are also continuities in planning practice being invested with expectations of particular normative spatial and social outcomes, by a select group of elites, whose visions, while being situated in the city, transcended to desires to build a modern Indian state.

In India, this group of urban elites was also part of the political elite in the Congress party post-independence and constituted the first professional body of planners in the country. Notions of spatial ordering and place-making emerged from their imaginations regarding cities as spaces for social engineering, albeit egalitarian ones, where scientific and ordered spaces would facilitate an ordered urban society comprising of a civil citizenry (Vidyarthi 2010). Moreover, planning was seen as a project to hold at bay haphazard urban growth tendencies, engendered either by the agency of an unbridled urban population, or an unbridled market. The ordered spaces of colonial 'White Towns' were spatial blueprints for early sanitary commissions and city improvement trusts, which introduced planning measures at the city scale. Greenfield administrative capitals such as New Delhi and Chandigarh translated state-led development visions of building a modern nation (Kalia 2006). Planning territorialised state power at the city scale through Eminent Domain enforced use-separated zones through planning permissions, developed transport networks and regulated land markets through development controls. Notions embedded in these exercises were normalised, with the third five-year plan mandating cities across the country to prepare Master Plans and the Delhi Master Plan, prepared by the Ford Foundation (Banerjee 2009), with its land use zoning and development controls, becoming the default template used unreflexively in cities across the country.

The planning experiences in the peri-urban areas of Kolkata and Hyderabad reveal continuities with these legacies, particularly in terms of planning mandates translating development visions of the state and local elites, through place-making initiatives. The contemporary inter-institutional conflicts and collaborations, which emerged between old and new planning authorities in Kolkata and Hyderabad, find historical parallels in the negotiations that marked interactions between planners during the planning of Chandigarh in the 1950s (Kalia 1999) and between Indian and international planners during Ford Foundation's preparation of Kolkata's Basic Development Plan (Banerjee 2009). Currently, many such conflicts between planners within planning practice have been rendered invisible, with ritualised making and enforcing of Master Plans. However, the disjunctures, are now clear in interactions with market and society, and in the difficulties in materialising and territorialising state power, without considering urban residents, actors in the land market, property rights, tenurial arrangements, etc.

Conclusion: planning 'failures' or emerging terrains of planning practice?

The experiences in Kolkata and Hyderabad reveal certain competencies on the part of planners navigating a complicated terrain. In both cities, planning mandates were determined by state-level development visions, which emerged in response to political and economic exigencies post-liberalisation. In both cities, planners were relatively successful in actualising their given mandates. In both cities, planners engaged, often within highly politicised terrains, with local and external players at scales that were more disaggregated than those typically associated with Master Plans. In both cities, planners learnt to engage with the logics of certain developers and investors.

These learning curves make it difficult to reduce the planning experiences of the two cities to planning 'failures', even though long-term normative outcomes traditionally associated with planning interventions are hard to find. The spatially fragmented value landscape of public–private investments, susceptible to local and external changes, reveal more than 'failure'. They highlight the ways in which planning practice has negotiated, collaborated and conflicted with a varied set of developer/investor/landowner-led urban practices. These highlight the need to go beyond teleological outcome-led development binaries such as success/failure, rule/exception, and planned/unplanned urban spaces as frames to analyse planning practice. They highlight how planning continues to be an evolving terrain of politics, where changes in state–society–market relationships are being actively constituted.

The exigencies shaping contemporary urbanisation demand something more than static planning ideals, which, as Chatterjee (2004) has argued, are conceived in 'universal urban space-time', i.e. place- and time-less imaginations of ordered and planned spaces beyond social and political dynamism. Producing more equitable urban spaces requires higher levels of dynamism in planning processes than is visible in the peri-urban areas of Kolkata and Hyderabad. Without incremental politics of planning operationalised and built over time with different creators and users of urban land value, across the tenure spectrum, incorporating knowledge networks that emerge as cities transform, planners are reduced to occupying a binary space of either enforcing a static Master Plan or facilitating highest and best use, dictated by powerful interest groups. Both roles are difficult to sustain within uneven and dynamic geographies of power in urban areas. Legacies of spatial control have disincentivised planners from building coalitions with diverse urban residents. Static, two-dimensional spatial planning regulations have made invisible much of the dynamism associated with production of urban space, creating a vacuum of data and imagination, and allowing colonisation by interest groups.

Deeper epistemological engagements with processes and instrumentalities shaping planning practice may in themselves be important to re-engage with normative expectations such as social and spatial equity. Advocacy and intervention, which acknowledge rather than dismiss politicised terrains of planning practice, offer more grounded practices of hope, to engage with planning in the face of power (Forester 1988), reimagine normative expectations from planning practice and construct potential pathways towards these expectations. While it may be naïve to have absolutist expectations from planners to hold and actualise larger normative values within a politicised urban realm, it may be imperative to devise relational practices of advocacy and intervention to lend weight to particular normative values. The contemporary challenge for planners is to conceive outcomes and instruments that engage with value-laden occupancy and development politics of residents and groups creating, capturing and using urban space and articulating urban land values, beyond dynamics of elite capture.

Note

1 City-level planning authorities in India are parastatals. Though they have city-level jurisdiction, they are not constituted through city-level elections. They comprise of state government appointees and report directly to the state government. As such, actions of city-level planning bodies often reveal clear instrumentalities of state governments.

References

Bandyopadhyay, D. (2003) "Land reforms and agriculture: The West Bengal experience", *Economic and Political Weekly* 38(9): 879–884.

Banerjee, T. (2009) "US planning expeditions to postcolonial India: From ideology to innovation in technical assistance", *Journal of the American Planning Association* 75(2): 193–208.

Basu, P.P. (2007) "'Brand Buddha' in India's West Bengal: The left reinvents itself", *Asian Survey* 47(2): 288–306.

Caldeira, T.P.R. (2016) "Peripheral urbanization: Autoconstruction, transversal logics, and politics in cities of the global south", *Environment and Planning D: Society and Space* 35(1): 3–20.

Chacko, E. (2007) "From brain drain to brain gain: Reverse migration to Bangalore and Hyderabad, India's globalizing high tech cities", *GeoJournal* 68(2–3): 131–140.

Chakravorty, S. (2000) "From colonial city to globalizing city? The-far-from-complete spatial transformation of Calcutta". In P. Marcuse and R. van Kempen (eds) *Globalizing Cities: A New Spatial Order?* Oxford: Blackwell, pp. 56–77.

Chatterjee, P. (2004) *The Politics of the Governed: Reflections on Popular Politics in Most of the World*, New York: Columbia University Press.

Dabla, A. (2004) "The role of information technology policies in promoting social and economic development: The case of the state of Andhra Pradesh, India", *The Electronic Journal of Information Systems in Developing Countries* 19(5): 1–21.

DuPont, V. and Sridharan, N. (eds) (2007) "Peri-urban dynamics: Case studies in Chennai, Hyderabad and Mumbai", CSH Occasional Paper No. 17. Available: https://ideas.repec.org/p/ess/wpaper/id974.html [accessed 15 January 2017].

Forester, J. (1988) *Planning in the Face of Power*, Berkeley, CA: University of California Press.

Harris, J. (1993) "What is happening to rural West Bengal?", *Economic and Political Weekly* 28(24): 1237–1247.

Healey, P. (2012) "The universal and the contingent: Some reflections on the transnational flow of planning ideas and practices", *Planning Theory* 11(2): 188–207.

Holston, J. (2008) *Insurgent Citizenship: Disjunctions of Democracy and Modernity in Brazil*, Princeton, NJ: Princeton University Press.

Kalia, R. (1999) *Chandigarh: The Making of an Indian City*. Oxford: Oxford University Press.

Kalia, R. (2006) "Modernism, modernization and post-colonial India: A reflective essay", *Planning Perspectives* 21(2): 133–156.

Kennedy, L. (2007) "Regional industrial policies driving peri-urban dynamics in Hyderabad, India", *Cities* 24(2): 95–109.

Kohli, A. (1988) "The NTR phenomenon in Andhra Pradesh: Political change in a south-Indian state", *Asian Survey* 28(10): 991–1017.

Miraftab, F. (2009) "Insurgent planning: Situating radical planning in the global South", *Planning Theory* 8(1): 32–50.

NASSCOM-McKinsey (2005) *Extending India's Leadership of the Global IT and BPO Industries*, New Delhi: McKinsey. Available: www.mckinsey.com/~/media/mckinsey%20offices/india/pdfs/extending_indias_leadership_of_the_global_it_and_bpo_industries.ashx [accessed 11 January 2010].

Ramachandraiah, C. (2003) "Information technology and social development", *Economic and Political Weekly* 38(12–13): 1192–1197.

Ramachandraiah, C. and Prasad, S. (2004) "Impact of urban growth on water bodies: The case of Hyderabad", Working Paper 60, Centre for Economic and Social Studies, Hyderabad: Centre for Economic and Social Studies. Available: www.cess.ac.in/cesshome/wp%5Cwp-60.pdf [accessed 24 October 2016].

Roy, A. (2003). *City requiem, Calcutta: Gender and the Politics of Poverty* (Vol. 10). Minneapolis, MN: University of Minnesota Press.

Sengupta, U. (2006) "Government intervention and public–private partnerships in housing delivery in Kolkata", *Habitat International* 30(3): 448–461.

Sinha, A. (2004) "The changing political economy of federalism in India: A historical institutionalist approach", *India Review* 3(1): 25–63.

The Hindu (2006) "ORR alignment changed 115 times: Devender Goud". Available: www.hindu.com/2006/08/31/stories/2006083111520400.htm [accessed 24 October 2016].

Vidyarthi, S. (2010) "Inappropriately appropriated or innovatively indigenized?: Neighborhood unit concept in post-independence India", *Journal of Planning History* 9(4): 260–276.
Wallerstein, I. (1991) "Braudel on capitalism, or everything upside down", *The Journal of Modern History* 63(2): 354–361.
West Bengal Electronics Industry Development Corporation (WEBEL) (n.d.). West Bengal's Policy on IT Enabled Services. Available: www.webel-india.com/pdf-download/west_bengal_ites_policy.pdf [accessed 3 February 2017].

Part II

Economy and economic actors

7

Urbanisation and development

Reinforcing the foundations

Ivan Turok

Introduction

The new conventional wisdom among global development organisations and think tanks is that cities function as powerful generators of national economic prosperity (OECD 2014; UN-Habitat 2012; World Bank 2009). Indeed, it has become a cliché that cities are 'engines of growth'. A corollary is that urbanisation drives economic progress by concentrating people in cities (GCEC 2014; McKinsey 2012). UN-Habitat's New Urban Agenda illustrates this well. The central theme of Habitat III has the bold title: "The transformative power of urbanisation" (UN-Habitat 2016). The website declares confidently: "Throughout modern history, urbanisation has been a major driver of development and poverty reduction . . . Urbanisation had become a driving force as well as a source of development with the power to change and improve lives" (UN-Habitat 2016).

Support for the idea that urbanisation is a force for growth and development has spread among politicians, professionals, planners and normally sceptical economists. For example: "Urbanisation is essential for (developing) countries to move from poverty to prosperity . . . Cities are one of the primary sources of economic growth in the developing world" (Glaeser and Sims 2015: 1, 7).

Yet the meaning of such pronouncements is ambiguous. There is patently an element of bullish rhetoric in challenging the anti-urbanism of many governments in the South and urging them to take cities more seriously. Yet for the sake of credibility and sound decision-making, the economic significance of an urbanising population should not be exaggerated. Does reliable evidence support the assumptions being made about the relationship between urbanisation and development, the time horizons under which it unfolds and the national context? Considering how important it is for these countries to raise household incomes and well-being, can one infer that urbanisation necessarily engenders development? Or does a positive connection depend on certain preconditions, which may require an active, enabling role for governments?

If urbanisation leads to prosperity, this has profound implications for countries wrestling with burgeoning urban populations. The palpable consequences seem overwhelmingly negative – chronic traffic congestion, overstretched public services, concentrated squalor, misery and social discontent (Buckley and Kallergis 2014; UN-Habitat 2014). These problems are generally most acute in sub-Saharan Africa, where urbanisation is happening at much lower levels of GDP per capita than occurred elsewhere (Jedwab *et al.* 2015; World Bank 2015). Three-quarters of governments are so concerned about the burden that they have policies to restrict rural–urban migration (UNPD 2012). Their developmental priorities reside in policies to bolster agriculture, mining, industrialisation, education or public health.

Yet if urbanisation can boost economic progress, this would justify a big shift in approach (Buckley and Simet 2016). It might mean that some nations are under-urbanised and would benefit from faster urban growth and bigger cities (Freire *et al.* 2014). The 2009 World Development Report argued that this applies to much of sub-Saharan Africa (World Bank 2009). Rwanda and Ethiopia have set targets to accelerate urbanisation, reflecting China's remarkable experience of urbanisation combining with industrialisation to lift some 500 million people out of poverty over the last three decades (OECD 2013; World Bank 2014).

If urbanisation can accelerate prosperity, government policy towards cities ought to become a priority. Urban policy would not be about redistributing resources to shift the benefits and burdens between urban and rural areas. It would be a means of generating additional resources and expanding livelihood opportunities. Planning city growth would transcend improving the liveability of poor communities. It would involve building the foundations to increase economic activity and mobilise tax revenues to plough back into human well-being. Investment in costly urban infrastructure would be justified because it can create decent work and provide lasting routes out of hardship, not simply because it relieves adversity and inconvenience for urban residents. In short, the new urban agenda would promote self-sustaining development.

The purpose of this chapter is to interrogate this proposition by exploring the relationship between urbanisation and development in the South. The first section outlines a strict interpretation, namely a direct causal link from urbanisation to growth. The second section summarises the available evidence, which actually challenges this perspective. The third section presents a broader view, namely a complex two-way relationship that is context-dependent. This suggests that urbanisation can foster development if suitable mediating conditions exist, but the connection is far from automatic. The fourth section of the chapter focuses on a key supporting condition that is often neglected, namely the urban land–infrastructure–coordination nexus. The final section concludes.

Could an urbanising population cause economic growth?

Quotations from two contrasting sources illustrate the view that urbanisation leads to growth. First: "Urbanisation is undoubtedly a key driver of development – cities provide the national platform for prosperity, job creation, and poverty reduction" (Glaeser and Joshi-Ghani 2015: 1). Second, in a well-publicised report on Africa's growth prospects, McKinsey wrote: "Africa's long-term growth also will increasingly reflect interrelated social and demographic trends that are creating new engines of domestic growth. Chief among these are urbanization and the rise of the middle-class African consumer" (2010: 3).

To unpick this relationship, two issues need clarification at the outset: (i) the definitions of urbanisation, growth and development; and (ii) the direction, magnitude and mechanisms of causality. Strictly speaking, urbanisation is the rising share of the national population living in cities. Growth is an increase in national economic activity, output or income. Development is a broader concept related to the changing character of the economy and linked with higher productivity, more equity or greater sustainability. The proposition that urbanisation drives growth implies that the urbanising population has a direct and ubiquitous causal effect on economic activity. It is an independent source of growth, and not merely a consequence of growth. The earlier statements that urbanisation is 'a key driver', 'essential' and 'a primary source' of growth also imply that it is a necessary condition for prosperity. Some authors have argued that the relationship between cities and growth is so regular and robust that it has the character of a rule or universal law, therefore diverse national and local contexts make little difference (Bettencourt and West 2010).

At first sight it seems implausible that an urbanising population would in itself bring about higher national output or average incomes (Storper *et al.* 2012). One mechanism might be through increased labour supply, but rural migrants in developing countries tend to be low skilled and the availability of such labour is rarely a constraint on business growth. Lack of labour demand is a much bigger problem, reflecting weak urban economies (de Haas 2012). Recent migrants are also unlikely to boost consumer demand because of their low incomes. Increased entrepreneurship is another possibility, but most new enterprises are informal and struggle to grow because of their limited business skills and access to capital (La Porta and Shleifer 2014). Consequently, major obstacles are likely to limit the economic impact of expanding urban populations (Buckley and Kallergis 2014; Fox 2014; Turok and McGranahan 2013).

In the longer run there are several ways in which the growing concentration of population might set in motion processes that improve economic performance. There is a large literature from the North that contains many theoretical possibilities, summarised as 'agglomeration economies'. These are well known and need little elaboration here beyond a brief outline. At the heart is the role of urban *density* (Glaeser and Sims 2015; Venables 2015; World Bank 2009). Jane Jacobs (1984) was one of the first to argue that people living in close proximity under conditions of relative density and social diversity give rise to intense *interaction*. This may be reinforced by skilled migrants from diverse origins offering new capabilities and insights, creating new opportunities and making new demands on host communities (Beall *et al.* 2010). The reciprocal influences take the form of transactions, negotiations, competition, comparison, co-operation and learning. The bustle, buzz and sheer pressure of human interaction encourage creativity, ingenuity, enterprise and innovation. The tangible and intangible results (such as accumulated knowledge) can be seen as *common-pool resources* or positive externalities that increase the efficiency and flexibility of firms, encourage investment, raise productivity and thereby spur growth in output and income (Glaeser 2011; Henderson 2003; Storper 2013; Turok 2016).

Allied to density is another fundamental issue of city size. Dense concentrations of population enable *economies of scale* in providing shared infrastructure and public services (Annez and Buckley 2009; Collier and Venables 2015). These reduce the costs of transport, energy, water and other resources for households and firms. The cost and availability of these services make living, working and investing in cities much more attractive than the countryside. Large cities offer superior opportunities than towns as places of reproduction and consumption because of the varied educational institutions, social amenities and housing options (Glaeser 2011). Economies of scale also operate in the labour market to improve the matching of workers and jobs, thereby improving efficiency and the choices available to employers (Duranton 2014b; Moretti 2014). Finally, large cities permit a division of labour between economic agents to organise themselves around particular tasks. This specialisation yields substantial efficiencies and enhanced skills, which also raise productivity and growth.

Many of these theories actually imply an interdependent relationship between urbanisation and growth – they are connected in both directions. Large compact cities enable interactions, efficiencies and shared assets that enhance business performance. However, the theories are often interpreted as implying that population growth comes first and gives impetus to the broader economic dynamics. Glaeser (2011), Florida (2003) and Moretti (2014) emphasise population movements leading to changes in firm location and performance. Yet this neglects the economic origins of cities and their subsequent development as local production systems (Storper and Scott 2009). It also overlooks how migration responds to opportunities – people need the possibility of earning a living to move from where they are settled (de Haas 2012). When urban economies are growing strongly, in-migration tends to accelerate and vice versa (Potts 2016).

What does the evidence show?

The most common forms of empirical evidence used to assess the relationship between urbanisation and growth are cross-country correlations (Annez and Buckley 2009; Duranton 2014a; Henderson 2010). They compare countries according to their rates of urbanisation and average incomes (Gross Domestic Product (GDP) per capita). The results show a robust statistical association: highly urbanised countries are much more likely to be prosperous. Henderson (2010) found a strong correlation ($R^2 = 0.57$) between the proportion urbanised and income per capita in 2004 for all countries around the world. Duranton (2014a) found an almost identical relationship for 189 countries using 2012 data. Each additional percentage point of urbanisation was associated with about five extra points of GDP per capita. Hence a modest increase in urbanisation yields a much greater increase in average incomes, indicating just how significant the urbanisation–development connection appears to be.

However, a statistical artefact does not constitute evidence of a causal relationship from urbanisation to growth. Urbanisation may be largely a consequence of economic development (uneven job opportunities) rather than a cause. Structural transformation from agriculture to manufacturing and services would inevitably involve a shift in economic activity from the countryside towards cities (Barrios *et al.* 2006; Henderson 2003). There is considerable historical evidence from around the world that people are pulled into cities by industrialisation and perhaps pushed out of rural areas by labour-saving technologies in agriculture (Henderson 2010; Scott and Storper 2015). It is also possible that a third (unknown) variable explains both GDP growth and urbanisation, such as improvements in education or technology. Duranton (2014a: 3) states frankly that

> what the relationship means is deeply unclear. What is cause? What is consequence? How much of that extra 5 percent of GDP per capita is a consequence of this extra percentage point in the rate of urbanization? … [W]e have almost no idea.

A superior approach than snapshot comparisons is to compare change in urbanisation and growth over time for different countries in order to explore which comes first. There are only two substantive studies of this kind. Henderson (2003) undertook a systematic analysis for 70 countries between 1960 and 1990. His conclusion was unequivocal: "the paper finds little support for the idea that urbanisation per se drives growth. Urbanisation is a by-product . . . as economic development proceeds, rather than a growth stimulus" (2003: 50). Duranton concurred that this "conclusion is unsurprising and confirms a broad consensus that urbanisation is a benign transition that to a large extent follows but does not profoundly affect the process of development" (2009: 102). Two of the leading economists in this field therefore believe that urbanisation is much more likely to be an effect of economic growth than a cause. This contrasts with the original assertions that urbanisation drives growth.

Brulhart and Sbergami (2009) undertook an even bigger study using data for 105 countries over the period 1960–2000. Unlike Henderson, they found some evidence of a positive association between urbanisation and subsequent economic growth, especially for low-income countries. The relationship also varied depending on the structure of the economy and its location, so it was far from homogeneous or ubiquitous. Yet they were very circumspect about their results and concluded that "our findings should be taken as suggestive at best" (Brulhart and Sbergami 2009: 60). Their methodology did not enable them to isolate the specific causal effect of urbanisation on economic growth. Consequently they did not say whether urbanisation had a strong or weak impact. Nevertheless, the possibility that urbanisation is intrinsically connected to growth in developing countries is an important finding.

A fundamental problem with these econometric techniques is that urbanisation and growth are so closely entwined, and surrounded by so much noise, that it may be impossible statistically to separate them and identify a causal effect in any particular direction (Storper and Scott 2009). An alternative method may be required that probes more deeply into the dynamics of industries and labour markets in different cities. The interdependent relationship between urbanisation and development is discussed in the next section. The fact that these processes are so strongly shaped by each other certainly calls into question the simple notion that an urbanising population necessarily leads to, and is an independent source of, economic growth. It seems that the confident pronouncements that urbanisation is an autonomous driver of progress do not have a robust evidential basis.

Another reason to doubt that urban growth drives prosperity is the indisputable evidence that many countries have urbanised rapidly without experiencing economic growth, sometimes over lengthy periods (Fay and Opal 2000; Henderson 2010; Jedwab and Vollrath 2015; Turok 2013). Many are in sub-Saharan Africa. 'Push' factors in rural areas seem to be the main reasons for urbanisation, including climate change, agricultural stress, civil conflicts and related disasters (Annez and Buckley 2009; Barrios *et al.* 2006; Henderson *et al.* 2014). This demonstrates that urbanisation is not always even associated with growth. At the very least there are some vital conditions in these countries inhibiting a positive relationship forming between urbanisation and growth. These mediating factors are barely understood.

Urbanisation and economic development go hand in hand

A broader perspective seems more plausible, involving a two-way relationship between urbanisation and development with extensive feedback effects. Urbanisation is neither simply a cause of growth, nor merely a symptom. By increasing density and proximity between human activities, it both enables growth and is an outcome of growth. These interactions also transform how economic growth itself evolves, so it is more accurately referred to as development. Urbanisation involves the spatial concentration of capital as well as labour. It is an integral part of economic development, and fuels continuing prosperity. The strength of the connection depends on the context – the relationship does not occur in a vacuum. Much depends on the form of urbanisation and the composition of the economy. Neither urbanisation nor growth is a straightforward or uniform process occurring in the same way in different places. These complexities make it extremely difficult to disentangle and measure the magnitude of the causal relationship between urbanisation and development.

There are multiple ways, or channels, through which urbanisation and development influence each other in a recursive and evolving manner. A positive relationship in any particular city or country is also not inevitable because it may be disrupted by other factors (Jedwab and Vollrath 2015). Hence urbanisation sometimes occurs without development, just as growth may occur in advanced economies without further urbanisation. A mutually reinforcing relationship requires urban environments that are economically functional. These preconditions don't emerge spontaneously through the operation of market forces (Collier and Venables 2015). They depend on collective action, i.e. planning, regulation, coordination and investment in infrastructure. Conducive conditions can spur the dynamics of agglomeration and establish a circular and cumulative process of increased output, greater efficiencies, learning new techniques, more specialisation, higher value production, higher incomes and further job creation.

The foundations of agglomeration dynamics stem from the clustering together of economic activities to save transport costs. Creating an interconnected system also enables a division of labour within production as firms organise themselves around particular products or tasks, which yields efficiencies and enhanced skills. Proximity enables them to perform complementary tasks, collaborate and trade with each other to generate synergies. The larger the town or city, the bigger firms can grow and the greater the economies of scale available to them in terms of their unit costs of production. There are additional efficiencies from proximity to other firms, such as access to specialised suppliers, customers, a large labour pool and shared infrastructure. Specialisation also applies at the level of the city, with benefits from focusing on a function and building a distinct advantage. Specialisation becomes more important as transportation technologies improve, external trade grows and competition intensifies.

The connection between urbanisation and development was strongest during the industrial revolution in the North, causing an unprecedented increase in productivity (Scott and Storper 2015). Mechanisation also enabled major advances in agriculture and removed the constraint of food insecurity that had restricted urban growth before. Design improvements in clean water supply and sanitation systems reduced the public health problems that had plagued dense settlements. A similar relationship between urbanisation and development has recurred across Asia over the last four or five decades (World Bank 2009; Overman and Venables 2010), despite the much faster rate of urbanisation. Accelerated industrialisation in countries such as Japan, Korea, Taiwan, China, Malaysia and Indonesia has underpinned the link between urbanisation and development.

Experience in sub-Saharan Africa has been different. The 1980s and 1990s were 'lost decades' when many countries experienced prolonged stagnation (Potts 2016; Turok 2013). The fortunes of many African economies have turned around since then, driven by natural resources. The exploitation of large stocks of oil, gas, metals and minerals such as diamonds and coal has boosted the exports of countries such as Nigeria, Angola, Ghana, Gabon, Mozambique and Tanzania. The problem for African cities is that these resources have been exported as primary commodities rather than processed capital and consumer goods. The rents have not been reinvested in building productive capacity and industrial diversification, but rather in importing consumer goods. There have been relatively few jobs created and limited human capital development. Many African cities have consumer economies (retail, transport and personal services) rather than producer economies (manufacturing, financial services and other tradeable industries). This has hindered increases in urban productivity, employment and incomes, and meant higher rates of poverty and slum housing than would otherwise have existed (Gollin *et al.* 2015).

It seems that urbanisation can reinforce development, but this is not assured. The economic structure also influences whether the outcome is broad-based prosperity. Industrialisation seems to go with urbanisation to generate higher productivity, more employment and rising incomes. Growth driven by natural resources tends to promote inequality rather than generalised well-being in urban areas.

The urban land–infrastructure–coordination nexus

Another fundamental reason why urbanisation may not contribute to development is because its form is inefficient and exclusionary. The physical fabric is the foundation of productive and liveable cities. Urban economies may be impaired because this is left to the vagaries of individual decisions and market forces (Glaeser and Sims 2015; Venables 2015; World Bank 2013). Haphazard and unstructured urban growth may give rise to gridlock, overloaded infrastructure, degraded environments, and higher costs of living and doing business (Buckley and Kallergis 2014; UN-Habitat 2014). This deters private investment, inhibits external trade and curbs entrepreneurial dynamism (Turok and McGranahan 2013). The diseconomies of agglomeration outweigh the advantages, creating a cycle of low investment, low productivity, informal employment, low taxes, deficient infrastructure, poverty and vulnerability. According to Scott and Storper, "these dysfunctionalities would unquestionably undermine the viability of the city, for market logic alone is congenitally incapable of regulating the urban commons in the interests of economic efficiency and social well-being" (2015: 8). There are four interrelated ways in which the physical structure of fast-growing cities can become disorderly and even chaotic.

First, urban land is frequently not allocated to appropriate uses at sufficient densities and building heights to generate the economic and social interactions that spur creativity, productivity and investment (Turok 2016; Venables 2015; World Bank 2013). Sprawling informal settlements disconnected from economic opportunities are frequently the result. This is partly because there is uncertainty over property rights and overlapping statutory and customary land tenure systems. Confusion over who has secure rights to particular land parcels causes drawn-out conflicts, inhibits investment, and makes it easier for exploitative and corrupt practices to thrive (Durand-Lasserve *et al.* 2015; Fox 2014; World Bank 2015). A related problem is the lack of a systematic method of assessing land values. This is required to assist land conversion from rural to urban uses, compensate farmers, make the incentives for property development more transparent, and raise property taxes fairly and efficiently. Inappropriate land-zoning schemes and building regulations inherited from colonial times can also prevent higher density development in well-located areas.

Second, coherent urban development is often impeded by fragmented land ownership and unfavourable topography that complicate the task of site assembly (Turok 2015). Extensive areas of land are required for integrated development schemes, new residential areas, major industrial sites, new transport connections or other infrastructure networks. Landowners are tempted to withhold their land from the market for speculative or symbolic reasons. Many city authorities lack the legal powers, political authority or resources to expropriate under-used land and consolidate sites for strategic projects. Some expropriate land in a heavy-handed way without compensation, which provokes discontent and conflict (Fox 2014). Alternative approaches, such as land readjustment, can be time-consuming because of the negotiations required to persuade all the landowners and occupiers to cooperate in redeveloping their areas (Ingram and Hong 2012; UN-Habitat 2013).

Third, efficient urban growth is frequently undermined by inadequate investment in infrastructure (Ingram and Brandt 2013). The exceptional requirements of cities are often overlooked by governments and international funders: "infrastructure needs and financing options at the sub-national level, especially for growing urban areas, have been largely ignored" (Gutman *et al.* 2015: 3). In 2009 the World Bank estimated that Africa needed USD93 billion per year to fill the glaring infrastructure gap. Urban infrastructure provides the spatial skeleton that supports the location of housing, industry, retailing and public services. Reliable energy and water supplies are vital for productive activity and household welfare. Transport networks are essential to get employees to work, material inputs to firms and products to consumers. Bottlenecks and breakdowns in such systems damage business performance and city-wide development.

The high cost of urban infrastructure and the low affordability of the population are serious obstacles. Fixed infrastructure disrupts settled communities if it means retrofitting after the land has already been occupied. The expense of public utilities is a massive challenge for cities with fragile economies and weak institutions. Spiralling cities make enormous demands on public services, but the resources aren't available to pay for them. There is a sequencing problem in that it is best to prepare ahead for urban growth by investing in serviced land in advance of settlement, but the tax base only grows after development (Angel 2016). Borrowing is the solution, but many municipalities lack the authority or capabilities to do this. Otherwise, the mutually reinforcing links between investment in infrastructure, economic development and rising land values create a compelling case for financing these fixed assets through land value capture. This can be done through property taxation, development charges and/or public–private partnerships, but these mechanisms tend to be weak in Southern cities (Ingram and Hong 2012).

Fourth, coordination of built environment decisions is vital to realise the advantages of urbanisation for productivity and liveability (Glaeser and Joshi-Ghani 2015; OECD 2014; Turok 2016). Well-configured, compact and connected cities generate positive value for firms and households in terms of access to services, income, information and jobs. These advantages do not emerge automatically as cities grow because firms, households and state entities make their investment decisions separately. Yet the utility of housing to households depends upon complementary investments by firms in nearby premises and by government in infrastructure (thereby providing basic facilities and links to other parts of the city) (Collier and Venables 2015). Similarly, the productivity of premises to firms depends upon corresponding investments in infrastructure and adequately housed workers and consumers nearby. Coordinating these reciprocal investments in housing and business premises requires collective action. This is typically achieved through a long-term spatial plan or framework that provides a shared vision of the future form and direction of the city's growth.

These four issues are connected. Cities that manage the urban land–infrastructure–coordination nexus are more likely to function better and prosper over time. Access to land for businesses and migrants will be simpler, enabling them to occupy productive positions in cities. Developing coherent land parcels and redeveloping well-located sites will be streamlined, permitting carefully structured layouts and higher density buildings. Infrastructure investment will improve mobility and liveability. Built environment coordination will improve the functional efficiency of the city and access to opportunities. Aligned actions by households, firms and government will create more valuable places that generate higher incomes and raise more taxes to be reinvested in improved public services and infrastructure (Turok 2016).

Conclusion

Urbanisation appears to have considerable potential to foster long-term economic development. This stems from the more intense interactions, resource efficiencies, learning and enhanced productivity achieved by concentrated activity. However, this is not assured. Urbanisation also causes congestion, pollution and conflict. It is misleading to say that urbanisation drives development as if it is an independent force for progress. It is more plausible to argue for interdependence – urbanisation is both shaped by and helps to shape development paths. Urbanisation is intimately bound up with the evolving dynamics of economic development, but is not a separate driver of change. The relationship is also context-dependent, rather than immutable through time and unvarying across different places.

One of the conditions for realising the potential of urbanisation is the composition of economic growth. Broad-based industrialisation seems more complementary to prosperity than exporting commodities. The employment intensity of manufacturing is greater and the economic linkages to local suppliers and service providers are stronger, so the multiplier effects are larger.

Manufacturing is more embedded in the regional economy than resource extraction. Therefore, urban policy needs support from industrial policy to help cities produce more of what they consume and to ensure the income from commodity exports is reinvested in urban infrastructure and productive activity. This may require interventions related to technology, international trade, human capital and financial support.

Another precondition for harnessing urbanisation's potential is improving the urban land–infrastructure nexus. Uncontrolled sprawl generates inefficiencies, injustices and degradations that undermine economic and human development. Continuing urbanisation in these circumstances simply reinforces existing patterns and problems. In more favourable conditions, urbanisation can accentuate positive outcomes and improve prosperity. Key elements requiring attention are systems of land allocation, land assembly, infrastructure investment and coordination of public, business and household investment decisions. This is a crucial agenda for government that cannot be left to the market.

Getting back to where the chapter started, a danger of arguing that urbanisation drives development is that urbanisation becomes an end in itself. It could imply that governments should set targets to accelerate rural–urban migration. But influencing the rate of urbanisation is probably not a worthwhile goal, and it may not be feasible if migration essentially reflects other factors. Enhancing the character or 'quality' of urbanisation would seem to be much more valuable than trying to accelerate or resist it. Governments should focus on improving how people and firms gain access to land and property in cities, and not try to exclude them. Most cities could do more to increase the availability of premises for new, incoming and expanding businesses. The same applies to housing, to prevent people living in squalor. Governments can do much to improve how informal settlements function as gateways to urban labour markets, schools, clinics and other public facilities. They could provide land, basic services, assist with upgrading shelter and livelihood creation and relax petty rules and regulations that frustrate people trying to gain a foothold in the city.

References

Angel, S. (2016) "Monitoring the share of land in streets". In G. McCarthy, G. Ingram and S. Moody (eds) *Land and the City*, Cambridge, MA: Lincoln Institute for Land Policy, pp. 62–101.

Annez, P. and Buckley, R. (2009) "Urbanisation and growth: Setting the context". In M. Spence, P. Annez and R. Buckley (eds) *Urbanisation and Growth*, Washington DC: The World Bank, pp. 1–46.

Barrios, S., Strobl, E. and Bertinelli, E. (2006) "Climactic change and rural–urban migration: The case of Sub-Saharan Africa", *Journal of Urban Economics* 60(3): 357–371.

Beall, J., Guha-Khasnobis, B. and Kanbur, R. (eds) (2010) *Urbanisation and Development: Multidisciplinary Perspectives*, Oxford: Oxford University Press.

Bettencourt, L. and West, G. (2010) "A unified theory of urban living", *Nature* 467: 912–913.

Brulhart, M. and Sbergami, F. (2009) "Agglomeration and growth: Cross-country evidence", *Journal of Urban Economics* 65(1): 48–63.

Buckley, R. and Kallergis, A. (2014) "Does African urban policy provide a platform for sustained economic growth?". In S. Parnell and S. Oldfield (eds) *The Routledge Handbook on Cities of the Global South*, London: Routledge, pp. 173–190.

Buckley, R. and Simet, L. (2016) "An agenda for Habitat III: Urban perestroika", *Environment and Urbanization* 28(1): 64–76.

Collier, P. and Venables, A.J. (2015) "Housing and urbanisation in Africa: Unleashing a formal market process". In E. Glaeser and A. Joshi-Ghani (eds) *The Urban Imperative: Towards Competitive Cities*, Oxford: Oxford University Press, pp. 413–436.

De Haas, H. (2012) "The migration and development pendulum: A critical view on research and policy", *Migration Review* 50(3): 8–25.

Durand-Lasserve, A., Durand-Lasserve, M. and Selod, H. (2015) *Land Delivery Systems in West African Cities*, Washington, DC: The World Bank.

Duranton, G. (2009) "Are cities engines of growth and prosperity for developing countries?". In M. Spence, P. Annez and R. Buckley (eds) *Urbanisation and Growth*, Washington DC: The World Bank, pp. 67–114.
Duranton, G. (2014a) "The urbanisation and development puzzle". In S. Yusuf (ed.) *The Buzz in Cities: New Economic Thinking*, Washington, DC: The Growth Dialogue, pp. 1–17. Available: www.growthdialogue.org/sites/default/files/publication/documents/Urbanization_web_10-25-14.pdf [accessed 25 January 2016].
Duranton, G. (2014b) "Growing through cities in developing countries", Policy Research Working Paper 6818, Washington, DC: The World Bank.
Fay, M. and Opal, C. (2000) "Urbanization without growth: A not so uncommon phenomenon", Working Paper No. 2412, Washington, DC: The World Bank.
Florida, R. (2003) *The Rise of the Creative Class*, New York: Basic Books.
Fox, S. (2014) "The political economy of slums: Theory and evidence from Sub-Saharan Africa", *World Development* 54: 191–203.
Freire, M.E., Lall, S. and Leipziger, D. (2014) *Africa's Urbanisation: Challenges and Opportunities*, Washington, DC: The Growth Dialogue. Available: www.dannyleipziger.com/documents/GD_WP7.pdf [accessed 25 January 2016].
Glaeser, E. (2011) *Triumph of the City: How our Greatest Invention makes us Richer, Smarter, Greener, Healthier, and Happier*, Basingstoke: Macmillan.
Glaeser, E. and A. Joshi-Ghani (eds) (2015) *The Urban Imperative: Towards Competitive Cities*, Oxford: Oxford University Press.
Glaeser, E. and Sims, H. (2015) "Contagion, crime and congestion: Overcoming the downsides of density", IGC Growth Brief Series 001, International Growth Centre, London. Available: www.theigc.org/wp-content/uploads/2015/05/Glaeser-Sims-2015-Growth-brief.pdf [accessed 25 January 2016].
Global Commission on the Economy and Climate (GCEC) (2014) *Better Growth, Better Climate: The New Climate Economy Report: Synthesis*, Washington. Available: www.newclimateeconomy.report [accessed 25 January 2015].
Gollin, D., Jedwab, R. and Vollrath, D. (2015) "Urbanization with and without industrialisation", *Journal of Economic Growth*. Available: http://home.gwu.edu/~jedwab/index.html [accessed 20 January 2016].
Gutman, J., Sy, A. and Chattopadhyay, S. (2015) *Financing African Infrastructure: Can the World Deliver?* Washington, DC: Brookings Institute.
Henderson, V. (2003) "The urbanization process and economic growth: The so-what question", *Journal of Economic Growth* 8(1): 47–71.
Henderson, V. (2010) "Cities and development", *Journal of Regional Science* 50(1): 515–540.
Henderson, V., Storeygard, A. and Deichmann, U. (2014) "50 Years of urbanisation in Africa: Examining the role of climate change", Policy Research Working Paper 6925, Washington, DC: The World Bank. Available: http://documents.worldbank.org/curated/en/2014/06/19679968/50-years-urbanization-africa-examining-role-climate-change [accessed 25 January 2016].
Ingram, G. and Brandt, K. (eds) (2013) *Infrastructure and Land Policies*, Cambridge, MA: Lincoln Institute for Land Policy.
Ingram, G. and Hong, Y. (eds) (2012) *Value Capture and Land Policies*, Cambridge, MA: Lincoln Institute for Land Policy.
Jacobs, J. (1984) *Cities and the Wealth of Nations: Principles of Economic Life*, New York: Random House.
Jedwab, R. and Vollrath, D. (2015) "Urbanization without growth in historical perspective", *Explorations in Economic History* 58(C): 1–21.
Jedwab, R., Christiaensen, L. and Gindelsky, M. (2015) "Demography, urbanization and development rural push, urban pull and … urban push?", Policy Research Working Paper 7333, Washington, DC: The World Bank. Available: http://documents.worldbank.org/curated/en/2015/06/24689971/demography-urbanization-development-rural-push-urban-pull-urban-push [accessed 25 January 2016].
La Porta, R. and Shleifer, A. (2014) "Informality and development", *Journal of Economic Perspectives* 28(3): 109–126.
McKinsey (2010) "Lions on the move: The progress and potential of African economies", McKinsey Global Institute. Available: www.mckinsey.com/insights/africa/lions_on_the_move [accessed 5 January 2016].
McKinsey (2012) "Urban world: Cities and the rise of the consuming class", McKinsey Global Institute. Available: www.mckinsey.com/insights/urbanization/urban_world_cities_and_the_rise_of_the_consuming_class [accessed 5 January 2016].

Moretti, E. (2014) "Cities and growth", IGC Working Paper. Available: www.theigc.org/wp-content/uploads/2014/09/IGCEvidencePaperCities.pdf [accessed 25 January 2016].
Organisation for Economic Cooperation and Development (OECD) (2013) *OECD Economic Surveys: China*, Paris: OECD.
Organisation for Economic Cooperation and Development (OECD) (2014) *OECD Regional Outlook 2014: Regions and Cities: Where Policies and People Meet*, Paris: OECD Publishing.
Overman, H.G. and Venables, A.J. (2010) "Evolving city systems". In J. Beall, B. Guha-Khasnobis and R. Kanbur (eds) *Urbanization and Development: Multidisciplinary Perspectives*, Oxford: Oxford University Press, pp. 103–123.
Potts, D. (2016) "Debates about African urbanisation, migration and economic growth: What can we learn from Zimbabwe and Zambia?", *The Geographical Journal* 182(3): 251–264.
Scott, A. and Storper, M. (2015) "The nature of cities: The scope and limits of urban theory", *International Journal of Urban and Regional Research* 39(1): 1–16.
Storper, M. (2013) *Keys to the City: How Economics, Institutions, Social Interaction, and Politics Shape Development*, Princeton, NJ: Princeton University Press.
Storper, M. and Scott, A. (2009) "Rethinking human capital, creativity and urban growth", *Journal of Economic Geography* 9: 147–167.
Storper, M., Marrevijk, C.V. and Oort, F. V. (2012) "Introduction: Processes of change in urban systems", *Journal of Regional Science*, 52(1): 1–9.
Turok, I. (2013) "Securing the resurgence of African cities", *Local Economy* 28(2): 142–157.
Turok, I. (2015) "Redundant and marginalized spaces". In T. Hutton and R. Paddison (eds) *The Economics of Cities*, London: Sage, pp. 74–92.
Turok, I. (2016) "Housing and the urban premium", *Habitat International* 54(3): 234–240.
Turok, I. and McGranahan, G. (2013) "Urbanisation and economic growth: The arguments and evidence for Africa and Asia", *Environment and Urbanisation* 25(2): 465–482.
United Nations Department of Economic and Social Affairs Population Division (UNPD) (2012) *World Urbanisation Prospects*, New York: UNPD.
UN-Habitat (2012) *State of the World's Cities*, Nairobi: UN-Habitat.
UN-Habitat (2013) *Urban Planning for City Leaders*, Nairobi: UN-Habitat.
UN-Habitat (2014) *State of African Cities Report*, Nairobi: UN-Habitat.
UN-Habitat (2016) *Habitat III: The New Urban Agenda*. Available: www.habitat3.org [accessed 6 January 2016].
Venables, T. (2015) "Making cities work for development", IGC Growth Brief Series 002, International Growth Centre, London. Available: www.theigc.org/wp-content/uploads/2015/05/IGCJ3155_Growth_Brief_2_WEB.pdf [accessed 25 January 2016].
World Bank (2009) *World Development Report: Reshaping Economic Geography*, Washington DC: The World Bank.
World Bank (2013) *Planning, Connecting and Financing Cities – Now*, Washington DC: The World Bank.
World Bank (2014) *Urban China: Towards Efficient, Inclusive and Sustainable Urbanisation*, Washington DC: The World Bank.
World Bank (2015) *Stocktaking of the Housing Sector in Sub-Saharan Africa: Challenges and Opportunities*, Washington DC: The World Bank.

8
Planning Special Economic Zones in China

Qianqi Shen

Introduction

A Development Zone, or Special Economic Zone in some international development literature, is a 'geographically delimited area' designated by the government that provides special policies applicable only to the businesses within the zone (FIAS 2008; Zeng 2010). Establishing such zones is deemed by some international development organisations, such as the World Bank, as an effective policy instrument for industrial and economic development especially for developing countries. China, top of the rankings for zone number and zone-related employment and FDI (Foreign Direct Investment) (FIAS 2008), became the exemplar for zone development.

In China, however, local governments experience a great deal of pressure from the central government to develop their economies, which also motivates the subnational governments to establish development zones. More importantly, development zones make it easier for the local governments to get authorisation from superior governments to requisition land and to improve their construction land quotas. China's loose land-management policies eventually led to at least two instances of 'zone fevers,' or development zone bubbles – one in the mid-1990s, and another in the early 2000s – in which large areas were enclosed and earmarked for development zones, but never actually utilised (Cartier 2001; Yang 1997; Yang 2005; Yang and Wang 2008).

The above description shows a different picture, one which is not aligned with the successful cases of development zones studied by international organisations such as the World Bank. In the following cases, I question what has led to the local variation in implementing the development zone policy in China. Viewing decentralisation as a dynamic process, which involves frequently changing relationships between the central and subnational governments, can reconcile the ideological split on the question of who is responsible for a country's economic growth and development. Development strategies at both levels change according to the relationships between interested parties. Therefore, I analyse here the economic development strategies and their effects against the backdrop of changing intergovernmental relationships in the course of decentralisation. Especially, my focus is on how inter-governmental negotiation worked as a key mechanism to engender development strategies that brought about different paths of development.

The reforms and changing planning

Since the 1980s, reform has been the theme in China's policy-making. Decentralisation reforms shifted the power balance on fiscal matters toward the subnational governments. Then, after 1994, the central government regained its fiscal power through tax-sharing reforms, which later, with reform in land development, triggered incidents of land-craving at the local level and tightening of land control from above.

Fiscal reform and administrative decentralisation, and the craving for land

Fiscal decentralisation in China has taken place in two distinct periods. In the first period, which began in the early 1980s, the central government allowed the majority of tax income to be kept at the provincial level to incentivise subnational economic development. Provincial governments were responsible for approving the budgeting plans of the lower level governments, including the municipal and county levels (Bahl and Linn 1992; Bahl and Wallich 1992). After a decade, the central government gradually found itself facing serious budgeting problems; moreover, it faced a loss of control of those provinces that were withholding most of the development dividends.

The fiscal challenges to the central government triggered the second period of fiscal decentralisation, which began with the fiscal reforms of 1994. Large tax items, such as income tax, were directly collected by the central government. As a result, local revenues from taxes decreased dramatically. The reform decreased the subnational governments' spending flexibility but strengthened the central government's ability to generate revenue and monitor taxation and, thus, reinforced its control over China's economy. The central government's share of budgetary revenue rose from about 20 per cent in 1993 to 55 per cent in 1994, as seen in Figure 8.1.

Parallel to this fiscal reform, the central government shifted more responsibility for public expenditures to the subnational governments. As seen in Figure 8.1, the local share of expenses had been increasing steadily from lower than 50 per cent in the early 1980s to almost 90 per cent in 2011. To ease the strained relationships and push forward both administrative and fiscal reforms, the central government allowed the local governments to keep fees from land sales at the local level, thus compensating them at least somewhat for the increases in administrative and other expenditures.

Subnational governments with unstable finances began to rely heavily on land sale fees under this system (Figure 8.2). Under the current Budget Law, local governments are not allowed to run deficits, nor are they permitted to issue local bonds. Land sales, then, have become the means for local officials to make ends meet. From 2000 to 2012, the urban built area increased by more than 100 per cent, and income from land sales increased 4500 per cent. In 2010, land sales accounted for 77 per cent of the local revenue income. In the first three quarters of 2013, the average price of land sales increased by 70 per cent compared to the same time period in 2012. Moreover, in the first three quarters of 2013, the total income from land sales in 300 cities was 2.111 trillion yuan (approximately 315 billion USD), compared to the previous year's 1.244 trillion yuan, an increase of 70 per cent (Li 2013).

The centre's tightening land policy and local discretion

Since implementing the previous round of fiscal reforms, the central government has tightened its regulatory control over land use. Under the current law, the Land Management Act (1998),

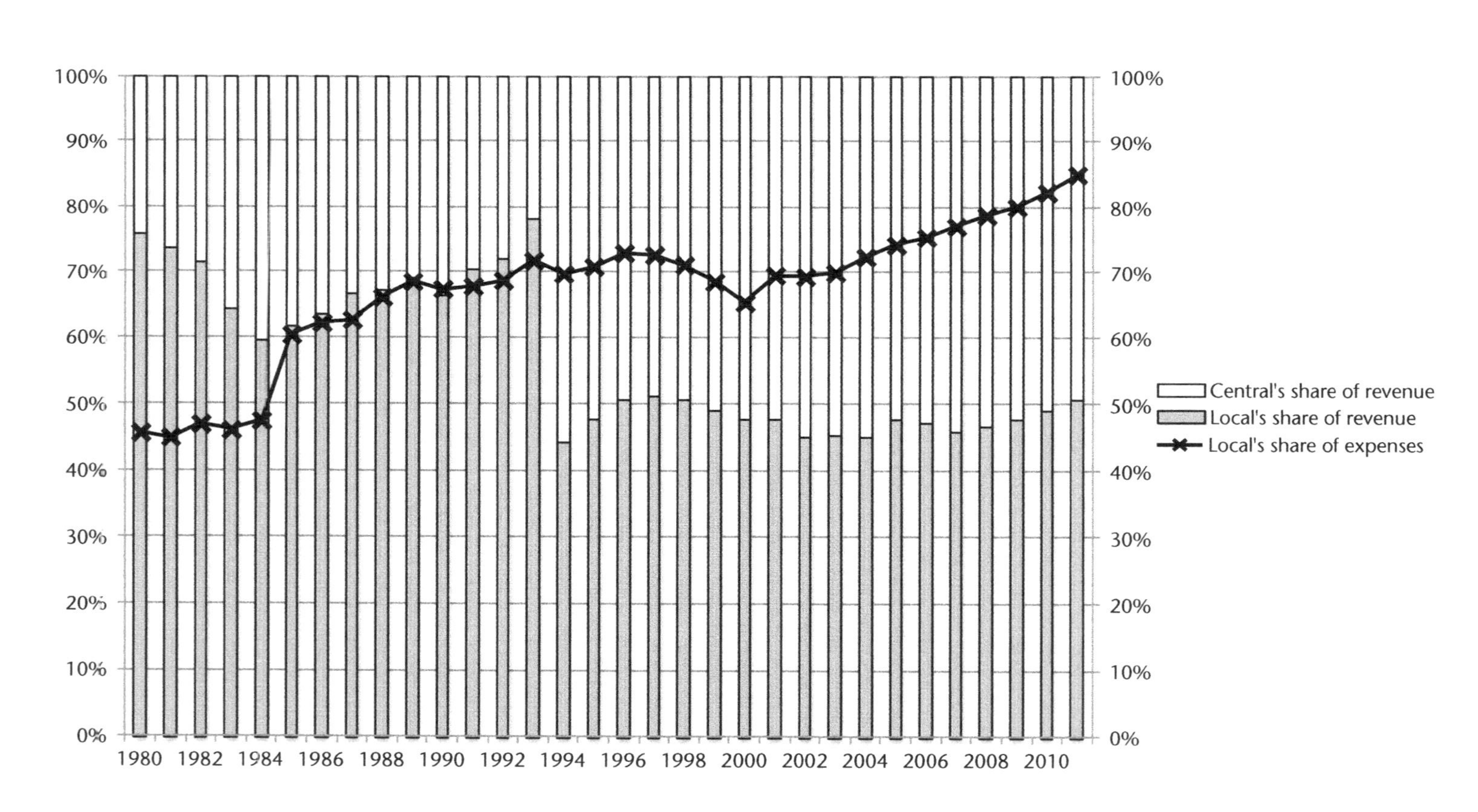

Figure 8.1 Central and local shares of revenues and expenses, 1980–2011

Source: data from the National Statistics Bureau (2014), http://data.stats.gov.cn.

each level of government is required to submit a Land Use Master Plan every 5–10 years for authorisation by the central government. In these plans, provision of land for construction projects is designated according to the assigned land development quotas passed down from a superior level of government. Without land quotas, local governments are not allowed to grant permission for new construction projects. Moreover, the monitoring system is also strict. The central government established nine regional bureaus of land supervision, which combine several provinces and cities to create jurisdictions; the bureaus are directly under the authority of the central Ministry of Land and Resources, which makes it difficult for any single subnational government to influence or corrupt the land supervision system (personal communication, July–September 2013).

However, in reality, local governments still use a variety of tactics to expand their land quotas and to avoid being punished for land use violations. For example, one common tactic is to set up more development zones and apply for a zone-level upgrade, moving from a local zone to a provincial or even national zone. When creating land use plans, development zones usually enjoy the top priority for construction quota assignments. The higher level the zone is, the easier for it to get land quota. As long as the local officials can prove that the outcome of economic development in their area is 'good', no official will be caught or punished for such bending of the rules (personal communication, September 2013). The consequences are, on the one hand, that the local governments develop the raw land and offer it at low prices to industrial businesses in order to attract industry to their region, while on the other hand, local governments also try to expand their development zones and sell housing and commercial land around or even within the zones at high prices to compensate for the deficits created by developing raw industrial land (*People's Daily* 2003).

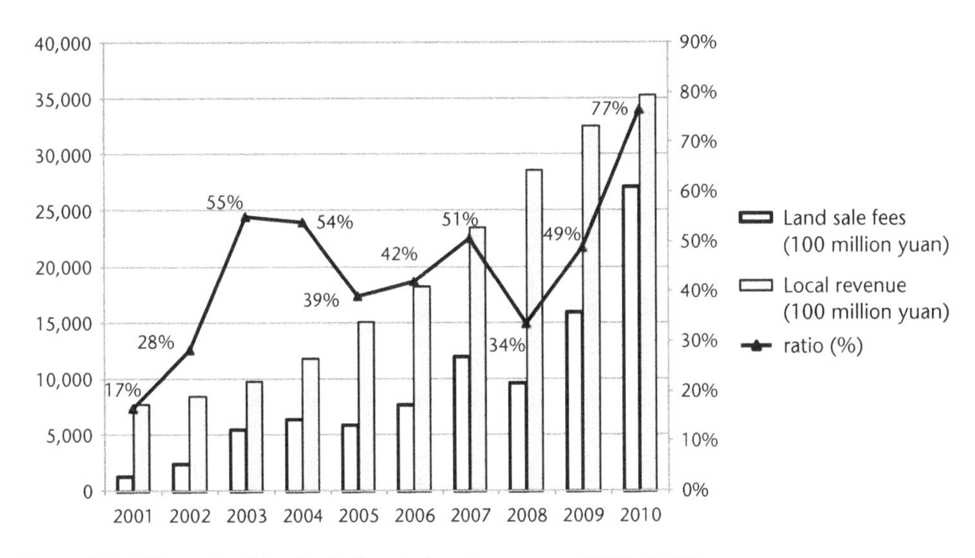

Figure 8.2 The ratio of land sale fees to local revenues, 2001–2010

Source: *China Land Resource Statistical Yearbook 2001–2010* and *China Statistical Yearbook 2001–2010*.

Note: According to standard accounting rules, land sale fees are counted as 'income beyond budget' and are not counted as part of local revenues. Thus, the ratio presented is a comparison of the quantity of the income under the two items and cannot be interpreted in the form of 'A takes x per cent of B'.

In addition, many subnational governments coordinate with the banking sector to secure infrastructure investments, promising to pay back these investment funds with revenue from future land sales. Large areas were enclosed and earmarked for development zones, but never actually utilised. Little of the land in these designated zones has actually been developed, leaving large swaths of land to lie fallow, reserved for future speculation by local governments and private companies (Cartier 2001; Yang 1997; Yang 2005; Yang and Wang 2008).

The changing role of planners and reform in the planning system

In the meantime, the planning system also experienced reform and the role of planners changed. Before the economic reform in the 1980s, city planners' main work was to coordinate industrial development and urban planning. However, since the economic reform, the planned economy ideology that grounded planning activities was shaken by the liberal ideology. Urban planning carried the stigma of a planned economy as a symbol of conservativism and anti-reform (Wu *et al.* 2007). The lack of coordination and regulation in urban space development called for a reform in the planning system.

In 1989, the City Planning Act was enacted. According to the Act, all development proposals are required to have approval from a planning department. However, in reality, the development control decisions are heavily interfered with by other governmental agencies. The master plan, the only statutory plan in most Chinese cities, is usually made after land parcels are leased out to users. Planners have to either revise their original plans or to seek compromise with developers (Xu 2001; Xu and Ng 1998). For most of the time, negotiations are carried out by the mayor, or other municipal officials, directly with developers, then the planning department and planners are notified afterwards to grant permission and rationalise and integrate the developer's plan into the city's plan. For example, the mayor of Guangzhou issued over 2,000 memos to the planning department to require special treatments to developers closely related to the government between 1992 and 1996 (Xu and Ng 1998). Negotiations on plans also take place among different levels of governments. For example, master plans are required to be submitted to upper-level government for approval. Especially for sub-provincial cities such as Guangzhou, and direct-controlled municipalities such as Shanghai, the master plan needs approval from the centre directly. This procedure creates yet another negotiating arena between the central and the local. For example, Guangzhou's 1996–2010 master plan was not able to get the centre's approval until 2007. By the time the master plan was approved, a lot of urban land development projects, legal and illegal (in terms of conforming to the plan or not), had taken place. The master plan therefore became an ineffective guidance and controlling tool for local urban development. The real negotiation is among the government of different levels, various governmental agencies, especially the economic development departments, and the developers and investors.

The three cases

In this section, I compare three cases of Special Economic Zones: Tianjin Economic Development Area (currently Binhai New Area), Guangzhou Development Zone (currently Luogang/Huangpu District), and Guangzhou Nansha Development Zone (currently Nansha New Area). The three zones were all set up under the national zone policy – the Tianjin and Guangzhou development zones were among the first set of zones established in the 1980s, and Nansha was created during the second round in the early 1990s. During my field study in China, I gradually came to the realisation that this national development zone project cannot be understood at a single scale; rather, the scales of how a zone is addressed depend on the powers and interests involved.

Tianjin: a coherent collaborative platform

The early stages of the development of Tianjin's Special Zone were predominantly characterised by explicit bargaining, as summarised by the strategy of 'waiting for, relying on, and requesting' (*deng, kao, yao*) activity from the centre. At the beginning of the establishment of the zones, it was hard to obtain authorisation for projects of varied scales from all different levels and departments of governments. How much support each zone actually received relied on the results of explicit bargaining between the local officials and the central government (personal communication, July 2013). With unconditional support from the higher levels of governments, the local zone officials were given flexibility in interpreting formal rules.

Entering the second phase beginning in the 1990s, the relationship between the central government and the municipality of Tianjin became more complicated as the city gained more independence in planning and implementation; the situation was especially fraught when there were conflicts with neighbouring Beijing's plans. The Tianjin municipality decided to modify its planning to place emphasis on the Binhai area in 1993. However, when this plan was sent to the central government for authorisation, the centre required planners to add the adjective 'important' in front of 'economic centre of Northern China,' reflecting the centre's reluctance to acknowledge Tianjin as rivalling Beijing as the economic centre of the northern area. In the meantime, the centre changed the expression 'Binhai New Area' to 'Binhai Town Centre,' seemingly to avoid having to acknowledge Binhai's national status.

Starting in 1994, a new fiscal and tax assignment system was adopted. Zone development became a highly localised economic development strategy. The zone now had to generate additional revenue to support development elsewhere in the municipality (Tianjin Municipal Archives 2008). The municipality government established local state-owned enterprises as financing apparatuses. From 1997 to 2005, the Binhai New Area Gross Domestic Product (GDP) grew from 38 billion yuan to 163 billion yuan, an average increase of 18.3 per cent annually. The proportion of the Binhai New Area's GDP to the city's GDP grew from 30 per cent to 42 per cent. The Binhai New Area is playing an increasingly important role in Tianjin's municipal economy.

In the early 2000s the Tianjin municipality began to seek support from members of the Chinese People's Political Consultative Conference (CPPCC), a powerful advisory body that connects to all high officials in the central government. The increasing support from the CPPCC after several rounds of discussions eventually led to the submission of a jointly authored report to the Central Secretary and the State Council Secretary in October 2004. This CPPCC report, entitled "Suggestions on Exerting the Effect of Tianjin Binhai New Area in Invigorating the Regional Economy of Bohai Gulf Area", suggested that the central leadership, including the party and the State Council, should support the continuing development of the Binhai area (Du *et al.* 2006). Premier Wen Jiabao subsequently commented on the report that "the planning and construction of Tianjin Binhai Area is not just about the long-term development of Tianjin but also will have an important impact on invigorating the Bohai gulf area's economy" (Liu 2008: 158).

At the end of 2005, the 11th National Five-Year Plan was published and approved; it officially acknowledged the Tianjin Binhai New Area as a national New Area. Five special policies were issued by the State Council, including making the Binhai area a comprehensive reform experiment district, establishing a free-trade port in the area, embarking on financial reform in the area, increasing the area's supply of building land, and lowering the tax rate for high-tech companies to 15 per cent (50 per cent lower than the normal tax rate). However, of the five special policies, the first three, regarding reforms and free trade, required additional research and discussions in order to be implemented; the preferential tax rate policy is applicable to high-tech companies across China. This left the policy increasing Tianjin's supply of building land as the single most important special policy for the Binhai New Area.

The local government put much focus on construction, including building and infrastructure, and attracting investment, both of which contributed to GDP and tax revenue. The total building area in Binhai New Area will be 14.5 million square metres, equal to 15 times the current supply of office space in Tianjin. However, this massive construction effort faces a significant problem: demand. According to a research report by CBRE Group, Inc. (one of the largest commercial real estate companies in the United States and major players in China), the office space in Tianjin would take Tianjin 60 years to saturate the market based on the current consumption (Jing, 2012). Construction projects began to face financing issues, delaying the completion of many projects. By 2013, a full 30 per cent of projects had stalled. We have yet to see any significant effects from financial reform in Tianjin, leaving as the only notable results from the area's development the CBD 'ghost cities' (abandoned or incomplete building projects) in Binhai.

Guangzhou development zone: a locally contained conflicting platform

Whereas Tianjin has constantly sought the centre's support and acknowledgement, Guangzhou intentionally broke away from any restrictions imposed by the centre. Guangzhou's leadership was explicit about being flexible in interpreting the rules and laws when such flexibility in implementation benefits the zone and the city economically. Compared to Tianjin, the minimal involvement by the central government made Guangzhou a less desirable site for large foreign investors. In order to attract large-scale businesses, the Guangzhou zone sometimes had to compromise its monetary interests in addition to implementing flexible policies, particularly in the early stage of zone development.

By the early 1990s, all the lots in the original planned area, which totalled 9.6 square kilometres, had been assigned to projects. Expanding the zone area became an urgent matter. The Guangzhou municipal government has led several major expansions of the development zone from 1990 to 2005 and the expansion efforts have accelerated during this time. The first period of expansion lasted from 1990 to 2000, when the zone expanded from its original 9.6 square kilometre area to an area of 78.9 square kilometres, more than eight times its original area. In this period, the city established new industrial zones around the original development zone area (Guangzhou Development Zone Chronicle Editor Committee 2004). In 2002, the Guangzhou government began to expand the development zone to create a larger administrative district, making the larger area an independent administrative district and a new sub-centre of the city. While retaining its focus on the automotive, chemical and high-tech industries, the city also began to assign more residential and commercial land quotas to the development zone area. In 2005, with approval from the State Council, the Guangzhou Development Zone and Luogang Administrative District merged, and the area of the development zone expanded to 393.22 square kilometres with a total population of 370,000. In the same year, a new master plan for the district was created by the city and district governments (Luogang District Government 2007). The new plan kept industrial development as its focus, but also included the real estate industry and tourism as new industries of interest. The ultimate goal was to transform the Guangzhou Development Zone from an economic zone into Guangzhou's new sub-centre town (Luogang Master Planning Department 2004). The zone's expansion, on the one hand, meets the area's needs for obtaining industrial land and attracting businesses and, on the other hand, meets the city's needs to expand toward the east.

In contrast to the situation in Tianjin, where the city and zone governments have formed stable partnerships, the Guangzhou city government has acted in a predatory manner toward the development zone (now district) and gave little financial help to the zone. After merging its administrative functions with the surrounding areas to become a district in 2005, the development zone not only has to pay for its own development, but also for the public affairs of the new villages and residents that did not belong to the zone originally. As a result, administrative expenses increased rapidly after the district merger. In addition, the district's infrastructure investment has increased substantially. The fiscal situation forced the district to rely more heavily on land sales to remain solvent. In 2013, the total disposable income of the district was 19 trillion yuan; of this, land sales income was 7.7 trillion yuan, more than 40 per cent of the district's total disposable income (personal communication, August 2013). The zone administration is struggling with the development of the zone, facing pressures from the city and competition from other districts in the city. Since it became a district, its contribution to the city's GDP has remained relatively low, at 12 per cent of the whole.

Nansha: a multi-scale conflicting platform

Nansha's development was initiated by a highly politically influential businessman from Hong Kong, Dr Henry Fok (Fok Ying Tung in Chinese) in the late 1970s. In the earliest phase, the Guangzhou municipal government did not actively participate in the area's planning and development. The change of Nansha's status in the city began in the late 1990s. For a long time, Guangzhou City's development was restricted by two factors: first, its geographic constraints on port industry development, and second, its lack of heavy industry. In the mid-1990s, the mayor of Guangzhou decided to build a deep-water port at the Pearl River's outlet to the sea, Nansha, but encountered significant opposition from Hong Kong, which saw the port as a competitor to its investment in Shenzhen, Yantian Port. Guangzhou city government, encountering postponement in the deep-water port plan, turned to advocate for establishing heavy industries in the Nansha zone area. This plan "if successful . . . would provide South China with the sort of heavy industrial base it has traditionally lacked" (Enright *et al.* 2003).

The Guangzhou government's strategy, shifting its focus from port development to comprehensive heavy industry development, successfully aligned with Guangdong's development strategy, gaining it provincial support. In addition, Guangzhou and Guangdong's plan for Nansha's development gradually gained support from the centre, especially from key officials such as the deputy director of the State Development and Reform Commission, Zhang Guobao, and the former prime minister, Zhu Rongji. Both administrative and financial support was soon forthcoming.

During the same period, the Guangzhou municipality drafted the Nansha Area Development Plan, which began in 2001 and was finally published in 2004. The plan outlined three focal industries: the port industry, high tech firms and the equipment industry. This focus was based on the firm belief that the new development zone would attract large projects to these areas. The planning agency also planned four clusters in Nansha, including an auto-making base, a ship-making base, a steel manufacturing base and a petrochemical base. In addition, in 2002, the Guangzhou government began investing heavily in Nansha's infrastructure to assure investors that the municipality was fully behind the area's development. From 2002 to 2005, the Guangzhou city government invested 11.7 trillion yuan in infrastructure development in Nansha, including building 95 roads and bridges, totalling 380 kilometres in length (personal communication, July 2013).

However, the plan was created without consulting with Nansha's neighbouring areas; as a result, Nansha's development plan caused increasing concern in the Pearl River Delta region, including Hong Kong. The most serious concern was the possible duplication of investments and the possible subsequent threat to their own economic development (Enright *et al.* 2003). These concerns finally led to a major project failure for Nansha when it was unable to attract a large petroleum industry investment from Kuwait in 2009. The project was opposed by the surrounding cities, including Hong Kong, on the grounds of environmental concerns. In 2009, the Guangdong governor, Wang Yang, decided to take the Kuwait project away from Guangzhou, moving it to Zhanjiang, another city in Guangdong.

Guangzhou's failure to secure the Kuwait project, as well as the later collapse of the Guangzhou Steel Manufacturing project, forced the city to change its plans for Nansha. In later rounds of planning, Guangzhou no longer aimed to be the heavy industrial centre of the Pearl River Delta.

In the 2008 Pearl River Delta Region Development Planning Guideline, the central government's Commission of Reform and Development proposed making Nansha a bridge for collaboration between Guangdong, Hong Kong, and Macau on developing the high-tech and modern device industries. Guangdong province also decided to make Nansha a model district for the Mainland and Hong Kong Closer Economic Partnership Arrangement (CEPA), a series of treaties between Hong Kong and the mainland signed in 2003. In 2011, the national 12th Five-Year Plan positioned Nansha as a "breakthrough to make a new Guangzhou", one that would further "strengthen Guangdong, Hong Kong, and Macao's cooperation" (Xinhua News 2011). Although Guangzhou adjusted its master plan for Nansha to address conflicts in the region and new guidance from the central government, it did not change its ambition to remain the economic and cultural centre of Southern China. In a meeting regarding the planning efforts hosted by the Guangzhou Planning Bureau and the various planning companies under contract, the Bureau's director stressed that Guangzhou needed to retain its primary status in Southern China and that it required Nansha to achieve this goal. On an industrial level, Nansha would retain its ship-making and auto-making industries, port-related industries, and large equipment industry. Nansha New Area would also be focused on high-tech companies and exporting services; it would act as a headquarters for outsourcing companies, emphasising intellectual capital and co-operation with overseas firms.

Nansha's growth is still very slow today. The Nansha District's GDP growth rate was ranked eighth in Guangzhou in 2013, even lower than Guangzhou's average rate, which put considerable pressure on local officials. Unsurprisingly, to facilitate Nansha's development, the Guangzhou and Nansha governments have been relying increasingly on real estate development and land sales. To allocate more land quota to Nansha, Guangzhou added three counties to the Nansha area, expanding it from 500 square kilometres to more than 800 square kilometres. From the fourth quarter of 2012 to the fourth quarter of 2013, the Nansha New Area sold 33 land lots; this made it the district that sold the most land in Guangzhou (Liu, 2013). Given the strong reliance on increasing land prices and significant land sale revenue, local officials often disobeyed the central government's orders about stabilising land prices and calming the residential market (personal communication, August 2013). Increasing land prices hurt the area's industrial development, however, creating a vicious cycle for the local government, which faced even less industrial activity and tax revenue, forcing it to rely still more on land sales.

Conclusion

The comparison of the three development zones in China has shown that the effectiveness of using delimited economic zones for economic development and industrial development needs more scrutiny. International organisations such as the World Bank have promoted spatially delimited zones, such as special economic zones and various types of development zones, as primary industrial development strategies across the world. In many of the research reports of these organisations, China was used as the example of how this policy brings growth, employment and technological innovation and upgrading by attracting foreign direct investment and state-led infrastructure investments. However, the different paths and outcomes of development of the three zones in China raise questions about this approach and the best practices related to this policy. This study demonstrates the importance of subnational politics in carrying out national economic development policy.

As for planners in China, providing alternatives for development paths for the zones does not seem to be an option when the development is heavily influenced by the will of local officials and intergovernmental politics. As discussed by Zhang (2002: 72), "to tell the truth to the power" is still the biggest challenge facing planners in China today. The purpose of planning from the government's perspective, in a growth-focused regime, is to provide information and direction in order to attract business and capital. Planning is subject to change if the desired businesses are not attracted. Under such circumstances, equality, as the core planning philosophy, seems to have to subsume itself to obedience.

However, planners do not have to be pessimistic. Today, many lawyers in China who used to be civil servants are now turning to the private sector. Among them, an emerging group of private lawyers have become rights-protection lawyers, who look for issues with a profound social impact or that affect the public interest; they advocate for clients in the face of more powerful interests (Fu and Cullen 2008). *Weiquan* lawyers are a small but influential group of legal practitioners in China, and the movement they initiated, weiquan (rights protection) law, has been the boldest initiative in legal practice in China in recent years.

Can planners form a group similar to that of the weiquan lawyers? Davidoff's (1965) "Advocacy and pluralism in planning" is certainly worth revisiting in this context. The advocate planner, as proposed by Davidoff (1965), does more than provide information and envision future options: he or she is also a "proponent of specific substantive solutions" (p. 333) – alternative plans – and must be responsible to his or her client, especially the underrepresented groups. Weiquan planners are the advocate planners confronting the authoritarian regimes of China and other developing countries that do not have open and welcoming environments for public participation. These planners must learn how to break the current political boundaries and help citizens escape from the "paradox of weiquan movements", in which the rights of the powerless weiquan participants are tightly restricted by the terms narrowly defined by the central authority even as they struggle to exercise those rights (Shih 2015: 154).

References

Bahl, R. and Linn, J. (1992) *Urban Public Finance in Developing Countries*, Oxford: Oxford University Press.

Bahl, R. and Wallich, C. (1992) "Fiscal intergovernmental relations in China", World Bank Policy Research Working Papers WPS 863. Available: http://documents.worldbank.org/curated/en/1992/02/699758/intergovernmental-fiscal-relations-china [accessed 2 February 2016].

Cartier, C. (2001) "'Zone fever,' the arable land debate, and real estate speculation: China's evolving land use regime and its geographical contradictions", *Journal of Contemporary China* 10(28): 445–469.

China City Planning Act (1989) Order of the President of the People's Republic of China (No.23). Available: www.china.com.cn/zhuanti2005/txt/2003-07/22/content_5370682.htm [accessed 23 February 2017].

Chinese Statistical Yearbook 2001–2010. Available: www.stats.gov.cn/tjsj/ndsj/ [accessed 26 May 2017].

Davidoff, P. (1965) "Advocacy and pluralism in planning", *Journal of the American Institute of Planners* 31(4): 331–338.

Du, Y., Gu, S. and Zhang, J. (2006) "Zheng Xie Diao Yan Bao Gao Zu Tui Tianjin Binhai Xin Qu Jia Kuai Jian She [The report of CPPCC accelerating Tianjin Binhai New Area's development]", *Guangming Daily*, 1 March. Available: www.gmw.cn/content/2006-03/01/content_381317.htm [accessed 1 February 2016].

Enright, M., Chang, K., Scott, E. and Zhu, W. (2003) "Hong Kong and the Pearl River Delta: The economic interaction", Hong Kong University Scholars Hub.

Foreign Investment Advisory Service (FIAS) (2008) *Special Economic Zones Performance, Lessons Learned, and Implications for Zone Development*, Washington, DC: The World Bank.

Fu, H. and Cullen, R. (2008). "Weiquan (rights protection) lawyering in an authoritarian state: Toward critical lawyering", *The China Journal* 59: 111–127.

Guangzhou Economic and Technology Development Zone Chronicle Editor Committee. (2004) *Chronicles of Guangzhou Development Zone 1991–2000*, Guangzhou: Guangdong People's Publisher.

Jing, B. (2012) "Oversupply of office in Tianjin", Sina Finance. Available: http://finance.sina.com.cn/roll/20120510/032812032600.shtml [accessed 1 February 2016].

Land and Resources China Statistical Yearbook 2001–2010, Beijing: Geological Press. Available: www.chinabookshop.net/land-resources-china-statistical-yearbook-2015-p-23163.html [accessed 26 May 2017].

Li, H. (2013) "National land selling revenues for the first three quarters exceeded 2,000 billion". Available: http://news.xinhuanet.com/finance/2013-10/15/c_125539684.htm [accessed 2 February 2016].

Liu, G. (2008). *Di San Ji: Tianjin Binhai Xinqu Fazhan Ji Shi* [*The Third Pole: Recording Tianjin Binhai New Area Development*] Tianjin: Tianjin People's Publisher.

Liu, Q. (2013) "Nansha new area sold 33 land lots in last year", *Southern Metropolis Daily*, 6 September. Available: http://gz.house.sina.com.cn/news/2013-09-06/06483546264.shtml [accessed 2 February 2016].

Luogang District Government (2007) *Luogang District Yearbook*. Guangzhou. Available: http://lgdfz.gdd.gov.cn/Sites/Main/HtmlBook.aspx?id=152 [accessed 23 February 2017].

Luogang District Planning Department (2004) Luogang Master Planning 2004–2020.

People's Daily (2003) "Why is it so hard to regulate development zones?", *People's Daily*. Available: www.china.com.cn/chinese/OP-c/446508.htm [accessed 23 February 2017].

Shih, M. (2015) "The paradox of weiquan movements in urban China: Social inequality and individual negotiation in land development". In F. Miraftab, K. Salo, and D. Wilson (eds) *Cities and Inequalities in a Global and Neoliberal World*, London: Routledge, pp. 153–169.

Tianjin Municipal Archives (2008) *Tianjin Binhai Xin Qu Jing Ji Fa Zhan Shi Lu Di Er Juan* [Tianjin Binhai New Area Economic Development Volume 2]. Tianjin: Tianjin Government.

Wu, F., Xu, J. and Yeh, A.G.O. (2007) *Urban Development in Post-Reform China: State, Market and Space*, London and New York: Routledge.

Xinhua News (2011) National 12th Five-Year Planning, China Net. Available: www.china.com.cn/policy/txt/2011-03/16/content_22156007.htm [accessed 23 February 2017].

Xu, J. (2001) "The role of land use planning in the land development process in China: the case of Guangzhou", *Third World Planning Review* 23(3): 229–248.

Xu, J. and Ng, M.K. (1998) "Socialist urban planning in transition: The case of Guangzhou, China", *Third World Planning Review* 20(1): 35–51.

Yang, D.L. (1997) *Beyond Beijing: Liberalization and the Regions in China*, London: Routledge.

Yang, D.L. (2005) *Remaking the Chinese Leviathan: Market Transition and the Politics of Governance in China*, Stanford, CA: Stanford University Press.

Yang, D.Y.R. and Wang, H.-K (2008) "Dilemmas of local governance under the development zone fever in China: A case study of the Suzhou Region", *Urban Studies*, 45(5–6): 1037–1054.

Zeng, D.Z. (2010) "Building engines for growth and competitiveness in China: Experience with Special Economic Zones and Industrial Clusters", *The World Bank*. Available: www.worldbank.org/en/news/feature/2010/12/19/building-engines-for-growth-and-competitiveness-in-china-experience-with-special-economic-zones-and-industrial-clusters [accessed 2 February 2016].

Zhang, T. (2002) "Challenges facing Chinese planners in transitional China", *Journal of Planning Education and Research* 22: 64–76.

9

Planning in the midst of informality

An application to youth employment programmes in Egypt

Ragui Assaad

Introduction

In the context of a pronounced youth bulge, a phenomenon associated with a stage in the demographic transition when the share of youth in the population rises substantially, creating productive employment for youth has risen to the very top of the policy agenda in Egypt as in many other developing countries. I argue in this chapter that the way in which policy makers and planners have been thinking about this issue is excessively focused on a single indicator, the unemployment rate. I argue that the unemployment rate means something quite different in a context where informality of employment is common, if not the norm, outside the public sector, than it does in contexts where informality is the exception. In a low informality context, unemployment is highest among the least employable, namely those with lower skills, youth entering the labour market for the first time, and members of marginalised groups. It rises and falls with the cyclical fluctuations of the economy and is an indicator of the adequacy of overall employment levels. In a high informality context, unemployment is a phenomenon that typically affects somewhat more privileged groups, namely those who have some probability of obtaining formal employment and are seeking such work. This tends to be a select group that is typically more educated, and generally more privileged. The less educated and the poor know that their chances of obtaining formal employment are negligible and therefore directly enter the informal economy by engaging in various livelihood strategies, including casual wage labour, unpaid work in a family enterprise or farm, or, if they are lucky, regular but informal wage employment. They simply cannot afford to remain without work for extended periods of time while searching for better jobs and are therefore rarely captured in unemployment statistics. In such high informality contexts, an excessive focus on the unemployment rate leads to an exclusive concern with the labour market problems of a relatively privileged group and ignores the plight of large swaths of the population who have to eke a living in the informal economy.

Egypt is an interesting context to study in terms of issues related to labour market informality. In the 1960s and early 1970s, Egypt went through a period of state-led development where the public sector, and therefore formal employment, grew rapidly (Assaad 1997). This development strategy soon proved unsustainable and, like many other developing countries, Egypt had to embark on various economic restructuring programmes to reduce the role of the public sector in the economy. As was the case in many developing economies where public sector retrenchment occurred as part of neo-liberal structural adjustment programmes, a formal private sector response never materialised and the slack had to be filled by a substantial expansion of the informal economy (Assaad and Krafft 2015b). Thus the Egyptian labour market has been characterised over the past four decades by increasing informality despite rapidly rising levels of educational attainment among labour market entrants.

In the context of widespread and growing informality, I argue that the typical approaches used in more developed countries to promote youth employment and to assist new labour market entrants in their transition from school to work are generally ineffective. These approaches can be broadly classified under the rubric of active labour market policies or programmes (ALMPs) and typically include job search assistance, skills and employability training, initial wage and training subsidies, among others.[1] Such policies and programmes often have an implicit or explicit assumption that they are assisting labour market entrants to obtain formal employment, either in the public or private sectors. As a result, they end up assisting a relatively small and fairly privileged segment of job seekers who qualify for such jobs. Alternative approaches would acknowledge the reality of informality and would seek to create a better business climate for small firms to encourage them to grow and eventually formalise, leading to organic growth of formal employment.

One issue that requires special attention when discussing labour market informality is the gendered nature of the phenomenon. In a socially conservative setting such as Egypt's, the informal economy has generally proven to be inhospitable to women. Unlike men for whom informal wage employment has served as the default option, women in Egypt, as in a number of other Middle Eastern and North African countries, find informal wage employment in small informal enterprises that are mostly owned and operated by men to be either a threat to their sexual and reputational safety prior to marriage or highly incompatible with their domestic responsibilities after marriage (Assaad 2014b; Assaad and El-Hamidi 2009; Hendy 2015). A few women are able to set up small home-based businesses or livelihood activities, especially after marriage, but these often fail to develop beyond a very basic survival level. Women are therefore often left with only two options: either queue for scarce public sector jobs and thus remain unemployed for long period of times, or simply withdraw from the labour force altogether. This results in both low female participation rates and very high unemployment rates among the women who do participate. Female labour force participation rates in Egypt were in the vicinity of 23 per cent in 2012, among the 20 lowest countries in the world and the female unemployment rate was 24 per cent, nearly six times the male rate (Assaad and Krafft 2015a).

A labour market structure characterised by high informality is the norm in much of the global South and thus much of what I argue here applies more generally to a wide range of settings in the developing world. Nevertheless, there are some specificities to the Middle Eastern and North African settings to which Egypt belongs. First, the co-existence of high informality and high unemployment is a feature that is characteristic of middle-income settings with a legacy of state-led development strategies that have resulted in highly dualistic labour market structures that encourage more privileged workers with a chance to access the formal economy to queue for such jobs by remaining unemployed (Assaad 2014a). The second specificity relates to the barriers that prevent women from entering even informal employment, which is the fallback position for men, and is typically the fallback position for women as well in other high informality settings.

An informalising labour market

In discussing informality in the context of the labour market, it is appropriate to focus on informal employment rather than simply the informal sector. Informal employment includes workers working on their own account and as unpaid family labour in informal firms, but also casual wage workers who float from employer to employer and informal wage workers who work without proper documentation or social insurance coverage in either formal or informal firms.[2] Formal employment is further subdivided into private and public wage employment and a small segment of formal own account workers (mostly professionals). To simplify presentation, I lump this small category of formal self-employed workers with the much larger category of the informal self-employed, keeping in mind that the resulting category is virtually all informal.

To illustrate the growing informalisation of the Egyptian labour market, I rely on analysis in Assaad and Krafft (2015b) who use information from the 2012 wave of the Egypt Labour Market Panel Survey to assess the changing nature of the labour market facing new entrants over time, starting in the early 1980s. They use retrospective data provided by respondents at various stages of their life cycles at the time of the survey in 2012 on the nature of the job they obtained when they first entered employment.[3] They examine the nature of first jobs over time for all labour market entrants by sex, and then separately for those with upper secondary degrees and higher.

As shown in the top left-hand panel of Figure 9.1, the largest category of first jobs for new entrants in the early 1980s was public sector employment, at just under 40 per cent of new entrants, followed closely by informal private wage work at about one third of new entrants, and by unpaid family workers, who constitute over 20 per cent of new entrants and are also entirely informal. Formal private wage work barely registered as an option at that time, at less than 5 per cent of first jobs. By 1985, informal private wage work had overtaken public sector employment as the largest category of first jobs and kept increasing in share until the mid-2000s when it reached about half of all first employment. The share of public sector employment declined sharply starting in the late 1980s to fall to less than 20 per cent of new entrants by the early 2000s. The share of formal private wage employment increased from its very low base, but stayed below about 15 per cent of first jobs. In total, formal employment constituted well below 35 per cent of first jobs throughout much of the period.

The increasing informalisation of employment is even more apparent among educated workers. As shown in the top right-hand panel of Figure 9.1, more than 60 per cent of workers with upper secondary degrees and higher obtained first jobs in the public sector in the early 1980s. That proportion declined by two-thirds to just 20 per cent by the mid-2000s. Again the increase in formal private wage employment for this group remained very anemic, going from around 10 per cent of first jobs in the early 1980s to no more than 18 per cent in the mid-2000s. The proportion of formal jobs for educated workers had thus declined from about 70 per cent of first jobs in the early 1980s to less than 40 per cent in the mid-2000s. Again the fastest growing segment of the labour market for secondary and post-secondary graduates was informal private wage employment, which increased from 20 per cent to over 40 per cent of first jobs.

Given the relatively low female employment rates, the pattern for males is almost identical to that of all workers and will therefore not be discussed further. Female new entrants were even more reliant on public sector employment at the beginning of the period than their male counterparts. As shown in the bottom right-hand side panel of Figure 9.1, over 80 per cent of educated female new entrants who got jobs went into the public sector in the early 1980s. That proportion fell steadily to just under 40 per cent by the mid-2000s, only to recover temporarily

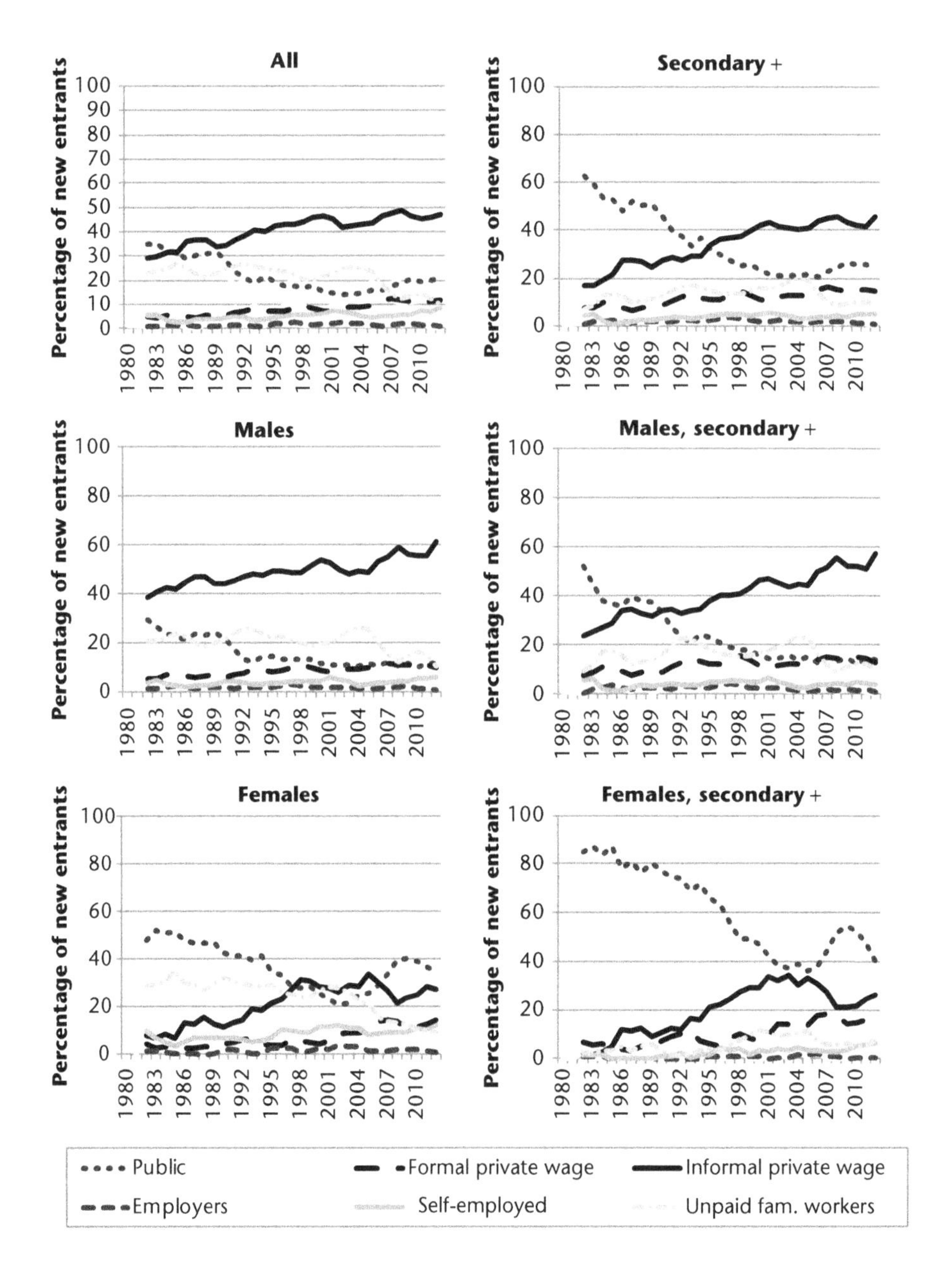

Figure 9.1 First jobs by type of job and year of first employment and sex, all new entrants and new entrants with secondary education and above, 3-period moving average, 1980–2011 (percentage)

Source: Assaad and Krafft (2015b), © by permission of Oxford University Press (www.oup.com).

and fall again by the late 2000s. Although private formal wage employment was growing modestly for this category of new entrants, the absorbing category was informal wage employment whose share grew explosively from nearly 5 per cent of educated female new entrants in the mid-1980s to nearly 40 per cent in the early 2000s. It should be kept in mind, however, that the decline in public sector job opportunities for educated female new entrants resulted in a similar long-term decline in educated female employment rates, as women who couldn't find public sector employment simply stayed out of the labour force rather than enter into informal employment (Assaad 2014a).

Access to formal employment is not only predicated on own educational attainment, but also increasingly on social class. Assaad and Krafft (2014c) compared the transition to first employment for two generations of labour market entrants in the Egyptian labour market, disaggregating by gender, own education and father's education, as a marker of social class. The main factor that determined access to formal jobs for males in the older generation (those aged 45–54 in 2012) was own education. While only 10 per cent of those with less than secondary education got formal jobs, nearly 60 per cent of those with higher education did, with little difference between those from a high and low social class. Males from a younger generation (25–34 in 2012) who were facing a much more privatised labour market, as we saw above, not only had a lower chance of entering into formal employment, but saw these chances increasingly being determined by social class. Men with higher education and with more educated fathers in this generation still had more than a 50 per cent chance of obtaining a formal job, but those with the same level of education but less educated fathers had only a 35 per cent chance. A similar difference across generations is observed for women, but in their case the fallback position is not informal employment, but non-participation or unemployment. Women with higher education from the older generation had a 68 per cent chance of getting a formal job, irrespective of social class, whereas women from the younger generation had a 42 per cent chance if their father was educated and a 26 per cent chance if their father was not. Less educated women of both generations, like less educated men, had virtually no chance to access formality, and ended up mostly outside the labour force.

The inadequacy of unemployment as an indicator of labour market health in the context of widespread informality

Open unemployment, defined as not working in excess of an hour in the reference week, while desiring to work, being available for work and actively searching for it, is a state that most young men in search of work in the developing world simply cannot afford to be in. A total lack of work means a lack of income in the absence of income support from other sources. The alternative to unemployment is the pursuit of livelihood strategies in the informal economy that combine between marginal self-employment, participation in family businesses or farms, casual wage work, and/or, if one is fortunate, regular but informal wage work. Unemployment is therefore a strategy that is pursued primarily by those with some probability of getting formal wage employment and who can afford to remain jobless while searching for such opportunities. Being unemployed allows those job seekers to dedicate more time and effort to searching for such jobs, while being supported by their families in the meantime.

For young women in a socially conservative setting that places a high value on modesty and sexual safety, engaging in the informal economy is often too risky a strategy. The predominant normative framework guiding gender roles is one that conceives of women as homemakers and men as breadwinners (Hoodfar 1997). Educated women will gladly pursue careers in the formal sector, particularly in the public sector, which is considered more family friendly, and will often remain unemployed for long periods of times to obtain such careers. Less educated women, who are simply not eligible for such employment, will often simply stay out of the labour force altogether rather than obtain any sort of work in the informal economy (Assaad and El-Hamidi 2009; Hendy 2015).

The unemployment profile by education and gender for youth described above is readily observed in Figure 9.2,[4] which is drawn from Krafft and Assaad (2014). The figure shows the distribution of employment states for young people aged 15–34 in Egypt, who are neither in school nor in mandatory military service, by level of education. Neither males nor females with less than secondary education bother to remain unemployed since they have almost no prospect of formal employment as we have seen above. Unemployment increases with education for both men and women as prospects for formal employment improve. For women, employment rates also rise with education, confirming that many women only enter employment if they can find formal employment.

Now that we have established that unemployment is a poor measure of labour market distress for disadvantaged youth, what would be better measures? Clearly some measure of income and its fluctuations over time would be ideal, but income data is notoriously difficult to collect in highly informalised economies. Some have suggested measures of vulnerable employment, which includes unpaid family workers (ILO 2013). Some measures proposed by Krafft and Assaad (2014) include the extent of irregular or casual employment and the degree of visible underemployment, defined as individuals working less than full-time and desiring to work more.[5] Krafft and Assaad (2014) show that both of these measures respond strongly to economic downturns and the phenomena they capture are much more likely to involve poorer people with lower levels of education.

Interventions to improve youth employment outcomes in the context of widespread informality

Addressing youth unemployment is an issue that tops public policy discourses and policy-makers' pronouncements in Egypt. With a long history of the government guaranteeing employment in the public sector for those who achieve a threshold level of education, there is an expectation among middle-class youth, and those aspiring to be in the middle class by virtue of their educational attainment, that solving their employment problem is a government responsibility,

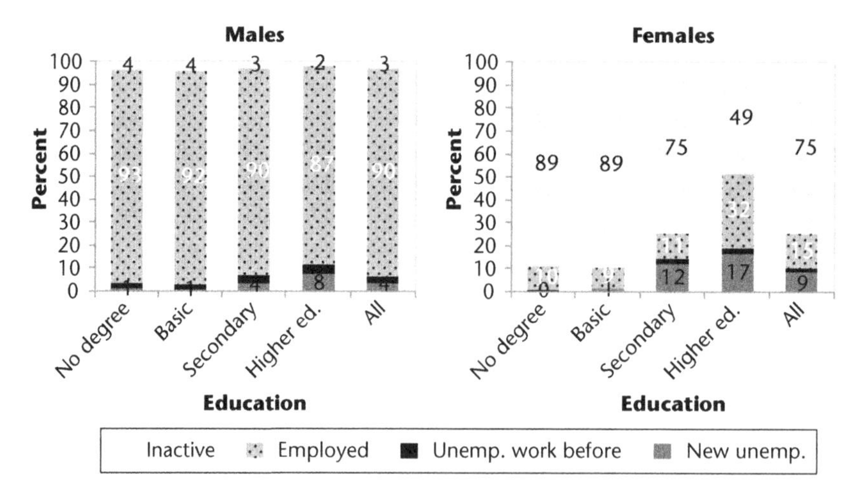

Figure 9.2 Labour market statuses by education and gender, ages 15–34, 2012

Source: Krafft and Assaad (2014).

Note: excludes students and males out of the labour force for military service.

and being relegated to the informal economy in precarious and unstable jobs is perceived by them with a deep sense of injustice (Barsoum 2016). Young people in Egypt continue to express strong preferences for government employment despite the perceived erosion of public sector wages and the vastly reduced hiring into the civil service (Barsoum 2015). This places enormous pressure on policy makers to respond, either by providing jobs in the public sector for educated youth, which they did in large numbers after the January 25 2011 revolution (Semlali and Angel-Urdinola 2012), or by instituting various programmes to assist unemployed youth under the general rubric of active labour market programmes or ALMPs.

ALMPs include programmes such as job search assistance and matching, wage and on-the-job training subsidies to private employers who hire unemployed new entrants, vocational training and soft-skills training and internship programmes (Card *et al.* 2010). In Egypt, these programmes are implemented by centralised bureaucracies, such as the Ministry of Manpower and Immigration or the Industrial Training Council of the Ministry of Trade and Industry, rather than by regional or local governments.[6] For the most part, these programmes assume that jobs do exist in the economy, but that various kinds of friction prevent youth from accessing them. Moreover, they almost always implicitly or explicitly target formal employers. In the context of highly informalised labour markets, such as the one I described above, these programmes invariably end up assisting the most privileged youth who meet the eligibility requirements for the scarce formal jobs and can afford to remain unemployed while searching for them. They do nothing to increase either the quantity or quality of jobs in the economy, which appears to be the binding constraint in a country such as Egypt.

What are policy makers and planners to do if they wish to improve employment prospects for young people? If the problem is anemic job creation and widespread informality, I follow Krafft and Assaad (2015) in arguing that the priority should be to create more and better jobs by improving the business and investment climate for micro and small firms, the predominant

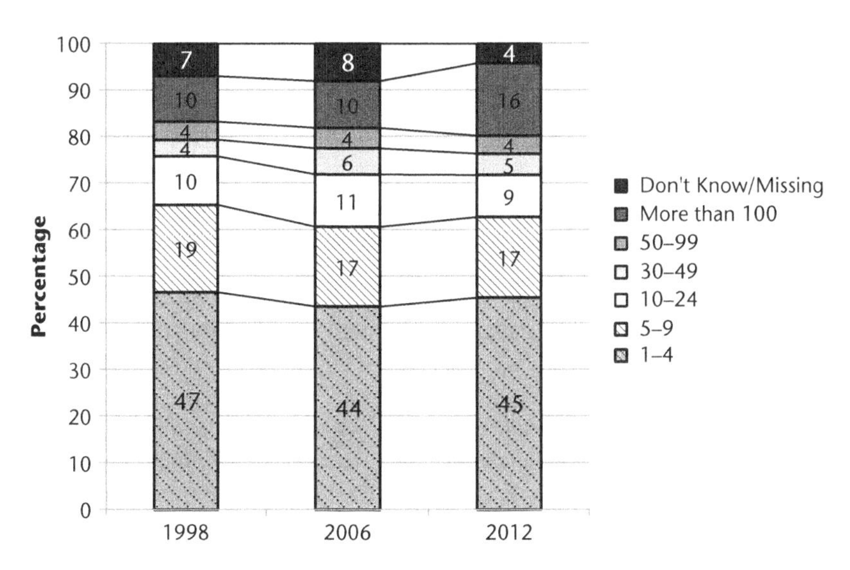

Figure 9.3 Distribution of private wage employment by firm size, workers aged 15–64, Egypt, 1998, 2006, 2012

Source: Krafft and Assaad (2015).

form of enterprises in the Egyptian context. As shown in Figure 9.3, just under half of private wage employment in Egypt is in firms of 1–4 workers and another 17 per cent is in firms of 5–9 workers. The vast majority of these firms are informal. Only 16 per cent of private wage employment is in large enterprises of 100 workers or more. These large firms have been the focus of most efforts to promote private sector development, and the more connected among them have captured the lion's share of resources, such as bank credit, land allocation, access to foreign exchange, export markets, and government contracts (Chekir and Diwan 2014; Diwan *et al.* 2014). The fact that so little employment is in mid-sized firms of 10–99 employees is evidence of a highly dualistic firm structure, a phenomenon often referred to as a "missing middle" (Krueger 2013).

A slew of factors constrains the growth of small and micro firms, some of which fall under the jurisdiction of national authorities and some are the purview of regional or municipal governments. Most important among these factors are ones that make the cost of formality prohibitively high for small and micro firms. National tax regulations not only impose a high monetary burden on small and micro firms, but also subject them to a great deal of uncertainty due to the inability of tax authorities to accurately assess production and revenues for small producers and hence their proclivity to impose arbitrary assessments. Complying with local licensing and registration rules imposes large fixed costs on firms, both in terms of time and money. Corrupt municipal officials often use the prohibitively high costs of compliance to strike deals that extract as much rent as possible from small business owners in return for less stringent enforcement. When firms are small, these fixed costs loom large in their cost structure and quickly become unaffordable, pushing them into informality.

The World Bank's *Doing Business* report attempts to ascertain some of these costs in areas such as starting a business, getting credit, trading across frontiers, paying taxes, registering property, enforcing contracts, dealing with construction permits, resolving insolvency among other areas (World Bank 2016). The *Doing Business* report ranks countries annually based on their distance from the frontier of best practices in each of these areas and also prepares an overall "doing business" rank. Egypt's overall rank in 2017 was 122 among the 190 countries included in the report. It obtained some of the poorest rankings in the sub-areas of paying taxes, trading across borders and enforcing contracts, but also fared poorly in registering property and resolving insolvency (World Bank 2016). The only major area where significant progress has been made in recent years is "starting a business", where Egypt now ranks 39 out of 190.

National legislation, such as the adoption of a comprehensive small and micro-enterprise law, is needed to streamline and dramatically simplify all the dealings small and micro enterprises have with the bureaucracy. Some steps have been taken in this direction with the enactment of the Small Enterprise Development Law (Law 141 of 2004), which improves the incentives to start small and micro enterprises and simplifies the procedures for registering enterprises. It also includes various provisions for facilitating access to capital through the Social Fund for Development (IBP Inc. 2007). The improvements in Egypt's ranking on the "starting a business" indicator of the *Doing Business* report is likely attributable to the enactment of this law. However, a great deal more must be done to reduce the cost burden and uncertainty small and micro firms face in their ongoing dealings with both the national and municipal bureaucracies, including in the determination and collection of tax payments, the enforcement of labour and social insurance regulations, the ease of compliance with municipal licensing and registration rules and health and safety regulations, among others.

A number of additional factors can substantially raise the costs of doing business for small firms in an urban setting. These include traffic congestion, which raises the cost of transactions, transport and commuting on the part of employees, zoning and land use constraints, which often raise the cost of land acquisition and end up pushing small and micro enterprises into informal areas that are less accessible and less well served by public transport and other infrastructure. The absence of specific infrastructure that supports small-scale economic activities, such as public markets and industrial zones designed to meet the needs of small and micro firms and that are well served with public transport, road networks, and other urban infrastructure, can also be an important obstacle to the growth of small and micro enterprises.

There have been various attempts since the 1970s in Egypt to create industrial towns and zones to encourage the deconcentration of industrial activities away from populated areas (Sims 2015). However, these new industrial towns have exclusively served large enterprises and have been characterised by a virtual absence of low-cost housing, forcing both the enterprises and their workers to incur heavy transport and commuting costs. The absence of low-cost housing in the new towns is essentially another manifestation of a failure of planning in the midst of informality. In their desire to keep these new towns modern and orderly, planning authorities were keen to make them free of informal housing, and have thus been far more diligent about enforcing housing regulations and preventing the emergence of informal settlements in these new towns. The reality however is that in the absence of widely accessible housing finance, informal settlements, where incremental housing construction is possible, are the primary mechanism through which the working classes in Egypt, as in many other developing countries, can avail themselves of affordable housing. Thus the intolerance of informality that is at the core of these modernist planning approaches is at the root of the failure observed in the housing market; a failure that is in turn constraining the creation of employment.

If the bureaucratic and regulatory burden on small and micro firms can be reduced sufficiently to encourage them to emerge from the shadows of informality, this could unleash significant growth of output and employment. At the moment, the cost of formality acts as a barrier that forces small firms to stay small and thus remain somewhat invisible to regulators, probably foregoing many opportunities for growth and expansion (Loewe *et al.* 2013).

Another set of interventions that could substantially improve employment prospects for youth are interventions designed to reduce barriers to employment for women. There is already considerable evidence that women strongly prefer to work in larger workplaces where other women are present and where exposure to sexual harassment by male bosses is less likely. While 47 per cent of male wage workers in the private sector in Egypt work in firms of fewer than five workers, only 27 per cent of female wage workers do (Assaad 2014b). Conversely, female workers are disproportionately represented in larger enterprises. Thus any interventions that allow small firms to grow and possibly formalise are also likely to create a more hospitable environment for female workers.

Given that an important barrier to women's employment in the Egyptian private sector is the possibility of being exposed to sexual harassment both at the workplace and on the way to work, any interventions that create safer public spaces and workplaces will go a long way in alleviating this problem (Nassar 2003). It is not only a question of passing the appropriate laws, which has been done recently, but to take practical steps to allow women to avail themselves of these laws through more welcoming reporting structures, support networks and confidentiality protections.[7]

A further barrier faced by married women in the private sector is the inability to reconcile their work with their family responsibilities. Long working hours, long commutes and heavy domestic burdens make it very difficult for women to remain employed in the private sector after marriage (Hendy 2015). Policy interventions that reduce the fixed cost of employment and that incentivise part-time work, job sharing, telecommuting and home production could go a long way toward lowering this barrier. For example, minimum wages in Egypt are currently specified on a monthly basis and are a significant barrier for employers to reduce work hours or to accommodate part-time work (Assaad 2014b).

With their double burden, women's ability to reconcile home and work crucially depends on the extent of accessibility to their workplaces. As mentioned above, the planning strategies that sought to reduce congestion in urban areas by relocating industries to new towns and industrial zones have in fact resulted in an increasing need for commuting, an issue that is particularly burdensome for women. Assaad and Arntz (2005) have shown that the increasing need to commute may have contributed to women's inability to work outside the home in Egypt. These well-meaning physical planning strategies have had the unintended consequence of not only being inhospitable to small and micro businesses, but also creating often insurmountable barriers for women to work outside the home.

Conclusion

I argued in this chapter that, in contexts characterised by widespread informality, standard approaches to promoting youth employment derived from the experience of developed countries are generally ineffective. Such policies tend to focus interventions on a group of relatively privileged youth who have some probability of obtaining formal jobs and are therefore willing and able to spend time and resources searching for them. I also argued that in high informality contexts, standard measures of labour market health, such as the unemployment rate, can be highly misleading. With their focus on the complete absence of work during a fairly long reference period, these measures tend to focus attention on that same group of relatively privileged youth who remain jobless while searching for formal employment opportunities. They ignore the labour market difficulties faced by poorer and less educated youth who are forced to engage in whatever livelihood strategies are available to them to make a living. Measures of the precariousness of employment and income rather than the total absence of employment are much more relevant for this group.

I also emphasised that the labour market experiences of youth are highly gendered, especially in socially conservative settings such as Egypt's. While the default state for young men is informal employment of various kinds, the default state for young women is non-participation in market work. If they are able to access formal employment, especially public sector employment, they will indeed participate and find ways to reconcile their job and household responsibilities. Those who are eligible for such employment but are unable to get it right away often state that they are available and searching for work and are therefore counted as unemployed. Non-working less educated women who have virtually negligible chances of getting formal employment simply state that they are not seeking employment and therefore tend to be enumerated as outside the labour force. As a result, unemployment for women, as it is for men, increases with rising educational attainment and socio-economic status.

Active labour market policies and programmes either implicitly or explicitly assume that formal jobs exist and that the main challenge is to match job seekers to these jobs and to prepare them adequately for them. In a context of widespread informality and anemic growth of formal employment, such programmes end up assisting the more privileged and educated youth who can compete for formal jobs. Instead, I advocate interventions that allow small and micro firms to grow and possibly formalise by substantially reducing the regulatory and bureaucratic burden on these firms, increasing their access to capital, supportive urban infrastructure and markets, and facilitating their participation in international trade. In contrast to strategies that focus exclusively on large formal firms, a focus on the growth of small firms can affect the broad base of the labour market. Modernist physical planning strategies that aim to deconcentrate economic activities by developing new industrial towns and zones often fail to make any accommodations for small and micro firms and thus contribute to the inhospitable investment climate for these firms.

Finally, a strategy to promote the growth of small and micro firms must be accompanied by specific interventions to make such firms more hospitable to women. At the moment, women find that work in these firms is too risky for their sexual and reputational safety and incompatible with their domestic responsibilities. Although measures to encourage the growth and formalisation of small firms will help make the private sector more hospitable to women, they must be complemented by measures that make both workplaces and public spaces safer, and measures that incentivise flexible work arrangements such as part-time work, job-sharing, telecommuting, and home-based production.

Notes

1 See Card *et al.* (2010) for an international review of such programmes.

2 See the International Labour Conference resolution on the informal economy at its 90th session meeting in 2002. Available: http://ilo.org/public/english/standards/relm/ilc/ilc90/pdf/pr-25res.pdf [accessed 11 January 2017].

3 For the purposes of this analysis, a first job is defined as the first job the individual engaged in that lasted at least six months. This avoids counting summer employment for students as a first job.

4 Earlier versions of Figures 9.2 and 9.3 were published in Economic Research Forum's (ERF) *Policy Perspectives*, numbers 14 and 15 respectively. This work was sponsored by the ERF and has benefited from both financial and intellectual support. The contents and recommendations do not necessarily reflect ERF's views.

5 See also discussion of an alternative framework for portraying youth employment in developing countries in ILO (2013), pp. 50–56.

6 Local governments in Egypt are appointed by central authorities, rather than being elected, and depend on centralised budget allocations for almost all of their funding. As a result, they have limited discretion in policy formulation and implementation.

7 A law amending article 306a of the penal code was passed in June 2014 to explicitly criminalise harassment for the first time in Egyptian legal history. The law criminalises harassment in the form of words, gestures and actions expressed in person or through other means of communication. See El-Rifaie (2014).

References

Assaad, R. (1997) "The effects of public sector hiring and compensation policies on the Egyptian labor market", *World Bank Economic Review* 11(1): 85–118.

Assaad, R. (2014a) "Making sense of Arab labor markets: The enduring legacy of dualism", *IZA Journal of Labor and Development* 3(6): 1–25.

Assaad, R. (2014b) "Women's participation in paid employment in Egypt is a matter of policy not simply ideology", *Egypt Network for Integrated Development Policy Brief No. 22*. Available: www.enid.org.eg/Uploads/PDF/PB22_women_employment_assaad.pdf [accessed 11 January 2017].

Assaad, R. and Arntz, M. (2005) "Constrained geographical mobility and gendered labor market outcomes under structural adjustment: Evidence from Egypt", *World Development* 33(3): 431–454.

Assaad, R. and El-Hamidi, F. (2009) "Women in the Egyptian labor market: An analysis of developments, 1988–2006". In R. Assaad (ed.) *The Egyptian Labor Market Revisited*, Cairo: American University in Cairo Press, pp. 219–257.

Assaad, R. and Krafft, C. (2015a) "The evolution of labor supply and unemployment in the Egyptian economy: 1988–2012". In R. Assaad and C. Krafft (eds) *The Egyptian Labor Market in an Era of Revolution*, Oxford: Oxford University Press, pp. 1–26.

Assaad, R. and Krafft, C. (2015b) "The structure and evolution of employment in Egypt: 1998–2012". In R. Assaad and C. Krafft (eds) *The Egyptian Labor Market in an Era of Revolution*, Oxford: Oxford University Press, pp. 27–51.

Assaad, R., and Krafft, C. (2014c) "Youth transitions in Egypt: School, work, and family formation in an era of changing opportunities", Silatech Working Paper, Silatech. Available: www.silatech.com/docs/default-source/publications-documents/youth-transitions-in-egypt-school-work-and-family-formation-in-an-era-of-changing-opportunities.pdf?sfvrsn=6 [accessed 11 January 2017].

Barsoum, G. (2015) "Young people's job aspirations in Egypt and the continued preference for a government job". In R. Assaad and C. Krafft (eds) *The Egyptian Labor Market in an Era of Revolution*, Oxford: Oxford University Press, pp. 108–126.

Barsoum, G. (2016) "Job opportunities for the youth: Competing and overlapping discourses on youth unemployment and work informality in Egypt", *Current Sociology* 64(3): 430–446.

Card, D., Kluve, J. and Weber, A. (2010) "Active labour market policy evaluations: A meta-analysis", *The Economic Journal*, 120: F452–F477.

Chekir, H. and Diwan, I. (2014) "Crony capitalism in Egypt", *Journal of Globalization and Development*, 5(2): 177–211, 452–477.

Diwan, I., Keefer, P. and Schiffbauer, M. (2014) *On Top of the Pyramids: Cronyism and Private Sector Growth in Egypt*, Washington, DC: The World Bank.

El-Rifaie, Y. (2014) "Egypt's sexual harassment law: An insufficient measure to end sexual violence", Middle East Institute, July 2014. Available: www.mei.edu/content/at/egypts-sexual-harassment-law-insufficient-measure-end-sexual-violence [accessed 11 January 2017].

Hendy, R. (2015) "Women's participation in the Egyptian labor market: 1998–2012". In R. Assaad and C. Krafft (eds) *The Egyptian Labor Market in an Era of Revolution*, Oxford: Oxford University Press, pp. 147–161.

Hoodfar, H. (1997) *Between Marriage and the Market: Intimate Politics and Survival in Cairo*, Berkeley, CA: University of California Press.

IBC Inc. (2007) *Egypt Business Intelligence Report: Practical Information, Opportunities, Contacts*. Washington, DC: International Business Publications.

International Labour Organization (2013) *Global Employment Trends for Youth: A Generation at Risk*, Geneva, Switzerland: ILO.

Krafft, C. and Assaad, R. (2014) "Why the unemployment rate is a misleading indicator of labor market health in Egypt", *Policy Perspective No.14*, Economic Research Forum, Cairo. Available: http://erf.org.eg/publications/why-the-unemployment-rate-is-a-misleading-indicator-of-labor-market-health-in-egypt/ [accessed 11 January 2017].

Krafft, C. and Assaad, R. (2015) "Promoting successful transitions to employment for Egyptian youth", *Policy Perspective No. 15*, Economic Research Forum, Cairo. Available: http://erf.org.eg/publications/promoting-successful-transitions-to-employment-for-egyptian-youth/ [accessed 11 January 2017].

Krueger, A.O. (2013) "The missing middle". In N.C. Hope, A. Kochar, R. Noll and T.N. Srinivasan (eds) *Economic Reform in India: Challenges, Prospects, and Lessons*, Cambridge: Cambridge University Press, pp. 299–318.

Loewe, M., Ayouty, I.A., Altpeter, A., Borbein, L., Chantelauze, M., Kern, M. and Reda, M. (2013) "Which factors determine the upgrading of small and medium-sized enterprises (SMEs)? The case of Egypt", *Studies*, Deutsches Institut for Entwicklungspolitik No. 76, Bonn. Available: www.die-gdi.de/uploads/media/Studies_76.pdf [accessed 11 January 2017].

Nassar, H. (2003) "Egypt: Structural adjustment and women's employment". In E.A. Doumato and M.P. Posusney (eds) *Women and Globalization in the Arab Middle East: Gender, Economy, and Society*, Boulder, CO and London: Lynne Rienner, pp. 95–118.

Semlali, A. and Angel-Urdinola, D.F. (2012) "Public employment services and publicly provided ALMPs in Egypt", *World Bank Other Operational Studies* No. 12993. Available: https://openknowledge.worldbank.org/handle/10986/12993 [accessed 11 January 2017].

Sims, D. (2015) *Egypt's Desert Dreams: Development or Disaster?* Oxford: Oxford University Press.

World Bank (2016) *Doing Business 2017*, Washington DC: The World Bank Group. Available: www.doingbusiness.org/reports/global-reports/doing-business-2017 [accessed 11 January 2017].

10

No global South in economic development

Smita Srinivas

Introduction: the diverse contexts of economic development

Is there a legitimate shared 'global South' in economic development? This chapter argues that we need a new set of terms and frameworks to look at economic development within public planning processes.

My aim is to deepen the conversation about the title of the book, *Routledge Companion to Planning in the Global South*, and make two points. First, to caution that the 'global South' label offers minimal analytical benefit as far as economic development plans are concerned, and what may appear to be the essential common empirical reality of developing countries is deceptive. This may help us better connect national contexts and development debates with what occurs in villages, towns and cities. Post-colonial institutional contexts and global market pressures may simply be insufficient common conditions to make wider claims on what planning should do.

The first section therefore focuses on early historical common ground on ideology but substantial economic plan variation between and within nations. I revisit the historical context of the Bandung Conference of 1955, and subsequent 'Third Worldism' challenges. Even at Bandung, it appears that national economic goals were quite distinct as part of a wider nation-building philosophy, and within groups of countries, the economic strategies deployed for similar goals were sufficiently different.

Precisely because of this diversity, we may need to tread a new path building on past critiques that developing, and especially industrialising, economies have insufficient autonomy to plan, with too much interference from a core of industrialised economies. There is certainly truth to this interference and to the critique, but the pendulum has arguably swung too far and development from 'within' and 'from here' can be much more strategically approached. The easy critique of development plans occurs when umbrella labels such as the 'global South' or 'Southern' is overused. It is easier conceptually and politically to criticise plans, and much harder (but perhaps more honest) to craft such plans in the service of the disadvantaged, and harder still to defend any single normative approach to individual or collective planning responses.

The second section of the chapter therefore discusses the practical context and some examples of economic development plans and the contradictions they contain. Because such plans often require dealing with multiple levels of economic governance in contexts where such governance capabilities are mixed at best, and for economic activity that often involves scale and scope, such contradictions may be inevitable. Such critique is often too easy because the alternatives are rarely specified. If anything, planners may need to hone their skills further to negotiate plans to develop a new lexicon and maximise gains for those worse off.

Above all, economic development involves productive structural change as well as institutional reform, and therefore 'plans' depending on context may include visioning processes, documents with goals and specific targets, policy analysis and design, as well as (usually hoped for) designated public administrative capacity, all of which require a political arena of power to exert some industrial and fiscal instruments. These economic development planners are invariably, by discipline and training, and institutional location, distinct from spatial and physical planners. The public process of economic development planning in which they sit is largely conceived and managed within public ministries, agencies and projects, sometimes with parastatals, multilateral agencies or development partnerships and sometimes through private firms and contracted-out services and missions. If anything, this diversity of means to accomplish public plans is evidence of the challenge of analytical taxonomies such as the global South.

Variation of economic plans between and within nations

Developing countries have historically marinated in a stew of multilateral policy advice and conditional lending to economically transform by following 'leader' countries in today's advanced industrialised economies. This advice has come alongside early industrial economies 'kicking away the ladder' for these latecomers and dismantling the institutional conditions for them to advance (Chang 2002). As we shall see, the notion of linear industrial progression may be too simple for the complex empirical national and sub-national realities of many developing economies, least of all because their sector differences and their responses are so diverse (Srinivas 2016). Many are indeed 'late' to the global trading contexts in specific sectors and technologies, yet 'early' relative to other indicators of advance (such as mobile phone use or solar deployment). Understanding this economic planning context thus requires substantial vigilance about national trading, standards setting and governance when considering the normative goals of change (Benería *et al.* 2012; Kabeer 2004; Srinivas 2012, 2015).

Yet structural change challenges conditions of unequal trading access, and the lack of specific production and knowledge situates many regions and towns in a dilemma about long-term economic strategies. Plans can isolate and crowd out crucial skills. Majority populations of underemployed and unemployed, piece-rate or contracted work, 'informal' workers and few opportunities for upgrading and learning, underscore that economic participation is uncomfortably squeezed between lack of local opportunity and severe international price and other pressures. Amsden (2001: 2) offers a version of human opportunity that reflects an emphasis on knowledge and structural change, with strategic choices required to move away from unskilled, low-entry work: "The transformation involves attracting capital, human and physical, out of rent seeking, commerce, and 'agriculture' (broadly defined), and into manufacturing, the heart of modern economic growth". Economic development in this broader sense requires attention to scale and scope. A wider ports and logistics development plan, or skills and training for SEZs and industrial zones would qualify, while real estate development in contrast – with nothing to build on except land value – would fail the wider economic development goal even if it raises money for the treasury. Additionally, important dual or triple aims can be satisfied within economic plans, such as the development of the health industry, soil remediation, waste processing or the creation of agriculture seed banks. The conversion of planning from these national and often normative aims to the practical manifestations of local transformation, is the force that shapes cities, towns and villages.

However, in the post-colonial context of several newly independent countries, the global South was equated primarily with nation-states. The first Asian–African conference to bring together mostly newly independent African and Asian nations was held in Bandung, Indonesia in 1955, over 60 years ago, setting in motion one important continuing conversation about whether a global South existed. Bandung was also an important precursor to the Non-Aligned Movement (NAM) in the Cold War period. Yet, a parallel tradition of town and city planning had also grown, focusing on physical investment and spatial outcomes. The intervention scale and built form varied. 'Town planning' often drew from British colonial traditions and priorities, with an emphasis on sanitation engineering and cantonment architecture, but rarely incorporating development realities such as informal work, migration, the need for jump-starting industrial scale and scope investments or evolving to respond to the cultural tensions of nation-building.

Given these different realities experienced by newly independent and other nations in both national and 'local' planning, it was perhaps unsurprising that policies even for ambitious import substitution industrialisation (ISI) resulted in different plans, instruments and urban outcomes. Making at home what could be bought abroad had been used to different degrees in industrialised economies, but the full-throttle use of it for late industrial economies captured the imagination of several political leaders and economists. Not all developing countries used this indigenisation approach. Those that did, however, hewed quite distinct paths. India, Brazil, Indonesia and South Korea approached this ISI challenge with different assumptions about urban and regional plans. In many cases, they used similar instruments in different ways: special export vehicles, logistics and regional development, shop-floor training priorities and domestic substitution for capital-intensive imports. 'Elite' demand and powerful interest groups shaped national priorities in production plans and policy design. Understandably, these worsened the domestic political legitimacy of several industrial choices and urban investments.

The 'rise' of Japan, China, South Korea (largely through domestic productive capabilities, then export market competitiveness), and to a much lesser degree India and Indonesia, created breakaway nations and cities and regions, signalling a successful shift in the productive structure in manufacturing, services, and extractive industries, and substantial regional income and trade gains. Yet sub-national administrative capacity to enact and implement plans varied hugely: Indian municipal and metropolitan state capacity may have been the weakest and least credible from a democratic standpoint; Indonesia and China made other, more centralised plan choices. While labour-intensive manufactured exports were the path of many, some gained manufacturing share and value added, but not always jobs. All developed shares outside agriculture and into manufacturing, construction, extractive industries and tech-intensive services, even where their specific urban or rural emphasis varied. Lall (1982, 1984, 1993) underscored the vital process of building technological capabilities including construction engineering, building plant capacity, to managerial capabilities in all sectors, organisational innovations and investments in project execution and governance. Amsden (1989) highlighted the crucial state capacity to monitor private manufacturing firms in the public interest which transformed Japan, South Korea and Taiwan, and which capitalised on selective market access opportunities.

Several developing country coalitions, whether termed Third World, global South or Non-Aligned, were thus geared towards diverse development and distributional goals. The political coalitions masked domestic diversity, but successfully, because they increased membership in the Security Council and United Nations Economic and Social Council (ECOSOC) to reflect their own needs. These included priorities of economic and technological advance, disarmament, reducing racial discrimination and decolonisation gains (Acharya and Tan 2008; Berger 2001, 2004; Nesadurai 2005). In addition, they increased the General Assembly's number of vice presidents. Arguably, as East Asian economies grew in influence, the existing coalitions ceased to be effective. It was hardly surprising therefore, that by the 1970s and with deep post-independence domestic discontent, alternative bilateral and regional arrangements emerged for access to export markets and technology transfer. The NAM's own imperatives to improve environment and human rights tempered its approach, and new and differentiated ties emerged among them with now industrialised Japan, the US, European nations (and former colonisers) and some selecting technology transfer from the Soviet Union (Korany 1994).

Given this emerging diversity by the 1970s, uniform 'Southern' qualities were clearly analytically insufficient to explain breakaway countries and regions (Amsden 2003: 37).

Today, such industrial plans rarely focus exclusively on manufacturing alone, but also include agricultural and services transformation with complex networks of tasks and locations. Because of this, although urban planning theories and urban economics may teach otherwise, there is no one-to-one correlation between economic development plans per se and their urban or rural built outcomes. This causes challenges of its own: local politicians and bureaucrats have neither the industrial or fiscal autonomy to fully respond, nor citizens the participatory power to resist detrimental growth or investment outcomes.

East Asia, the most dramatic twentieth-century story of public planning resulting in economic prosperity, challenged any idea of Southern dependency on an industrial core of nations. Furthermore, the 'East Asian Tigers' as a group had substantial variety in the composition of their public and industrial plans. Between larger nations (South Korea) and smaller, homogeneous nations (Taiwan) or practically city-states (Hong Kong and Singapore), they responded with dramatic differences based on land reform, labour controls, investments by lead firms and original equipment manufacturers and widely different approaches to the informal economy and investment. Table 10.1 shows stylised characteristics of the 'East Asian Miracle'.

These industrial employment plans in turn played decisive roles in the spatial and physical footprint of their cities (and city-states). Singapore invested in public housing but placed strict controls on migrant labour; Hong Kong relied on private housing and expanding global trade

Table 10.1 Diversity in 'East Asian' development

Country	*Industrial organisation*
Hong Kong (pre-PRC)	Middle-size informal economy; plan/policy emphasis on MNCs and OEMs; private firms manage labour; strong ties to global suppliers and retailers
Taiwan	Large informal economy; plan/policy emphasis on OEMs, small firms; public and private firms manage labour controls; strong ties to domestic supplier base and small firm networks
Singapore	Small informal economy; plan/policy emphasis on large MNCs; state manages labour controls; strong ties to international suppliers
South Korea	Small informal economy, plan/policy emphasis on large domestic conglomerates and state-owned firms; state manages labour controls; strong ties to building domestic supplier base

Source: adapted from Amsden (1989, 2001) and Cheng and Gereffi (1994).

ties to drive their economy; Taiwan privileged science and technology parks and small supplier networks in knowledge-intensive sectors; South Korea did the same, but emphasised big business *chaebol* (conglomerates) and delayed any dedicated regional plans, thus leaving substantial sub-national skew to its economic and infrastructure outcomes. South Korea and Taiwan had perhaps the most interventionist and credible planning strategies in building indigenous firms and long-term monitoring (Amsden 1989, 2001). The plans, building on geopolitical and trade opportunities, privileged proprietary knowledge acquisition and skills, using industrial instruments such as local content requirements and cherry-picked regional investments (sometimes the two processes merged in privileged quotas and investments for original equipment manufacturing (OEM) supplier networks). 'East Asian' industrial welfare also points to analytical diversity, from industrial SEZs and housing dormitories (in China), public housing (Singapore), and substantial consolidation of localised community group risk-mitigation strategies, with the enormous social insurance and health challenges of micro, small and medium firms alongside large ones (Srinivas 2010).

As the diverse East Asian plans demonstrate, the confusion generated by the term 'the global South' rests in no small part on its multiple loose usage. The global South can in turn refer to Third World, developing countries, even post-colonial territories, even simply 'poor', or 'informal' whether 'informal labour' or 'informal settlements', both with different connotations and assumptions. The post-colonial heritage of the global South was evident neither in Brazil, Russia, India, China or South Africa (BRICS) (not least because China and Russia were not colonised), nor did all ex-colonies provide unequivocal support for the NAM, nor even constitute a cohesive club for the G77. Moreover, only some nations self-identified with agencies such as the United Nations Conference on Trade and Development (UNCTAD) or the United Nations Industrial Development Organisation (UNIDO), which were explicitly focused on aiding their economic transformation. Some nations, Malta for instance, were members of the largely industrialised later European Union (EU), while retaining membership of the G77, yet also philosophically allied to the NAM (Berger 2001, 2004; Korany 1994; Nesadurai 2005). The Southern taxonomy was further weakened by new economic and climate realities of the late twentieth century. Emerging tensions between these nations came into sharp relief with complex new cross-jurisdictional spatial challenges: controls on service professionals and refugee migrations, environmental degradation, water conflict and cross-national ecosystems and poverty responses.

Furthermore, the South had no noteworthy sub-national cohesion or urban pattern, as witnessed in East Asian 'Tiger' diversity and larger federal and quasi-federal contexts such as India or Brazil. While labour informality, uneven physical environment and weak governance characterised some of the commonalities, industry successes fissured these nations further, building on growth in pharmaceuticals, biotech, energy, steel, automotive, garments, textiles, leather, furniture and electronics.

Ultimately, any existing cohesion rapidly dissipated. Many former NAM and G77 members 'graduated' economically into the Organisation for Economic Co-operation and Development (OECD), G20, or other memberships, with their towns and cities selectively benefitting from these memberships and trade opportunities, and their national plans intent on augmenting this divergence. Others tied their economic fortunes to tight political controls dominated by defence-led economic priorities. Only some were historically part of NAM, and only Chile, member of both NAM and OECD, remained a G77 member.

Nesadurai (2005: 68–69), for example, underlines the tensions of global ambition set against a backdrop of dependency concerns yet dissent among members. The New International

Table 10.2 Political economy norms and clubs

Country	*'Origin world' of national development*	*Select trade bloc/club memberships*
Cuba	Second World (Marxist-Leninist, single party) NAM	G77
India	Third World, NAM (socialist (mixed market) republic, parliamentary democracy)	G20, G77, BRICS, emerging economy, 'enhanced engagement' country EEC/OECD
Brazil	Third World (NAM observer only)	G20, G77, BRICS, emerging economy, EEC/OECD
Thailand	Third World (constitutional monarchy, and former military leader elected prime minister), NAM	G20, G77, BRICS, emerging economy, OECD, ASEAN
Indonesia	Third World (military led), NAM	G20, G77, BRICS, emerging economy, EEC/OECD ASEAN
Egypt	Third World (military led, later democracy) NAM	G77
South Korea	Third World (capitalist; military led, later democracy)	G20, OECD, ASEAN
Chile	Third World (capitalist; presidential democracy now), NAM	OECD, G20, G77 (only OECD member exception)

Source: author.

Note: some memberships are fluid.

Economic Order (NIEO) was thus ostensibly both a call to solidarity among disparate post-colonial and newly independent nations, but also a platform for projected domestic planning. Processes were arguably poorly defined and unevenly executed, through which the new Order would come about.

The variety in actual framing and adoption of plans reflected a localised pragmatism. Some were set against Gandhian, Fabian and other economic perspectives (India), less or more militant forms of Marxism and Maoism (China, Cuba, Vietnam), an ideology of nationalising assets (Venezuela), public (sometimes military) controls of national resources and rents (Indonesia under Sukarno), pragmatic approaches to global mediation (Malaysia) and selective supports for foreign and domestic investments (Tito in Yugoslavia, Nasser in Egypt and Nkrumah in Ghana). ISI policies were critical planning junctures that attempted to reclaim value in the entire supply chain for homemade goods, but there were dissimilar strategies deployed. South Korea, Brazil and India offered three national (and also urban-regional) approaches to ISI and then moved at different stages into unique strategies for export promotion, further shaping their regions and cities in distinct ways. Yugoslavian cooperatives offered another ideological, local organisational approach, driving its own spatial and land logic, public sector ownership and variegated success of jobs. Breakaway sectors and regions offered both employment and/or high value addition: garment and apparel sectors (Bangladesh), ICTs (India, the Philippines, Brazil); aviation and aerospace (India, Brazil), optoelectronics (Japan, Brazil); pharmaceuticals (to different degrees, India, Nigeria, Brazil, Kenya, China, South Africa); early electronics and often semi-conductors, steel and automotive (Japan, South Korea, Taiwan, Hong Kong, China, Malaysia).

Economic development plans: contradictions, critique, future challenges

At best, there now remain multiple 'Southern' groupings of economic realities requiring an updated lexicon. All NAM nations (other than Belarus and Uzbekistan) constitute 114 of 134 members of the G77. Others from Brazil to Haiti and Uruguay may be termed 'Southern', but their economic planning institutions and forms of geopolitical strategy are dramatically different. New G20 groupings and strategy are also pulling members in different directions: Turkey (waitlisted EU member, but OECD founder, has just been the G20 President); China (Association of Southeast Asian Nations ASEAN) and pivotal member of the contentious new Trans-Pacific Partnership Agreement, and India (neither ASEAN nor TPP) have grown in very different forms, sculpting their towns and cities. Even single industry sectors may not provide shared strategies. Nations pull in single directions at the World Trade Organisation (WTO) on access and affordability issues based on sector and regional criteria protections (e.g. geographic recognitions), but in markedly opposing directions on intellectual property, subsidies or compliance on local content requirements and national mission priorities. From local building materials and solar panels, to municipal write-offs to attract industries, the trade and industry repercussions have been felt in most cities. Agriculture and food plans in villages and cities are today closely tied to G20 and WTO dialogue on farm subsidies.

The traded context is thus vital to understanding the domestic political argument for planning autonomy. Arguably, the Bandung Conference was more about structural dependency of the time but also moral and ideological obligations to domestic development planning (Acharya and Tan 2008). The development autonomy was more clearly expressed in the Cairo meeting in 1962 (Nesadurai 2005; Srivastava 2001), while the dependency arguments were theorised and organisationally embedded by the 1964 Geneva UNCTAD meeting. These culminated (via Raul Prebisch and others) in a shared package of market access and a 'one nation, one vote' system, but with considerable dissent on items such as postcolonial restitution and nationalisation of assets. NAM members forged their own national and sub-national paths. The institutional contexts and political and social arrangements of national trajectories (socialist, Maoist, cooperatives, chaebols), inevitably created domestic industrial organisation patterns with urban and regional path dependencies.

Nations with industrial and structural change ambitions sought selective coalitions to counter the juggernaut of the Washington Consensus' conditional lending and structural adjustment plans. However, precisely because the East Asian nations pulled away on many dimensions, the fragility of a shared front for the others was revealed. Even large, unwieldy nations such as India and Indonesia managed to boost medium- and high-tech activities to almost a third of industrial output between 2006 and 2011, while other emerging industrial economies pushed manufactured export shares up further (22–27 per cent) and manufacturing value added (24–31 per cent) from 2007 to 2011 (UNIDO 2013). This was in sharp contrast to those whose manufactured export shares and value added stagnated at 2–3 per cent in the same periods (UNIDO 2013). Indonesia of course benefitted more substantially from oil prices than India. However, even these struggled to extend the learning gains to wider developmental outcomes. For much smaller less developed countries (LDCs), the technology focus generated some sector successes. Yet, dependent on aid and changing terms of trade, their fiscal and administrative autonomy remained far less secure (UNCTAD 2009).

There is a new momentum for a return to national development plans. The United Nations Economic Commission for Africa (UNECA) and NEPAD are increasingly supporting the adoption of national development plans for jobs and growth, but embedded in African multilateralism. These new cross-national treaties rest on mutual recognition from travel and trade to accreditation and exchange in education, and build investment corridors for enhancing trade and mobility.

If indeed the 'global South' today provides a poor descriptive and analytical umbrella for economic development planning, where do we take this critique next? Consider that most dominant theories of economic development see planning processes as the main mechanism by which to bring about institutional reform, or even that planning *is* itself institutional reform, including developing strong ethical norms in governance, monitoring systems and accountability standards. More challenging, the principles of 'similarity and difference' underpinning development can always be too easily criticised:

> Development studies is an unusual enterprise. It is committed both to the principle of difference (the Third World is different, hence the need for a separate field of studies) and to the principle of similarity (it is the job of development policy to make 'them' more like 'us').
>
> *(Corbridge 2007: 179)*

These questions of inequality and distributional justice between and within groups of nations (and certainly within each) are hardly easy to resolve. 'Post-development' analysis thus would have to move beyond critique alone (Corbridge 2007). 'Planning from here' can indeed extract some public benefit from larger economic processes underway. However, the 'morality of critique' (Corbridge 2007) needs to be made explicit because the economic and institutional reform choices are not simple, and even small power shifts to reduce inequality, which planners surely aspire to but that are embedded in ethical responses, are necessary and urgent.

Yet if there is no easy 'us' or 'them' in terms of 'South' and 'North', the mantle for planning action 'from here' falls to highly uneven sub-national governance and institutional reform. Planning foresight, administrative capabilities, accountability, ethical norms and transparency, and political legitimacy become paramount. Because the goals of economic development are wide (income strategies, jobs, technological choices, women's participation, etc.), such planning would be constantly buoyed by national and global economic winds, and remain without an easy single normative goal or strategy. Planners will have to focus on crafting the 'big-picture' plans, but also expend energy on connecting seats of power and control from national ministries to local development communities (see also dilemmas discussed in Benería *et al.* 2012; Kabeer 2004; Srinivas 2012).

More promisingly, 'planning from here' can identify dual or multiple goals for plans. Demand and delivery concerns, or attention to learning and innovation can converge, resulting in employment, skills and productivity, but also better customisation of products and processes to local conditions (Kaplinsky 2011; Lall 1982, 1993; Srinivas and Sutz 2008). Social policies and environmental demand factors offer productive expansion (Albuquerque 2007; Arocena and Sutz 2000; Srinivas 2012, 2014, 2017), making industrial welfare an essential element of this economic strategy. Plan instruments such as procurement and local content requirements can be deployed for urban public services from health, transport and water and sanitation. Fuel-efficient and modular public bus networks across South America have spawned a bus body works industry in Brazil; municipal and national health programmes have created India's formidable vaccine industries; ironically, the miserable state of Indian clean water availability has also generated an immense private water filter industry. As I have argued elsewhere (Srinivas 2012, 2014, 2017), industry transformations and technological advances, even when done well, complicate, rather than simplify, domestic redistributive politics. 'Planning from here' requires substantial public reform. This reform includes building shared ethical norms and citizens and residents insisting on transparent public finance, process monitoring and third party accountability systems to improve municipal and regional government actions.

Only some places have benefitted from global production networks and 'value chains', mostly due to the capabilities they developed through slow, ambitious, multi-organisational investments and political reforms (UNCTAD 2009; UNIDO 2013). Urbanisation when not accompanied by such industry reform, job growth, or social protections brings its own deep insecurities. However, there are several learning and employment opportunities within the domestic, largely untraded, economy. If an economic governance process is dynamic, these islands can converge; if not, plans can create troubling sector and product enclaves of their own (Albuquerque 2007; Kaplinsky 2011; Srinivas 2017; Srinivas and Sutz 2008).

While they might have generated similar concerns of inclusion, shared colonial experiences did not create similar plans: India and Ghana offer different Anglophone industrial and urban outcomes, as did Brazil and Mozambique in the Lusophone context. Similarly, Brazil and India, often grouped together for BRICS, analytically break apart on economic development, with Brazil arguably doing much more than India to improve social policy commitments (health affordability, access and housing are examples). The Petrobras economic governance scandal of mismanagement of the success of the oil and gas industry is now reshaping these commitments in cities across Brazil. Even NAM Chair in the 2000s, Malaysia, broke from global partnerships to its own version of labour-intensive exports, followed its own ideological piper and responded primarily to its domestic religious and ethnic considerations.

Economic development is not only 'traditional' industrial policy of 'targeting' key technologies (although these persist), but it can frame the priority setting for specific sectors and investments that directly galvanise the economics of cities: solar panels, wind generation, electronics, freight and distribution logistics, ports, waste processing or construction. Dramatic urban shifts across Africa appear to be lending further appeal for national plans. Ethiopia's extraordinary economic growth and a target of 200,000 jobs annually has meant industrial parks with far greater focus on 'eco-industrial' development, close ties to provincial governments and cities, and much greater emphasis on fiscal incentives. Its Growth and Transformation Plan from 2011 to 2015 uses similar ambitious language.[1]

National plans however, are not substitutes for regional ones. Urban politicians often promise outcomes for which they lack jurisdiction. Indian mayors of even powerful metropolises lack the authority to plan industries without regional government interventions. South American mayors in contrast wield greater power, often moving to the national political stage, but they too are limited. In South Africa, the African National Congress (ANC) has supported mayor Danny Jordaan's bid to be re-elected and boost the ANC's lagging fortunes. Press coverage on his campaign has been focused on five-year plans for Port Elizabeth and pivots on two key industrial outcomes: education and youth employment in response to the Eastern Cape's youth unemployment rates ranging from 37 to 41 per cent between 2008 and 2015 (StatsSA 2015); and a focus on economic diversification away from automotive investments and towards tourism and ports development.

Sub-national urban and regional infrastructure and investment is not only connected to the credibility of such political devolution or democracy, but can reduce costs of trade. High transport costs make agricultural produce more expensive for both farmers and consumers alike, and make it more likely that nations such as Ethiopia look to manufacturing to achieve economic prosperity, despite approximately 40+ per cent of their GDP originating in agriculture. The Maghreb Highway has experienced several delays because of its cross-national ambition and budgetary uncertainties. If completed, Libya, Tunisia, Algeria, Morocco and Mauritania, with approximately 55 cities and 22 international airports, rail and port connectivity, will be able to create large cultural exchanges and travel, lower costs of imports that affect healthcare products, basic infrastructure, food, and of course hoped-for jobs (Mohamed-Chérif and Ducruet 2016).[2] Djibouti is also expanding toward plans for a ports and logistics hub for Africa. At its Red Sea location, it enjoys 30 per cent of global maritime trade. The associated new oil terminal and container terminals will be helped by the Chinese-built 750 km electric railway connecting Addis

Ababa and Djibouti, and 75 per cent cargo from Addis. This forms a plan to build 5,000 km of rail by 2020 to link Addis to Sudan, South Sudan, and Kenya (*Guardian* 2016).

It is evident that the fissured 'global South' reflects the highly uneven ability of governments to seize economic opportunities, develop economic governance norms with citizens and bring prosperity (Korany 1994; Morphet 2004; Srivastava 2001). These divisions were visible within Asia and between Asians and Africans of the original Bandung meeting (Nesadurai 2005). The new trade pacts of the twenty-first century such as the Trans-Pacific Partnership (TPP) also establish new governance frameworks for nations in unexpected 'North' and 'South' configurations; the recently withdrawn US, to Peru, Japan to Canada and Vietnam to Mexico. From agriculture to financial services, and pharmaceuticals to automotive issues, there are complex existing multilateral and new bilateral agreements emerging. For generic drug producers such as India, and for patients around the world, there remain important concerns about access to medicines with the scope of intellectual property, data protections and other tariff barriers. The Regional Comprehensive Economic Partnership (RCEP) has also expanded in the Asia Pacific region, making more challenging the ability of sub-national governments to fully flex their economic muscles or plan jobs or investment strategies with any certainty. While foreign direct investment affects these outcomes, so do intellectual property rules, accreditation for labour and phytosanitary standards, the laboratory and technical capabilities for such verification and the storage, freight, fabrication, process and project execution capabilities to build. These include credible rules for land acquisition, re-allocation of land and relocation of people.

Conclusions: what now for planning?

Bandung offered a symbolic site of crucial changes in the articulation of planning autonomy, economic solidarity and cultural difference, and led, 60 years later, to economic dialogue but also divergence. Bandung nevertheless generated a vital historical break by claiming a prior economic order, and an important new one that offered advantageous terms of trade for nations integrating into world markets, but that required domestic commitments to planning.

Power consequently rested neither exclusively within the 'North' (industrialised nations), nor only within 'Souths' within 'Norths' (the United States' poor communities for instance), or even, within 'Souths within Souths' (e.g. lower- versus middle-income nations, some prospering cities versus others, billionaires among the poor). If 'planning from here' points to location-specific planning, then there is something surely to be gained by the good practice of customised plans and policy design suitable to the location. If 'for here' means to respond to some shared reality of institutions and organisations in a presumed global South, or to assume that what is suitable for Egyptian or Thai towns or regions is inevitably useful for Indian or South African ones, we should proceed with extreme caution.

Because nation-states have been a recent formulation of history, and their economic zones often contested, the actually deployed instruments and strategies in specific industry sectors became more vital to any explanation of planning outcomes. In turn, this focus on interconnected industries and sub-sectors challenged urbanisation and local planning frameworks that pictured cities with cohesive political or jurisdictional power to design, physically build or bootstrap themselves out of poverty. Even in non-traded sectors, the learning and social integration challenges were immense to create meaningful and well-paying work opportunities, and to finance critical redistributive processes.

Planning – whether economic or spatial and physical – forces projections and conjectures, reveals normative assumptions of the better society, but also manifests in a hard, multi-year slog initiated by the state under problematic political conditions. It leaves no easy space within development debates of taking a single moral or normative higher ground for the locus of planning. The cohesion of the 'global South' might have seemed important as a reaction to a seemingly dominant core of industrialised nations. But as we have seen in this chapter, this would be a selective reading of history, glossing over important regional, sectoral and sub-national divergences. Contradictions and conflict about the economy can reflect a healthy degree of debate and dissent. However, the evidence from many countries and cities on basic law and order, stability, social services and life quality showed a disturbing lack of credible state capacity, often triggering persistent (often state-induced) violence and violations. If this breakdown in social norms and law and order is a point of solidarity among the so-called South, it is a bitter tribe to belong to, only strengthening the claims of conservative economic hawks that these nations cannot plan or govern themselves.

At the same time, traditional disciplinary social science terminology of 'hegemony', 'neoliberal', and 'Southern' has not provided the degree of facility needed to customise a domestic planning process or underscore difference. These terms straitjacket the conversation. The irony of the 'global South' is a selective reifying of the nation-state whenever it suits the analytical arguments of solidarity, yet retreating to local contexts in critiquing national or regional government effects of planning. 'Planning from here' cannot romanticise self-help, because truly customising a local planning response requires some wider vision of economic dynamism and innovation. Climate change alone requires us to imagine new economic sectors and jobs, skills toward improving our towns, forests, farms, waterways, air and soil.

Professional planners – including policy designers, public administrators and national and regional professionals – with imagination and tenacity are required. Disciplinary critique will not be sufficient to respond to these development challenges. At the same time, economic development planners require the essential knowledge of evolutionary, open-ended change in order to anticipate how specific locations might be better embedded in regional, national and international economic ties. This requires planners trained in different ways, working in teams and at different scales. It requires them to claim more, not less, ambition for non-determinist approaches to plans and willing to take on substantial experimentation in economic plans.

Notes

1 [T]o build an economy which has a modern and productive agricultural sector with enhanced technology and an industrial sector that plays a leading role in the economy; to sustain economic development and secure social justice; and, increase per capita income of citizens so that it reaches at the level of those in middle-income countries.

(Federal Democratic Republic of Ethiopia 2010: 21)

2 See www.gefco.net/en/news/maghreb-highway-strong-axis-of-development-international-trade [accessed 4 February 2017].

References

Acharya, A. and Tan, S.S. (eds) (2008) *Bandung Revisited: The Legacy of the 1955 Asian–African Conference for International Order*, Singapore: NUS Press.

Albuquerque, E.M. (2007) "Inadequacy of technology and innovation systems at the periphery", *Cambridge Journal of Economics* 31: 669–690.

Amsden, A.H. (1989) *Asia's Next Giant: South Korea and Late Industrialization*, New York: Oxford University Press.

Amsden, A.H. (2001) *The Rise of the Rest: Challenges to the West from Late-Industrialization Economies*, Oxford: Oxford University Press.

Amsden, A.H. (2003) "Comment: Good-bye dependency theory, hello dependency theory", *Studies in Comparative International Development* 38(1): 32–38.

Arocena, R. and Sutz, J. (2000) "Looking at national systems of innovation from the South", *Industry and Innovation* 7: 55–75.

Benería, L., Deere, C. and Kabeer, N. (2012) "Gender and international migration: Globalization, development, and governance", *Feminist Economics* 18(2): 1–33.

Berger, M.T. (2001) "The nation-state and the challenge of global capitalism", *Third World Quarterly* 22(6): 889–907.

Berger, M.T. (2004) "After the Third World", *Third World Quarterly* 25 (1): 9–39.

Chang, H. (2002) *Kicking Away the Ladder: Development Strategy in Historical Perspective*, London and New York: Anthem Press.

Cheng, L.-L. and Gereffi, G. (1994) "The informal economy in East Asian development", *International Journal of Urban and Regional Research* 18(2): 194–219.

Corbridge, S. (2007) "The (im)possibility of development studies", *Economy and Society* 36(2): 179–211.

The Federal Democratic Republic of Ethiopia (2010) "Growth and Transformation Plan (GTP) 2010/11–2014/15", Ministry of Finance and Economic Development (MoFED), September, Addis Ababa, Draft. Available: www.ethiopians.com/Ethiopia_GTP_2015.pdf [accessed 25 February 2017].

Guardian (2016) "Next stop the Red Sea: Ethiopia opens Chinese-built railway to Djibouti". Available: www.theguardian.com/world/2016/oct/06/next-stop-the-red-sea-ethiopia-opens-chinese-built-railway-to-djibouti [accessed 23 February 2017].

Kabeer, N. (2004) "Globalization, labor standards, and women's rights: Dilemmas of collective (in)action in an interdependent world", *Feminist Economics* 10(1): 3–35.

Kaplinsky, R. (2011) "Schumacher meets Schumpeter: Appropriate technology below the radar", *Research Policy* 40(2): 193–203.

Korany, B. (1994) "End of history, or its continuation and accentuation? The global South and the 'new transformation' literature", *Third World Quarterly* 15(1): 7–15.

Lall, S. (1982) "Technological learning in the Third World: Some implications of technology exports". In F. Stewart and J. James (eds) *The Economics of new Technology in Developing Countries*, London: Frances Pinter, pp. 165–186.

Lall, S. (1984) "Exports of technology by newly industrializing countries: an overview", *World Development* 12(5–6): 471–480.

Lall, S. (1993) "Understanding technology development", *Development and Change* 24(4): 719–753.

Mohamed-Chérif, F. and Ducruet, C. (2016) "Regional integration and maritime connectivity across the Maghreb seaport system", *Journal of Transport Geography*, 51: 280–293.

Morphet, S. (2004) "Multilateralism and the non-aligned movement: What is the global South doing and where is it going?", *Global Governance* 10: 517–537.

Nesadurai, H.S. (2005) *Bandung and the Political Economy of North–South Relations: Sowing the Seeds for Revisioning International Society*, Singapore: Institute of Defence and Strategic Studies, Nanyang Technological University.

Srinivas, S. (2010) "Industrial welfare and the state: nation and city reconsidered", *Theory and Society* 39(3–4): 451–470.

Srinivas, S. (2012) *Market Menagerie: Health and Development in Late Industrial States*, Stanford, CA: Stanford University Press.

Srinivas, S. (2014) "Demand and innovation: Paths to inclusive development". In S.V. Ramani (ed.) *Combining Economic Growth with Inclusive Development*, Cambridge: Cambridge University Press, pp. 78–106.

Srinivas, S. (2015) "Healthy industries, unhealthy populations: Lessons from Indian problem-solving". In M. Mackintosh, G. Banda, W. Wamae and P. Tibandebage (eds) *Making Medicines in Africa: The Political Economy of Industrializing for Local Health*, Basingstoke: Palgrave Macmillan.

Srinivas, S. (2016) "Knowledge assets in late industrial development: Amsden's challenge and challenge for Amsden", unpublished draft (available from author on request).

Srinivas, S. (2017) "Evolutionary demand, innovation, and development". In D. Nathan, S. Sarkar and M. Tewari (eds) *Upgrading and Innovation in GVCs in Asia*, Cambridge: Cambridge University Press.

Srinivas, S. and Sutz, J. (2008) "Developing countries and innovation: Searching for a new analytical approach", *Technology in Society* 30: 129–140.

Srivastava, P. (ed.) (2001) *Non-Aligned Movement: Extending Frontiers*, New Delhi: Kanishka Publishers.
Statistics South Africa (StatsSA) (2015) "National and provincial labour market: Youth". Available: www.statssa.gov.za/publications/P02114.2/P02114.22015.pdf [accessed 23 February 2017].
UNCTAD (2009) "The state and development governance". In *Least Developed Countries Report 2009*, Geneva: UNCTAD, pp. 141–175.
UNIDO (2013) "Sustaining employment growth: The role of manufacturing and structural change". In *Industrial Development Report 2013*, Vienna: UNIDO, p. 208, Table A6.3.

11

The informal economy in cities of the global South

Challenges to the planning lexicon

Caroline Skinner and Vanessa Watson

Introduction

In most cities of the global South, the majority of the working population earns a living in the informal economy. Street and home trading, car guarding, small-scale tailoring, mechanical repairing, waste collecting and sorting, water selling and many other forms of income generation are found everywhere. Yet most fail to meet state regulations and find themselves classified as illegal. Informal employment comprises more than one-half of non-agricultural employment in most regions of the global South and can therefore be considered the norm rather than the exception. The dominance and persistence of these activities pose a particular challenge to urban policy and planning as they are often seen as antithetical to those with ambitions for 'modern' or 'world-class' cities. Moreover, much of the thinking on how to support the informal economy has its origins in the field of economic policy, which rarely engages with the urban planning field and is remarkably space-blind in terms of policy development. A central purpose of this chapter is to argue the importance of understanding the informal economy from an urban and spatial planning perspective, as planning regulations and controls in many cities of the global South are a major factor constraining and marginalising the informal economy. Finding an approach to urban planning that acknowledges and supports the informal economy, while managing its relationship to the many other activities in cities, is one of the most intractable, but important, issues facing planning practitioners and theorists. A critical starting point is a thorough understanding of why different informal workers choose specific work locations and how they currently use space, and letting this inform spatial and infrastructural inventions that enhance their productivity and income earning potential.

In setting out this argument we need to clarify the scope of this chapter. We are defining urban planning as the discipline and profession concerned with the collective, or societal, effort to re-imagine towns, cities and regions and to translate the result into priorities for area investment, strategic infrastructure investment, conservation and mitigation measures and principles of land and space use regulation. This activity can happen at national, regional and local institutional levels. Our brief review of shifts in informal economy policy (which usually remains institutionally and intellectually separate from spatial planning) is unable to include here significant policy initiatives relating to social protection that also impact on the informal economy (but see Lund and Srinivas 2000; Srinivas 2009, 2010).

The first part of the chapter examines the statistical evidence on the size and nature of informal work, as well as major factors that impact on its growth and performance, such as global economic shifts. The huge scale of the urban informal economy has elicited a wide range of policy and theoretical positions, and the next part of the chapter considers this variation. The last part of the chapter explores the implications for planning of recognising informal work in both the homes and public spaces of cities, and the fundamental revision of planning concepts that this would require. This is supported by examples of successful cases of inclusion of informal work.

The size and contribution of the informal economy

Most people who work outside of the agricultural sector in global South regions are in the informal economy,[1] although there is significant regional diversity (see Table 11.1) with South Asia showing a clear dominance of informal employment.

As Vanek *et al.* (2014: 1) point out, while regional estimates provide a useful overview, they hide the diversity that exists within a region. In sub-Saharan Africa, for example, informal employment accounts for a smaller share of non-agricultural employment in southern Africa (e.g. 33 per cent in South Africa and 44 per cent in Namibia) relative to countries in other regions (e.g. 82 per cent in Mali and 76 per cent in Tanzania). Regional estimates also hide the significant role of women in the informal economy: in three out of six regions, informal employment is a greater source of non-agricultural employment for women than for men. The difference is highest in sub-Saharan Africa where 74 per cent of women are in informal work compared to 61 per cent of men (Vanek *et al.* 2014: 2). Given country specificity as to what is considered urban, non-agricultural work is used as a proxy for urban. While the focus of this analysis is on the global South, informal employment is also a growing phenomenon in the global North (see Vanek *et al.* 2014: 15–31 for details).

Due to the way national statistical agencies collect labour force survey data, city-level labour market estimates are rare. Women in Informal Employment Globalising and Organising (WIEGO), however, drawing on a variety of sources, constructed city-level estimates which are reflected in Table 11.2. Again this demonstrates significant variations between cities but does highlight the predominance of informal work and its importance as a source of employment for women. What these statistics demonstrate is that informal work is the norm rather than the exception in these contexts.

Although the individual incomes of informal workers are often low, cumulatively their activities contribute to gross domestic product (GDP). A compilation of the contribution to GDP of the informal sector (see Table 11.3) suggests that rather than marginal, these activities in many countries are a central part of the economy.

Table 11.1 Informal employment as a proportion of non-agricultural employment

Region	*Percentage (%)*
South Asia	82
East and South East Asia	65
Sub-Saharan Africa	66
Middle East and North Africa	45
Latin America	51
China (estimates based on six cities)	33

Source: adapted from Vanek *et al.* (2014: 7).

Table 11.2 Informal employment as a percentage of total employment in a selection of cities

Country	*City*	*Year*	*Informal employment as % of total employment, all*	*Informal employment as % of total employment, women*	*Informal employment as % of total employment, men*
Benin	Cotonou	2001–2003	83	89	73
Burkina Faso	Ouagadougou	2001–2003	81	87	77
China	Composite of six cities – Fuzhou, Guangzhou, Shanghai, Shenyang, Wuhan and Xi-an	2010	32	36	30
Côte d'Ivoire	Abidjan	2001–2003	79	90	70
India	Ahmedabad	2011–2012	85	86	85
	Chennai	2011–2012	66	62	67
	Delhi	2011–2012	71	61	73
	Kolkata	2011–2012	85	87	85
	Mumbai	2011–2012	80	85	78
	Pune	2011–2012	58	57	59
Madagascar	Antananarivo	2007	64	69	61
Mali	Bamako	2001–2003	83	92	76
Mexico	Mexico City	2015	49	50	49
Niger	Niamey	2001–2003	77	84	74
Peru	Lima	2014	55	60	51
Senegal	Dakar	2001–2003	80	88	74
Togo	Lome	2001–2003	83	90	76
Vietnam	Hanoi	2007	53	57	50
	Ho Chi Minh City	2007	55	56	54

Source: adapted from http://wiego.org/dashboard/statistics.

The notion of the informal economy disguises significant heterogeneity. Three worker groups that are of particular relevance to planners are home-based workers, street vendors and waste pickers. There have been recent advances in quantifying these different worker groups (Vanek *et al.* 2012; ILO 2013: 45–47). What these worker-group statistics show is the diversity

Table 11.3 Percentage contribution of the informal sector (excluding agriculture) to GDP in select developing countries

Sub-Saharan Africa		*Middle East and North Africa*		*Asia*	
Benin (2000)	61.8	Algeria (2003)	30.4	India	46.3
Burkina Faso (2000)	36.2	Egypt (2008)	16.9	*Latin America*	
Cameroon (2003)	46.3	Iran (2007)	31.1	Colombia (2006)	32.2
Niger (2009)	51.5	Tunisia (2004)	34.1	Guatemala (2006)	34.0
Senegal (2000)	48.8	Palestine (2007)	33.4	Honduras (2006)	18.1
Togo (2000)	56.4			Venezuela (2006)	16.3

Source: adapted from ILO (2013: 22).

of workplaces. Particularly relevant for planning is that homes are often doubling up as places of work (hence contravening planning land use zoning and building regulations) and public space is an important workplace (although it is not designated as such in spatial plans).

There has been debate in the economics literature about the role the informal economy plays in economic growth in general (Elgin and Birinci 2016) and particularly during times of economic crisis (Fallon and Lucas 2002; Rogan 2016; Verick 2012). Research conducted through the global research policy network WIEGO in 10 cities across the global South in the immediate aftermath of the global economic recession suggested that informal workers were impacted in much the same way as their formal sector counterparts through price fluctuations, reduced demand for goods and services and the related increase in competition due to shrinking aggregate demand (Horn 2009, 2010).

Informality: policy and theoretical positions

The urban informal economy has been a field of enquiry for some time. Keith Hart (1973), through his analysis of the subsistence activities of the urban poor in Ghana, coined the term 'informal sector' and countered the commonly held view that these 'traditional' activities would disappear by being absorbed into the modern capitalist economy with industrialisation. Hart's pioneering study sparked further research and policy on the topic.

Chen (2012: 4–6) crystallises the academic and policy debates on the informal economy since Hart's study into four schools of thought, as follows. The 'Dualist' school, exemplified by Hart, but then widely promoted by the International Labour Organisation (ILO), sees the informal sector of the economy as comprising marginal activities – distinct from and not related to the formal sector – that provide income for the poor and a safety net in times of crisis. The 'Structuralist' school, a critique from the left (Moser 1978; Portes *et al.* 1989) views the informal economy as subordinated economic units and workers that serve to reduce input and labour costs and, thereby, increase the competitiveness of large capitalist firms. The 'Legalist' school, exemplified by de Soto (1989, 2000), sees the informal sector as comprised of 'plucky' micro-entrepreneurs who choose to operate informally in order to avoid the costs, time and effort of formal registration and who need property rights to convert their assets into legally recognised assets. A World Bank team (Perry *et al.* 2007) has more recently posited a variant on the legalist school which Chen refers to as the 'Voluntarist' school. This view holds that the informal economy is comprised of informal entrepreneurs who volunteer to work informally, not due to cumbersome regulations but as a strategic choice. Chen (2012: 6) concludes that, given the heterogeneity of the informal economy, there is merit to each of these perspectives, noting that each school reflects one or another "slice of the (informal) pie". This literature, derived largely from development studies, makes little or no reference to the spatial, nor is there an explicit or even implicit spatial or urban imagination.

Across the policy and planning spectrum there appears to be growing interest in regularising and formalising informal work, inspired by a recognition that it is neither disappearing nor formalising on its own. This approach was strongly echoed in the ILO's International Labour Conferences of 2014 and 2015 and the resulting Recommendation on "the transition from the informal economy to the formal economy", to which all member states are obligated to give effect (ILO 2015). The ILO discourse on the informal economy has shifted from largely regarding them as 'tax evaders' in the early 2000s, to the present Recommendation, which sees them as vulnerable workers and economic units needing protection and policies that ensure decent work for all within a rights-based approach to formalisation. The Recommendation states that during the transition, existing livelihoods should be preserved and improved and that public space is a workplace to which informal workers should have regulated access as well as access to public natural resources. The potential ambiguity of this approach emerges in some of the recent theoretical debates on urban informality as well as in discussions of rights-based urban planning approaches, below.

The policy positions above in the development studies field stand in stark contrast to conceptualisations of informality that have been emerging from the field of urban and planning theory. These theorists suggest that informality is not separable from formality, and must rather be seen as an organising logic or system of norms that governs urban transformation (Roy 2005); or, as Porter (2011: 116) puts it: informality is not "outside" of formal systems but instead is "produced by formal structures and always intimately related to them". Therefore, for Roy (2009: 81) the state itself can be regarded as an "informalised" entity that "actively utilizes informality as an instrument of both accumulation and authority". For this reason, she argues, engagement with informality has always been particularly difficult for planners as "to deal with informality therefore partly means confronting how the apparatus of planning produces the unplanned and unplannable" (Roy 2005: 155–156).

Yiftachel has also argued that the state, and planning, is deeply implicated in the production and maintenance of informality. As such, informality can be seen as a tactic and strategy of governments to contain the "ungovernable" and politically manipulate urban space (Yiftachel 2009: 89). Along with Roy, Yiftachel states that the "informality of the powerful" is often authorised by the state, while alternative forms of informality remain indefinitely "gray" or are officially "blackened". Urban planning is implicated here as well as plans that "design the city's 'white' spaces which usually create little or no opening for inclusion/recognition of most informal localities and population, while their discourse continuously condemns them as a chaotic danger to the city" (2009: 94).

Given these understandings of the role of the state and planning in urban informality, it is tempting to be sceptical of the hope by organisations such as the ILO that governments can be charged with protective regularisation and supporting workers' rights to public space and resources. Nonetheless, it is possible to find instances where this has been successfully achieved, indicating the importance of identifying such gaps and successes and feeding them back into policy debates. The following section of the chapter draws on such cases and also identifies some key conceptual shifts that need to inform a more nuanced approach to new and developing policy and planning.

Planning concepts and practice

Urban planning institutions, regulations and practices are highly variable across the globe, but it is difficult to find a city where planning and other forms of land use and business regulation do not in some way attempt to constrain and control the informal economy. Governments are more likely to achieve this in regions of the world where local institutions are strong and well resourced, but in rapidly growing and poorly resourced cities of the global South, the extent to which this is achieved is more likely to be the outcome of various local political factors. As Lindell (2010) has argued, some governments have taken restrictive and even violent measures against segments of the informal workforce, but at other times politicians have seen them as a potential 'vote bank' and their presence has been tolerated or encouraged.

Pressure from the urban elite or private sector land developers has also at times resulted in planning being mobilised to remove or curtail informality. The Warwick Junction area in inner-city Durban (South Africa) is an example of this kind of collusion between state and developers. In the immediate post-apartheid period, a prime piece of real estate, owned by the local authority and located at the main transport interchange, was upgraded to accommodate large numbers of informal traders. In this unique period of political transition from apartheid to democracy, the flux occurring in bureaucracies, policies and laws allowed this highly innovate and pro-poor intervention to 'slip under the radar'. By 2000 the project was able to accommodate some 6,000 street traders (Dobson and Skinner 2009). However, by 2009 the political landscape had changed in part due to the hosting of the Soccer World Cup, and collusion between local government and a property developer wanting this prime site for a shopping mall resulted in threatened demolition of the market. Court challenges mounted by the informal traders and NGOs held off this demolition, but the implication that informal trading is an inferior use of a well-located urban site has not disappeared.

The use of urban planning in these opportunistic ways to constrain and regulate the informal economy is facilitated in those global South regions that have a colonial history of planning legislation. For example, there are many countries in Africa and South Asia that were once British colonies and subject to British planning laws and urban spatial models in use in the UK at the time. Land use laws aimed at the 'orderly' development of towns and cities, single-function land zoning schemes, minimum standards for plot sizes and building materials and efficient and car-dominated traffic movement, were all intended to produce cities that would function to meet the ideal image of a modern British town. In addition, spatial concepts such as Garden Cities, neighbourhood units and urban park-ways, contained in detailed urban master plans, all served to promote the idea of towns without an informal economy. In many of these post-colonial countries planning laws have hardly changed up to the present, although informal work and shelter are now a dominant phenomenon with informal workers using both the public space and their homes as places from which to generate incomes.

Across the global South the instruments of planning are now also used to reinforce 'world-class city' and 'global city' ambitions. With growing demands for well-located urban land by property developers, new urban plans are emerging that promise to remake cities in the image of claimed 'global cities' such as Dubai and Singapore (Bhan 2014; Watson 2014). The graphics that accompany these new visions and plans demonstrate high levels of similarity: glass-box skyscraper buildings, landscaped lawns and public spaces and car-oriented movement routes all devoid of any informality. In these future city visions the presence of informality is considered non-modern; the global city has managed to 'wipe away' poverty, unemployment, congestion and dirt and with these, all informality. This raises the question: could these new plans be re-imagined so fundamentally as to accommodate informality, or does this lie beyond the very concept of planning? We hold that a different planning is possible and cases below support this.

The rhetoric of 'global cities' is more often than not strongly supported by local politicians and business interests hoping for either direct profit or what has been called "symbolic power" (Acuto 2010) – narratives of city importance addressed to other urban elites and foreign investors. World city-making is also directly supported through state practices: Goldman (2011) has termed this "speculative urbanism", drawing on the case of Bangalore where the main business of government has become that of land speculation and dispossession of those living on land earmarked for private development. As Bhan (2014) has argued, there is more to these processes than just profit-making, and the supportive involvement of governments in most of these projects suggests a desire for new forms of 'semi-authoritarian' governance where higher levels of political control and order match these qualities in the built environment.

Clearly planning can and does play a central role in the ability of informal workers to generate livelihoods. In most parts of the world this role has been a negative and constraining one, yet a realisation of the importance and persistence of the urban informal economy demands that planning responds differently. Reconsidering the role of planning in relation to the informal economy needs to happen conceptually as well as in practice. There are a number of 'conventional wisdoms' in planning regarding attitudes to informality which can be challenged.

Rights to work in public space

Do informal economic actors have a right to work in public spaces? Mexico, Colombia and India all recognise the constitutional right of people to work and court decisions have affirmed the right to work on the street. In March 2014, the Indian Parliament ratified the Protection of Livelihood and Regulation of Street Vending Act, providing legal protection for street traders and affirming that street hawking is a fundamental right when carried out in designated spaces. The new Act also requires recognition of "natural markets" where street traders have congregated in response to dynamic local demand. The Act is a result of many years of lobbying from activists, most notably the Self Employed Women's Association (SEWA) and the National Alliance of Street Vendors. Court decisions in all three of these countries essentially take a regulatory position, which both allows and constrains street economic activity, with regulatory controls (licences, demarcated spaces, etc.) often significantly limiting possibilities of street trade. The entrenchment of these constitutional rights represents an advance on the far more prevalent view that any form of street vending is against the law, but they do fall short of the position that work is a right and prohibition of street trading is a denial of that right (Meneses-Reyes and Caballero-Juárez 2014).

Brown (2015) develops this principle by arguing that urban public space should be considered as a common resource offering open access to those attempting to secure an income. Secure tenure for livelihoods demands as much recognition as it does for housing and it is quite possible to extend civil and common law traditions in many parts of Africa to accommodate a broader definition of these rights. In her African case studies, Brown (2015) shows that in the absence of constitutional measures, collective action on the part of street traders can also open up space for dialogue on rights to public space, and strong trader organisations are critical here.

These arguments on access to streets and public spaces for work have significant implications for urban planning, as it is usually through these channels that conflicting claims are mediated and details of regulatory measures are formulated. Not only street traders but also pedestrians, vehicles and recreational activities lay claim to public space and some kind of mutual accommodation needs to be found. Frequently, land use zoning schemes are based on single-use categories, which assume that commercial and other uses can be formally assigned to particular land parcels. The idea of mixed, flexible land use and less rigid approaches to use management are more recent ideas usually absent in planning schemes. Further, street traders are not a homogenous group: they range in degrees of permanence, ability to pay for space, infrastructure requirements and the nature of goods traded, and patterns of demand and use of space usually fluctuate over a day, week and year. Innovative planning approaches are needed to accommodate flexibility: pedestrian routes that share space with traders; transport hubs that also make space for pedestrian flows and fluctuating trader presence; streets that can carry traffic during the day and become markets at night, and so on.

Once access to public space has been secured, there is a critical need for basic infrastructure such as running water and toilets and services such as rubbish removal. These are often to the advantage of all users of public space. In addition, there is a need for work-related infrastructure. For street traders priorities are shelter from the elements, tables to display goods and storage facilities (for example, see findings from interviews with street traders in five cities in

Roever 2014). In Warwick Junction, when small kiosks with electricity were provided, traders started more lucrative and value-adding trades – carpentering, sewing, catering and computer and cell phone repairs (Dobson and Skinner 2009: 104). For waste pickers, sorting and storage facilities can transform their work environment and earning capacity (Dias 2016; Dias and Samson 2016).

The home as workspace

A second conceptual shift that needs to inform livelihood-supporting planning approaches is the acknowledgement that housing also has an economic function. Poor households all over the world use their homes to generate income through making and/or selling from these sites, or renting out accommodation, and increasingly this is common in wealthier cities and neighbourhoods as well. Yet single-use zoning schemes and infrastructure planning, as well as many state housing policies, fail to recognise this important role of housing and can end up criminalising such activity and adding to economic costs and inefficiencies.

The Mahila Housing Trust's (MHT) slum upgrading work in four states of India effectively incorporates the economic function of housing. MHT was promoted by SEWA in recognition that homes of their now nearly two million members also doubled up as workplaces. Obino (2013: 4) notes that in the vast majority of cases housing improvements were linked to a desire to expand economic productivity:

> For home-based workers, better housing and services directly translates into more time at productive work, easier access to water, safer storage for stocks and better equipment. An individual electrical connection, for example can make it possible to use an electric sewing machine, while a water connection quickens the production of food to be sold on the street.

In a post-upgrade evaluation of their Ahmedabad programme, over a quarter of respondents felt that the new infrastructure had improved their incomes and employment patterns (Rusling 2010).

Studies have also found that improved infrastructure can stimulate investment in housing, but often this is also dependent on loans from financial institutions that require improved tenure security. Formal loans are beyond the reach of those living and working informally, but MHT developed a concept of transitional tenure, which is a 10–15 year non-eviction guarantee. This in turn required the co-operation of the local authorities and a check against urban planning regulations to gain an assurance that a settlement was not being threatened with demolition.

The informal economy, urban nutrition and health

Planning needs to shift from the usual assumption that informal food-making and vending is a health threat and should be removed to an acknowledgement of the positive health benefits of this sector if it is correctly supported and managed.

Urban food security is a key concern worldwide, especially in poorer communities in rapidly growing and under-resourced cities, and the informal economy is a centrally important conduit through which the poor gain access to food (https://consumingurbanpoverty.wordpress.com/). For example, a 2009 survey across 11 cities in southern Africa found that 76 per cent of poor urban households experience some form of food insecurity. While there is variation between and within cities, overall the survey found that many poor households depend on informal sources for their food needs and as food insecurity increases, so the reliance on informal outlets increases.

> Some 70 per cent of households in the AFSUN survey normally sourced food from informal outlets ... while 32 per cent of households patronised the informal food economy almost every day and 59 per cent did so at least once a week.
>
> *(Crush and Frayne 2011: 798)*

Abrahams' (2010) detailed study of the changing food system with the greater penetration of supermarkets in Lusaka, Zambia, found that the informal food economy remained resilient and competitive, despite the growth of supermarkets.

Yet planning and health regulations frequently make it extremely difficult for food vendors to operate either in public spaces or in residential areas and this undermines access of households to cheap and nutritious food. Moreover, when food vendors are subject to large-scale eviction measures, as happened in Blantyre (Malawi) in 2006, then the "geography of urban poverty is reshaped" and households no longer able to access these cheaper outlets suffer worsened food insecurity (Riley 2014).

Where informal traders are operating in public spaces and markets it is a key task for the municipality to provide services and facilities to ensure a healthy selling environment. In the case of the Warwick Junction in Durban a partnership between the traders and the municipality, along with the careful design of facilities, ensured that a healthy environment was achieved. Water points were designed and located to allow for their multiple use for bathing, washing, vegetable cleaning, cooking and taxi washing. Much time was spent on designing toilets that were easy to maintain, and were arranged in smaller blocks distributed throughout the market. Municipal waste removal was supplemented by volunteer market cleaners and intermittent 'cleaning blitzes' along with high-pressure surface hosing and a check on storm water drainage (Dobson and Skinner 2009: 114–117).

The informal economy and sustainability

The contribution of the informal economy to urban sustainability is rarely recognised. It is more often cast as a direct polluter of the environment, and planning, health and environmental regulations are mobilised to remove and repress it. Yet informal traders often source locally and make less use of polluting plastic packaging, and municipal schemes that use informal recyclers, as has been proved in Colombia, emit far fewer greenhouse gases (Sintana *et al.* 2015).

In a number of countries, waste picker movements have managed to negotiate with governments to secure integration into municipal recycling schemes. Peru and Brazil have both passed progressive national laws that support the formalisation of waste picking and encourage cooperatives. Peru's (2010) law introduced a series of incentives to achieve this, and Brazilian law mandates the inclusion of waste picker associations into solid waste-management systems (Dias 2011). Belo Horizonte (Brazil), Bogotá (Colombia) and Pune (India) offer examples of successful integration of waste pickers into municipal waste management schemes. Dias (2016: 379–382) details the different approaches. What is common to all of them is that there are strong co-operatives of waste pickers who have lobbied for the right to access waste (in the case of Bogotá through the courts) and have secured formal agreements with local authorities. In Belo Horizonte the council has established facilities to sort and process waste, which the co-operatives manage, and have recently received a financial incentive for the services they provide. In Pune the waste picker cooperative is more of an independent service provider with agreed performance indicators doing door-to-door collection receiving user fees. The Council does, however, provide equipment, working space and technical training. In Bogotá, in response to a Constitutional Court ruling granting waste pickers status as subjects of special protection, the mayor in 2012 created a public cleansing company to replace private contractors and introduced a new programme that incorporated waste pickers into collection, transportation and recuperation of recyclables. In 2013, the municipality also launched a payment scheme, financed through the service fees, to remunerate waste pickers for the services they provide.

Implications for planning practice

There is a significant gap between the needs of those working informally and the expertise and training of municipal officials charged with the planning and management of informal workplaces and spaces, even in contexts where authorities have been persuaded (or pressurised) to accommodate the informal economy. It is rare that planning officials understand the highly specific and differentiated locational and service requirements of informal workers or where their activities fit into local, national and even international value chains. For example, many global South cities can show formally provided market infrastructure that is wrongly located and over-designed and hence not in use (see Bromley and Mackie 2009; Donovan 2008). More often than not planners and architects draw on training biased towards the requirements of the formal retail sector, which are very different from those of informal traders. It is therefore essential that any urban intervention aimed at supporting the informal economy (whether public space or home based) involves extensive negotiation and participatory planning processes, as well as professionals willing and able to consider new forms of infrastructure provision and new or adapted rules to manage their use and servicing.

Key to supportive interventions is understanding economic dynamics. In Warwick Junction, for example, understanding the specific economic dynamics within different segments of the informal economy was crucial. Through observation, consultation and many one-on-one discussions it became clear that interventions to enhance the livelihoods of, for example, traditional medicine traders, were different from interventions needed for bovine head cookers, which was different again from supporting waste pickers operating in the area. The process of gleaning these insights involved municipal staff spending much time on the streets observing how space was used; one-on-one discussions about backward and forward linkages of particular trades with individual traders, their suppliers and customers; and carefully designed participatory processes with groups of traders using role play, among other techniques (Dobson and Skinner 2009). Planners also need to understand the impact of value chains. For example, for waste pickers the price of recycled metal is a global one determined by the needs of the big players in China and the East, and this directly affects ability to survive economically.

Common to the approach in these cases is that individual informal workers (particularly women) and their organisations are integrally involved in the process; there is recognition of informal workers as knowledgeable and legitimate partners; and planners thus function as facilitators rather than all-knowing experts. In addition, these cases show the importance of an incremental approach to upgrading, making use of experimentation through 'pilots', rather than end-state and inflexible master planning.

A central issue, however, is the necessary balance between regulation and control on the one hand, and inclusive and supportive planning approaches on the other. Where there are competing and conflicting claims on public space (traders, pedestrians, social uses of space), or home-based work that negatively affects neighbours, agreements need to be in place, and institutionalised, to mediate wider public impacts. Such arrangements will inevitably be highly context specific: there are no models of planning law that can work across all places. The importance of consultative approaches and collective action on the part of informal workers has been emphasised above, allowing agreements (even in conflictual situations) to be worked out and reinforced in wider forums. An important precursor for planners here is that there is an understanding of the informal economy as an urban asset rather than a liability and a willingness to engage with often quite complex economic drivers, which dictate the survival of these activities.

Conclusion

This chapter has argued that informal work forms a dominant element in urban economies in many parts of the world, and especially in global South cities. Moreover, assumptions that it is declining over time as informal workers become formal, along with 'modernising' economies, are not supported by evidence and informality will be contributing to urban economies well into the future. With urban policy and planning regulations and practices in most parts of the world still heavily biased towards control and containment, there is a growing disconnect between these and theoretical advances in the field that point to the highly embedded role of informality in urban political economies. Yet policy discourses and practices do shift (unevenly and for complex reasons), and this chapter highlights these and key planning assumptions regarding informality that can be challenged. The need for planners to understand the economic informants of these activities suggests that urban spatial planners would do well to engage with development studies debates and vice versa.

Note

1 This data uses international statistical norms according to which the 'informal sector' refers to employment and production that takes place in unincorporated, small or unregistered enterprises, while 'informal employment' refers to employment without social protection through work both inside and outside the informal sector. The 'informal economy' refers to all units, activities and workers so defined and the output from them.

References

Abrahams, C. (2010) "Transforming the region: Supermarkets and the local food economy", *African Affairs* 109(434): 115–134.

Acuto, M. (2010) "High-rise Dubai urban entrepreneurialism and the technology of symbolic power", *Cities* 27: 272–284.

Bhan, G. (2014) "The real lives of urban fantasies", *Environment and Urbanization* 26(1): 232–235.

Bromley, R. and Mackie, P. (2009) "Displacement and the new spaces for informal trade in the Latin American city centre", *Urban Studies* 46(7): 1485–1506.

Brown, A. (2015) "Claiming the streets: Property rights and legal empowerment in the urban informal economy", *World Development* 76: 238–248.

Chen, M. (2012) "The informal economy: Definitions, theories and policies", WIEGO Working Paper No. 1. Available: http://wiego.org/sites/wiego.org/files/publications/files/Chen_WIEGO_WP1.pdf [accessed 1 October 2015].

Crush, J. and Frayne, B. (2011) "Supermarket expansion and the informal food economy in southern African cities: Implications for urban food security", *Journal of Southern African Studies* 37(4): 781–807.

De Soto, H. (1989) *The Other Path: The Economic Answer to Terrorism*, New York: Basic Books.

De Soto, H. (2000) *The Mystery of Capital: Why Capitalism Triumphs in the West and Fails Everywhere Else*, London: Bantam Press.

Dias, S. (2011) "Overview of the legal framework for inclusion of informal recyclers in solid waste management in Brazil", WIEGO Policy Brief (Urban Policies) No. 6. Available: http://wiego.org/sites/wiego.org/files/publications/files/Dias_WIEGO_PB6.pdf [accessed 1 October 2015].

Dias, S. (2016) "Waste pickers and cities", *Environment and Urbanization* 28(2): 375–390.

Dias, S. and Samson, M. (2016) "Informal economy monitoring study: Waste sector report". Available: http://wiego.org/sites/wiego.org/files/publications/files/Dias-Samson-IEMS-Waste-Picker-Sector-Report.pdf [accessed 10 January 2017].
Dobson, R. and Skinner, C. (2009) *Working in Warwick: Including Street Traders in Urban Plans*, Durban: University of KwaZulu-Natal, School of Development Studies.
Donovan, M. (2008) "Informal cities and the contestation of public space: The case of Bogotá's street vendors, 1988–2003", *Urban Studies* 45(1): 29–51.
Elgin, C. and Birinci, S. (2016) "Growth and informality: A comprehensive panel data analysis", *Journal of Applied Economics* XIX(2): 271–292.
Fallon, P. and Lucas, R. (2002) "The impact of financial crises on labor markets, household incomes, and poverty: A review of evidence", *World Bank Research Observer* 17(1, 2): 1–46.
Goldman, M. (2011) "Speculative urbanism and the making of the next world city", *International Journal of Urban and Regional Research* 35(3): 555–581.
Hart, K. (1973) "Income opportunities and urban employment in Ghana", *The Journal of Modern African Studies* 11(1): 61–89.
Horn, Z. (2009) "No cushion to fall back on: The global economic crisis and informal workers", WIEGO Research Report. Available: http://wiego.org/sites/wiego.org/files/publications/files/Horn-Global-Economic-Crisis-1.pdf [accessed 9 January 2017].
Horn, Z. (2010) "Coping with crisis: Lingering recession, rising inflation, and the informal workforce", WIEGO Research Report. Available: http://wiego.org/sites/wiego.org/files/publications/files/Horn-Global-Economic-Crisis-2_0.pdf [accessed 9 January 2017].
ILO (2013) *Women and Men in the Informal Economy: A Statistical Picture. Second Edition*, Geneva: International Labour Office. Available: www.ilo.org/stat/Publications/WCMS_234413/lang--en/index.htm [accessed 1 October 2015].
ILO (2015) "Recommendation No. 204 concerning the Transition from the informal to the formal economy". Available: www.ilo.org/wcmsp5/groups/public/---ed_norm/---relconf/documents/meetingdocument/wcms_377774.pdf [accessed 10 January 2017].
Lindell, I. (2010) "Introduction: The changing politics of informality – collective organizing, alliances and scales of engagement". In I. Lindell (ed.) *Africa's Informal Workers: Collective Agency, Alliances and Transnational Organizing in Urban Africa*, London and New York: Zed Books, pp. 1–30.
Lund, F. and Srinivas, S. (2000) *Learning from Experience: A Gendered Approach to Social Protection for Workers in the Informal Economy*, Geneva: International Labour Organization.
Meneses-Reyes, R. and Caballero-Juárez, J. (2014) "The right to work on the street: Public space and constitutional rights", *Planning Theory*, 13(4): 370–386.
Moser, C. (1978) "The informal sector or petty commodity production: Dualism or dependence in urban development?", *World Development* 6(9): 1041–1064.
Obino, F. (2013) "Housing finance for poor working women: Innovations of the Self-Employed Women's Association in India", WIEGO Policy Brief (Urban Policies) No. 14. Available: http://wiego.org/sites/wiego.org/files/publications/files/Obino-Home-Based-Workers-India-SEWA-Housing-WIEGO-PB14.pdf [accessed 1 October 2015].
Perry, G.E., Maloney, W.F., Arias, O.S., Fajnzylber, P., Mason, A.D. and Saavedra-Chanduvi, J. (2007) *Informality: Exit and Exclusion*, Washington, DC: The World Bank.
Porter, L. (2011) "Informality, the commons and the paradoxes for planning: Concepts and debates for informality and planning", *Planning, Theory and Practice* 12(1): 115–120.
Portes, A., Castells, M. and Benton, L. (1989) *The Informal Economy: Studies in Advanced and Less Developed Countries*, Baltimore, MD: Johns Hopkins University Press.
Riley, L. (2014) "Operation Dongosolo and the geographies of urban poverty in Malawi", *Journal of Southern African Studies* 40(3): 443–458.
Roever, S. (2014) "Informal economy monitoring study: Street vendor sector report". Available: http://wiego.org/sites/wiego.org/files/publications/files/IEMS-Sector-Full-Report-Street-Vendors.pdf [accessed 10 January 2017].
Rogan, M. (2016) "Informal employment and the global financial crisis in a middle-income country". Available: http://wiego.org/blog/informal-employment-and-global-financial-crisis-middle-income-country [accessed 6 January 2017].
Roy, A. (2005) "Urban informality: Towards an epistemology of planning", *Journal of the American Planning Association* 71(2): 147–158.

Roy, A. (2009) "Why India cannot plan its cities: Informality, insurgence and the idiom of urbanization", *Planning Theory* 8(1): 76–87.

Rusling, S. (2010) "Approaches to basic service delivery for the working poor: Assessing the impact of the Parivartan slum upgrading programme in Ahmedabad, India", WIEGO Policy Brief (Urban Policies) No. 1. Available: http://wiego.org/sites/wiego.org/files/publications/files/Rusling_WIEGO_PB1.pdf [accessed 1 October 2015].

Sintana, V., Damgaard, A. and Gomez, D. (2015) "The efficiency of informality: Quantifying greenhouse gas reductions from informal recycling in Bogota, Colombia", *Journal of Industrial Ecology* 1: 1–13.

Srinivas, S. (2009) "Cost, risk and labour markets: The state and sticky institutions in global production networks", *Indian Journal of Labour Economics* 52(4): 583–605.

Srinivas, S. (2010) "Industrial welfare and the state: Nation and city reconsidered", *Theory and Society* 39(3): 451–470.

Vanek, J., Chen, M. and Raveendran, G. (2012) "A guide to obtaining data on types of informal workers in official statistics: Domestic workers, home-based workers, street vendors and waste pickers", WIEGO Statistical Brief No. 8. Available: www.inclusivecities.org/wp-content/uploads/2012/07/Vanek_WIEGO_SB8.pdf [accessed 1 October 2015].

Vanek, J., Chen, M.A., Carre, F., Heintz, J. and Hussmanns, R. (2014) "Statistics on the informal economy: Definitions, regional estimates and challenges", WIEGO Working Paper (Statistics) No. 2. Available: http://wiego.org/sites/wiego.org/files/publications/files/Vanek-Statistics-IE-WIEGO-WP2.pdf [accessed 1 October 2015].

Verick, S. (2012) "Giving up job search during a recession: The impact of the global financial crisis on the South African labour market", *Journal of African Economies* 21(3): 373–408.

Watson V. (2014) "African urban fantasies: Dreams or nightmares?", *Environment and Urbanization* 26(1): 215–231.

Yiftachel, O. (2009) "Theoretical notes on 'gray cities': The coming of urban apartheid?", *Planning Theory* 8(1): 88–100.

12
Urban finance
Strengthening an overlooked foundation of urban planning

Paul Smoke

Introduction

Finance is a neglected element of urban development planning. With rapid urbanisation and advancing decentralisation in many countries in the global South, the international community is devoting substantial effort to financing sustainable development post-2015. Urban governments have been assigned a prominent role in the process (Ingram *et al.* 2013; UCLG 2010, 2014; UN 2015). This is fitting because there is an increasingly wide consensus on the need to unlock the developmental potential of urban areas and the bodies that govern them.

For finance to empower urban governments to develop and implement robust plans, the role it plays must be understood and its future potential determined. The link between planning and finance seems clear in theory, but it has been inadequately prioritised in practice. Urban governments operate in diverse settings that complicate their ability to act developmentally. They are often embedded in several levels of government and administration involving relationships that are complex and poorly defined in law and practice. At each level, there are actors who influence planning and finance. This chapter examines the nature of the planning–finance relationship and how finance could be better tapped by urban governments.

This next section outlines contextual factors that affect the intersection of urban planning and finance. This is followed by a review of the critical role of finance in developing and implementing plans and a synopsis of key financial mechanisms commonly available, briefly characterising their use in the global South. The chapter then considers challenges faced by countries of the global South in improving their fiscal position and planning–finance linkages. The chapter closes with thoughts about how to improve finance to support urban planning.

Setting the context

If steps are to be taken to better link urban planning and finance to improve development outcomes, this cannot be done in a mechanical or piecemeal way, as has often been the case. There are different types of planning, some based on formal techniques, while others focus more on processes and actors. Urban finance has long been dominated by public finance specialists promoting normative approaches to generating and allocating public revenues. Urban planning and finance, however, occur in a world of cultural, political and institutional diversity, shaping what is desirable and feasible. It is impossible to do justice to all relevant variations in context, but a few introductory comments on salient factors are provided here.

Framing urban planning

Planning has different meanings (Connel 2010; Healey 1997, 2012; MacDonald *et al.* 2014; Rodwin and Sanyal 2000; Sanyal *et al.* 2012). There has been a historical divide in many countries in the global South between physical (land use and physical infrastructure) and development (socio-economic) planning. Another distinction is between more comprehensive approaches and sectoral (e.g. transport) planning. Of course, these dimensions are inherently related and over time there have been efforts to connect them through adoption of integrated strategic development planning. Systems also vary in scope, but many countries in the global South have moved towards broader planning approaches.

Another key consideration is who is involved in the process. Early planning was conceived largely as an objective, technical exercise conducted by experts, but there has long been movement towards recognising it as a political process that involves multiple actors at the urban level. Although the planning–finance nexus has essential technical dimensions, intergovernmental relations, interactions between elected urban bodies and the staff executing urban functions and linkages between elected bodies, planning and budgeting entities and citizens who live in urban areas are critical considerations in bridging planning and finance.

Understanding institutional realities

Urban governments do not exist in a vacuum (Romeo and Smoke 2016; Smoke 2013). Most countries have multiple levels of administration and government, and decentralisation policies and intergovernmental relations vary. Some countries have semi-autonomous elected local governments (devolution), while in other cases lower levels report to the central government (deconcentration). In some cases, there is a mix of devolution and deconcentration across levels, with great variation in functional and revenue authority allowed to each level and the autonomy they enjoy in exercising these powers.

There are also varying relationships among levels – in some countries, such as Indonesia, at least one level is comparatively independent, while in other cases the relationships are hierarchical, such that lower levels are subject to supervision or approval of plans and budgets by higher levels, as in Uganda. In federal systems, states or provinces may have more regular control over local governments than the central government, as in India and Pakistan.

Urban governments may have distinct legal status in countries of the global South. Capital cities commonly have a special designation. Other urban areas may also have special status (as with metropolitan municipalities in South Africa) or may be legally equivalent to other local governments (as in Indonesia). Even without special status, urban governments may be more empowered than other local governments (as in Ghana). In virtually all cases, however, urban governments are subject to formal mandates or informal influences from higher levels.

An emerging trend in decentralisation is to think beyond functional assignment to subnational governments, emphasising a broader mandate of urban governments to provide for the overall development and welfare of their territories. This demands more holistic territorial planning than the often fragmented sectoral planning of traditional approaches and requires more urban autonomy to tailor plans and budgets to local conditions than is often the case in the global South (Commonwealth Local Government Forum 2013; Romeo 2013).

Differences in intergovernmental structures and processes have great implications for urban planning and finance. Integrated territorial planning is complicated if urban governments face interference from one or more higher levels in general or in specific key sectors. Even with relatively independent plans, urban resources may largely come from higher levels. If intergovernmental transfers are highly conditional, or if there are restrictions on urban revenue generation or borrowing for infrastructure by creditworthy urban governments, the best devised plans may not be implementable. Such challenges pervade the global South.

Navigating accountability

Urban governments need to face the discipline of accountability, including in developing, financing and executing urban plans (Brinkerhoff and Azfar 2010; Boex and Yilmaz 2010; Faguet 2014). In devolved systems, downward accountability is prioritised through elections and other means that can offer citizens more routine and consequential interface with urban governments. Transparency and access to information on urban processes and decisions – through managerial mechanisms (budgeting, financial management, audit, etc.) and freedom of information laws – also support robust downward accountability.

Even in decentralised systems, upward accountability – a framework for urban government planning, finance and staffing, as well as standards for basic service delivery – is needed. Assessment occurs through urban government financial and physical reports, performance reviews and external audits, among others. These provide information to citizens and central agencies, although oversight mechanisms can unduly constrain urban governments and undermine downward accountability if too inflexible or inconsistently applied.

Horizontal accountability – between elected urban councils and staff who deliver local services – is also vital. Elected officials need to be able to exert some control over staff. In traditionally centralised countries, staff may look to higher levels for direction, hindering urban councils' ability to take action to be responsive to their constituents.

In short, multiple accountability relationships are essential for developing and implementing urban plans and financing arrangements, but they will necessarily vary with context. A core challenge is how to negotiate a workable balance between upward and downward accountability, which can evolve as urban government capacity grows.

The critical role of finance in urban planning

Although urban finance is essential for urban planning, the topic has not been a focus of the planning literature. The bulk of work on this topic has been in the public finance and fiscal decentralisation fields (Ahmad and Brosio 2014; McClure and Martinez-Vazquez 2004). There has, however, been some attention in planning to instruments directly tied to development projects, such as capital grants, land value capture and borrowing (more below).

A highly simplified diagram of the conventional planning cycle is presented in Figure 12.1. Urban development plans are, in the first instance, often prepared relatively independently of resource availability and the budgeting process because they are focusing primarily on crafting a larger shared vision for improving a city. The basic cycle involves six steps.

The first step is to review past plan performance and underlying policy objectives, so that goals can be modified as needed for a new plan. At this stage there may be preliminary attention to finance, but only in step 3 is the plan framed in operational terms and tied to budget constraints. Implementation in step 4 includes action to generate capital and recurrent funding to build,

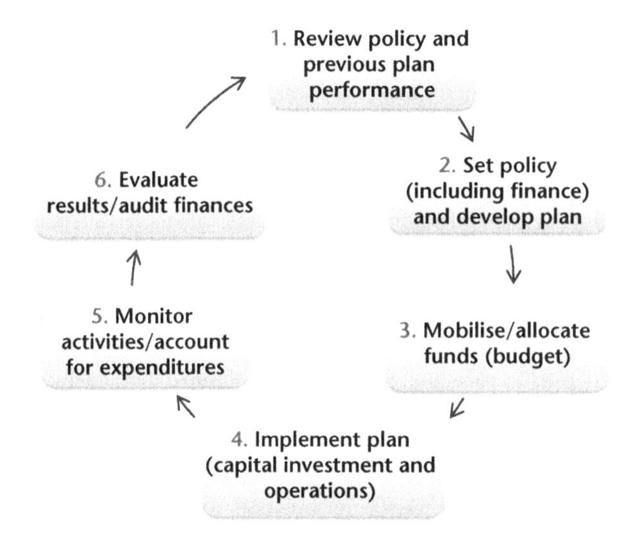

Figure 12.1 The basic planning and implementation process

operate and maintain planned facilities and services. As implementation unfolds, monitoring of plan activities and finances is conducted. Finally, step 6 involves initial attempts to evaluate the attainment of the plan and audit the financial transactions involved.

This is, of course, a very simplified version of a process that has been elaborated in more detail elsewhere (Asian Development Bank 2001; Rajaram *et al.* 2010; World Bank 1998). The main concern here is to note the relationship between urban planning and budgeting. In the global South, development planning has commonly been based on creating a broad vision without adequate attention to how to implement and resource it (Romeo and Smoke 2016).

Following the lead of national public financial management reforms, there has been some movement in the global South to adopt local medium-term planning and expenditure frameworks. This involves maintaining aspirational strategic plans that cover a longer period of 5–10 years, but also linking them directly to operational instruments that bridge the priority goals for the longer time frame with the financial planning needed for implementation. An illustration of such a mechanism is presented in Figure 12.2.

In this version, a five-year strategic plan provides a foundation for expenditure prioritisation and resource allocation. This plan is operationalised in three-year rolling local investment programmes and annual programme/performance budgets. These must be linked to fiscal (own-source revenues, intergovernmental transfers and others), political (e.g. elected urban councils in devolved systems) and civic engagement (e.g. participatory planning/budgeting) mechanisms. Three-year rolling investment plans can be revised based on previous performance and new information, while annual budgets are obviously issued each year.

These processes and diagrams are highly stylised and simplified. Their effective use is dependent on myriad assumptions about certain systems, attitudes, behaviours and other conditions being in place. They do, however, provide a basic schematic that can be interpreted and applied as appropriate and feasible in real contexts.

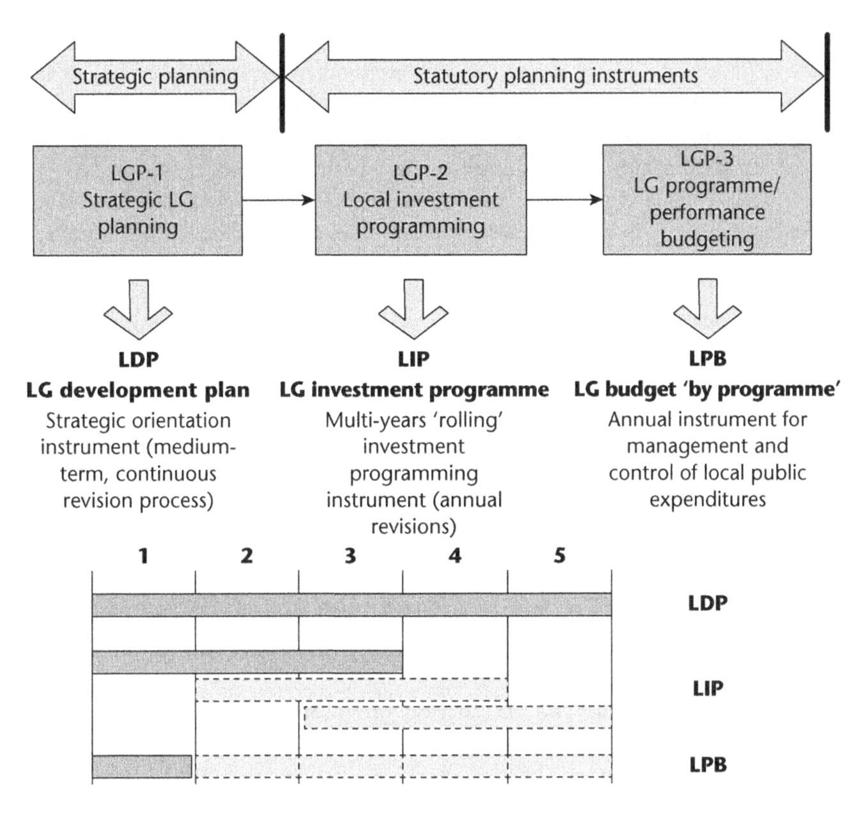

Figure 12.2 Urban planning, investment programming and budgeting framework

The landscape of urban/local finance instruments

As noted above, urban finance operates in a framework of intergovernmental relationships that collectively affect its use. Urban governments need a mix of financial instruments to play a stronger role in inclusive and sustainable development (Bird and Slack 2013; Martinez-Vazquez 2015; UCLG 2010). Although some already have fiscal capacity and may even be creditworthy, this is not the norm. Few countries in the global South give urban governments adequate access to resources, and available options are often not used productively.

This section outlines a range of urban revenue sources for recurrent and development purposes essential for development planning. The mix of sources and the extent to which they can be used must be tailored to country context. Revenue reforms will often need to be phased in and supplemented by governance, technical and capacity development efforts.

Recurrent sources of urban/local revenue

The foundation for urban finance is regularised revenues authorised by the centre and raised by urban governments. This has not been an area of strength in the global South, where locally raised revenues are often modest – e.g. property taxes account for over 2 per cent of GDP in OECD countries, but only 0.5 per cent in developing countries.

Subnational, especially urban, governments have certain advantages in providing a range of public services (due to their local nature and site-specific needs and preferences), while central governments have inherent advantages in generating funds due to the nature of major revenue bases and administrative scale. Thus, intergovernmental transfers are typically imperative, and how they are allocated to urban governments is a key policy concern.

Despite central revenue superiority, urban governments should raise as much revenue as they can. This reduces pressure on the central budget and frees up funds for redistribution to poorer areas. Urban revenues also decrease dependence on the centre, link benefits and costs of urban public services and help to service loans needed to finance urban capital investment.

Urban own-source revenues

Common urban revenue sources include property taxation, fees and charges, licenses, certain business taxes, and occasionally motor vehicle taxes and licenses and turnover taxes (Bahl and Bird 2008; Martinez-Vazquez 2013). Not all sources, however, are purely local. Urban governments may request higher levels to assist them in collecting certain revenues, such as a local business tax for which payees are already registered with a central system. Some countries permit adding a local surcharge to a higher level source, such as VAT or natural resource taxation, but this mostly benefits regional governments in the global South. Central government concerns with allowing stronger urban revenue powers include possible tax competition, administrative challenges and limiting benefits to wealthier areas.

Urban own-source revenue and tax sharing provisions in practice are highly diverse (Martinez-Vazquez 2015; UCLG 2010). Full local autonomy over any tax is unusual, but there can be discretion over rates. User charges are often regulated, but there can be flexibility, including to effect redistribution. On balance, many central governments in the global South decentralise fewer revenue sources than warranted by fiscal logic, urban needs and country context.

Intergovernmental fiscal transfers

Most countries – from the OECD to the global South – use intergovernmental transfers (Bird and Smart 2002; Shah 2013) due to the above-noted asymmetry between expenditure and revenue decentralisation. Urban areas are commonly less dependent on the centre than other local governments, but in many countries transfers remain essential for them. The focus here is on general transfers, with more discussion of development planning grants below.

Urban governments are affected by multiple aspects of the transfer system, including the predictability of resource flows and discretion over transfer use. Predictability depends on stability of the resource flow and how it is allocated. Countries in the global South long avoided committing specific shares of national resources to transfers, leaving the decision to the annual budget. Central governments need flexibility to respond to economic conditions, but urban governments need predictability to plan for service delivery and revenue generation. Many countries in the global South have taken steps to define a rule-based transfer pool (e.g. a specific share of national resources), with some exceptions, such as South Africa and Uganda.

The specific criteria/formulas to allocate transfers among local jurisdictions will depend on the desired goals, including redistribution. All transfers should use objective criteria so that urban governments understand why they receive a certain amount. In addition, transfer formulas should not create incentives that undermine urban revenue generation activity required to cover operating expenditures or debt service related to important infrastructure expenditures.

Unconditional transfers raise urban government autonomy to spend in accordance with constituents' needs. Some countries (e.g. Indonesia and the Philippines) prioritise unconditional transfers. Reasonable conditions to ensure provision of priority basic services are legitimate, but in some cases they can be very restrictive (e.g. Uganda). Other countries (e.g. South Africa) have a balanced mix of unconditional and conditional transfers. To plan effectively, urban governments must be aware of how much flexibility they have in using transfers.

Long-term urban/local development financing

Urban governments in many countries rely heavily on transfers for long-term development activities. Urban infrastructure in some OECD countries has long been financed with loans from capital markets, which are mostly limited to large cities in selected countries of the global South. Such access must expand to meet growing urban needs, but this requires actions from national and urban governments to reform fiscal frameworks and to build creditworthiness.

Intergovernmental development transfers

Subnational governments account globally for nearly two-thirds of total public infrastructure investment, and a third of this is financed by grants (Martinez-Vazquez and Timofeev 2012). Grants dominate urban government development budgets in the global South, supporting key infrastructure functions, such as schools, health facilities, housing, roads, etc.

Development transfers can be ad hoc, application-based (for a specific project) or formula-based. There are a limited number of grant programmes targeted specifically to urban governments, such as for urban transport or housing, and more recently 'Smart City' programmes have emerged. Many capital grants use matching requirements, such that the recipient government must finance a percentage of the investment. For this to work, however, urban governments must be able to raise funds for the match from recurrent revenue sources.

Urban borrowing

Increasing urban and local government access to loans for development projects in the global South has centered on two mechanisms – a special bank or fund offering subsidised loans and borrowing frameworks to allow market access where feasible. Special credit facilities often underperform due to poor design. Generally, such facilities have been closely tied to the central government and their operations were politicised (Friere and Petersen 2004; Peterson 2000).

Efforts to improve urban development finance are growing. These include new or enhanced borrowing/fiscal responsibility frameworks (e.g. Argentina, Brazil, South Africa), restructured municipal banks operating on more market-based principles (e.g. Colombia, Czech Republic) and direct market access for creditworthy urban governments (e.g. Brazil, India, Mexico, Philippines, South Africa) (Ingram *et al.* 2013; Smoke 2013). Expanded market access mostly benefits large urban areas (UCLG 2010, 2015). Borrowing can also be facilitated with risk mitigation strategies, such as credit guarantees, co-financing, bond banks and credit pooling (Alam 2010; Kehew *et al.* 2005; Martinez-Vazquez 2015; Petersen 2006; Platz 2009).

Land value capture

A potentially productive but underutilised urban revenue base is the increment in land value generated by urban infrastructure, including roads, sewerage, transit and water, among others (Ingram and Hong 2012; Peterson 2009; Suzuki *et al.* 2015; Walters 2012; World Economic Forum 2014). Options include betterment levies and special assessments (lump-sum levies on developers or property owners to finance improvements that raise property values), tax increment financing (surtaxes on properties redeveloped and financed from bonds issued against anticipated property tax increases) and land readjustment (pooling land with a share sold to raise resources to finance new infrastructure). Various countries, including Brazil, China, Colombia, India and several OECD countries, have used variations of these options.

There are debates about how to measure land value to recover infrastructure costs because other factors influence land value. There is, however, general agreement that land value capture is a reasonable and potentially fruitful urban revenue source. If it is to be more productively tapped, however, countries in the global South with weaker institutions will need to improve property rights, land valuation systems, information and capacity. In many cases, stronger political credibility must be cultivated with citizens expected to foot the bill, and efforts to improve general urban finances needed to operate new infrastructure are essential.

Public–private partnerships

Public–private partnerships (PPPs) can support urban governments to secure resources and expertise they need to pursue development (Brinkerhoff and Brinkerhoff 2011; Ingram *et al.* 2013; Marin 2009). Much PPP activity has been through central governments, but there have been advances with urban PPPs in recent years, particularly in certain sectors, such as transport (e.g. Beijing and Rio de Janiero) and water (e.g. Cartagena). Generally speaking, engaging the private sector has been challenging for urban governments in the global South.

A key constraint is that PPPs require a solid legal framework and urban government capacity. A framework takes time to develop, demands good urban fiscal practices and often requires central government support. PPPs do hold promise for urban governments, but they need to be started in an appropriate manner and expanded/adjusted on the basis of experience.

Challenges to improving planning–finance linkages

Having outlined key dimensions and mechanisms of urban finance, this section turns to some common but underappreciated constraints on improving linkages between urban planning and finance. These include national political factors, national bureaucratic dynamics, urban political/ institutional context and implementation considerations.

National political economy

The structure of intergovernmental systems and the role that different actors play must be understood in terms of historical trajectories and national politics (Eaton *et al.* 2011; Smoke 2014). The role of urban governments is a function of various factors, including governance traditions, external/colonial influences, social divisions and political rivalries, among others.

Such considerations may preclude a stronger and more independent role for urban planning and finance even if warranted on other grounds (Bahl *et al.* 2013; Connerley *et al.* 2010; UCLG 2010). Urban governments, for example, may be kept weak because of pressure from regional governments that risk losing control of resources or because they are dominated by opposition parties. If underlying forces preclude greater urban empowerment, prospective reformers must try to identify feasible alternatives under prevailing conditions.

National bureaucratic dynamics

Although political and historical factors shape broad intergovernmental system parameters, policy details and execution fall primarily under national agencies operating in fractious bureaucratic environments (Eaton *et al.* 2011). Multiple agencies are often mandated to develop and/or oversee specific policies relevant for urban planning and finance. Individual agencies with divergent views of the system and their role in it may independently develop inconsistent policies. If, for example, planning is devolved to urban governments but the finance ministry limits resources or line ministries place conditions on sectoral transfers that weaken urban government expenditure choice, integrated urban planning can suffer.

In aid-dependent countries, donor agencies that partner with specific national agencies can also sway policy (Development Partner Working Group on Decentralisation and Local Governance 2011; Dickovick 2014). Many donors have supported urban planning and finance, but often in fragmented ways (including at a sectoral level). Despite global aid agreements, many of them continue to use unsustainable arrangements (such as separately managed funds to finance urban infrastructure) and to compete with each other, potentially reinforcing policy contradictions of their partner government agencies.

Urban political and institutional dynamics

Even countries with strong empowerment of and support for urban planning and finance may face local challenges (Brinkerhoff and Azfar 2010; Faguet 2014; Grindle 2013; Yilmaz *et al.* 2010). How urban governments use powers depends on political conditions on the ground. The relative influence of economic elites, political parties, ethnic groups, labour unions, civil society groups and others shape the local environment. These factors can enhance urban accountability, but elite capture and corruption can undermine urban planning and finance.

Elections are central for urban governance, but their impact depends on the factors outlined above (Bland 2010). They are also a blunt instrument for discerning what citizens want. Other mechanisms that provide more regular involvement in urban government decisions or feedback on results include participatory planning and budgeting, town meetings, oversight boards, complaint bureaus, report cards, social auditing, etc. Their effects are heterogeneous (Blair 2013; Boulding and Wampler 2010; Brinkerhoff and Azfar 2010; Cheema 2013). Effectiveness depends on whether people feel empowered to engage, who participates, how fairly the mechanisms are used and if the results influence urban decision-making.

Multiple accountability channels can also compromise urban planning and finance. Urban governments may co-exist with local offices of national agencies and service delivery bodies. Functional boundaries may be blurred, and national offices often have superior staffing and resources. Under such conditions, citizens may be confused about which actor is responsible for which services, delivery gaps may occur and inequities may be created. If these entities were coordinated, they could jointly mobilise their resources for urban development.

Finally, governance of large urban areas is often a concern (Slack 2015). With a unified metropolitan government (e.g. Cape Town), opportunities for problems are reduced, although challenges with other levels of government remain. If there is jurisdictional fragmentation and a weak metropolitan development authority (e.g. Manila), service delivery and financing can be compromised. Individual jurisdictions tend to focus on their own needs, even if they could deliver services more effectively and raise more funds with collaboration. In still other cases, different types of jurisdictions with dissimilar powers co-exist in an urban area (e.g. the mix of governorates and 'new cities' in Cairo). Each of these arrangements is based on political and institutional incentives that influence urban government performance.

Implementation

Strong commitment and careful attention to good design of planning and fiscal systems is not enough – implementation is key but often receives inadequate attention. Needed reforms in the global South can be extensive and involve major operational and behavioural changes, so it may not be feasible for urban governments to assume new financial instruments to improve the quality and sustainability of development plans quickly. There is growing recognition of the need to consider implementation strategy (see, for example, Bahl and Martinez-Vazquez 2006; Eaton *et al.* 2011; Shah and Thompson 2004; Smoke 2010; World Bank 2008).

Implementation is relevant at both central and local levels. The central government, for example, can use criteria to determine if an urban government is ready to use new revenue sources and provide relevant incentives and support. Urban governments also face challenges when undertaking reforms to improve planning and finance. Even if relatively capable, they may need to strategically roll out reforms that require new processes and skills and place new demands on their constituents.

Looking forward: implications for planning–finance linkages

There is ample room to improve linkages between urban planning and finance throughout the global South. The first step is to document the current situation in a specific country or urban government. Many planning systems suffer from well-known problems that need to be remedied. But even if planning systems are well designed, improving urban finance to support their realisation has often proven challenging and merits dedicated attention.

Poor urban resource mobilisation can result from various mixes of issues discussed above. These include excessive central controls on urban revenue authority (including borrowing), inadequate urban capacity to generate revenue and/or deliver services, revenue generation disincentives embedded in improperly designed intergovernmental transfers, or weak accountability to urban residents and businesses, resulting in lacklustre revenue compliance and limited trust in urban governments.

Certain remedial actions are clearly in the central government realm. Urban governments often need additional authority to act more autonomously and developmentally, including broader revenue access. Some institutional challenges noted above, including blurry functional boundaries and relationships among actors at different levels and the structure of accountability, cannot be fully resolved locally. Urban governments can, however, often take steps to improve revenue generation and use, including better links to planning. Reforms requiring major changes, however, must often be pursued gradually and strategically so as to contain national political and bureaucratic resistance, be consistent with urban capacity, and reduce the risk of sabotage by local political dynamics.

Although the immediate link between planning and financing is on the development side of the budget, new facilities developed under urban plans require funds to be operated and maintained. Stronger recurrent revenues can also demonstrate the fiscal capacity required for urban governments to access long-term finance for key public investments. Since the sources of revenue available to urban governments must work together to support development, they need to be considered collectively rather than as isolated reform targets.

With respect to own-source revenues, urban governments can often improve the structure and administration of existing revenues, such as property taxes and user fees. They may also have the option to piggyback on productive revenues collected at higher levels or to adopt new sources. As noted above, a potentially productive but underutilised urban revenue base is the increment in land value generated by infrastructure development, although it is only one component of effective revenue system enhancement.

Raising current or adopting new revenues face logistical and political constraints, but these can be alleviated by strategic incrementalism and flexibility. If an urban government wishes to raise low/stagnant property valuations, increases could be phased in. Likewise, new or higher user charges could build progressively towards cost recovery. Such strategies can limit political opposition and adverse equity effects compared to what might occur with swift major increases. Enhanced payment convenience can also improve compliance, especially if payments are large, as with betterment assessments or infrastructure connection charges.

More generally, linking revenue increases more closely to better services through public awareness raising and consultation can improve accountability. If beneficiaries can be identified, urban governments might negotiate payment increases with service users, neighbourhood groups, business associations, etc. Participatory mechanisms noted above can aid urban governments in gathering input on proposed and past expenditure priorities and revenue changes. Urban governments could publicise new revenue rules to make sure their basis is transparent and perceived as fair. They can also create citizen appeal and grievances mechanisms and make efforts to safeguard consistent enforcement.

Intergovernmental transfer reforms must generally be undertaken by the central government. Some common reforms that help urban governments, such as ensuring the stability of the transfer pool and using objective allocation criteria that do not create distortions or disincentives, were outlined earlier. Increasing dedicated transfers for urban public investment can directly support planning and use of performance-based transfers can create incentives for urban local governments to behave responsibly (Steffensen 2010).

Despite the central government lead, urban governments can play a role in transfers. It may be productive, for example, to negotiate specific transfer performance objectives. This can reduce centralistic paternalism, placing responsibility on urban governments to realise targets they consent to. Urban governments can also use public consultation to help set and monitor targets and programme transfers. Such measures may enhance urban government credibility.

Beyond own-source revenues and transfers, there has been a growing movement to improve urban government access to private development finance, which is very relevant for urban planning. There are several elements – developing more robust borrowing/fiscal responsibility frameworks, facilitating market opportunities and adopting/reforming public lending mechanisms that operate on more market-based principles than past versions but still provide funds access to urban governments not yet able to borrow directly.

An intergovernmental fiscal system needs to offer a range of investment finance options, from grants and subsidised loans for fiscally weaker urban governments and non-self-financing projects, to loan mechanisms for more fiscally robust governments and self-financing projects. Other innovations and risk mitigation strategies noted above will also benefit urban finances. Sources and mechanisms vary considerably across countries, and planners need to know what is available for their urban government to consider how to assemble an appropriate package of options. For many urban governments in the global South, creditworthiness remains a challenge, and dedicated efforts are needed to build it.

It is important to reiterate that any of these illustrative finance reforms and how they can best be blended to improve urban planning–finance linkages must be framed in terms of broader institutional, territorial and political structures and incentives, as well as logistical considerations. These of course differ across countries and even across urban governments in a specific country. Not everything can be done at once and various actors will need to learn to behave in different ways, reinforcing the value of collaboration, partnership and carefully devised implementation strategies. The challenge for each country and urban government is to seek – within relevant contextual factors and constraints – pragmatic ways to improve planning–finance linkages that allow urban governments to play a stronger role in sustainable and inclusive development.

References

Ahmad, E. and Brosio, G. (eds) (2014) *Handbook of Fiscal Federalism and Multilevel Finance*, Cheltenham: Edward Elgar.

Alam, M. (ed.) (2010) *Municipal Infrastructure Financing: Innovative Practices from Developing Countries*. London: Commonwealth Secretariat.

Asian Development Bank (ADB) (2001) "Linking planning and budgeting: The medium-term expenditure framework", *The Governance Brief* 2. Available: https://think-asia.org/bitstream/handle/11540/3451/governancebrief02.pdf?sequence=1 [accessed 8 November 2016].

Bahl, R. and Bird, R. (2008) "Subnational taxes in developing countries: The way forward", *Public Budgeting and Finance* 28(4): 1–25.

Bahl, R. and Martinez-Vazquez, J. (2006) "Sequencing fiscal decentralization", Policy Research Working Paper No. 3914, Washington, DC: The World Bank.

Bahl, R., Linn, J. and Wetzel, D. (eds) (2013) *Metropolitan Finance in Developing Countries*, Cambridge, MA: Lincoln Institute of Land Policy.

Bird, R. and Slack, E. (2013) "Metropolitan public finance: An overview". In R. Bahl, J. Linn and D. Wetzel (eds) (2013) *Metropolitan Finance in Developing Countries*, Cambridge, MA: Lincoln Institute of Land Policy, pp. 135–158.

Bird, R. and Smart, M. (2002) "Intergovernmental fiscal transfers: International lessons for developing countries", *World Development* 30(6): 899–912.

Blair, H. (2013) "Participatory budgeting and local governance". In J. Ojendal and A. Dellnas, *The Imperative of Good Local Governance: Challenges for the Next Decade of Decentralization*, Tokyo: United Nations University Press, pp. 145–178.

Bland, G. (2010) "Elections and the development of local democracy". In E. Connerley, K. Eaton and P. Smoke (eds) *Making Decentralization Work: Democracy, Development and Security*, Boulder, CO: Lynne Rienner, pp. 47–80.

Boex, J. and Yilmaz, S. (2010) "An analytical framework for assessing decentralized local governance", IDG Working paper No. 2010-6, Washington, DC: The Urban Institute. Available: www.urban.org/sites/default/files/alfresco/publication-pdfs/412279-An-Analytical-Framework-for-Assessing-Decentralized-Local-Governance-and-the-Local-Public-Sector.PDF [accessed 8 November 2016].

Boulding, C. and Wampler, B. (2010) "Voice, votes and resources: Evaluating the effect of participatory democracy on well-being", *World Development* 38(1): 125–135.

Brinkerhoff, D. and Azfar, O. (2010) "Decentralization and community empowerment". In E. Connerley, K. Eaton and P. Smoke (eds) *Making Decentralization Work: Democracy, Development and Security*, Boulder, CO: Lynne Rienner, pp. 81–114.

Brinkerhoff, D. and Brinkerhoff, J. (2011) "Public–private partnerships: Perspectives on purposes, publicness and good governance", *Public Administration and Development* 31(1): 2–14.

Cheema, G.S. (2013) "Engaging civil society to promote local governance: Emerging trends in local democracy in Asia". In J. Ojendal and A. Dellnas, *The Imperative of Good Local Governance: Challenges for the Next Decade of Decentralization*, Tokyo: United Nations University Press, pp. 233–254.

Commonwealth Local Government Forum (2013) *Developmental Local Government: Putting Local Government at the Heart of Development*, London: Commonwealth Local Government Forum.

Connel, D. (2010) "Schools of planning thought: Exploring differences through similarities", *International Planning Studies* 15(1): 269–280.

Connerley, E., Eaton, K. and Smoke, P. (eds) (2010) *Making Decentralization Work: Democracy, Development and Security*, Boulder, CO: Lynne Rienner.
Development Partner Working Group on Decentralization and Local Governance (2011) "Busan and beyond: Localizing Paris principles for more effective support to decentralization and local governance reforms", Bonn: GIZ. Available: www.delog.org/cms/upload/pdf/DeLoG_Busan_and_beyond.pdf [accessed 8 November 2016].
Dickovick, T. (2014) "Foreign aid and decentralization: Limitations on impact in autonomy and responsiveness", *Public Administration and Development* 34(1): 193–205.
Eaton, K., Kaiser, K. and Smoke, P. (2011) *The Political Economy of Decentralization: Implications for Aid Effectiveness*, Washington, DC: The World Bank.
Faguet, J.-P. (2014) "Decentralization and governance", *World Development* 53(1): 2–13.
Friere, M. and Petersen, J. (eds) (2004) *Subnational Capital Markets in Developing Countries: From Theory to Practice*, Oxford: Oxford University Press.
Grindle, M. (2013) "Sanctions, benefits and rights: Three faces of accountability". In J. Ojendal and A. Dellnas (eds) *The Imperative of Good Local Governance: Challenges for the Next Decade of Decentralization*, Tokyo: United Nations University Press, pp. 207– 232.
Healey, P. (1997) *Collaborative Planning*, Vancouver: University of British Columbia Press.
Healey, P. (2012) "The universal and the contingent: Some reflections on the transnational flow of planning ideas and practices", *Planning Theory* 11(1): 188–207.
Ingram, G. and Hong, Y.-H. (2012) *Value Capture and Land Policies*, Cambridge, MA: Lincoln Institute of Land Policy.
Ingram, G., Liu, Z. and Brandt, K. (2013) "Metropolitan infrastructure and capital finance". In R. Bahl, J. Linn and D. Wetzel (eds) *Metropolitan Finance in Developing Countries*, Cambridge, MA: Lincoln Institute of Land Policy, pp. 339–366.
Kehew, R., Matsukawa, T. and Petersen, J. (2005) "Local financing for sub-sovereign infrastructure in developing countries: Case studies of innovative domestic credit enhancement entities and techniques", Discussion Paper No. 1: Infrastructure, Economics and Finance Department, Washington, DC: The World Bank. Available: http://siteresources.worldbank.org/INTGUARANTEES/Resources/Local_Financing_for_Sub-sovereign_infrastructure.pdf [accessed 8 November 2016].
McClure, C. and Martinez-Vasquez, J. (2004) "The assignment of revenues and expenditures in intergovernmental fiscal relations", Washington, DC: The World Bank. Available: www1.worldbank.org/publicsector/pe/PEAMMarch2005/AssignmentRevenues.pdf [accessed 8 November 2016].
MacDonald, K., Sanyal, B., Silver, M., Ng, M.K., Head, P., Williams, K., Watson, V. and Campbell, H. (2014) "Challenging theory, changing practice: Critical perspectives on the past and potential of professional planning", *Planning Theory and Practice* 15(1): 95–122.
Marin, P. (2009) *Public–Private Partnerships for Urban Water: A Review of Experiences in Developing Countries*, Washington, DC: The World Bank.
Martinez-Vazquez, J. (2013) "Local non-property revenues". In R. Bahl, J. Linn and D. Wetzel (eds) *Metropolitan Finance in Developing Countries*, Cambridge, MA: Lincoln Institute of Land Policy, pp. 182–212.
Martinez-Vazquez, J. (2015) "Mobilizing financial resources for public service delivery and urban development". In *The Challenge of Local Government Financing in Developing Countries*, Nairobi: UN-Habitat, pp. 15–36. Available: http://unhabitat.org/the-challenge-of-local-government-financing-in-developing-countries/ [accessed 8 November 2016].
Martinez-Vazquez, J. and Timofeev, A. (2012) *Propensity to Invest and the Additionality of Capital Transfers: A Country Panel Perspective*, Atlanta, GA: Andrew Young School of Policy Studies, Georgia State University.
Petersen, J. (2006) "Municipal funding arrangements: Global experience and lessons learned", International Workshop on Strengthening Local Infrastructure Financing, Washington, DC: The World Bank.
Peterson, G. (2000) "Building local credit institutions", Urban and Local Government Background Series, No. 3. Washington, DC: The World Bank.
Peterson, G. (2009) *Unlocking Land Values to Finance Urban Infrastructure*, Washington, DC: The World Bank.
Platz, D. (2009) "Infrastructure finance in developing countries: The potential of sub-sovereign bonds", DESA Working Paper No. 76. New York: UN Department of Economic and Social Affairs.
Rajaram, A., Le, T.M., Biletska, A. and Brumby, J. (2010) *A Diagnostic Framework for Assessing Public Investment Management*, Washington, DC: World Bank.

Rodwin, L. and Sanyal, B. (eds) (2000) *The Profession of City Planning: Changes, Images and Challenges*, New Brunswick, NJ: Rutgers Center for Urban Policy Research.

Romeo, L. (2013) "Decentralizing for development: The developmental potential of local autonomy and the limits of politics-driven decentralization reform". In J. Ojendal and A. Dellnas (eds) *The Imperative of Good Local Governance: Challenges for the Next Decade of Decentralization*, Tokyo: United Nations University Press, pp. 60–90.

Romeo, L. and Smoke, P. (2016) "The political economy of local infrastructure planning". In J. Frank and J. Martinez-Vazquez (eds) *Decentralization and Infrastructure in the Global Economy*, Abingdon: Routledge, pp. 381–417.

Sanyal, B., Vale, L. and Rosen, C. (eds) (2012) *Planning Ideas that Matter: Livability, Territoriality, Governance and Reflective Practice*, Cambridge, MA: MIT Press.

Shah, A. (2013) "Grant financing of metropolitan areas: A review of principles". In R. Bahl, J. Linn and D. Wetzel (eds) *Metropolitan Finance in Developing Countries*, Cambridge, MA: Lincoln Institute of Land Policy, pp. 213–242.

Shah, A. and Thompson, T. (2004) "Implementing decentralized local governance: a treacherous road with potholes, detours, and road closures", Policy Research Working Paper No. 3353, Washington, DC: The World Bank. Available: http://documents.worldbank.org/curated/en/349441468761683659/pdf/wps3353.pdf [accessed 8 November 2016].

Slack, E. (2015) "Innovative governance approaches in metropolitan areas of developing countries". In *The Challenge of Local Government Financing in Developing Countries*, Nairobi: UN-Habitat, pp. 61–82.

Smoke, P. (2010) "Implementing decentralization: Meeting neglected challenges". In E. Connerley, K. Eaton and P. Smoke (eds) *Making Decentralization Work: Democracy, Development and Security*, Boulder, CO: Lynne Rienner, pp. 191–217.

Smoke, P. (2013) "Metropolitan cities in the national fiscal structure". In R. Bahl, J. Linn and D. Wetzel (eds) *Metropolitan Finance in Developing Countries*, Cambridge, MA: Lincoln Institute of Land Policy, pp. 57–84.

Smoke, P. (2014) "Why theory and practice are different: The gap between principles and reality in subnational revenue systems". In R. Bird and J. Martinez-Vazquez (eds) *Taxation and Development: The Weakest Link*, Cheltenham: Edward Elgar, pp. 287–325.

Steffensen, J. (2010) *Performance-Based Grants: Concepts and International Experience*, New York: United Nations Capital Development Fund.

Suzuki, H., Murakami, J., Hong, Y.-H. and Tamayose, B. (2015) *Financing Transit Oriented Development with Land Values: Adapting Land Value Capture in Developing Countries*, Washington, DC: The World Bank.

United Cities and Local Governments (UCLG) (2010) *Local Government Finances: The Challenges of the 21st Century*, Barcelona: UCLG.

UCLG (2014) *Access to Basic Services*, Barcelona: UCLG.

UCLG (2015) *Financing Urban and Local Development: The Missing Link in Sustainable Development Finance*, Barcelona: UCLG.

UN (2015) *Transforming Our World: The 2030 Agenda for Sustainable Development*, New York: United Nations.

Walters, L. (2012) *Land Value Capture in Policy and Practice*, Provo, UT: Romney Institute, Brigham Young University.

World Bank (1998) *Public Expenditure Management Handbook*, Washington, DC: The World Bank.

World Bank (2008) *Decentralization in Client Countries: An Evaluation of World Bank Support 1990–2007*, Washington, DC: Independent Evaluation Group, World Bank.

World Economic Forum (2014) *Accelerating Infrastructure Delivery: New Evidence from International Financial Institutions*, Geneva: World Economic Forum.

Yilmaz, S., Beris, Y. and Serrano-Berthet, R. (2010) "Linking local government discretion and accountability in decentralization", *Development Policy Review* 28(3): 259–293.

Part III

New drivers of change

Ecology, infrastructure and technology

13

Urban climate adaptation in the global South

Justice and inclusive development in a new planning domain

Eric Chu, Isabelle Anguelovski and Debra Roberts

Introduction

Cities across the global South are increasingly responding to climate change by pursuing mitigation, adaptation and other environmental sustainability actions. While many cities have long promoted broad waste reduction and urban greening programmes, dedicated climate change actions have only gained ground recently. This chapter focuses on emerging climate change adaptation efforts, that is, processes of adjusting to actual or expected climate impacts in order to moderate or avoid harm (Field *et al.* 2014), and which range from raising early awareness and assessing risks to implementing adaptation in sectoral work plans (Anguelovski and Carmin 2011; Carmin *et al.* 2013). Adaptation is important for cities in the global South because of their disproportionate exposure to impacts, lower capacity to respond and relative concentration of poor populations (Bicknell *et al.* 2009).

Many cities are protecting housing, infrastructure, public services and other economic assets against climate impacts. However, there is little understanding of how municipalities balance climate adaptation needs with urban development priorities, especially those related to economic growth and poverty reduction, as well as navigate relevant institutional structures and actors who have conflicting priorities and practices. This chapter asks: through what processes do urban adaptation planning approaches promote more equitable forms of development? To answer this, we describe adaptation actions in Medellín (Colombia), Durban (South Africa) and Indore (India), and assess emerging experiences of integrating adaptation priorities into urban development agendas. Specifically, we examine Medellín's actions to reduce climate risks through spatial planning and green infrastructure projects; Durban's plans to integrate adaptation into biodiversity protection programmes; and Indore's approach to bringing climate resilience into community development projects. We compare these approaches and highlight sources of planning tension and the ways they reshape subsequent adaptation interventions.

We find that when advanced with a deliberate focus on alignment with development, emerging adaptation agendas have a better chance of taking root in urban programmes and practices. The integration of adaptation into development can also promote stronger leadership, resource support and agenda awareness across municipal departments. This increased embeddedness between issues facilitates showcase or pilot adaptation projects that can lead to greater municipal buy-in and, more importantly, trigger dialectical relationships among different municipal actors that catalyse more inclusive development outcomes.

Theories of climate adaptation planning

In this chapter, we refer to development as processes of economic growth, livelihood improvement and poverty reduction, though we acknowledge the contentious and context-dependent manifestations of development plans and policies on the ground. Within these parameters, climate adaptation and development are closely related through four conceptual links. First, climate change is the direct result of unsustainable development. Second, sustainable development reduces climate vulnerability and mitigates future carbon emissions (Bizikova *et al.* 2007). Third, climate change erodes past gains in reducing poverty (Ayers and Dodman 2010; Carmin *et al.* 2015). Finally, climate adaptation can transform unsustainable development pathways if conceived in ways that fundamentally shift future development paradigms (Shi *et al.* 2016).

Climate impacts can exacerbate poverty (Ayers and Dodman 2010), so scholars have introduced the idea of mainstreaming adaptation into development – broadly defined – to better articulate how adaptation can contribute to peoples' capacities to deal with these impacts (Halsnæs and Trærup 2009). Programmatically, mainstreaming takes place across a wide range of policy arenas, including disaster risk reduction (Serrao-Neumann *et al.* 2015; Solecki *et al.* 2011), ecosystem protection (Roberts *et al.* 2012), spatial and land use planning (Anguelovski *et al.* 2016), public health (Bowen and Ebi 2015) and poverty reduction (Jerneck and Olsson 2008). Procedurally, adaptation can be mainstreamed into environmental management, asset procurement and public finance mechanisms (Carmin *et al.* 2012). Some scholars argue that effective adaptation planning must also build on co-operation with civil society organisations to improve awareness and knowledge transfer (Chu *et al.* 2015). Therefore the hallmark of climate adaptation across the global South is a pursuit of diverse planning strategies, reliance on cross-sectoral tools and experimentation with different participatory arrangements.

Local governments are often most attuned to the context-specific nature of risks and vulnerabilities (Bulkeley 2010; Hunt and Watkiss 2011), and many also oversee responsibility for managing infrastructure and services that are essential for good living standards (Anguelovski *et al.* 2014; Dodman and Satterthwaite 2009). Cities often find greater policy traction when adaptation priorities are linked with existing development concerns rather than being framed as issues of hazard risk avoidance. However, for many cities in the global South, mainstreaming adaptation into different development agendas is a practical challenge because it requires them to bridge deficits in finance, staffing capacity, information, local leadership and supportive cultural values (Carmin *et al.* 2013). The lack of capacity also constrains how municipal planners, engineers and service delivery professionals are able to anticipate or cope with increasingly severe climate impacts (Carmin *et al.* 2012).

Even though there is broad consensus that cities must adapt to climate impacts and that such efforts must be integrated into ongoing development agendas (Leichenko 2011), there is little understanding of the extent to which adaptation planning processes can promote more equitable forms of development. Equity and justice are important parameters for assessing adaptation outcomes due to the uneven distribution of power in current development processes (Paavola 2008; Schlosberg 2012), which in turn affects the degree to which adaptation projects can truly improve capacities on the ground, redress structural forms of poverty, as well as facilitate more sustainable development pathways (Shi *et al.* 2016).

To address this gap, we assess three different approaches to mainstreaming adaptation into development in Medellín, Durban and Indore to distil lessons for enhancing municipal integration and coordination, promoting inclusion of vulnerable groups, and facilitating new pathways of urban development. These lessons contribute to identifying tools necessary for assessing the quality of urban adaptation programmes and the extent to which inclusion and justice priorities are recognised on the ground.

Experiences of integrating climate adaptation into urban development

This section evaluates different approaches to integrating adaptation and development planning in Medellín, Durban and Indore. These cities are emblematic cases of early adopters of adaptation in the global South and, in recent years, have pursued a wide array of strategies to combat climate impacts that recognise the needs of vulnerable communities in the process. As Table 13.1 summarises, adaptation strategies range from large-scale ecosystem measures to community-based livelihood technologies. Through documenting these interventions, this section highlights the intricacies of planning approaches – as well as implications for urban inclusion and justice – across the different contexts.

Medellín: a case of green infrastructure and territorial planning

Medellín faces increasing frequencies of extreme rainfall events, extended dry periods and increasing temperatures affecting the Andean ecosystems that contribute to the urban water supply (Anguelovski *et al.* 2016). Today, due to inadequately constructed buildings along hillsides, 180,000 households in Medellín are at risk of mudslides.

Table 13.1 Examples of climate adaptation and development planning tools

City	*Select examples*
Medellín	*Spatial and territorial planning:* containing urban growth; managing landslides risks; and protecting ecosystems through Green Belt
	Zoning and building codes: delineating urban land use zones; improving structural quality of public housing
	Demonstration projects: improving access to public transportation; upgrading marginalised neighbourhoods; designing new public spaces.
Durban	*Project evaluation and assessment:* developing tools to prioritise beneficial outcomes of adaptation options
	Urban management: drafting policies at the adaptation and mitigation interface; establishing multi-stakeholder networks
	Ecosystem protection: protecting ecological infrastructure; increasing water security; conserving river basins; managing coastal erosion; protecting biodiversity and ecosystem services
Indore	*Slum management:* supporting microfinance and women's groups; improving housing quality; providing access to public health services
	Water conservation: protecting water bodies; promoting waste-water-harvesting technologies and water recycling facilities
	Municipal finance: offering property tax incentives for water-harvesting technologies; creating a new municipal budget line-item for climate action

In response to uncontained urban growth and increasing climate risks, in 2012 former Mayor Aníbal Gaviria announced the construction of a 46-mile-long Metropolitan Green Belt (*Jardín Circunvalar*). The USD249 million ring of protected natural space was conceived to integrate emerging adaptation needs – especially water protection and heat management – with other urban greening, congestion alleviation and property development priorities. As one of Mayor Gaviria's 31 flagship projects, the Green Belt builds upon Medellín's tradition of urban rebranding, spatial planning and entrepreneurialism to address pervasive social problems.

At an elevation of 1,600 metres, the Green Belt connects three distinct projects. First, a 'protection zone' preserves natural hillside ecosystems. Next is a 'transition zone' that includes new parks, bike paths and risk-management infrastructures. This area also coincides with the highest concentration of low-income neighbourhoods that lack basic services and amenities. Finally, beyond the Green Belt is the 'consolidation zone', which, in the words of Mayor Gaviria, is designed to "re-conquer the valley" with new parks, multi-family housing and multi-modal transportation networks (Municipio de Medellín 2014). These projects are designed to help Medellín achieve sustainable and climate resilient development, but may displace approximately 230,000 residents who live in high-risk areas or on future Green Belt protected land.

All Green Belt projects are planned in ways that can be implemented in concert with existing or upcoming spatial development plans. For example, the city's new Territorial Organisation Plan (POT) focuses on protecting water resources, densifying the urban core and increasing access to public transportation (Municipio de Medellín 2014). Medellín is also introducing new Integral Urban Projects (PUI) as part of the city's planned expansion into the north-western region. PUIs are addressing adaptation needs by bringing housing units in line with building codes, strengthening existing infrastructure, improving the quality of public spaces and creating new transport links in low-income neighbourhoods. The Plan Bio 2030 articulates the expansion of the Green Belt into the metropolitan valley and focuses on environmental services, disaster prevention and urban growth containment (Municipio de Medellín 2011). The connection between the Green Belt and sectoral planning highlights the municipality's commitment to integrating emerging adaptation needs with existing development objectives and institutional arrangements.

Yet, in the midst of these projects, the Green Belt is raising equity implications for low-income communities. Mimicking historical patterns of spatial development, the municipality has started to relocate communities living on land that is deemed unstable or risky. Notably, in Communa 8 – from where the municipality is relocating 6,600 households – residents are opposed to resettlement in faraway public housing. That said, this new public housing can be more comfortable and safe if constructed using participatory designs that compensate for residents' loss of social networks and livelihoods. However, these relocation projects will not actually solve the problem of growing low-income housing demand across the region, and will instead promote expansion of new low-quality settlements along fragile hillsides.

Controversies over eviction and relocation highlight the contentious politics surrounding how climate risks are defined and communicated to different communities. In particular, there are discrepancies between different climate risk assessments – including those produced by the city's Risk Zone Maps, the Geological Suitability Map and different resident-produced estimates (Municipio de Medellín 2014) – about the specific number of households located in 'non-recoverable risk' areas. Rather than benefitting from onsite retrofitting and upgrading, low-income residents are simply relocated. In contrast, there are no plans to move higher income residents in the neighbourhoods of El Poblado, Cedro Verde and Alto de las Palmas. Rich neighbourhoods also seem to be unrestrained in their expansion up the hills in the south-east of Medellín, which correspond to areas that the municipality previously deemed at high risk of landslides. Finally, gated communities located next to native forest reserves, such as in the case of Alto de Escobero, are permitted to expand without any mandates for resettlement. Incidences of displacement highlight how adaptation processes may actually facilitate the entrenchment of historic power differentials, and that a focus on adaptation as an environmental good may only benefit the urban elite (Anguelovski *et al.* 2016).

Through the Green Belt, low-income residents are also losing access to green space and agricultural land around the Pan de Azúcar Mountain, which historically has housed those displaced by the Colombian armed conflict. The Green Belt converted this area into ecological corridors, recreational spaces, playgrounds and educational centres. In response, displaced communities argued that the planning process lacked meaningful engagement of vulnerable communities and disregarded traditional land uses. Again in Communa 8, residents prepared a community development plan highlighting their own development vision against the perceived non-transparency of the Green Belt planning process. Residents also advocated for formal housing as well as construction of new rental units in anticipation of general urban growth needs. The municipality has, in turn, proposed alternative strategies to better preserve existing housing, protect livelihoods and broaden access to urban agriculture. Such tensions highlight dilemmas between climate and development agendas, as well as the tendency of city governments to designate protected natural areas through imposing new spatial control tools at the expense of local residents' access to traditional lands.

In summary, Medellín's Green Belt project shows that despite commitments to integrate new green infrastructure into development plans in socially conscious ways, climate adaptation efforts can at times accelerate cultural, economic and physical displacement of vulnerable residents (Anguelovski *et al.* 2016). Under this scenario, the distribution of adaptation benefits continues to be rooted in historic power inequities that characterised the city's development in the past. Although Medellín has achieved some recent success in containing, beautifying and protecting against development stresses and climate impacts, the city is now facing new concerns of economic loss, social disintegration and 'green' gentrification among low-income communities.

Durban: a case of ecosystem-based infrastructure management

Durban is the largest port on the east coast of Africa and the third largest of South Africa's metropolitan areas. Among South Africa's major cities, Durban has the highest percentage of people living in poverty and has high levels of inequality. The city also has considerable backlogs in infrastructure and basic services. In response to these challenges, Durban's Integrated Development Plan established a vision of a caring, equitable and liveable city where poverty is addressed through infrastructure-led growth and job creation (eThekwini Municipality 2015). Climate projections show that Durban will experience increases in temperature, more variable rainfall, sea level rise and storm surge (Golder Associates 2011). These impacts put Durban at risk from sudden as well as slow onset disasters, ranging from flash floods and droughts to coastal erosion exacerbated by sea level rise, currently calculated to be 2.7 mm per year (Mather *et al.* 2009).

Even though South African cities have no formal climate change planning obligations, the Environmental Planning and Climate Protection Department (EPCPD) of eThekwini Municipality – the local government responsible for managing Durban – nonetheless initiated the Municipal Climate Protection Programme (MCPP) in 2004. The MCPP's adaptation workstream began following a climate impact assessment, and is comprised of several components. First, adaptation plans focused on three strategic sectors: municipal water, health and disaster management functions. Second, there is a focus on improving capacity of local communities through community-based adaptation. Third, there is a strong ecosystem-based component in all interventions. Fourth, there has been a series of urban management interventions that address specific challenges, such as urban heat islands and increased storm water runoff. Fifth, there is an attempt to develop locally appropriate climate change tools, particularly in the form of sea level rise and cost-benefit models focusing on human benefit and ecological integrity. Finally, the municipality has taken actions to mainstream climate protection, such as through mega-event greening, to raise the profile of climate change and institutional restructuring (Diederichs and Roberts 2016).

Durban's adaptation approach has been phased and opportunistic because of limited precedents, interest, leadership, institutional support and resources (Roberts and O'Donoghue 2013). Knowledge gained through early interventions has shaped and refined subsequent actions, while the focus has been on 'no-regrets' projects that are beneficial under a range of climate change scenarios. These have helped build constantly evolving planning pathways composed of manageable and adjustable steps over time, each triggered by a change in available resources, knowledge and response to unexpected opportunities (Roberts *et al.* 2016). Both the resulting successes and failures have generated a cycle of continuous learning that has helped build a clearer understanding of the complexity of adaptation action.

Durban has relied on cultivating institutional champions who have deep sectoral knowledge, and who can then identify points of integrative action and overlapping spheres of influence (Sutherland *et al.* 2014). This is a particularly strategic approach since adaptation in Durban is an unfunded mandate and is dependent on strong leadership. These champions have been critical in re-casting climate change as a key development issue, and minimising the marginalisation associated with environmental programmes. For example, the municipality helped establish a multi-stakeholder partnership addressing the role of ecological infrastructure in increasing water security and adaptive capacity in the uMgeni River catchment. This reflects a shift towards a 'socio-ecological systems approach' to managing water, biodiversity, climate and poverty challenges (Sutherland *et al.* 2014). This strong alignment of agendas has led the city to adopt adaptation policies early on, and has generated numerous tools to assist with integrating adaptation priorities into urban planning. Still, these tools have so far yielded mixed results (Cartwright *et al.* 2013; Walsh *et al.* 2013) and have often proved expensive and time consuming.

The developmental needs of the city, its high climate risk profile and limited capacities make adaptation a priority for the foreseeable future. As a result, Durban has begun to combine local adaptation and mitigation agendas into a unified climate change approach. For example, three large-scale community reforestation projects – initiated to offset the carbon footprints of the FIFA Football World Cup™ in 2010 and the COP17 meeting in 2011 (Diederichs and Roberts 2016) – created new carbon sinks and delivered multiple adaptation co-benefits, such as biodiversity conservation and improved ecosystem services. Socioeconomic co-benefits, particularly in terms of job creation and skills development, are important for encouraging and sustaining local climate action among vulnerable communities, where risk acceptance levels are generally higher and where climate risks are secondary to livelihood concerns. The expansion of adaptation to include equitable transitions to a low-carbon future represents a clearer link to local developmental objectives. Finally, these projects have also catalysed a broader discussion around how the urban climate change agenda itself should be framed. For example, in 2015 Durban established a mayoral climate change committee.

In summary, Durban's experience highlights the complexity of adaptation planning processes, which often necessitates a portfolio of incremental solutions that emerge from opportunistic and experimental approaches. Many of these actions are conceived with specific economic development and poverty reduction co-benefits in mind. Plans are coordinated, synergised and constantly re-evaluated to guard against maladaptation and inequitable consequences. The relationship between adaptation and development, therefore, is determined by conscious decisions to find these synergistic priorities, minimise institutional contestation and highlight points of integration with the city's overall development agenda.

Indore: a case of community-led service delivery projects

Indore, with a population of more than 2.2 million, is the commercial capital of Madhya Pradesh in India. The city has experienced 40 per cent decadal population growth and approximately 6.5 per cent annual economic growth over the past several decades (Indore Municipal Corporation 2006). One third of the population lives in slum settlements and a significant proportion of these settlements are prone to flooding, waterlogging and vector-borne diseases (Bhat *et al.* 2012). Rapid urbanisation has also led to traffic congestion, high rates of solid waste generation, inadequate public services and general environmental degradation. Even though Indore is not directly exposed to many natural hazards, the city does have a history of addressing slow-onset risks such as droughts.

With support from the Rockefeller Foundation's Asian Cities Climate Change Resilience Network (ACCCRN), adaptation planning in Indore began in 2009, which culminated in the release of the Indore City Resilience Strategy in 2012. The adaptation planning process began with a series of scenario-building workshops to help raise awareness of climate impacts in relation to current socioeconomic development needs (Kernaghan and da Silva 2014). The Indore City Resilience Strategy identified water, public health and human settlements sectors as most vulnerable to climate change and, therefore, proposed pilot projects for addressing these key impacts (Bhat *et al.* 2012). Since the local government – in the form of the Indore Municipal Corporation (IMC) – is resource constrained, the planning process prioritised engagement with civil society actors for capacity support.

Water scarcity and supply consistency have been perennial issues impacting development in Indore (Bhat *et al.* 2012). Much of the urban poor depend on public or community water sources such as standpipes, hand pumps and wells. Currently, approximately 80 per cent of the city's water comes from the Narmada River located more than 70 kilometres away. This is supplemented by two municipal water tanks and more than 2,000 tube-wells distributed across the city. To anticipate increased water shortages under climate change, adaptation projects have facilitated a renewed local focus on water conservation and protection as critical urban development priorities.

Between 2010 and 2013, the city experimented with different community-based water conservation technologies and devised new decentralised wastewater management models. One example is in Rahul Gandhinagar, which is a settlement of 5,000 residents without piped water, where a reverse osmosis plant was built to improve the quality of drinking water. The facility is managed by a women's group dedicated to championing the benefits of reverse osmosis-treated water, which include reducing gastrointestinal disease infection rates and improving community health (Chu 2016). Although the municipality did not directly finance the reverse osmosis facility, the municipality did subsidise electricity rates for the facility's operational needs. In a second community, Ganeshnagar, the municipality launched a water-harvesting programme to provide water to households without access to public pipelines. The programme initiated a local system of collecting and storing rainwater, filtering this water through drums consisting of coal, sand and brick fragments, and, finally, collecting water through common-access taps. Profits from both the reverse osmosis plant and community water-harvesting system were subsequently used to complement existing community economic development and livelihood improvement projects.

Apart from improving water access in slum communities, the municipality also made use of ACCCRN support to rehabilitate and conserve existing urban water bodies. Many urban lakes across the city have been degraded due to development pressures and pollution. So, targeting communities near Khajrana Talab, Lasudiya Talab, Talavali Chanda Talab and Rangvasa Talab, the municipality conducted biodiversity and household surveys, drafted water quality protection plans, and constructed community sewage treatment plants. These municipal efforts are supported by various ward committees and community welfare associations (Chu 2016), which fostered co-operation between community beneficiaries and different local waste management utilities.

Many of Indore's community-based projects have facilitated a renewed local focus on water conservation as critical urban development priorities and have catalysed some institutional change in the local government itself. For example, the municipality is integrating wastewater management mandates into revisions of City Development Plans, prioritising adequate storm water drainage for new road developments, and introducing financial incentives for household water harvesting technologies. Also, in recent annual budgets, the municipality has included a line item entitled 'climate change safety expenses', which earmarks approximately USD75,000 per year for urban climate change programmes. All of these actions prioritise and promote co-benefits between climate adaptation needs and urban development priorities, many of which target slum communities.

However, due to the city's existing capacity constraints, much of the adaptation process exists outside of formal municipal decision-making and is driven strongly by civil society interventions, such as the Rockefeller Foundation and various community-based organisations. Those strategies that are relatable, easy and cheap to implement receive priority. Although many of these programmes focus on community interventions – and thus have clear poverty reduction implications – there is no strategy for scaling-up these incremental projects to facilitate equitable development across the city (Chu 2015). Unlike in Durban, the lack of institutional champions who can bring together high-level planners and managers might be affecting Indore's ability to visibly and structurally institutionalise climate adaptation priorities. Without institutionalisation, adaptation interventions will only target discrete sectors, actors and locations, rather than help in building overall urban adaptive capacity.

The dialectics of adaptation and development: implications for planning in the global South

This chapter described different approaches to integrating emerging climate adaptation priorities into urban development planning, which range from constructing large-scale greenbelt infrastructures in Medellín, to mainstreaming adaptation into sectoral policies in Durban, to harnessing community-based strategies in Indore. These examples highlight how current adaptation plans are often not framed as completely new agenda items, but rather as extensions of existing sustainable development, spatial planning or economic growth strategies. The sheer number of actors and institutions involved in climate adaptation often means that particular interventions are politically and socially contested, as well as laden with multi-scalar implications for equity and justice.

Climate adaptation presents new opportunities – as well as challenges – for planning in the global South. On the one hand, increased embeddedness between adaptation and development can promote institutional leadership, resource support and agenda awareness. Contestations around showcase or pilot projects such as the Green Belt in Medellín or the reverse osmosis plant in Indore may then catalyse more just social outcomes. These planning pathways can yield important insights into facilitating poverty reduction co-benefits. On the other hand, a reliance on existing planning strategies may be indicative of structural constraints that inhibit the ability of cities in the global South to drastically redistribute resources to tackle new, multi-scalar environmental problems, of which climate change is a prime example.

Our analysis reveals two broad sources of planning contention in how adaptation is framed within a particular development context. The first source of tension comes from within local government, and is attributed to differences in internal priorities, strategies and capacities. Scholars note the importance of champions, leader departments and clear strategies (Anguelovski and Carmin 2011), but many of these champions find difficulty building partnerships across different sectors. For example, in Durban, the EPCPD experiences difficulty with generating buy-in from certain departments because of historically conflicting priorities, even though stronger partnerships have emerged more recently. Municipalities are testing ideas and quantifying specific co-benefits across sectors and between adaptation and mitigation (Anguelovski *et al.* 2014), which facilitates an incremental approach to navigating local governance constraints and opportunities. For example, water management pilot projects in Indore involved different community organisations, municipal actors and donors at different points in time. The experimental approaches taken by both Durban and Indore are contributing to more effective, legitimate and inclusive inter-institutional engagements around climate adaptation.

The second source of tension is between local governments and communities who are trying to influence adaptation by advocating against inequities from the bottom up. Grassroots actions contesting municipal adaptation plans create opportunities for addressing equity and social justice priorities. For example, in Medellín, local mobilisations against particular Green Belt projects have refocused municipal priorities from infrastructure development to broader issues of poverty reduction, environmental justice and livelihoods security. In Durban, vulnerable communities are increasingly benefiting from poverty reduction outcomes of local ecosystem protection projects. Here, equity has emerged as a by-product of the creation of new socio-ecological systems and development of a restoration economy. However in Indore, where there are institutional constraints to scaling-up community projects, there exists a gap between bottom-up advocacy and genuine improvements to structural poverty and inequality experienced by slum residents. These indirect social costs threaten the long-term sustainability of adaptation and prevent inclusive and equitable adaptation plans. Powerful urban residents or investors may end up reaping adaptation benefits at the expense of marginalised communities that are already vulnerable to climate impacts.

The presence of different planning tensions is prompting cities to identify more transformative adaptation strategies that remedy structural patterns of unjust or unsustainable development. Such tensions are even more acute in the global South, where resource constraints and capacity limitations pose additional challenges for adaptation planning. Cities like Medellín, Durban and Indore are making progress towards integrating adaptation into poverty reduction and other development needs, but future research must also critically reflect on the means and ends of current unsustainable urban development practices. Adaptation can be a double-edged sword, and plans should not entrench neoliberal processes of urban growth and production. Rather, cities should experiment with building new cross-sectoral partnerships and civil society networks to support inclusive, sustainable and pro-poor adaptation approaches. A pursuit of innovative and transformative development pathways in the context of future climate impacts is what is substantively – and paradigmatically – unique about adaptation.

References

Anguelovski, I. and Carmin, J. (2011) "Something borrowed, everything new: Innovation and institutionalization in urban climate governance", *Current Opinion in Environmental Sustainability* 3(3): 169–175.

Anguelovski, I., Chu, E. and Carmin, J. (2014) "Variations in approaches to urban climate adaptation: Experiences and experimentation from the global south", *Global Environmental Change* 27: 156–167.

Anguelovski, I., Shi, L., Chu, E., Gallagher, D., Goh, K., Lamb, Z., Reeve, K. and Teicher, H. (2016) "Equity impacts of urban land use planning for climate adaptation: Critical perspectives from the global north and south", *Journal of Planning Education and Research* 36(3): 333–348.

Ayers, J. and Dodman, D. (2010) "Climate change adaptation and development I: The state of the debate", *Progress in Development Studies* 10(2): 161–168.

Bhat, G.K., Kulshreshtha, V.P., Bhonde, U.A., Rajasekar, U., Karanth, A.K. and Burvey, M.K. (eds) (2012) "Indore city resilience strategy for changing climate scenarios", Indore, India: TARU Leading Edge.

Bicknell, J., Dodman, D. and Satterthwaite, D. (eds) (2009) *Adapting Cities to Climate Change: Understanding and Addressing the Development Challenges*, New York: Earthscan.

Bizikova, L., Robinson, J. and Cohen, S. (2007) "Linking climate change and sustainable development at the local level", *Climate Policy* 7(4): 271–277.

Bowen, K.J. and Ebi, K.I. (2015) "Governing the health risks of climate change: Towards multi-sector responses", *Current Opinion in Environmental Sustainability* 12: 80–85.

Bulkeley, H. (2010) "Cities and the governing of climate change", *Annual Review of Environment and Resources* 35(1): 229–253.

Carmin, J., Anguelovski, I. and Roberts, D. (2012) "Urban climate adaptation in the global South: Planning in an emerging policy domain", *Journal of Planning Education and Research* 32(1): 18–32.

Carmin, J., Dodman, D. and Chu, E. (2013) "Urban climate adaptation and leadership: From conceptual to practical understanding", 2013/26, OECD Regional Development Working Paper, Paris: Organisation for Economic Co-operation and Development (OECD).

Carmin, J., Tierney, K., Chu, E., Hunter, L.M., Roberts, J.T. and Shi, L. (2015) "Adaptation to climate change". In R.E. Dunlap and R.J. Brulle (eds) *Climate Change and Society*, Oxford and New York: Oxford University Press, pp. 164–198.

Cartwright, A., Blignaut, J., De Wit, M., Goldberg, K., Mander, M., O'Donoghue, S. and Roberts, D. (2013) "Economics of climate change adaptation at the local scale under conditions of uncertainty and resource constraints: The case of Durban, South Africa", *Environment and Urbanization* 25(1): 139–156.

Chu, E. (2015) "Urban development and climate adaptation: implications for policymaking and governance in Indian cities". In A. Garland (ed.) *Urban Opportunities: Perspectives on Climate Change, Resilience, Inclusion, and the Informal Economy*, Washington, DC: The Woodrow Wilson Center Press, pp. 6–29.

Chu, E. (2016) "Mobilising adaptation: Community knowledge and urban governance innovations in Indore, India". In M. Roy, S. Cawood, M. Hordijk and M. Hulme (eds) *Urban Poverty and Climate Change: Life in the Slums of Asia, Africa and Latin America*, London and New York: Routledge, pp. 238–254.

Chu, E., Anguelovski, I. and Carmin, J. (2015) "Inclusive approaches to urban climate adaptation planning and implementation in the global South", *Climate Policy* 16(3): 372–392.

Diederichs, N. and Roberts, D. (2016) "Climate protection in mega-event greening: The 2010 FIFA™ World Cup and COP17/CMP7 experiences in Durban, South Africa", *Climate and Development* 8(4): 376–384.

Dodman, D. and Satterthwaite, D. (2009) "Institutional capacity, climate change adaptation and the urban poor", *IDS Bulletin* 39(4): 67–74.

eThekwini Municipality (2015) "Integrated Development Plan 2015/2016", Durban, South Africa.

Field, C.B., Barros, V.R., Dokken, D.J., Mach, K.J. Mastrandrea, M.D., Bilir T.E., Chatterjee, M., Ebi, K.L., Estrada, Y.O., Genova, R.C., Girma, B., Kissel, E.S., Levy, A.N., MacCracken, S., Mastrandrea, P.R. and White, L.L. (2014) *Climate Change 2014: Impacts, Adaptation, and Vulnerability. Contribution of Working Group II to the Fifth Assessment Report of the Intergovernmental Panel on Climate Change*, Cambridge and New York: Cambridge University Press.

Golder Associates (2011) *Community-Based Adaptation to Climate Change in Durban*. Report Number 10290-9743-13, Durban, South Africa: Golder Associates.

Halsnæs, K. and Trærup, S.L.M. (2009) "Development and climate change: A mainstreaming approach for assessing economic, social, and environmental impacts of adaptation measures", *Environmental Management* 43(5): 765–778.

Hunt, A. and Watkiss, P. (2011). "Climate change impacts and adaptation in cities: A review of the literature", *Climatic Change* 104(1): 13–49.

Indore Municipal Corporation (2006) "Indore City Development Plan", Indore, India.

Jerneck, A. and Olsson, L. (2008) "Adaptation and the poor: Development, resilience and transition", *Climate Policy* 8(2): 170–182.

Kernaghan, S. and da Silva, J. (2014) "Initiating and sustaining action: Experiences building resilience to climate change in Asian cities", *Urban Climate* 7: 47–63.

Leichenko, R. (2011) "Climate change and urban resilience", *Current Opinion in Environmental Sustainability* 3(3): 164–168.

Mather, A.A., Garland, G.G. and Stretch, D.D. (2009) "Southern African sea levels: Corrections, influences and trends", *African Journal of Marine Science* 31(2): 145–156.

Municipio de Medellín (2011) "Bio 2030 Plan Director Medellín, Valle de Aburrá", Medellín, Colombia.

Municipio de Medellín (2014) "Plan de Ordenamiento Territorial – POT", Medellín, Colombia.

Paavola, J. (2008) "Science and social justice in the governance of adaptation to climate change", *Environmental Politics* 17(4): 644–659.

Roberts, D., Boon, R., Diederichs, N., Douwes, E., Govender, N., Mcinnes, A., Mclean, C., O'Donoghue, S. and Spires, M. (2012) "Exploring ecosystem-based adaptation in Durban, South Africa: 'Learning-by-standing' at the local government coal face", *Environment and Urbanization* 24(1): 167–195.

Roberts, D. and O'Donoghue, S. (2013) "Urban environmental challenges and climate change action in Durban, South Africa", *Environment and Urbanization* 25(2): 299–319.

Roberts, D., Morgan, D., O'Donoghue, S., Guastella, L., Hlongwa, N. and Price, P. (2016) "Durban, South Africa: Working towards transformative adaptation in an African city". In S. Bartlett and D. Satterthwaite (eds) *Cities in a Finite Planet: Towards Transformative Responses to Climate Change*, London: Routledge.

Schlosberg, D. (2012) "Climate justice and capabilities: A framework for adaptation policy", *Ethics and International Affairs* 26(4): 445–461.

Serrao-Neumann, S., Crick, F., Harman, B., Schuch, G. and Choy, D.L. (2015) "Maximising synergies between disaster risk reduction and climate change adaptation: Potential enablers for improved planning outcomes", *Environmental Science and Policy* 50: 46–61.

Shi, L., Chu, E., Anguelovski, I., Aylett, A., Debats, J., Goh, K., Schenk, T., Seto, K.C., Dodman, D., Roberts, D., Roberts, J.T. and VanDeveer, S.A. (2016) "Roadmap towards justice in urban climate adaptation research", *Nature Climate Change* 6(2): 131–137.

Solecki, W.D., Leichenko, R. and O'Brien, K. (2011) "Climate change adaptation strategies and disaster risk reduction in cities: Connections, contentions, and synergies", *Current Opinion in Environmental Sustainability* 3(3): 135–141.

Sutherland, C., Hordijk, M., Lewis, B., Meyer, C. and Buthelezi, S. (2014) "Water and sanitation provision in eThekwini Municipality: A spatially differentiated approach", *Environment and Urbanization* 26(2): 469–488.

Walsh, C., Roberts, D., Dawson, R., Hall, J., Nickson, A. and Hounsome, R. (2013) "Experiences of integrated assessment of climate impacts, adaptation and mitigation modelling in London and Durban", *Environment and Urbanization* 25(2): 361–380.

14

Social-environmental dilemmas of planning an 'ecological civilisation' in China

Jia-Ching Chen

Introduction

Over the past three decades, China has undergone social and environmental changes of astonishing speed and scale. Rapid industrialisation and urbanisation have contributed to dramatic economic growth and reduction of poverty. However, these successes have been accompanied by sharply increasing social inequality, starkly uneven development and widespread environmental degradation. In response, the party-state has promoted "the construction of an ecological civilisation" as a pillar of its ideology of socialist development, "merging comprehensively with national economic construction, political construction, cultural construction and all aspects of constructing society" (Xinhua News Agency 2012). At the centre of this undertaking is China's national programme of integrated rural transformation and urbanisation. In major policy statements, the Central Committee describes this programme as the simultaneous implementation of policies of urbanisation to stabilise economic growth and improve living standards, and environmental governance to curb pollution and to optimise the use of rural land and natural resources. Together, these policies and plans for 'ecological construction' are aimed at and ideologically justified as unifying efforts to transform and govern the economy, society and environment for sustainable (and nominally 'socialist') development on a coordinated national scale. In this context, spatial planning in contemporary China continues to pursue goals and techniques centred on economic growth and the zoning and regulation of land markets. However, it is distinctive as an emergent mode of territorialisation and as a regime for producing and governing a national spatial structure and nationally scaled environmental values.

These policies are most prominently drawn together in master planned construction of new towns and cities as models for remaking China's development in its pursuit of an ecological civilisation (State Council 2014; Xinhua News Agency 2015). As such model ecological construction projects entail widespread dispossession of village lands, and the eviction and demolition of millions of households each year, the party-state must also constantly justify its normative visions for change. Within the context of China's single-party government structure, I use the term 'party-state' to refer to the combined social, cultural and political apparatuses of the Communist Party of China (CPC) and the government at all administrative levels. In this context, phrases evoking 'the state' in relationship to power or authority must be read as extending beyond the formal government bureaucracy and into the mostly opaque domain of the party, whose leadership is legally enshrined in the constitution. Cartier (2015) highlights the salience of the term 'party-state' in analysing questions of state scale and territoriality in processes of urbanisation and socio-spatial transformation in China. Cartier (2015) argues that the party-state shapes processes of 'territorial urbanisation' as integral to a process of state building across scales. Here, by highlighting the party-state's efforts to optimise national spatial structure and to construct and govern environmental value at a national scale – efforts that I argue are at the centre of the ideological conception of an ecological civilisation – I complement Hoffman's (2011: 55) approach to 'urban modelling' as a "governmental practice that shapes, disciplines, and produces particular kinds of spaces and subjects". In this chapter's discussion of spatial planning within such processes of modelling, this conceptual approach facilitates a combined analysis of the party's authoritative discourse, municipal government and expert planning practices.[1] By analysing model urbanisation within a larger programme of national ecological construction, I argue that the party-state sees urbanisation as a mode of governing larger dynamic processes, and that China's contemporary spatial transformation cannot be fully grasped through analyses that privilege urbanisation (as multi-scalar or isolated place-based processes) per se (cf. Brenner 2013; Scott and Storper 2015).[2]

In this chapter I argue that the party-state's overarching paradigm of ecological civilisation and the contemporary process of ecological construction hinge critically on spatial planning, and analyse how authoritative discourse and expert practices of spatial planning work to define environmental value at a national scale. Utilising an analysis of national policy discourse on urban and rural transformation, this chapter demonstrates that (1) the Chinese party-state is promoting spatial planning as a key to maintaining long-term social stability and political legitimacy; and (2) that planning for spatial optimisation is conceived of in broadly environmental terms, with focus on constructing and governing nationally scaled environmental resource values. In the following sections the chapter outlines the context for the enrolment of eco-urbanisation planning as a key expression of the policy and practice of ecological construction, and then analyses this enrolment in discourse and policy. The chapter then turns to the implementation of this programme of national spatial restructuring, which is taking place in hundreds of master planned eco-urbanisation projects around the country. Planning practice in a model eco-city project in Yixing, Jiangsu Province (see Figure 14.1) is analysed in the context of national policy and

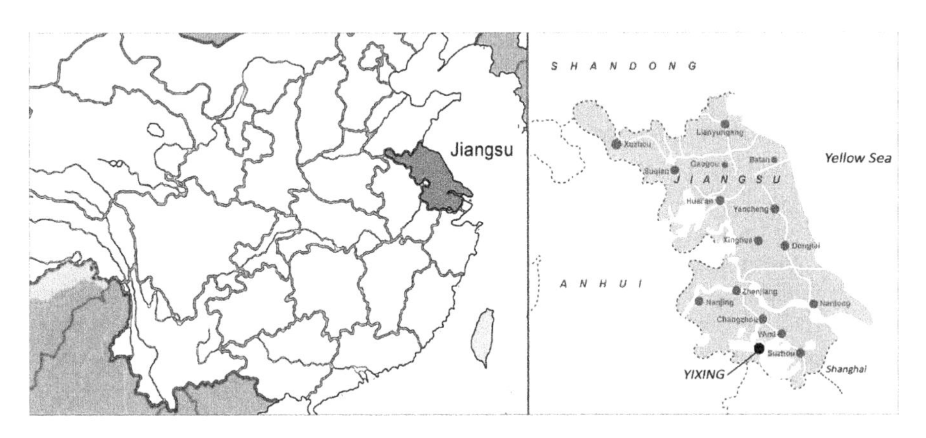

Figure 14.1 Locations of Jiangsu Province and Yixing City

Source: Chen (2012: 83).

authoritative discourse. The chapter argues that, driven by the political goals of the party-state, this model constitutes an ideological conflation of eco-urbanisation with ecological construction and its historically established notions of national development and modernisation. Through the case study, the chapter documents how model eco-urbanisation produces metonymies of national environmental rationality to justify the massive dispossession of rural residents and wholesale transformation of rural landscapes. In conclusion, the chapter examines the implications for planning and argues that these processes and outcomes produce fundamental dilemmas of social-environmental injustice across localities and scales.

The crisis context of China's eco-urbanisation paradigm

China is confronting the social and environmental limits of over three decades of growth-above-all development policy. Industrial pollution in city centres and rural hinterlands has gone under-regulated since the beginning of the reform period, resulting in severe contamination of water supplies in 75 per cent of the rivers and lakes and 90 per cent of urban groundwater supplies (Shapiro 2012: 8). For decades, unregulated industrial effluents and agricultural runoffs have fed massive hazardous blooms of cyanobacteria, leading to frequent shutdowns of factories and municipal water systems in several coastal provinces. Air and water pollution and resulting negative health and economic impacts are increasing causes for social unrest, with tens of thousands of 'mass incidents' occurring each year (Economy 2004; Jing 2000; Stern 2010). This is also a crisis in China's human environmental health. Liu (2010: 8) has counted 459 "cancer villages" documented in official and unofficial reports across 29 of China's 33 provincial-level administrative divisions. In 2007 a joint report by the Ministry of Environmental Protection (MEP) and the World Bank reported that the economic cost of air pollution in premature deaths and health care costs in 2003 was conservatively estimated at 157.3 billion CNY (World Bank and SEPA 2007).[3] In 2013 Chinese officials estimated that outdoor air pollution contributes to up to 500,000 premature deaths every year, and analysis based on the 2010 Global Burden of Diseases Study estimated 1.2 million premature deaths in 2010 alone (Health Effects Institute 2013).[4]

Meanwhile, China's national economy has been slowing down with nominal growth in GDP estimated at 5.3 per cent in 2015 compared to an average of over 10 per cent from 1978 through 2011 (IMF 2016). As a share of GDP, exports are down by more than 10 per cent over the past decade (IMF 2016). This decline presents serious challenges to the local governments that are tasked with generating revenue for infrastructural modernisation and social development targets. Following fiscal recentralisation in 1994 and rural tax reforms in the subsequent decade, local governments have increasingly relied on revenues derived from land and real estate development to pay for an expanding share of social services and infrastructure (Hsing 2010; Lin 2009; Lin *et al.* 2006). The local political economy of land generates 70 per cent or more of local government revenues (Cai 2012; Hsing 2010). Moreover, this has led to the steady enclosure of rural land and the corresponding dispossession of villagers by township and municipal governments (Hsing 2010). Cai (2012) has shown that since constitutional reforms created the leasehold market in the 1980s, the growth of the overall economy has closely tracked the conversion of arable land to other uses. In sum, the industrialisation and urbanisation of rural land has been a fundamental component of China's economic growth over the past three decades, with urban real estate speculation driving a disproportionate share of economic growth (e.g. Hsing 2010). With rural localities holding over 37 per cent of national debt, there is an even greater pressure to utilise urbanisation as a tool for government revenue generation, and ultimately for capital accumulation (Cai 2012).[5]

Local officials are tasked with broadening access to quality housing, transportation and basic physical and social infrastructural services. Moreover, the pressing need for improving rural access to health care and education is further heightened as under-employment and impoverishment are sharpened by the declining economy as well as by rural dispossession for urban modernisation projects. Caught in this pincer of massive rural dispossession and agrarian transition, rural-to-urban migration continues to feed unplanned urban population growth (Chuang 2015). These processes are producing multiple dynamics of social stratification, economic class differentiation and forms of socio-spatial segregation (Chen 2013a; Chuang 2014; Hsing 2010). Land dispossession is the leading cause of social unrest with tens of thousands of protests – amounting to 65 per cent of officially recorded "mass conflicts" each year (He 2010; Yu 2005). Between 1980 and 2006 Yu (2006) estimates that up to 66 million farmers were dispossessed of housing or land. Master planning is largely justified as a correction to these decades of illegal seizures and conversions of village lands for development that proceeded with little central oversight over compensation and resettlement practices.

Spatial planning as foundational development: institutionalising eco-urbanisation in discourse and policy

In the face of these myriad environmental, economic and social challenges, the party-state is pursuing master planned spatial optimisation as the foundational *sine qua non* for development. Discursive practices of modelling eco-urbanisation link spatial planning at local-regional scales to the construction of optimised national spatial structures. Hoffman (2011) highlights urban modelling as a mode of shaping what cities and their inhabitants are supposed to be. This section highlights modelling from another tack to analyse its role in national state building. In particular, this section analyses spatial planning for eco-urbanisation at the local-state level as a mode of governing – and thereby constituting – national environmental value and territorial space.

International media have reported frequently on China's urbanisation push (e.g. Johnson 2013). However, contrary to widespread impressions and dystopian imaginaries of seemingly endless urban agglomeration, the broader process of urbanisation is one of social and spatial differentiation. Rather than homogenisation under a uniform blanket of urban fabric, China's state-led urbanisation is producing a variety of new towns and peri-urban districts in addition to expanding secondary and tertiary cities and its vast metropolitan areas. These phenomena are more accurately conceptualised as master planned urban *and* rural transformation as opposed to a total subsumption of the rural by a singular urban form. Moreover, the urbanisation narrative de-emphasises the broader phenomenon of environmental resource governance and the maintenance of distinct rural spaces and populations (Chen *et al.* 2017).

Spatial planning policy for model eco-urbanisation frequently refers to "urban–rural integration" (e.g. State Council 2014); a concept illustrated by a plethora of slogans that include "without industry, no wealth; without agriculture, no stability" and "industry nourishes agriculture, the city supports the countryside". By combining such myriad goals, these contemporary practices of spatial planning must be analysed as a mode of regulation linking environmental, social and economic planning domains. The ability of contemporary spatial planning to bring such domains together depends upon a generalised rationale of development – a regime of development-truth – and the consolidation of its disciplinary status in this broadened domain.[6] The Major Functional Zones Plan, a key policy encapsulation of this development-truth is subtitled, "the construction of a high efficiency, coordinated, and sustainable structure for national territorial spatial development". The plan details the coordination of spatial planning at all administrative levels and a strategic agenda for technological, informational and expert disciplinary development. This expansive ambit for spatial planning thus extends what is commonly thought of as a practice of making urban places to one of optimising national territorial space as the "institutional materiality of the state" (Poulantzas 1978).

The past decade has seen official discourse place major importance upon 'scientific' spatial planning practices. As the first 'green' national development plan, the Eleventh Five-Year Plan (2006–2010) represented a significant reorientation of China's development goals and principles. Since then, the five-year plans have placed considerable emphasis on state planning authority and practice as foundational to sustainable overall social and economic development (NPC 2006, 2011; Xinhua News Agency 2015). The approach is characterised by a push to plan optimised functional zones of more homogenous and spatially efficient patterns of land use for urban development, industrial and agricultural production and ecological services. This emphasis follows on decades of Malthusian preoccupation with agricultural land supply and food security (Chen 2013b). Rooted in the party-state's overall goal of social stability, much of the environmental policy concern of the past decade addresses questions of the human environment and environmental quality. The National Plan for a New Model of Urbanization (State Council 2014), National Ecological Functional Zones Plan (MEP 2008) and National Major Functional Zones Plan (State Council 2010) are emblematic of this approach. Accordingly, these policies enrol spatial planning into addressing industrial location, pollution, environmental health, social development, economic growth, agricultural development and overall land supply across urban and rural locales.

This explicitly spatial strategy has been institutionalised as the integrated and comprehensive approach to ecological construction and 'sustainable development'. Beginning in the early 2000s, China pioneered efforts in master planned 'eco-city' and 'eco-industry' planning and construction on rural greenfield sites.[7] The regime of model eco-urbanisation rapidly expanded to include systems of standards, pilot programmes, new legislation and the official development policies set forth by the Eleventh and Twelfth Five-Year Plans (2006–2010 and 2011–2015, commonly referred to as the '11-5' and '12-5' plans). The 11-5 Plan set targets for energy intensity (the amount of energy used per unit of GDP) and renewable energy generation, and introduced a vocabulary of environmental governance into China's official development lexicon. Key central government development policies and initiatives by ministry-level agencies have defined master planned eco-urbanisation as including: comprehensive urban–rural spatial planning; the integration of social services, labour and land markets; the promotion of strategic energy and environmental industries; improved physical infrastructure; relocation and elimination of polluting rural industries; concentration of housing for rural populations; professionalisation and scientific industrialisation of agricultural production; preservation and economic development of ecological resources (MOST 2007, 2012; NDRC 2012; NPC 2011).

The Twelfth Five-Year Plan and the Eighteenth National Congress emphasised the "optimization of national spatial structure as the vehicle for the construction of ecological civilization" (Hu 2012). The National Development and Reform Commission, the Ministry of Housing, Urban and Rural Development, the Ministry of Environmental Protection and the Ministry of Science and Technology have produced complementary policies and standards for establishing model eco-city and sustainable development zone projects. According to the Ministry of Environmental Protection over 97 per cent of prefectural-level cities (284 of 293 total) and 80 per cent of county-level cities (288 of 363 total) now have state designated eco-city and low-carbon city projects. Together, this irruption of model eco-urbanisation discourse and policy constitutes a paradigm of Chinese ecological modernisation that centres on scientific–rational spatial planning as the method and the self-referential justification for social-environmental transformation and governance. Ecologically rational space qua sustainable development is a discursive syllogism produced through the evocation of science and state planning authority (e.g., He 2007; State Council 2014).[8]

In 2014, the State Council approved a plan to concentrate 60 per cent of the national population into urban areas by 2020 (State Council 2014).[9] By 2025, China is projected to have 221 cities with over one million inhabitants. The construction of model eco-urbanisation is distinguished within this broader process of state-led urbanisation in that it reflexively (in its official state and technical discourses) addresses the shortcomings of prevalent practices of building urban space through speculative economic bubbles centred on the construction sector itself. While these problems plague eco-urbanisation just the same, it is still important to note that the political, expert scientific and technical work that proceeds under its banner is distinct in arguing for a high degree of state intervention into integrating strategic industrial development planning, rural social transformation through dispossession and the optimisation of land and environmental resources *across all scales of territorial jurisdiction*.

Metonymies of eco-urbanisation: producing locales of national environmental rationality

Yixing is a site of what might be termed 'commonplace greening' in China. Its projects are not internationally known and reported. Rather, it is a site where green development ideas have attained authority in shaping municipal government processes of yoking economic development to spatial planning. It is a place where it has become common sense to join eco-city construction to the pursuit of rapidly expanding global markets in environmental commodities as levers for overall social-environmental transformation. Yixing is not simply a 'model' in the sense of a leading example. Rather, it also provides a paradigmatic standard – now one of the hundreds as mentioned above – a devised display of what is understood to be 'correct'. Hoffman (2011) argues that such models emphasise attainability and replicability. This section examines how Yixing's model of commonplace greening is embedded in the making of ecological civilisation and the particular social–environmental relationships it enacts across urban, regional and national scales.

Like other cities in the Yangzi delta region, Yixing changed rapidly as urbanisation and industrialisation accelerated through the Chinese countryside. Unlike its neighbours, Yixing has been an object of intense environmental policy attention and has received numerous citations from the central government for its work under official rubrics for ecological protection and sustainable development (see Table 14.1). In 1993, the Yixing Industrial Park for Environmental Science and Technology was designated as the research and development centre for China's Rio Declaration Agenda 21 Program for environmental protection. By 1998, Yixing generated 18 per cent of the national total value added in the environmental industry (Zhang 2002: 62). In 2006, the National Yixing Economic and Technological Development Zone (hereafter, 'the Zone') was established. The Zone was created to focus on 'greentech' and 'cleantech' industries, especially solar photovoltaics and optoelectronics, to support – and capitalise on – national targets for the production of commodities in these strategic sectors. These green industry projects formed the economic motor for what is conceived of in Yixing's master plan as a larger project of eco-urbanisation and rural resource integration. This regional vision was prominently articulated by a national Party Central Committee member, Li Yuanchao, who extolled Yixing as a national model "eco-city" and called for its integration of rural areas extending from the western shore of Taihu Lake across Yixing's chain of lakes (see Figure 14.2). In 2008, to achieve this vision as a model for simultaneous industrial upgrading, urbanisation, environmental protection and rural development, the Zone's planning authority was extended to 98.3 square kilometres. Its designation as a National Sustainable

Development Experimental Zone in 2009 recognised its strategic incorporation of projects of industrial development in key industries (especially environmental protection industries, solar energy, batteries for electric vehicles, and optical electronics), "high standard" agricultural land, ecological set asides, and village land consolidation for conversion to other uses. Now with projects occupying over 330 square kilometres of rural land dispossessed from hundreds of villages, Yixing's model development lies explicitly at the intersection of national land management policies and the urban–rural transformation processes of ecological construction at the local level (see Table 14.2).

Yixing's eco-urbanisation projects are predicated on large-scale enclosures of village land for direct land use allocation under its municipal authority. Dispossessed rural land has formed the basis for municipal development finance of eco-urbanisation including free facilities as incentives for strategic green industries. The Yixing eco-city master plan's "conservative

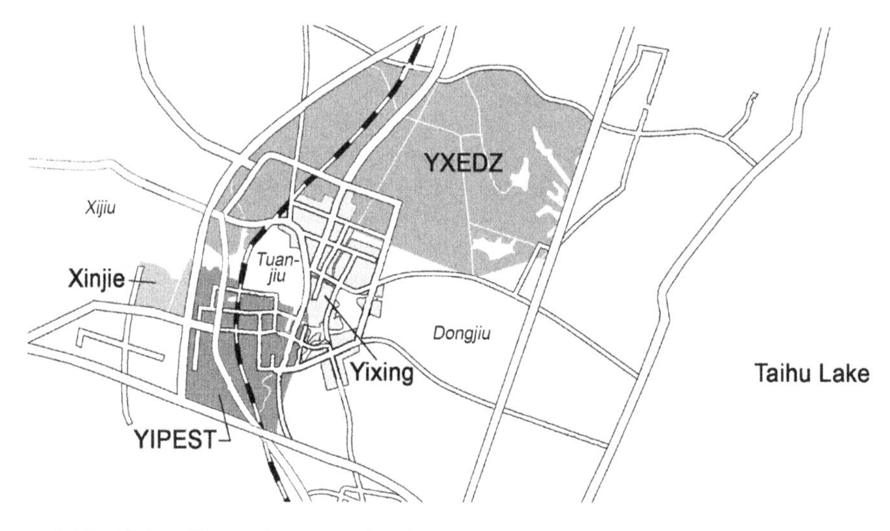

Figure 14.2 Yixing City major green development projects

Source: Chen (2012: 84).

Note: This map shows the central area of the Yixing City-region with its major green development projects. YXEDZ, YIPEST and the area of Xinjie annexed by YIPEST are all shaded in darker grey.

Table 14.1 State designations for model green development in Yixing

Year	*Designation*	*Agency*
2005	National garden city	Ministry of Urban–Rural Development
2006	National environmental protection model city	State Environmental Protection Agency*
2006	State ecological model zone	State Environmental Protection Agency*
2008	Jiangsu provincial sustainable development experimental zone	Jiangsu Department of Science and Technology
2009	National model experimental sustainable development zone	Ministry of Science and Technology
2011	National (state) ecological city	Ministry of Environmental Protection

Source: author.

*The State Environmental Protection Agency was restructured as the Ministry of Environmental Protection in 2008.

Table 14.2 Yixing City environmentalised land enclosures

Name	*Year*	*Project type/justification*	*Area (km²)*
National Yixing Industrial Park for Environmental Science and Technology (YIPEST)	1992	National R&D, Economic Development Zone	4
YIPEST	1993	Park expansion	11
Jiangsu Yixing Economic Development Zone (YXEDZ)	2006	Provincial economic development zone; national solar and new energy development land base	54
Jiangsu Yixing Economic Development Zone (YXEDZ)	2006	Integrated green urban and industrial development; ecological preservation	30
Taihu Greenbelt	2007	Ecological preservation	133
Scientific Innovation New City (under YXEDZ authority)	2009	Eco-city with 'green solar valley' R&D and manufacturing base	22
Environmental Science and Technology New City (under YIPEST planning authority)	2011	Eco-city and park-district integration; environmental industry R&D	87
Gaocheng New Town (under YIPEST planning authority)	2012	'World-class city' urban–rural integration of the YIPEST eco-industrial zone	110
Total			**331**

Source: author.

and intensive land use" entails the planned dispossession and resettlement of 100,000 village residents between 2010 and 2020 (Yixing BLR 2010).[10] As a consequence, Yixing's model eco-urbanisation extends state practices of rural dispossession and reconstitutes patterns of urban-rural inequality (Chen 2013a). As ecological construction has become unequivocally linked to spatial restructuring, the discourse and practices of planning represent predominant patterns of rural society, economy and landscapes as environmentally irrational.

Under the rubric of urban–rural integration, local governments are expanding their direct planning authority over the rural villages within their jurisdictional boundaries. This rural land is held directly by village collectives. Because law stipulates that the state conducts direct allocation of land for non-agricultural purposes only in urban areas at the county level and higher, the Yixing government has reorganised and eliminated rural town, township and village administrations in order to facilitate the state-rationalisation of rural land under its master plan (e.g. Jiangsu BLR 2006). Since 2006, these activities have increased pace and scope to integrate an expanding array of rural land uses into municipal planning. In urban–rural integration planning, a major objective is the consolidation of village land resources for centralised reallocation for urban development, higher intensity agriculture, and environmental governance. However, the conversion of agricultural land to other uses is accounted for and regulated through a national quota system (Cai 2012). In this regard, rural land is constituted as national space and through spatial planning as the infrastructural foundation for ecological construction. In contrast to earlier forms of practice, where planning was limited to state-owned land in urban districts, ecological construction endeavours to encompass all rural land in the cohesion of a national state scaled environmental space and territory.

The implications for planning

As spatial planning for ecological construction increases its efforts at totality, it blunts any analytical purchase in narrative distinctions of top-down versus bottom-up drivers of socio-spatial transformation in China. Despite its historical singularity, Yixing yielded practical models for its neighbours and instantiations of the party-state's ideological conception of ecological civilisation. For the former, the enclosure of rural land as a source of financing and incentives for the production environmental commodities like solar panels became so widespread that the global solar industry experienced a three-year oversupply crisis, with hundreds of millions of dollars lost to bankruptcies and millions of tonnes of CO_2-equivalent benefit lost in dismantled production facilities. In terms of the latter, by centring so much on governing and optimising rural land supply, practices of ecological construction makes the *unit area* of rural land into a fungible environmental value and spatial planning into a simultaneous practice of speculative place-making, state territorialisation and intense socio-spatial transformation. These processes must be seen in a relational context, simultaneously in the locale of Yixing, the political reality of the party-state, and an imaginary national space that is produced and represented in the spreadsheets of land managers rather than in official maps.

Acting upon space and the environment through planning is in many ways seen as a basic state function (e.g., Holston 1989; Rabinow 1989; Scott 1998). Graham and Marvin (2001: 43) describe the modernist "infrastructural ideal" as a unitary and total (nation-scaled) network producing "coherent urban relations" among the urban totality of people, institutions, places and the environment. However, with such a broad ambit of ideological justification and accelerated relevance, spatial planning for ecological construction is bound to fail, to continue to create unevenness and dialectics of conflict and social and environmental devaluation, with fundamental dilemmas of social-environmental injustice across localities and scales.

By insisting that environmental value can and should be planned, the party-state constructs environmental value as something that is inherently tied to landscapes, but is also paradoxically fungible and divisible from any particular place and the people who might inhabit them. In his discussion of the construction of biodiversity as an object of environmental governance, Bowker (2005) introduces the concept of "coinage" to highlight the erasures necessary in capital and aggregate environmental metrics alike. A totalising logic of planning that is justified by a universalised notion of environmental value and utility harkens Foucault's (2008: 259–271) brief discussion of a mode of "governmentality which will act on the environment and systematically modify its variables" as the basis for "the generalization of the 'enterprise form in the social field". The 'environmentality' of ecological civilisation, then, seems to depart in important ways from previous regimes of rural development, yielding a process of subjectivation as an epiphenomenon rather than as a central goal. This paradox of planned social-environmental value is a clear illustration of what Timothy Morton (2016) describes as the "Mobius strip" of the Human/Nature dichotomy; a spiralling "retroactive fantasy construct" (Morton 2007) that seeks to resolve previous modernist failures with ever broader efforts at totality.[11] As the social-environmental fabric of ecological civilisation takes shape through increasingly abstracted planning, it is not clear how its long-term effects will unfold, or how any potential counter-movement in policy or social resistance might take form. However, critical, reflexive and socially engaged practices will undoubtedly be needed to assert alternative forms of scientific, sustainable and ecological construction.

Notes

1 While it is fair to characterise this chapter as a 'state-centred' approach to analysing planning in China (where many non-state and transnational actors exist), it also does not presume that such empirical objects are mere units within a single monolithic entity. Rather, the chapter highlights planning as one domain of the party-state's ongoing construction of ideology and practice. In this regard, the use of 'party-state' also refers to the historical debate over the relationship between the vanguard party and the government apparatus in theory and in practice under one-party state of Bolshevik rule and subsequent

Marxist-Leninist regimes, including the CPC and People's Republic of China (e.g. James and Grimshaw 1992; LeBlanc 2014).

2 In the comparative context of this volume, the analysis of processes of planning as extending beyond the confines of government bureaucracies might serve to highlight the limits of conventional distinctions of state-society relationships, while putting the state 'in its place'. The practices of modelling discussed here as a domain of politics are also comparable with cases of 'world-class' and eco-city building elsewhere.

3 Approximately USD25.6 billion at current exchange rates. The State Environmental Protection Administration (SEPA) was expanded and strengthened as the Ministry of Environmental Protection in 2008.

4 The Global Burden of Diseases Study (GBD) is published by *The Lancet* (available at: www.thelancet.com/global-burden-of-disease). Chen Zhu, former Minister of the Environment, and members of the Chinese Academy for Environmental Planning reported the 2010 annual estimate in a *Lancet* commentary on the GBD (Chen *et al.* 2013).

5 In an update to its series on national debts, McKinsey estimates that nearly half of China's government, household and non-financial corporation debt is in real estate. At approximately 9.5 trillion dollars, this debt is nearly equivalent to China's national GDP in 2015 (MGI 2015).

6 This conceptualisation draws upon the work of Foucault and Latour. In *Truth and Power*, Foucault (1980 [1977]: 131–133) discusses "regimes of truth" as a "general politics" or "political economy" that in modern society consist of the modes through which truth can be discerned, produced and made to function; and that are centred on scientific discourse, mass circulation and production through politically and economically elite institutions; and that engender political incitement and social and political struggles. In the essay "Visualization and Cognition", Latour (1986: 5) examines practices of representation and the agonistic contexts in which they "make a difference".

7 Among such efforts was the internationally lauded plan for Dongtan Eco-City by Arup, the most renowned engineering firm in the world, for the Shanghai government. Although the plan failed to be implemented, Arup has continued to utilise the project as a means of marketing its eco-city expertise.

8 Such processes of making new domains of scientific expertise and political rationales are not new in China. In her definitive account the so-called One Child Policy, Greenhalgh (2008) reveals the evolution of state concerns over agricultural capacity into a biopolitical governmentality that ruled human reproduction through mathematical models and village level cultural transformations.

9 In the National Plan for a New Model of Urbanization (State Council 2014), the approximate 60 per cent figure refers to permanent residence in towns and cities. The urban household registration rate is projected at 45 per cent.

10 This figure is over 23 per cent of the 2010 registered village population, which does not account for migrant residents with household registration status in other locales.

11 Morton's image of the Mobius strip is also resonant with Rosalind Williams' (2008: 1) description of environment and technology as forming "not a dichotomy but a continuum".

References

Bowker, G. (2005) "Time, money, and biodiversity". In A. Ong and S. Collier (eds) *Global Assemblages: Technology, Politics, and Ethics as Anthropological Problems*, Malden MA: Blackwell, pp. 107–123.

Brenner, N. (2013) "Theses on urbanization", *Public Culture* 25(1): 85–114.

Cai, M. (2012) "Land-locked development: The local political economy of institutional change in China", PhD Dissertation, Political Science, University of Wisconsin.

Cartier, C. (2015) "Territorial urbanization and the party-state in China", *Territory, Politics, Governance* 2(3): 294–320.

Chen, J.-C. 2012. "Greening dispossession: Environmental governance and sociospatial transformation in Yixing, China". In T. Samara, S. He and G. Chen (eds) *Locating Right to the City in the Global South*, New York: Routledge, pp. 81–104.

Chen, J.-C. (2013a) "Greening dispossession: Environmental governance and sociospatial transformation in Yixing, China". In T. Samara, S. He and G. Chen (eds) *Locating Right to the City in the Global South*, New York: Routledge, pp. 81–104.

Chen, J.-C. (2013b) "Sustainable territories: Rural dispossession, land enclosures and the construction of environmental resources in China", *Human Geography* 6(1): 102–118.

Chen, J.-C., Zinda, J.A. and Yeh, E.T. (2017) "Recasting the rural: State, society and environment in contemporary China", *Geoforum* 78: 83–88.

Chen, Z., Wang, J.N., Ma, G.-X. and Zhang, Y.-S. (2013) "China tackles the health effects of air pollution", *The Lancet* 382(9909): 1959–1960.

Chuang, J. (2014) "China's rural land politics: Bureaucratic absorption and the muting of rightful resistance", *The China Quarterly* 219: 649–669.

Chuang, J. (2015) "Urbanization through dispossession: Survival and stratification in China's new townships", *Journal of Peasant Studies* 42(2): 275–294.

Economy, E. (2004) *The River Runs Black: The Environmental Challenge to China's Future*, Ithaca, NY: Cornell University Press.

Foucault, M. (1980) "Truth and power". In C. Gordon (ed.) *Power/Knowledge: Selected Interviews and Other Writings, 1972–1977*, New York: Pantheon, pp. 109–133.

Foucault, M. (2008) *The Birth of Biopolitics: Lectures at the Collège de France 1978–1979*, translated by Graham Burchell, New York: Palgrave Macmillan.

Graham, S. and Marvin, S. (2001) *Splintering Urbanism: Networked Infrastructures, Technological Mobilities and the Urban Condition*, New York: Routledge.

Greenhalgh, S. (2008) *Just One Child: Science and Policy in Deng's China*, Berkeley, CA: University of California Press.

He, C.Q. (2007) *China Modernization Report 2007: Ecological Modernization Research* [Chinese], Beijing: Beijing University Press.

He, D. (2010) "Land battles most dire rural issue: Report", *China Daily*, 16 December, Available: www.chinadaily.com.cn/china/2010-12/16/content_11709564.htm [accessed 16 May 2016].

Health Effects Institute (2013) "Ambient air pollution among top global health risks in 2010: Risks especially high in China and other developing countries of Asia (Summary of methods)". Available: www.healtheffects.org/International/HEI-GBD-MethodsSummary-033113.pdf [accessed 16 May 2016].

Hoffman, L. (2011) "Urban modeling and contemporary technologies of city-building in China: The production of regimes of green urbanisms". In A. Roy and A. Ong (eds) *Worlding Cities*, Malden, MA: Blackwell, pp. 55–76.

Holston, J. (1989) *The Modernist City: An Anthropological Critique of Brasilia*, Chicago, IL: University of Chicago Press.

Hsing, Y. (2010) *The Great Urban Transformation: Politics of Land and Property in China*, Oxford: Oxford University Press.

Hu, J. (2012) "Full text of Hu Jintao's Report to the 18th National Congress of the Communist Party of China", Xinhua News Agency. Available: http://news.xinhuanet.com/18cpcnc/2012-11/17/c_113711665.htm [accessed 16 May 2016].

International Monetary Fund (IMF) (2016) *World Economic Outlook Database*, International Monetary Fund, April 2016 edition. Available: www.imf.org/external/pubs/ft/weo/2016/01/weodata/index.aspx [accessed 10 June 2016].

James, C.L.R. and Grimshaw, A. (1992) *The C.L.R. James Reader*, Oxford: Blackwell.

Jiangsu Bureau of Land and Resources (BLR) (2006) "Yixing City 2006 urban construction land use levies [various releases]", Nanjing: Jiangsu Bureau of Land and Resources.

Jing, J. (2000) "Environmental protests in rural China". In E.J. Perry and M. Selden (eds) *Chinese Society: Chang, Conflict and Resistance*, New York: Routledge, pp. 204–222.

Johnson, I. (2013) "China's great uprooting: Moving 250 million into cities", *The New York Times*, 16 June. Available: www.nytimes.com/2013/06/16/world/asia/chinas-great-uprooting-moving-250-million-into-cities.html [accessed 16 May 2016].

Latour, B. (1986) "Visualisation and cognition: Thinking with eyes and hands". In H. Kuklick (ed.) *Knowledge and Society: Studies in the Sociology of Culture Past and Present Vol. 6*, Greenwich: JAI Press, pp. 1–40.

LeBlanc, P. (2014) *Marx, Lenin, and the Revolutionary Experience: Studies of Communism and Radicalism in an Age of Globalization*, New York: Routledge.

Lin, G.C.S. (2009) *Developing China: Land, Politics and Social Conditions*, New York: Routledge.

Lin, J.Y., Tao, R. and Liu, M. (2006) "Decentralization and local governance in China's economic transition". In P. Bardhan and D. Mookherjee (eds) *Decentralization and Local Governance in Developing Countries: A Comparative Perspective*, Cambridge, MA: MIT Press, pp. 305–327.

Liu, L. (2010) "Made in China: Cancer villages", *Environment: Science and Policy for Sustainable Development* 52 (2):8–21.

McKinsey Global Institute (MGI) (2015) "Debt and (not much) deleveraging". Available: www.mckinsey.com/global-themes/employment-and-growth/debt-and-not-much-deleveraging [accessed 16 May 2016].

Ministry of Environmental Protection (MEP) (2008) "National ecological functional zones plan". Available: www.mep.gov.cn/gkml/hbb/bgg/200910/W020080801436237505174.pdf [accessed 5 May 2015].
Ministry of Science and Technology (MOST) (2007) *National Sustainable Development Experimental Zone Management Measures*, Beijing: MOST.
Ministry of Science and Technology (MOST) (2012) 863 Program "Large scale grid-connected solar photovoltaic systems design integration technology research and equipment development", Major Project Opening Held in Beijing, Beijing: MOST. Available: www.most.gov.cn/kjbgz/201112/t20111215_91447.htm [accessed 8 June 2014].
Morton, T. (2007) *Ecology Without Nature: Rethinking Environmental Aesthetics*, Cambridge, MA: Harvard University Press.
Morton, T. (2016) *Dark Ecology*, New York: Columbia University Press.
National Development and Reform Commission (NDRC) (2012) *Twelfth Five-Year Plan Energy Efficiency and Environmental Protection Industrial Development Plan (2012–2020)*, Beijing: National Development and Reform Commission, PRC.
National People's Congress (NPC) (2006) *The Eleventh Five-Year Plan for Economic and Social Development of the People's Republic of China*, Beijing: National People's Congress, PRC.
National People's Congress (NPC) (2011) *The Twelfth Five-Year Plan for Economic and Social Development of the People's Republic of China*, Beijing: NPC, PROC.
Poulantzas, N. (1978) *State, Power, Socialism*, London: New Left Books.
Rabinow, P. (1989) *French Modern: Norms and Forms of the Social Environment*, Cambridge: MIT Press.
Scott, A.J. and Storper, M. (2015) "The nature of cities: The scope and limits of urban theory", *International Journal of Urban and Regional Research* 39(1):1–15.
Scott, J.C. (1998) *Seeing Like a State: How Certain Schemes to Improve the Human Condition Have Failed*, New Haven, CT: Yale University Press.
Shapiro, J. (2012) *China's Environmental Challenges*, Malden, MA: Polity.
State Council (2010) "The national major functional zones plan", Beijing: Information Office of the State Council of the People's Republic of China. Available: www.gov.cn/zwgk/2011-06/08/content_1879180.htm [accessed 9 September 2012].
State Council (2014) "National plan for a new model of urbanization (2014–2020)", Xinhua News Agency. Available: www.gov.cn/zhengce/2014-03/16/content_2640075.htm [accessed 16 March 2016].
Stern, R.E. (2010) "On the frontlines: Making decisions in Chinese civil environmental lawsuits", *Law and Policy* 32(1):79–103.
Williams, R. (2008) *Notes on the Underground: An Essay on Technology, Society, and the Imagination*, Cambridge, MA: MIT Press.
World Bank and State Environmental Protection Administration (SEPA) (2007) *Cost of Pollution in China: Economic Estimates of Physical Damages*, edited by Natural Resources and Environment Management Unit Rural Development, East Asia and Pacific Region, Washington, DC: The World Bank.
Xinhua News Agency (2012) "Full text of Hu Jintao's report to the 18th National Congress of the Communist Party of China". Available: http://news.xinhuanet.com/18cpcnc/2012-11/17/c_113711665.htm [accessed 22 February 2014].
Xinhua News Agency (2015) "Central Committee of the CCP Recommendations for the Thirteenth Five-Year Plan for National Economic and Social Development". Available: http://news.xinhuanet.com/fortune/2015-11/03/c_1117027676.htm [accessed 12 December 2015].
Yixing BLR (Bureau of Land and Resources) (2010) *2006–2020 Yixing Land Use Master Plan* [Chinese]. Yixing: Yixing People's Government.
Yu, J. (2005) "Land problems have already become the focus for farmers' protests to uphold their rights: An investigation of the current situation in China's rural communities", *The World of Survey and Research* 18 (3): 22–23.
Yu, J. (2006) "Urbanization in China and the protection of peasant land rights", speech given at the 21st-Century Strategic Development of China Urbanization Forum, Beijing, 17–19 February. Available: www.china.com.cn/chinese/jingji/1127514.htm [accessed 20 December 2010].
Zhang, L. (2002) "Ecologizing industrialization in Chinese small towns", PhD dissertation, Environmental Sociology, Wageningen University.

15

Open space provision and environmental preservation strategies

A case study in Brazil

Mônica A. Haddad

Introduction[1]

A majority of cities across the globe are confronting threats to their natural environment because of worrying trends in indicators such as urban population, PM2.5 air pollution, and CO2 emissions (World Bank 2016). Consequently, several strategies are being used in urban areas to minimise the negative impact of urbanisation on the natural environment. Examples of such strategies are increasing provision of open space and preserving areas that are characterised by high environmental importance. These issues may be less of a problem in public sector practice in the global North compared to the global South, because its public institutions generally have more human capital, information and resources to deal with them. For instance, public employees in the global North have better access to accurate and timely data (Arsanjani *et al.* 2016; Musakwa and van Niekerk 2015), which produces more effective policies.

Few studies indirectly address differences in public sector practice between the global North and the global South. The first difference, pointed out by Yazdani and Dola (2013), concerns sustainability. They advocate that sustainability should be addressed in ways that are relevant to the context of the global South, as the concept and practices used in the global North do not necessarily fit the Southern reality. Information pertaining to sustainable practices are widespread in the global North and lacking in the global South (Yazdani and Dola 2013: 39). The second difference, pointed out by Miraftab (2009), is about relationships between citizens and governments. She argues that in the North, neoliberalism has a negative effect on civil liberties and the public sphere, but the effect of neoliberalism is more damaging in the South. Citizens' rights in the South have expanded, but their ability to participate in the public sector has decreased (Miraftab 2009).

In this chapter I present a case study of a region, located in the global South, to investigate the performance of the public sector regarding open space provision and environmental preservation issues. I am specifically interested in understanding public sector practice, by analysing the spatial relationship between urban growth and environment-related planning tools.

This chapter has two main objectives: 1) to describe the spatial patterns of urbanisation in a region of Brazil from 2000 to 2015; and 2) to examine the current use of two urban planning tools related to open space provision and environmental preservation in this region.

Based on my results, I argue that the private and public sectors in Brazil are equipped with a variety of urban planning tools, but there is an urgent need to improve their implementation. Towards that end, the most important issue that needs to be addressed is the role of political influence in public sector practice. I offer recommendations to prioritise 'combating administrative corruption', invest in the automation of land and property databases, educate citizens about sustainable practices, and expand electronic-participatory budget.

Urban planning tools

When compared to other Latin American countries, Brazil should be ahead of the game of promoting urban planning practices because of its recognised success of enacting the 2001 urban planning law, Estatuto da Cidade (henceforth City Statute). This law, internationally acclaimed, has its basis on the 1988 Federal Constitution, and includes many urban planning tools to help shape urban areas. In theory, with this law, municipalities would have more power over the urbanisation process and would be able to manage land use more effectively (Rolnik and Klink 2011: 104). Additionally, based on the law, municipalities would promote public participation during urban planning processes. Participatory budgeting, even though not mandatory in the law, is an example of public participation being used by many Brazilian municipalities interested in making the democratic process stronger.

In this chapter, I examine the current use of two City Statute tools related to open space provision and environmental preservation: the right to preferential purchase (direito de preempção) and land use and zoning regulations. The right to preferential purchase (RPP) gives priority to municipal governments to buy properties that are in the real estate market, being sold by private owners. According to the City Statute, RPP should be included in municipal comprehensive plans, where RPP's uses and its period of validity should be determined. There is, however, a dichotomy in the literature concerning the public financial health of Brazilian municipalities, which directly relates to the ability to use RPP. Hübner *et al.* (2006) argue that there is a lack of RPP implementation in cities, which may be related to weak public financial health. On the other hand, Fernandes and Maldonado Copello (2009: 18) state that some municipalities are generating "impressive financial resources as a result of urban operations [derived from the City Statute]", which should allow municipalities to use RPP with these funds. Because of this dichotomy, and the lack of comprehensive studies in the literature focusing on RPP, in this chapter I carefully examine the use of this tool.

The City Statute lists several urban uses for which the public sector should have priority to buy private properties, in case there is a need for land acquisition for a public cause such as historical preservation (Presidência da República do Brasil 2001). Other listed uses of RPP are related to the natural environment. As Prieto (2006) points out, RPP could become a powerful tool to improve the quality of the environment for all citizens. This chapter focuses on three types of RPP's uses, included in the City Statute: land for development of open space, land for development of conservation areas that have high environmental importance, and land for installing environmental conservation units.

The second urban planning tool is land use and zoning regulations. Even though land use and zoning regulations are commonly used in urban planning practices in countries from the global

North, it only started to be more present in Brazilian practices after 2001, with the enactment of the City Statute. In fact, the time lag is quite shocking as the 1915–1930 period was characterised as "the technical age of zoning ordinances" for cities in the global North, according to Silver (2014: 104).

With the enactment of the City Statute, comprehensive plans became mandatory to all municipalities with 20,000 inhabitants or more. According to IBGE (2016) in 2005, 14.5 per cent of all municipalities with 20,000 inhabitants or more had comprehensive plans, and in 2015, this percentage increased to 89.2. Concerning land use and zoning regulations, in 2009, 38.3 per cent of municipalities had zoning regulations (IBGE 2010), and in 2015, this amount increased to 58.6 per cent (IBGE 2016). These statistics are all welcome news for urban planning practices, especially given the short time frame, but there is still much progress to be made concerning implementation.

The study area

The study area, depicted in Figure 15.1, is a sub-region of the Belo Horizonte Metropolitan Region (BHMR), located in the State of Minas Gerais, in Brazil; its total area is 1,405 km^2. BHMR is composed of 34 municipalities, and its core is the municipality of Belo Horizonte (BH). The sub-region is composed of eight municipalities (BH, Confins, Lagoa Santa, Pedro Leopoldo, São Jose da Lapa, Ribeirão das Neves, Santa Luzia and Vespasiano) that are part of the regional 'North' urban expansion axis. The study area encompasses 64 per cent of the 2010 total regional population, which was 4.8 million inhabitants. Of the top seven BHMR densest municipalities in 2010, four were in the study area.

This sub-region was selected because of its urban history and its continuous urban growth. Interestingly, the 'North' urban expansion axis started in the 1970s as bedroom communities for low-income people who were working in BH but could not afford housing there (de Andrade and de Medonça 2010: 170). Nowadays, as Costa and Magalhães (2011) describe, there are several state initiatives located in the 'North' axis, all contributing to BHMR urban expansion. Additionally, the presence of two state highways (i.e., MG 010 and MG 424) in the sub-region explain its attractiveness for growth.

Spatial analysis

In order to examine spatial patterns of urbanisation in the study area, land cover maps for 2000 and 2015 were created based on remotely sensed data. Two main sets of analysis were carried out: Land Use Land Cover (LULC) change and Normalised Difference Vegetation Index (NDVI). For LULC, first I detected the area under 'urban' and 'non-urban' land cover classes for 2000 and 2015. I identified areas where 'new development' occurred between 2000 and 2015, and results of LULC are shown in Figure 15.2: (A) represents the 2000 classified land cover, (B) represents the 2015 classified land cover, and (C) represents the 'new development' that occurred between 2000 and 2015. The total urbanised area was approximately 100 km^2, showing an increase of 7.2 per cent of urbanised area when compared to the whole study area, and a 23 per cent increase when compared to the existing urbanised area. Figure 15.2 also shows that BH is the most urbanised municipality in both years, having only a few vacant areas, most of which are located in the mountains that surround its south-west border. These mountains are preserved from any construction, designated under the environmental preservation category in the municipal zoning regulation.

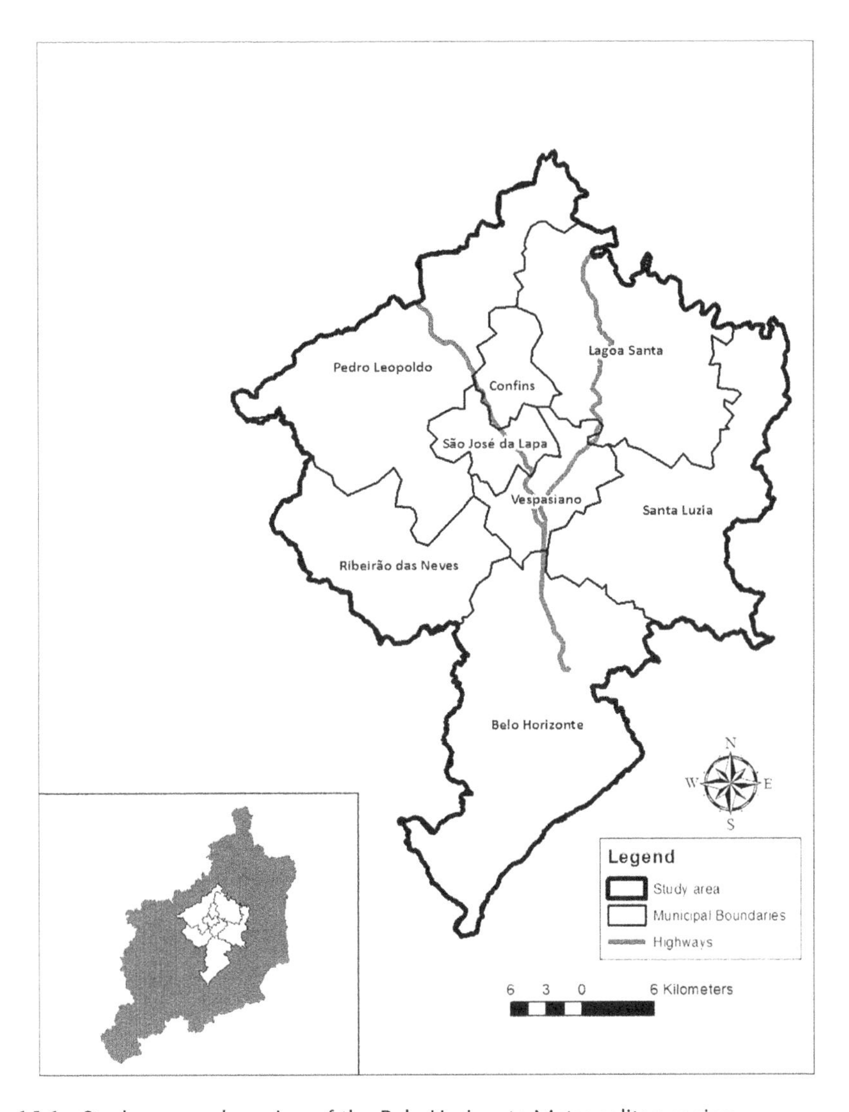

Figure 15.1 Study area: sub-region of the Belo Horizonte Metropolitan region

Table 15.1 summarises how 'new development' happened in each of the eight municipalities. In absolute terms, from the total 100 km^2, BH had the largest area of 'new development' (18.74 km^2). BH also experienced the highest level of urbanisation, in relative terms: around 25 per cent of its total vacant land in 2000 was urbanised by 2015. In addition to 'new development', it is important to highlight that BH was also experiencing redevelopment in the form of verticalisation, even though the spatial analysis presented here does not capture that. By demolishing old houses and/or three-floor apartment buildings, developers were constructing high-rise buildings in BH. The municipality of Ribeirão das Neves also witnessed high urban growth, about 15 per cent of its total vacant land in 2000 was urbanised by 2015.

Finally, to better understand the urbanisation process, I evaluated changes in the vegetation health from 2000–2015. I used the most common vegetation index – the Normalised Difference

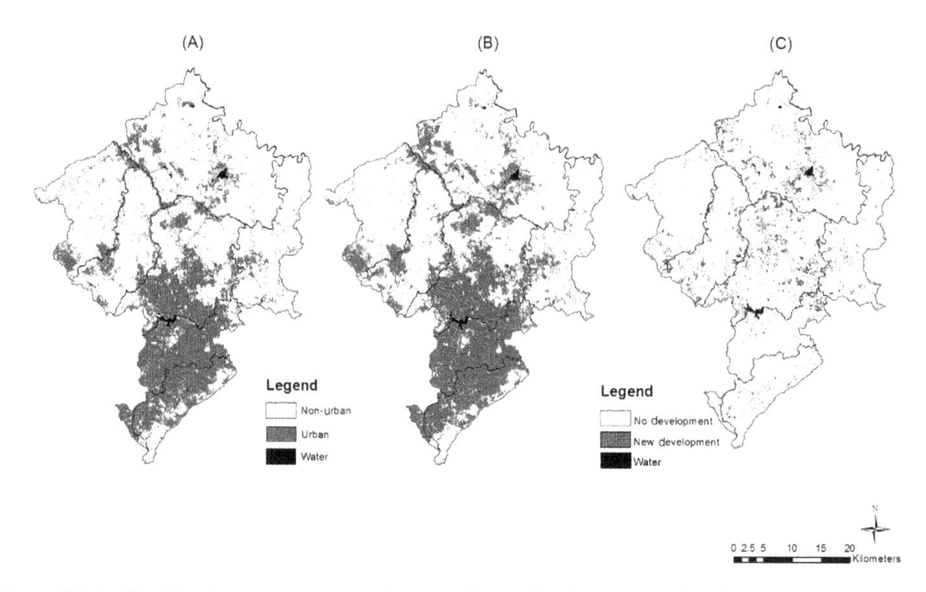

Figure 15.2 Final land cover maps, and map of new development identified

Vegetation Index (NDVI) – which is calculated from red and near-IR reflectance data. NDVI values can range from +1 to −1, with healthy green vegetation having values closer to +1 and unhealthy vegetation and non-vegetated features having values nearer −1.

Using Holben's (1986) study as a reference for NDVI classification categories, NDVI values ranging from −1 to 0.1 were included in the non-vegetated category, NVDI values ranging from 0.1 to 0.3 were included in the moderate vegetation category, and NDVI values above 0.3 were included in the high vegetation category. Table 15.2 compares the results of these three categories.

Table 15.1 Summary of new development among municipalities

Municipality	*New development (km^2) 2000–2015*	*Municipal total area (km^2)*	*Vacant land (km^2) 2000*	*Percentage of vacant land that was urbanised from 2000–2015*
Belo Horizonte	18.74	330.93	73.76	25.41
Confins	4.04	42.22	34.78	11.62
Lagoa Santa	15.78	230.58	199.19	7.92
Pedro Leopoldo	14.77	291.49	251.33	5.88
Ribeirão das Neves	15.9	154.58	105.31	15.10
Santa Luzia	14.27	233.63	193.66	7.37
São José da Lapa	6.51	48.62	36.22	17.97
Vespasiano	9.66	69.78	50.65	19.07

Table 15.2 Summary of NDVI classification categories

NDVI categories	*NDVI value*	*2000 area km^2*	*2015 area km^2*	*Percentage change*
Non-vegetated	0.1 or less	403	514	27.54
Moderate vegetation	0.1–0.3	567	615	8.47
High vegetation	0.3 or high	435	276	-36.55

One can observe that there was a decrease in 'high vegetation' of around 36 per cent, and the non-vegetated category increased by approximately 27 per cent. Both statistics indicate that the vegetation health was getting worse during the study period.

Examining urban planning tools

Right of preferential purchase (RPP)

A content analysis of municipal comprehensive plans was conducted to understand how RPP was being used by the municipal governments that encompass the study area. This research process took place between August and September 2016. I was looking for answers for the following questions: Is RPP included in the comprehensive plan? Is there a broad list of RPP's uses present in the plan? Are RPP's uses copied verbatim from the City Statute? Are RPP's procedures copied verbatim from the City Statute? Is there a specific list of RPP's uses present in the plan? Is the City Statute included as the primary regulatory law? Is there any additional specific law for RPP mentioned in the plan? Table 15.3 summarises the findings of the content analysis.

Seven out of the eight comprehensive plans included RPP in their scope.[2] Only the municipality of São José da Lapa did not mention RPP in its comprehensive plan. Lagoa Santa included RPP in its comprehensive plan, but did not mention the City Statute as a regulatory law for the tool, nor did it describe the broad list of RPP's uses. On one hand, Belo Horizonte and Vespasiano included RPP in their plans and referred to the City Statute as the primary and only law that could regulate the tool. On the other hand, although Confins, Pedro Leopoldo, Ribeirão das Neves and Santa Luzia refer to the City Statute as the primary law that could regulate the plan, they all require that the RPP be the object of a specific municipal law. In the Ribeirão das Neves comprehensive plan, despite the fact that the RPP text closely resembled the original City Statute text, nowhere in the comprehensive plan was the 2001 law mentioned. Finally, BH and Confins were the only municipalities to detail RPP uses.

In summary, even though RPP was included in seven municipal comprehensive plans, the implementation of such a tool was not found in any municipality. This finding reinforces what Santos Júnior and Montandon (2011) explained about RPP: it is one of the City Statute tools

Table 15.3 Summary of content analysis for RPP in the municipal comprehensive plans

Municipality	*RPP included*	*Broad list of RPP's uses present*	*RPP's uses copied verbatim from City Statute*	*RPP's procedures copied verbatim from City Statute*	*Specific list of RPP's uses presented*	*City Statute as the primary regulatory law*	*Additional specific law for RPP mentioned*
Belo Horizonte	Yes	Yes	Yes (copied from)	No	Yes	Yes	No
Confins	Yes	Yes	No	No	Yes	Yes	Yes
Lagoa Santa	Yes	No	NA	NA	No	No	No
Pedro Leopoldo	Yes	Yes	No	No	Yes	Yes	Yes
Ribeirão das Neves	Yes	Yes	No	No	No	Yes	Yes
Santa Luzia	Yes	Yes	Yes	No	No	Yes	Yes
São José da Lapa	No	NA	NA	NA	NA	NA	NA
Vespasiano	Yes	Yes	Yes	No	No	Yes	No

most often included in municipal comprehensive plans, but it is almost never regulated, making its implementation unlikely to happen. Even with this knowledge, I found it surprising that BH was not implementing RPP during the time of this study. BH residents have a very small amount of open space per capita, and yearly issues with flooding are part of residents' reality. Having such a tool remain only on paper is an example of what Klink and Denaldi (2016: 404) describe as "disappointing progress" in the effort toward Brazilian urban reform.

Land use and zoning

When examining land use and zoning regulations, I only selected zoning categories related to provision of open space and environmental preservation. I was able to gather zoning regulations of seven out of the eight municipalities, having adoption dates ranging from 2006 to 2012. Only the municipality of São José da Lapa did not have zoning regulations, or did not make them available to the public, at that time. Using a GIS software, I overlaid 'new development' with selected zoning categories related to the provision of open space and/or environmental preservation.

For this analysis, I only selected the areas of 'new development' that were equal to or greater than the mean patch size (i.e. 4,318 m^2). This decision was based on the fact that my zoning analyses required robust 'new development' data, and by not considering the smaller areas, I made 'new development' more robust. Out of the 100 km^2 of total 'new development', this examination includes 74 km^2, corresponding to 74 per cent of the total urbanisation that occurred between 2000 and 2015.

Table 15.4 displays the summary of the overlay between 'new development' and the zoning categories. Based on the description of all these zoning categories, development should not have been occurring in these zones, which are all related to open space and/or environmental preservation. Five municipalities had two types of zoning scales, macro-zoning and zoning. I made sure to include both categories when analysing these municipalities. When observing Table 15.4, the two most unexpected results are: 14 per cent of new development took place in areas where it should not have happened, and 43 per cent of new development in BH took place in areas with environment-related zoning categories. The municipality of Santa Luzia also had a similar result to BH, with 35 per cent of its new development taking place in areas within environment-related zoning categories. The situation in São José da Lapa was unclear because of unavailable zoning regulations.

These regulations were dated between 2006 and 2012. This is an important piece of information because my analysis covers 2000–2015, and one may argue that the results presented here may be biased. In other words, municipalities may have zoned these areas after development took place there. While this is possible, the high likelihood is that these areas that were characterised by high environmental importance were already present in these zones. For instance, an existing neighbourhood may be adjacent to a forest, and new housing units may have been built there in 2003. After some evaluation, the city decided to preserve the rest of the remaining forest, and a new zoning for preservation may have been enacted in 2006. Situations like this may affect the magnitude of the percentage of new development that coincided with environment-related zoning categories presented in Table 15.4, i.e., the numbers may have been smaller. To illustrate, instead of having 43 per cent of new development in areas with environment-related zoning categories, BH may have a smaller percentage. Its zoning regulations are from 2010, and part of the 11.63 km^2 of 'new development' may have taken place before that. Therefore, I cannot make my conclusion with a very high level of certainty.

Table 15.4 Summary of the overlay of 'new development' and environment-related zoning categories

Municipality	*New development with areas => 4,318 m^2 (in km^2)*	*New development coincide environment-related zoning (in km^2)*	*% areas coincide in relation total area new development*	*Macro-zoning regulation*	*Zoning regulation/year*
Belo Horizonte	11.63	5.01	43.04	no	yes/2010
Confins	2.96	0.31	10.48	no	yes/2009
Lagoa Santa	12.34	0.78	6.36	yes	yes/2006
Pedro Leopoldo	10.12	0.00	0.02	yes	yes/2006
Ribeirão das Neves	11.79	0.74	6.25	yes	yes/2012
Santa Luzia	9.83	3.49	35.49	yes	yes/2008
São José da Lapa	6.93	NA	NA	no	no
Vespasiano	8.57	0.04	0.50	yes	yes/2006
New development	**74.18**	**10.37**			

Nonetheless, I argue that it is likely that new development may be happening in areas where it should not have, even after the land use and zoning regulations have been passed. This argument is due to the fact that in Brazil, "plans have not been properly implemented, and many forms of disrespect for the legal order have been left unquestioned" (Fernandes and Maldonado Copello 2009: 16). The other issue that I cannot be certain about is related to the legality of these new developments. I presume that some of the new developments are legal, while others are illegal. To illustrate my point, a well-known illegal settlement that took place circa 2013 is the Isidoro/Izidora settlement, located in an area in BH that was supposed to be devoted to environmental preservation.

Regardless of legality and/or illegality, in my opinion, this disrespect towards land use and zoning regulations happens because of three main factors: the fragility of policies supporting urban planning processes in Brazil, the lack of practical experience in implementation of urban planning tools in many municipal governments, and the national political process that facilitates the formation of bureaucratic links between developers' interests and public administration. This systematic disrespect towards land use and zoning regulations is due to these three factors and is not a problem related to lack of information nor the technical capacity of public employees.

Discussion

My findings point to the fact that there is a lack of implementation for the City Statute tools in the sub-region of BHMR. Friendly (2013) also found this lack of implementation in her case study about the city of Niterói. There are two main reasons that may help us understand my findings for the BHMR sub-region.

The first reason is that in Brazilian cities, political influence drives economic interest for development over social interest for provision of public goods (Rolnik 2011). As she explained, developers have an economic interest in furthering their own agendas by influencing politicians. These politicians, in turn, rather than implementing policy that is focused on the public good, support development for profit (Rolnik 2011: 244–247). I believe this type of behaviour may be related to the lack of RPP implementation for environment-related uses because developers are interested in using the land for corporations rather than goods and services for the public.

Marques (2010) had similar insights, when studying urban governance in São Paulo. He concluded that political influence has a profound effect on policy implementation in Brazil. For instance, many mayors and municipal officials have vested interests in their own companies outside of government, and sometimes they change land regulation to suit their own needs rather than the needs of the public (Marques 2010: 11–18). Moreover, my findings echo what Roy (2009) describes about urban governance in India: urban governance operates based on 'unmapping' processes, allowing "considerable territorialised flexibility to alter land use" (Roy 2009: 81) and zoning regulations.

The second reason is related to policy implementation per se. Policy implementation may be extremely limited if municipal governmental entities do not have funds, personnel and/or power to enact the changes a policy may demand (Milio 2010). When a policy is implemented, oversight and accountability must be taken into consideration. Ideally, a specific group would be responsible for overseeing the process so that it is implemented as intended, providing accountability during the completion and implementation of the policy (May 2003: 228–230). Additionally, if more than one organisation is involved, inter-organisational cooperation is imperative. Each entity that is involved in the implementation process must understand that they all share a common goal, and must facilitate cooperation to proceed (O'Toole 2003: 235–240).

In Brazil, the way the public sector operates does not generally help policy implementation to be successful. According to Arellano *et al.* (2013: 583), the Brazilian public sector "has no commitment to effectiveness and efficiency". The Brazilian public sector has been affected by politicians, who have had the power to place their own allies in positions of power, causing issues in efficacy and accountability. On the other hand, Rhodes *et al.* (2012) observed that there are systems in place to monitor the public sector performance, allowing opportunities to improve accountability. To illustrate, as they described, the Ministry of Planning has released a guide for analysing performance success, providing program administrators with a method to assess their outcomes.

When facing these two reasons, I believe the most important issue that needs to be addressed is the role of political influence in public sector practice. Traditional politics seep into the process of implementation. Municipal politicians begin re-interpreting urban legislation, bribing fiscal supervisors, and ignoring the rights of groups with low representation. As a consequence, the implementation of urban planning tools is affected adversely. This can be illustrated by Caldeira and Holston's (2015) article analysing participatory planning during São Paulo's 2007 Master Plan revision. They described that "the Electoral Justice Tribunal considered accusations of corruption affecting over 50% of the members of the City Council" (Caldeira and Holston 2015: 2012). The members were accused of receiving illegal donations from a developer's organisation interested in changing the Plan.

Recommendations and conclusion

Some recommendations should be addressed to ensure that strategies, such as RPP and environment-related zoning categories, are indeed being implemented. I have four main recommendations. The first recommendation is about the public administrative aspects of planning. I recommend that the public sector makes 'combating administrative corruption' the number one priority in its agenda in order to improve institutional practice. Second, if Brazilian municipalities decide to implement RPP, municipalities should invest in the automation of their land and property database. By having updated and easily accessible land and property registries, planners will be able to have a solid foundation for implementation of RPP, which will likely lead to a successful conclusion, as Braga (2001) asserts.

The final two recommendations are related to public participation, i.e., increase citizens' ability to participate in the public sector. Looking at the history of the City Statute, Friendly (2013: 162) explained that from 1988 to 2001, "an intense negotiation process" took place between many actors, including NGOs, environmental organisations and private citizens. Nowadays, however, "the renovation of [these] processes" (Fernandes and Maldonado Copello 2009: 19) is one of the central factors needed to improve implementation of the City Statute.

One bright spot regarding public participation is found in the municipality of BH. It has received special attention in the literature because of its successful implementation of participatory budgeting (Wood and Murray 2007). This success provides optimism for the future. However, there is cause for concern. In the 2015/2016 BH participatory budgeting, 7 out of the 116 approved projects were allocated to the environment-related theme, representing only 6 per cent of the total budget.[3] My immediate recommendation would be to educate citizens about sustainable practices and the importance of prioritising topics related to the natural environment. The goal would be for an increase in the percentage of the total budget devoted to this issue.

The final recommendation is about the electronic version of BH's participatory budgeting. As technology is more and more ubiquitous, electronic-participatory budgeting should be expanded: public employees should organise topics for discussion and be active in participating and responding online; political representatives should take part in online forums; and online forum per se should be used to evaluate the participatory process (Barros and Sampaio 2016: 295–308).

There are important opportunities for future research on the topic of public sector practice in the global South. One idea could be to interview public employees to better understand their perspectives on the role of political influence in public sector practice, and on policy implementation per se.

Notes

1 I would like to acknowledge Marina Magalhães de Castro, Devanshi Mehta and Morgan Bradley for their research assistance, and Jayme Wilken for her editorial suggestions.
2 A list of comprehensive plans, laws and orders follow: São José da Lapa, Plano Diretor, Law 881/2015; Lagoa Santa, Plano Diretor, Law 2.633/2006; Belo Horizonte, Plano Diretor, Law 7.166/1996 Secretaria Municipal de Serviços Urbanos/Secretária Municipal Ajunta de Regulação Urbana; Vespasiano, Plano Diretor, Law 002/2006 and Complementary Law 003/2007; Confins, Plano Diretor, Complementary Law 012/2009; Pedro Leopoldo, Plano Diretor, Law 3.034/2008; Ribeirão das Neves, Plano Diretor, Complementary law 002/2006; Santa Luzia, Plano Diretor, Law 2.699/2010; and Ribeirão das Neves, Plano Diretor, Law 036/2006.
3 Prefeitura Municipal de Belo Horizonte, Secretaria Municipal de Governo and Secretaria Municipal Adjunta de Gestão Compartilhada 2015, Plano Regional de Empreendimentos: Orçamento Participativo 201/2016 and Orçamento Participativo Digital 2013, Belo Horizonte, Brazil.

References

Arellano, E.B., Wakamatsu, A. and Ribas, R. (2013) "Organizational values in the Brazilian public sector: An analysis based on the tri-axial model", *Cross Cultural Management* 20(4): 578–585.

Arsanjani, J.J., Tayyebi, A. and Vaz, E. (2016) "GlobeLand30 as an alternative fine-scale global land cover map: Challenges, possibilities, and implications for developing countries", *Habitat International* 55: 25–31.

Barros, S.A. and Sampaio, R.C. (2016) "Do citizens trust electronic participatory budgeting? Public expression in online forums as an evaluation method in Belo Horizonte", *Policy and Internet* 8(3): 292–312.

Braga, R. (2001) "Gestão ambiental no Estatuto da Cidade: alguns comentários". In P.F. Carvalho and R. Braga (org) *Perspectivas de Gestão Ambiental em Cidades Médias*, Rio Claro: LPM-UNESP, pp. 95–109.

Caldeira, T. and Holston, J. (2015) "Participatory urban planning in Brazil", *Urban Studies* 52(11): 2001–2017.

Costa, G.M. and Magalhães, F.N.C. (2011) "Processos socioespaciais de metrópoles de países de industrialização periférica: Reflexões sobre a produção do espaço metropolitano de Belo Horizonte, Brasil", *Revista Brasileira de Estudos Urbanos e Regionais* 13(1): 9–25.

de Andrade, L.T. and de Mendonça, J.G. (2010) "Explorando as consequências da segregação metropolitana em dois contextos socioespaciais", *Cadernos Metrópole* 12(23): 169–188.

Fernandes, E. and Maldonado Copello, M.M. (2009) "Law and land policy in Latin America: Shifting paradigms and possibilities for action", *Land Lines* 21(3): 14–19.

Friendly, A. (2013) "The right to the city: theory and practice in Brazil", *Planning Theory and Practice* 14(2): 158–179.

Holben, B.N. (1986) "Characteristics of maximum-value composite images from temporal AVHRR data", *International Journal of Remote Sensing* 7(11): 1417–1434.

Hübner, C.E, Pinto, J.F. and Oliveira, F.H. (2006) "Cartografia Cadastral urbana e o direito de preempção", *Anais do Congresso Brasileiro de Cadastro Técnico Multifinalitário* 1(10): 1–7.

IBGE (2016) "Pesquisa de Informações Básicas Municipais: Perfil dos Municípios Brasileiros – 2015", Rio de Janeiro. Available: http://loja.ibge.gov.br/perfil-dos-municipios-brasileiros-2015.html [accessed 13 December 2016].

Instituto Brasileiro de de Geografia e Estatística (IBGE) (2010) "Pesquisa de Informações Básicas Municipais: Perfil dos Municípios Brasileiros – 2009", Rio de Janeiro. Available: http://biblioteca.ibge.gov.br/visualizacao/livros/liv44692.pdf [accessed 13 December 2016].
Klink, J. and Denaldi, R. (2016) "On urban reform, rights and planning challenges in the Brazilian metropolis", *Planning Theory* 15(4): 402–417.
Marques, E. (2010) "Government, political actors and governance in urban policies in Brazil and São Paulo: Concepts for a future research agenda", *Brazilian Political Science Review* 7(3): 8–35.
May, P. (2003) "Policy design and implementation". In B. Peters and J. Pierre (eds) *Handbook of Public Administration*, Trowbridge: Cromwell Press, pp. 223–233.
Milio, S. (2010) *From Policy to Implementation in the European Union: The Challenge of a Multi-Level Governance System*, London and New York: Tauris Academic Studies.
Miraftab, F. (2009) "Insurgent planning: Situating radical planning in the global South", *Planning Theory* 8(1): 32–50.
Musakwa, W. and Van Niekerk, A. (2015) "Earth observation for sustainable urban planning in developing countries: Needs, trends, and future directions", *Journal of Planning Literature* 30(2): 149–160.
O'Toole, L.J. Jr (2003) "Interorganizational relations in implementation". In B.G. Peters and J. Pierre (eds) *Handbook of Public Administration*, Trowbridge: Cromwell Press, pp. 234–244.
Presidência da República do Brasil (2001) "Lei federal no. 10.257, de 10 de julho de 2001". Available: www.planalto.gov.br/ccivil_03/leis/LEIS_2001/L10257.htm [accessed 13 August 2016].
Prieto, E.C. (2006) "O estatuto da cidade e o meio ambiente", *Anais do Congresso Brasileiro de Direito Urbanístico* 4(12): 1–20.
Rhodes, M.L., Biondi, L., Gomes, R., Melo, A.I., Ohemeng, F., Perez-Lopez, G., Rossi, A. and Sutiyono, W. (2012) "Current state of public sector performance management in seven selected countries", *International Journal of Productivity and Performance Management* 61(3): 235–271.
Rolnik, R. (2011) "Democracy on the edge: Limits and possibilities in the implementation of an urban reform agenda in Brazil", *International Journal of Urban and Regional Research* 35(2): 239–255.
Rolnik, R. and Klink, J. (2011) "Crescimento economico e desenvolvimento urbano: Por que nossas cidades continuam tão precárias?", *Novos Estudos—CEBRAP* (89): 89–109.
Roy, A. (2009) "Why India cannot plan its cities: Informality, insurgence and the idiom of urbanization", *Planning Theory* 8(1): 76–87.
Santos Júnior, O.A. and Montandon, D.T. (eds) (2011) *Os Planos Diretores Municipais Pós-estatuto da Cidade: Balanço Crítico e Perspectivas*, Rio de Janeiro: Letras Capital.
Silver, M. (2014) "The role of planning in the twenty-first century and beyond", *Planning Theory and Practice* 15(1): 103–106.
Wood, T. and Murray, W.E. (2007) "Participatory democracy in Brazil and local geographies: Porto Alegre and Belo Horizonte compared", *European Review of Latin American and Caribbean Studies* 83: 19–41.
World Bank (2016) "Indicators". Available: http://data.worldbank.org/indicator [accessed 13 December 2016].
Yazdani, S. and Dola, K. (2013) "Sustainable city priorities in global North versus global South", *Journal of Sustainable Development* 6(7): 38–47.

16

Cities, planning and urban food poverty in Africa

Jane Battersby

Introduction

Food insecurity has historically been viewed as predominantly rural in location and hunger-based in manifestation. As such, it has been understood as being largely outside of the realm of urban planning. However, this view will need to change. Sub-Saharan Africa is undergoing rapid urbanisation. As a result, the locus of poverty and therefore food insecurity is shifting towards urban areas (Crush *et al.* 2012). The increasing urbanisation of food insecurity has been identified as a challenge throughout the global South (Fundación *Alternativas* undated; Reardon and Timmer 2014). This urbanisation is precipitating changes in the food system and in the nature of rural–urban linkages. The nature of food insecurity is shifting with hunger-based malnutrition co-existing with increasing rates of obesity and diet-related diseases, a trend most evident in urban areas (Doak *et al.* 2005; Steyn and Mchiza 2014).

There is a pressing need to develop new ways to address the food and nutrition security challenges facing African cities. This chapter argues that there is a new role for urban planning (which in the African context takes the form of municipal spatial and land use planning) as a discipline to meet this challenge. The argument is premised on the potential role of urban planning in advancing food security, as defined by the FAO (1996): "Food security is . . . the situation that exists when all people, at all times, have physical and economic access to sufficient safe and nutritious food that meets their dietary needs and food preferences for an active and healthy life". The achievement of food security is based on four pillars: availability, accessibility, utilisation and stability. This chapter proposes that all four pillars can be advanced by urban planning. Availability is not simply about caloric sufficiency, but also the relative balance of types of food made available within the food system and why particular kinds of food are more available than others. This raises the question of the fundamental drivers of the food system. Accessibility must consider both economic and spatial accessibility of affordable and nutritious foods. Spatial accessibility must be considered in the context of where food sources are located relative to where people work, live and commute. Economic access extends beyond food price control, but also considers the impact of other costs incurred by urban households. For example, increases in the costs of energy or transport may change the sources of food, nutritional quality and frequency of food consumed. Effective utilisation is dependent on access to clean water, sanitation, storage, refrigeration and health care, all of which have urban planning components. Finally, urban planning can play a vital role in creating an urban environment and food system that is resilient to sudden shocks (economic, environmental or political) that disrupt households' abilities to access food (SACN 2015: 11).

Writing from the North American perspective, Pothukuchi and Kaufman (2000: 118) have noted:

> Air, water, food and shelter are among the essentials of life. Planners have been involved in efforts to improve the quality of air and water through pollution control programs and more comprehensively in shelter planning. But the fourth essential, food, has been virtually ignored by planners.

In recent years there has been an upsurge in interest in urban food issues and food systems governance in North America and Europe, focussing on food policy and planning for food justice, addressing food deserts and obesogenic environments, and food system sustainability (Halliday 2015). This chapter argues that there are important contextual differences that need to be considered by African urban planners in order to address food insecurity.

David Maxwell (1999) has suggested that in the African context, the absence of concerted policy thinking about the urban food security challenge can largely be attributed to three reasons. First, urban policy makers and practitioners have limited budgets and capacity and therefore give priority to "more urgently visible problems" (Maxwell 1999: 1940), such as housing and sanitation. Second, food insecurity in urban areas largely manifests at the household scale and households employ a range of coping strategies, which effectively render food insecurity invisible. Finally, the long-standing perception of food insecurity as a rural issue makes policy makers less likely to acknowledge urban food insecurity.

More recently, Battersby (2015) has highlighted the problem of lack of a clear food security mandate for sub-national or urban government in many African countries as a barrier to concerted planning for food security. This can be attributable to the persistent framing of food insecurity as a predominantly rural problem that can be addressed either through support for small-scale subsistence farmers, or as a national problem that can be addressed by increasing the productivity of large-scale commercial agriculture, which will reduce the cost of food and therefore reduce food insecurity. In both cases, the solutions are rural and production-oriented, leaving urban policy makers with little power or direct funding to address the problem.

However, in many African countries there has been a recent devolution of powers to local government and an increased focus on developing national urban policies (UN-Habitat 2015, e.g. Kenya, Zimbabwe and Zambia). This transition provides an opportunity for new roles for urban planning and new approaches to food security to be developed.

The following section outlines the limitations of the current dominant approaches to addressing food insecurity in the urban African context, and provides cases studies of production, processing and retail to discuss the impact of these dominant approaches. This is followed by a suggestion of a set of new opportunities for spatial planning to address the burgeoning food challenge.

Limitations of current approaches

There has been little overt attention paid to urban food security by planners in Africa. This chapter proposes that in the absence of urban food security planning, two dominant discourses have led to a series of inappropriate policy and planning responses. These are 1) the existing rural framing of food insecurity, and 2) the vision of the 'modern' African city. Although apparently distinct, these have common origins in Africa's post-colonial development pathways, their underlying logics and dissonances evident through Africa's history of postcolonial developmentalism, structural adjustment with attendant decentralisation and NGO-ification, and the current donor model of the private sector as partners in development (Chachage and Chachage 2004; Rizzo 2009; Young 2004).

Current approaches to food insecurity are dominated by what has been termed the 'twin-track approach' of

> (a) direct interventions and social investments to address the immediate needs of the poor and hungry (food aid, social safety nets, and so on) and (b) development programmes to enhance the performance of the productive sectors ... create employment and increase the value of assets held by the poor.
>
> *(CFS 2006: 16, cited in Crush and Frayne 2011: 529)*

This approach has focused almost exclusively on rural areas on the basis that "about 75 percent of the poor and hungry in developing countries living in rural areas" (FAO/IFAD/WFP 2005: 7). Under this approach there is little overt focus on the urban or scope for considering mitigation strategies that extend beyond the household or community scale. Informed by this dominant framing, the response by development practitioners and the state to increased awareness of urban food insecurity has been increased support for urban agriculture, which will be discussed at greater length below.

The second discourse is that of the 'modern' African city, which is usually mobilised through the imposition of laws, zoning ordinances and adherence to a master planning approach inherited from the global North. Although its roots are in the colonial era, post-independence governments have "tended to reinforce and entrench colonial plans and land management tools, sometimes in even more rigid form than colonial governments" (UN-Habitat 2009: 55). Within this planning approach, there is an implicit and explicit anti-informality in all its forms, be it housing, employment or food system (Kamete and Lindell 2010; Watson and Agbola 2013). In recent years, this over-arching trend has been modified reflecting two shifts. The first is the new narrative of Africa Rising and the re-visioning of African cities as 'future cities', reflected in new master plans referred to by Watson (2014) as 'new urban fantasy plans' that bear little resemblance to urban realities. The second is related to the first. This is the increased role of large private sector actors, increasingly viewed as 'partners in development' by the state, in shaping city visions and funding new developments shaped by an economic growth agenda (UN-Habitat 2015).

These framings of food security and the city have led to food security and the food system being officially largely absent from planning, and yet planners having a profound impact on the food system and therefore food security. This chapter provides three brief examples of the ways in which the conjunction of these dominant discourses on food security and the African city translates into impacts of planning on the food system and therefore food security. In each case (urban agriculture, food processing and food retail) there is a disjuncture between the vision of the city represented in planning practice and the lived reality of the needs of the food insecure. These three cases pay particular attention to Kitwe (Zambia), Kisumu (Kenya) and Epworth/Harare (Zimbabwe), the three sites of the ESRC/DFID-funded Consuming Urban Poverty project (https://consumingurbanpoverty.wordpress.com).

Urban agriculture

Municipal responses to urban agriculture have shifted over time from repression to being the default policy and programmatic response to urban food insecurity. The initial repression had its roots in the dominant framings described above. In the wake of structural adjustments and ongoing concerns about urban bias, urban agriculture was repressed on the grounds that urban production could undermine domestic demand for nationally produced agricultural products, when agriculture was viewed as a crucial engine of economic growth (World Bank 1993: 130, cited in Maxwell 1999: 1948). This political imperative bolstered the existing perception of production being a rural activity and has informed the current rurally focused twin track approach to food insecurity.

At the same time, urban agriculture was repressed by local governments on the basis that it did not reflect the ideal, modern city. Smith (1998: 214) argues that "within the city itself, illegal cultivation has, like illegal retailing, proven to be anathema to most city authorities, tarnishing as it does the modernizing image that they wish to pursue". It was commonly repressed on the grounds of public health concerns, such as spreading malaria, cholera and dysentery (Simatele and Binns 2008). While there is a need to mitigate public health risks, it is essential to note that in African cities, public health has been used by local government to repress urban agriculture, informal housing and the informal economy. Rogerson (2016: 233) states that in urban Africa "the pathologisation of informality is routine and pervasive".

However, in the last 15 years or so there has been a marked shift in municipal approaches to urban agriculture. As the extent of urban food insecurity has become more apparent, major development agencies and NGOs have begun to work in urban areas. Informed by previous experiences and food security discourses, their primary entry point has been through urban agriculture. In light of the advocacy by groups like RUAF (Resource Centres on Urban Agriculture and Food Security), municipalities have begun to accommodate urban agriculture. Municipalities also became more open to urban agriculture in response to the price shocks in the wake of structural adjustment, which removed subsidies and increased food prices for the urban poor. Urban agriculture was therefore permitted to retard the social protest that emerged in response to these austerity measures. Maxwell argues that the rapid growth of urban agriculture is the outcome of the "shifting of responsibility for bearing the costs of economic adjustment from the state or other employers to the poor themselves" (Maxwell 1999: 46) and to prevent declining urban living standards becoming a political problem (Maxwell 1999: 47). The increased advocacy for urban agriculture should not be viewed as politically progressive, but as a continuation of the neoliberal call for "personal responsibility" in a context of market liberalisation and a rolling back of social safety nets (McClintock 2014). This merely puts pressure on the most vulnerable to resolve crises not of their making.

The outcome of this shift has been that urban agriculture is increasingly incorporated into city policy and programmes, but in often contradictory ways. For example, in Kitwe, agriculture is not officially recognised as an urban land use and common agricultural practices, such as small livestock and poultry rearing, are illegal. However, the Department of Community Development, as part of its Food Security Pack programme, provides loans to women's groups, many of which it recognises are being used to conduct livestock and poultry farming (Muyoba 2015). In the development of an Urban Agriculture Policy for the neighbouring city of Ndola, RUAF identified a critical challenge to the policy being conflicting pieces of legislation across sectors (RUAF/MDP undated). This lack of legislative cohesion is in part the result of changing perspectives on urban agriculture, but is more fundamentally because of this lack of policy focus on urban food issues. These regulatory ambivalences make urban food production a precarious livelihood with farmers beholden to the political will of local government officials and politicians.

It is essential to note that while there has been considerable focus on urban agriculture by policy makers and NGOs, there is very little evidence to support the assertion that it is the most viable food security solution (Zezza and Tasciotti 2010). Indeed, the findings of the AFSUN survey conducted in 11 cities in 9 southern African countries in 2008/2009 found that urban agriculture was practiced more by wealthier households than poorer households and that households did not use urban agriculture as a frequent source of food (Frayne *et al.* 2014). The vast majority of households source most of their food through formal and informal retail market sources. Planning's intersection with the food system that delivers food to the market is therefore an essential element of food security.

Milling

Maize meal is a staple food for the urban poor in many parts of Africa. However, many households are unable to buy maize meal in the unit sizes sold by retailers. They therefore take small quantities of unmilled maize, which they have either bought, grown or received from rural relatives, to small neighbourhood hammer mills that will grind the maize for a limited cost. These are an essential part of the urban food system that is responsive to the needs of the urban poor. However, these are not recognised within formal planning and zoning regulations. In the case of Harare, milling is only legally permissible within industrially zoned areas (Jayne and Rubey 1993). This implies a vision of a food system that is highly formalised and operating at an industrial scale. As a result, most of the city's hammer mills are 'illegal' and the millers are vulnerable to being closed down, fined or even arrested should there be a shift in political perspective on the informal sector.

The vision of the food system evident in the zoning and planning regulations for food processing is at odds with the lived experience of urban residents. The absence of recognition of hammer mills is indicative of negative perceptions of informality and a set of planning norms and standards inappropriately imported from the global North. However, it is also essential to view this in the context of a legacy of an imagined set of relationships between rural producers and urban consumers that emerged in the era of colonial and postcolonial developmentalism. The history of maize in Africa is one of encouragement of maize production to feed an urbanising population. Maize marketing boards were therefore introduced as part of the post-colonial developmentalism project, which led to the development of large-scale, concentrated grain milling industries (Smale and Jayne 2003). These were underpinned by an imagining of a modern and industrial food system that had no need for small, informal processors. While market liberalisation has broken the monopolitistic power of the large mills, and hammer mills have become important sources components of the food system feeding the poor (Seshamani 1998), they remain absent from planning codes. In the case of milling, the parallel logics of the modern city and the modern agricultural sector have led again to a mismatch between the lived and planned realities of poor households and the food system.

Food retail

The mismatching of urban vision to lived urban realities is most evident in the treatment of food retail, where attempts are repeatedly made to remove or formalise informal food vendors and market places, and large shopping malls are increasingly viewed as indicators of development and modernization. In Kisumu, Kenya, for example, there are plans to remove street vendors from main streets in town and place them in a Vendors' Mall (Omollo 2015). Although there is evidence that the location of the street traders is essential to their businesses' survival and that the food sold by these vendors is cheaper than formal retailers and is fresh (Onyango *et al.* 2012: 112), there is an articulated need to move the traders out of sight as they "make the town disorganised" (Omollo 2015). The town planner also spoke of concerns about health and safety associated with street traders (Otieno, personal communication 14 July 2015). This articulation of disorder and disease is not unique to Kisumu and is a common discursive practice to argue for the removal of street traders throughout Africa, as is evident by recent efforts to move traders in Harare (Kadirere and Kwaramba 2015). These efforts to move street traders reflect a fundamental weakness in the ways in which local governments and planners view marketing. They fail to understand the economic and geographical rationality of why marketeers locate where they do. It has been repeatedly demonstrated that the viability of markets is reduced when they are moved as they locate in response to consumer needs.

The second component of food retail currently affected by planning is the promotion of formal retailing, of which food retail is an important component. Across Africa the shopping mall is emerging as a symbol of urban development and cosmopolitanism. At the opening of the Mukuba Mall in Kitwe, President Lungu identified the mall as serving the needs of a "growing segment of our urban citizenry that has evolved and become more cosmopolitan and global in their exposure and preferences and have even more disposable income and options" (Lungu 2015). Likewise, the proposed Lake Basin Development Authority Mall in Kisumu is argued to serve to attract further businesses and accelerate local economic development (Ochiel 2015).

Within this desire for a modern retailing sector, there is no reflection of the impact of this on the existing food retail sector. However, directly and indirectly, the emergence of these malls is reshaping the local food system. Each of these malls has at least one large supermarket as an anchor tenant. Planning decisions about retail space are fundamentally shaped by a vision of a good city, characterised by a modern, formal food system. Given the centrality of the supermarket as the embodiment of this food system, it is essential at this point to consider what the expansion of supermarkets in African cities might mean for food security. Although it has been argued that supermarkets may prove to be an "urban food security boon" because of their capacity to lower food prices (Reardon and Minten 2011), there is a need to consider what changes in the food system they bring.

Evidence from South Africa suggests that when large malls enter into urban areas there is an erosion of existing local businesses, many of which are food retailers (Ligthelm 2008). While this is often viewed as the work of the invisible hand of the market, there is considerable evidence that large supermarket chains put pressure on local government to control informal trade near to the supermarkets (Potts 2007). In the case of Kitwe, the Minister of Sport, Youth and Child Development articulated the need to regulate street trade as follows: "Shoprite is one company whose business is being negatively affected and yet it contributes a lot of money to Government in taxes" (*Zambia Daily Mail* 2012). Likewise in Kisumu, the City Manager argued that street trading needed to be controlled because "legitimate traders had raised concerns over insecurity, unfair competition and unfair treatment by the hawkers who sell similar goods right at their door steps" (Omollo 2015).

However, informal traders play a vital role in food security through selling food in affordable unit sizes that allow frequent purchases to overcome storage and refrigeration limitations, and by selling food on credit to poor consumers (SACN 2015). The loss of these businesses will have a negative impact on urban food security. Additionally, although supermarkets and shopping malls may bring cheaper, more diverse food to consumers, there are concerns that they are accelerating the nutrition transition, characterised by increased consumption of highly processed, nutritionally poor foods (Popkin *et al.* 2012: 6), and the rise of obesity and diet-related non-communicable diseases.

Although local government ostensibly has little to do with food security, there are clearly a number of points of intersection of planning with the food system that have a direct impact on food and nutrition security. As Pothukuchi notes, "if planners are not conscious [of food issues], then their impact is negative, not just neutral" (Pothukuchi 2000, cited in Roberts 2001). The following section therefore suggests a set of possible opportunities for planners to proactively intervene in the food system and the wider urban system to create a generative food security environment.

Envisioning a food security role for planners

In the wake of Habitat III's New Urban Agenda (UN-Habitat 2016) and its prioritisation of decentralisation and devolution, there is scope for more purposive food system planning in African cities. This will, however, require a cognitive shift for both food security practitioners and urban planners. There is a need to acknowledge that although food insecurity manifests at the household scale, its causes and therefore solutions extend into the food system and wider urban system.

As suggested above, there are already many urban planning functions that shape the food system and therefore food security, whether or not they are considered food system planning or food security interventions. This section therefore provides some opportunities for planners to engage in food systems planning, which work within existing mandates and develop appropriate, indigenous solutions. In order to achieve this, planners need to consider how households actually access and utilise food and develop planning responses that acknowledge these realities.

There has been a recent surge in interest in understanding urban food systems and moving beyond the household approach. The IIED, for example, is doing innovative work in Nairobi, Kenya examining urban food security's connection to the local food system and urban form (Ahmed *et al.* 2015). Likewise Consuming Urban Poverty, working in Kenya, Zambia and Zimbabwe, is seeking to understand the connections between the urban food system and its governance, and food security. These projects call for a more concerted focus on the role of planning in ensuring food security.

In order to do this, there is a need to consider how the food system interacts with other systems and what planning opportunities might exist to enhance access to safe, affordable, nutritious food for the urban poor. There are a number of practices within the largely informal food sector that are responsive to the food needs of the poor that can be harnessed to generate such a food system. There is a role for planning to work with these systems and not against them. So, for example, informal food vendors tend to cluster around transport interchanges, as individuals purchase food on the way home from work. Despite this, transport infrastructure planning rarely considers these transport/trade intersections. There are opportunities to provide better trading infrastructure in these spaces and, following the example of cities like Belo Horizonte, provide preferential trading space for retailers selling healthier foods (Rocha and Lessa 2009). In the Philippines there have been a number of local food ordinances passed that ban the sale of junk food near sites of education (see, for example, City of General Santos 2013).

Beyond these food system specific interventions there is a need to further think how food security is affected by other kinds of planning decisions. In Cape Town, the promulgation of the City's 2012 Single Zoning scheme, which sought to 'modernise' and 'standardise' the city's zoning has been labelled as anti-poor as it renders 70 per cent of the house shops (or spaza stores) illegal. The underlying logic of this has been questioned by activists who ask "whether these By-Laws were designed to benefit the Corporate Retailers who are increasingly encroaching on the townships with the proliferation of shopping malls" (Western Cape Informal Traders Coalition 2013). Whether intentional or not, this seemingly non-food related planning tool has significant power in shaping the food system. It is therefore important to consider the ways in which non-food decisions shape food security.

Conclusion

Food insecurity is increasingly an urban problem in Africa and the global South. Responses thus far have focused at the household level and largely neglected the relationships between food security, the food system and the urban system. This chapter has argued that there is a need to develop new responses that engage the role of planning in shaping the urban food system. This chapter argues that urban planners have played a significant role in shaping the food system through decisions made without considering their food system impact. Many of these have been driven by a modernisation agenda that has marginalised and rendered illegal many components of the food system that feed the urban poor. This chapter has focused on just three examples of this within the wider food system.

There are many direct food system interventions available to planners, such as ensuring that zoning regulations are not exclusionary to informal food vendors, planning food processing sites for small-scale processors, designing municipal market precincts to be responsive to transportation and sanitation needs of traders, integration of space for small-scale farming within municipal plans. However, there is a wider need for food to be considered in non-food planning decisions that affect the food system. These include the planning of transport infrastructure and nodes, the provision of water and sanitation, and the location and design of residential areas, all of which impact on the structure of the food system and ultimately food security. Building on the example of Water Sensitive Urban Design, this chapter calls for the promotion of Food Sensitive Planning and Urban Design (FSPUD), which requires the food system be considered in wider planning decisions. Its creators present its rationale as follows:

> FSPUD means thinking about 'and' opportunities rather than 'or'. By planning and designing food-sensitive places, we have the opportunity to create jobs, build communities and transform, for the better, the environmental sustainability of our settlements and the environmental welfare enjoyed by their inhabitants. Planners and designers can use food to simultaneously address multiple objectives, creating diverse opportunities for people to meet their needs.
>
> *(Donovan* et al. *2011: 13)*

I am aware that asserting the planners should not only start to consider specifically food-based planning interventions, but also incorporate consideration of the food system impacts of non-food planning decisions, may be viewed as naïve and overly ambitious. However, it is clear that the food system plays a major role in the economic, social and environmental life of cities, and if proactively managed can address issues of justice, poverty and health (Moragues *et al.* 2013). There are many components of food systems planning that fall out of the jurisdiction of municipal planners, and food strategies acting across jurisdictions should be developed. Ultimately, planning for food security can provide new entry points for municipalities to meet their economic, social, environmental, health and other objectives, without extending beyond their existing mandates.

References

Ahmed, S., Simiyu, E., Githiri, G., Sverdlik, A. and Mbaka, S. (2015) "Cooking up a storm: Community-led mapping and advocacy with food vendors in Nairobi's informal settlements", IIED Working Paper, June 2015, IIED: London.

Battersby, J. (2015) "Food insecurity amongst urban households". In S. Fukuda-Parr and V. Taylor (eds) *Food Security in South Africa: Human Rights and Entitlement Perspectives*, UCT Press: Cape Town, pp. 97–121.

Chachage, C.S. and Chachage, C.S.L. (2004) "Nyere: Nationalism and post-colonial developmentalism", *African Sociological Review/Revue Africaine de Sociologie* 8(2): 158–179.

City of General Santos (2013) "An ordinance providing for a comprehensive policy for the respect, promotion, fulfillment and protection of children's rights, enhancement of support system and mechanism, and for other purposes", Ordinance No. 07 Series of 2013. Available: http://spgensantos.ph/2013/11/ordinance-07-series-of-2013/ [accessed 30 September 2015].

CFS (Committee on World Food Security) (2006) *Mid-Term Review of Achieving the World Food Summit Target*, Rome: Food and Agriculture Organisation of the United Nations.

Crush, J.S. and Frayne, G.B. (2011) "Urban food insecurity and the new international food security agenda", *Development Southern Africa* 28(4): 527–544.

Crush, J., Frayne, B. and Pendleton, W. (2012) "The crisis of food insecurity in African cities", *Journal of Hunger and Environmental Nutrition* 7(2–3): 271–292.

Doak, C.M., Adair, L.S., Bentley, M., Monteiro, C. and Popkin, B.M. (2005) "The dual burden household and the nutrition transition paradox", *International Journal of Obesity* 29(1): 129–136.

Donovan, J., Larsen, K. and McWhinnie, J. (2011) *Food-Sensitive Planning and Urban Design: A Conceptual Framework for Achieving a Sustainable and Healthy Food System*, Melbourne: Report commissioned by the National Heart Foundation of Australia (Victorian Division).

FAO (1996) "Declaration on World Food Security", Rome: World Food Summit. Available: www.fao.org/docrep/003/w3613e/w3613e00.htm [accessed 27 June 2016].

FAO/IFAD/WFP (2005) "Background note on roundtable 1: Dialogue on the eradication of poverty and hunger, ECOSOC 2005 High Level Segment, 29 June–1 July 2005". Available: www.un.org/en/ecosoc/meetings/2005/hl2005/rt1_bg.pdf [accessed 30 September 2015].

Frayne, B., McCordic, C. and Shilomboleni, H. (2014) "Growing out of poverty: Does urban agriculture contribute to household food security in Southern African cities?", *Urban Forum* 25(2): 177–189.

Fundación *Alternativas* (undated) "Who we are". Available: http://alternativascc.org/sites/default/files/FundacionAlternativas_WhoWeAreWhatWeDo.PDF [accessed 27 June 2016].

Halliday, J.J. (2015) "A new institutionalist analysis of local level food policy in England between 2012 and 2014", Unpublished PhD Thesis, Centre for Food Policy, City University London.

Jayne, T.S. and Rubey, L. (1993) "Maize milling, market reform and urban food security: The case of Zimbabwe", *World Development* 21(6): 975–987.

Kadirere, H. and Kwaramba, F. (2015) "Questions mount over vendors war", *Daily News*, 13 July. Available: www.dailynews.co.zw/articles/2015/07/13/questions-mount-over-vendors-war [accessed 30 September 2015].

Kamete, A.Y. and Lindell, I. (2010) "The politics of 'non-planning' interventions in African cities: Unravelling the international and local dimensions in Harare and Maputo", *Journal of Southern African Studies* 36(4): 889–912.

Ligthelm, A.A. (2008) "The impact of shopping mall development on small township retailers", *South African Journal of Economic and Management Sciences* 11(1): 37–53.

Lungu, E. (2015) "Speech at the opening of Mukuba Mall, Kitwe", 23 April 2015. Available: http://lusakavoice.com/2015/04/23/president-lungu-opens-mukuba-mall-in-kitwe/ [accessed 30 September 2015].

McClintock, N. (2014) "Radical, reformist, and garden-variety neoliberal: Coming to terms with urban agriculture's contradictions", *Local Environment* 19(2): 147–171.

Maxwell, D. (1999) "The political economy of urban food security in Sub-Saharan Africa", *World Development* 27(11): 1939–1953.

Moragues, A., Morgan, K., Moschitz, H., Neimane, I., Nilsson, H., Pinto, M., Rohracher, H., Ruiz, R., Thuswald, M., Tisenkopfs, T. and Halliday, J. (2013) "Urban food strategies: The rough guide to sustainable food systems", document developed in the framework of the FP7 project FOODLINKS (GA No. 265287).

Muyoba, M. (2015) Department of Community Development: Presentation at the Consuming Urban Poverty Stakeholder Meeting in Kitwe, 23 June.

Ochiel, H. (2015) "Kisumu basks in glory as investors troop in", *Standard Digital*, 15 April. Available: www.standardmedia.co.ke/business/article/2000158423/kisumu-basks-in-glory-as-investors-troop-in?pageNo=2 [accessed 30 September 2015].

Omollo, K. (2015) "Construction of 3 hawkers: Markets begins in Kisumu", *Standard Digital*, 26 August. Available: http://212.100.244.246/business/article/2000174190/construction-of-3-hawkers-markets-begin-in-kisumu [accessed 28 September 2015].

Onyango, J.O., Olima, W.L.A. and Onyango, L. (2012) "Dynamics of street vending phenomenon in the Kisumu Municipality, Kenya", *International Journal of Arts and Commerce* 1(4): 107–120.

Popkin, B.M., Adair, L.S. and Ng, S.W. (2012) "Global nutrition transition and the pandemic of obesity in developing countries", *Nutrition Reviews* 70(1): 3–21.

Pothukuchi, K. (2000) "Community food mapping", address to Ontario Public Health Association workshop on community food mapping, Toronto, September 11, 2000.

Pothukuchi, K. and Kaufman, J.L. (2000) "The food system: A stranger to the planning field", *Journal of the American Planning Association* 66(2): 113–124.

Potts, D. (2007) "The state and the informal in sub-Saharan African urban economies: Revisiting debates on dualism", Crisis States Research Centre Working Paper, 18.

Reardon, T. and Minten, B. (2011) "The quiet revolution in India's food supply chains: Vol. 1115", IFPRI Discussion Paper.

Reardon, T. and Timmer, C.P. (2014) "Five inter-linked transformations in the Asian agrifood economy: Food security implications", *Global Food Security* 3(2): 108–117.

Rizzo, M. (2009) "The struggle for alternatives: NGOs' responses to the World Development Report 2008", *Journal of Agrarian Change* 9(2): 277–290.

Roberts, W. (2001) "The way to a city's heart is through its stomach: Putting food security on the urban planning menu", Toronto Food Policy Council, Crackerbarrel Philosophy Series. Available: www.toronto.ca/health/tfpc_hs_report.pdf [accessed September 2015].

Rocha, C. and Lessa, I. (2009) "Urban governance for food security: The alternative food system in Belo Horizonte, Brazil", *International Planning Studies* 14(4): 389–400.

Rogerson, C.M. (2016) "Responding to informality in urban Africa: Street trading in Harare, Zimbabwe", *Urban Forum* 27 (2): 229-251.

RUAF/MDP (RUAF Foundation and Municipal Development Partnership) (undated) "Ndola urban agriculture programme: Policy analysis narrative report". Available: www.ruaf.org/sites/default/files/Ndola%20Policy%20analysis%20on%20urban%20agriculture%20report.pdf [accessed 30 September 2015].

South African Cities Network (SACN) (2015) "A study on the current and future realities for urban food security", South African Cities Network, Johannesburg. Available: www.sacities.net/wp-content/uploads/2015/11/Urban-Food-Security-Report.pdf [accessed 27 June 2016].

Seshamani, V. (1998) "The impact of market liberalisation on food security in Zambia", *Food Policy* 23(6): 539–551.

Simatele, D.M. and Binns, T. (2008) "Motivation and marginalization in African urban agriculture: the case of Lusaka, Zambia", *Urban Forum* 19(1): 1–21.

Smale, M. and Jayne, T. (2003) *Maize in Eastern and Southern Africa: "Seeds" of Success in Retrospect*, Washington, DC: Environment and Production Technology Division, International Food Policy Research Institute.

Smith, D.W. (1998) "Urban food systems and the poor in developing countries", *Transactions of the Institute of British Geographers* 23(2): 207–219.

Steyn, N.P. and Mchiza, Z.J. (2014) "Obesity and the nutrition transition in sub-Saharan Africa", *Annals of the New York Academy of Sciences* 1311(1): 88–101.

UN-Habitat (2009) *Global Report on Human Settlements 2009: Planning Sustainable Cities*, London: Earthscan, for UN-Habitat.

UN-Habitat (2015) *Towards an African Urban Agenda*, UN-Habitat: Nairobi.

UN-Habitat (2016) Habitat III: New urban agenda: Draft outcome document for adoption in Quito, October 2016. Available: www2.habitat3.org/bitcache/97ced11dcecef85d41f74043195e5472836f6291?vid=588897&disposition=inline&op=view [accessed 2 November 2016].

Watson, V. and Agbola, B. (2013) *Who Will Plan Africa's Cities?* London: Africa Research Institute.

Watson, V. (2014) "African urban fantasies: dreams or nightmares?", *Environment and Urbanization* 26(1): 215–231.
Western Cape Informal Traders Coalition, The Somali Association of South Africa, The National Consumer Forum, COSATU Western Cape, PASSOP, The South African Council of Churches, The Scalabrini Foundation and The Black Business Chamber (2013) "Press Release: Implementation of new by-laws affecting spaza shops". Available: www.passop.co.za/news/the-implementation-of-new-by-laws-could-force-up-to-70-of-spaza-shops-in-cape-town-to-close [accessed 27 June 2016].
World Bank (1993) *Uganda: Agriculture*, Washington, DC: The World Bank.
Young, C. (2004) "The end of the post-colonial state in Africa? Reflections on changing African political dynamics", *African Affairs* 103(410): 23–49.
Zambia Daily Mail (2012) "Street vending needs to be controlled – Kambwili", *Zambia Daily Mail*, 11 March. Available: www.lusakatimes.com/2012/03/11/street-vending-controlledkambwili/ [accessed 30 September 2015].
Zezza, A. and Tasciotti, L. (2010) "Urban agriculture, poverty, and food security: Empirical evidence from a sample of developing countries", *Food Policy* 35(4): 265–273.

17

Technology and spatial governance in cities of the global South

Nancy Odendaal

Introduction

The 'smart city' discourse has become a container for many dimensions of the relationship between technology and cities. Throughout urban planning history, the relationship between the city and technology has assumed that technological innovation enables efficiency and progress. Planners are key to formulating well-articulated spatial responses where spatial planning and infrastructure investment co-evolve within a comprehensive frame. This changed under neo-liberalism as corporatisation of infrastructure and unbundling, together with the loss of faith in comprehensive planning, led to fragmented delivery of infrastructure, often outside the realm of spatial planning. The manifestation of landscapes of premium networked spaces and forgotten, 'in-between', areas where infrastructure does not necessarily connect, but divides, is a feature of increasing 'splintering urbanism' (Graham and Marvin 2001). A version of this is discernable in the smart city visions touted for African cities, referred to by Watson (2014) as the African urban fantasy phenomenon where designs for new satellite cities in Nairobi and Kigali bear no resemblance to the 'real' city.

This chapter explores ways through which the relationship between contemporary technology and spatial governance could manifest in the urban global South in an empowering and inclusive way. Rather than focus on the smart city visual narrative, it takes a more finely grained approach to technology appropriation, given its many articulations with urbanity at varying scales. Exponential technical innovation and a reordering of the relationship between society and ICT means that "the lifecycle of technology is now measured in months" (Townsend 2014: 134). The intensification of social media activity has led to reciprocity between consumer and service provider that has delivered beyond the initial expectations of Web 2.0.[1] Smart phone growth in developing countries has surpassed the global North in value of handsets sold. In 2014, 3 of the top 10 smart phone markets were in emerging economies (China, Brazil and India). Favourable pricing and a saturation of the smart phone market in the global North led to a shift in 2015 where all the top 10 markets were in the global South (Davidson 2014). We now carry communications infrastructures in our pockets. Not only can our phones connect to the Internet, but urban hardware from household appliances to electricity grids are fitted with networked sensors that can measure light, sound and moisture, as well as microprocessors that are optimised to process limited data effectively and speedily. The Internet of people has transitioned to the Internet of things, and more objects are now connected to the Internet than people (Townsend 2014). The generation of data by objects connected to city grids and information systems, and people that engage on social media together with the usual forms of information gathering contribute to big data accumulation, leading to multinational firms such as IBM and Cisco now playing dominant roles in cities. Furthermore, the use of ICT to enable access to health care, policing and basic local government services in remote rural areas, for example, show the power of technology to enable resource optimisation in developing contexts. The use of smart meters for water and electricity consumption, for example, facilitates household budget prioritisation, while ensuring basic services access. Smart inventions therefore cover a broad spectrum of methods of technology appropriation. Despite work on local innovations, there is surprisingly little theoretical work on the notion of smart in urban planning and urban studies in the global South (see Luque-Ayala and Marvin 2015).

My theoretical position is one that rejects the objectification of technology as an outside force that liberates or constrains, hence the departure from the corporate smart city narrative. The myth of "transcendence" (Graham 2008) that presupposes technology can overcome distance barriers is deeply dominant. The notion that technology can overcome physical barriers and remove the need for face-to-face contact (the "death of distance" hypothesis referred to in Hall 2002), goes back to the innovation of the telegraph and the telephone. A similar logic applies in development thinking where technology is seen as the key to unlocking socio-economic potential. Both assumptions are invalid. Technological innovation has developed in tandem with, and in many ways has contributed to, urban agglomeration and growth (Graham and Marvin 2001; Hall 2002; Townsend 2014). Socio-economic development requires more than a technology fix. The multifaceted nature of the digital divide continues to manifest in cases where such projects underestimate the importance of human and social systems in technology adoption and use (Warschauer 2004). The review of literature, therefore, will focus on the relational nature of technology adoption, rather than a comprehensive review of all that is technology and urban.

The city is taken as representing the outcome of continuous socio-technical processes that manifest spatially as the relationships between the material and human agency evolve. This represents a "dance" between technological innovation and appropriation through social action and livelihood imperatives (Odendaal 2011). This is an opportunity for planning, as a form of spatial governance where the actions of multiple actors intentionally combine to enable spatial change, to be effective. A renewed understanding of the relationship between technology and space is necessary. I explore three dimensions of this relationship here: the first refers to grassroots efforts to render the margins more visible through online mapping and geo-referenced data capture, thereby focusing on 'for whom' we plan. The second dimension then explores how this information can empower and enable local communities to engage powerful stakeholders in the management of place, exploring ways through which planning can be more participatory, with the third exploring the extent to which this makes input into spatial governance in general.

ICT, cities and planning in the global South

While eschewing the idea that technology is a panacea, I would argue that it nevertheless does provide unique opportunities for innovation in situations where resource access is uneven. This review seeks to show how this is mediated by context and livelihood choices. The first part of the review, therefore, is selective in its consideration of literature on the developmental potential of ICT and works towards a lens that entwines human and material agency.

The relationship between technology and space is one that is informed by context and human agency. It continuously evolves and follows a network configuration that is dependent on the strength of connections. The second part of this review considers this relationship in more detail. The third part of the review focuses on the evolution of the contemporary reading of smart cities in the global South.

Technology, livelihoods and governance

The spatial imagination that accompanies technological export is associated with the assumption that technologies are likely to spread everywhere with the same impact, regardless of context (Thrift 1996). A second related evolutionary view is that technologies progressively replace one another. Socio-technical change is not linear however. One of the features of technology adoption is how 'older' forms of communication, such as analogue radio, coexist with digital forms in ways that best represent local conditions. Mytton (2000: 2) estimates that the million radio sets in Africa in the mid-twentieth century increased to approximately 100 million by 2000. Community and amateur stations broadened listenership while also allowing for growth in specific sectors such as university campuses and churches. Radio is a highly effective mass medium, given the lower cost of production and transmission (Paterson 1998). In Cape Town it continues to play an important role in soliciting opinion on public issues (Bosch 2011). New media is beginning to be recognised as a tool for social organisation and sharing in Uganda (Javuru 2013), but radio remains a core part of political mobilisation (Mwesige 2009). Technological appropriation is closely tied to livelihoods and lifestyles (Odendaal 2011). Processes of technical innovation are informed by capacity, access, need and cost.

The premise of technology-led development initiatives is that ICT is critical to economic production and distribution and shares some assumptions with modernisation theory: the notion that technology enables a progression away from more traditional forms of communication and media, progressing towards more complex structures (Schech 2002). It also speaks to a temporal linearity that is deemed inevitable, where developing countries cannot be 'left behind' (Graham 2008). Neo-liberal expectations of technological diffusion influence policy implementation (Kleine 2009: 181). Contemporary iterations of the ICT for Development (ICT4D) paradigm have been more mindful of the interface between capacity and objectives however. Moving from a supply-driven model to more demand-centred approaches frames the poor as potential innovators, not just passive consumers (Heeks 2008) with a focus on local practices (Krishna and Walsham 2005).

Examination of the efficacy of a systematic application of ICT in local governance shows a high degree of failure in e-governance projects in developing countries, due to the lack of contextualisation and local engagement. Aims of e-governance projects include using digital connectivity as a democracy enabler, improving the efficiency and efficacy of bureaucratic processes and enhancing service delivery. E-governance projects in the global South are often driven by government departments and funded by external agencies (and more recently private players), raising questions about long-term sustainability and maintenance (Dada 2006). Despite developmental intentions in using ICT towards deepening democracy and enabling last-mile access, it is tempting for cash-strapped local governments to get carried away by the 'shine' of technology-led interventions (Odendaal 2011). The entry of private firms in the smart city market has broadened the terrain of actors involved in governance with much of the necessary technical and processing capacity vested in the private sector. Where local government capacity is constrained, using public resources to address basic needs trumps technology-intensive processes that do not necessarily yield quick results.

Entrenched institutional cultures are not necessarily supportive of innovation (Krishna and Walsham 2005). Research in Andhra Pradesh in India, for example, shows that a participatory approach that understands the service and informational needs of a range of stakeholder groups, which also seeks diverse inputs, extending innovation to the organisational sphere and consistent political and organisational leadership, are success factors for sustainability (Krishna and Walsham 2005).

E-governance and e-democracy are often assumed to enhance one another, but as Netchaeva (2002: 471) succinctly states: "To participate in the social, political and democratic life of the country through information technologies people must have access to them, know how to use them and want to do it". Heeks (2003) identifies three sets of issues here. The first relates to the gaps between technology and social systems into which automated governance systems aspire to be embedded. Organisational cultures and societal demands are deeply contextual. The second refers to the disjuncture between private and public sector demands and the third to unequal access.

The digital divide manifests in various dimensions. Spatially, there are divides between the South and North illustrated by the distribution of Internet penetration, for example. The UK and US and much of Europe show Internet user penetration percentages of between 89.8 per cent (the UK) to 96.5 per cent (Iceland). By contrast, the range in the South is wider, with Argentina sitting at 59.9 per cent and the Democratic Republic of Congo at 2.2 per cent (The Internet Society 2016). Mobile broadband population penetration alone shows a less predictable pattern where countries such as Botswana reflect 74.1 per cent and Portugal 36.7 per cent coverage. Rural–urban digital divides must also be considered (Furuholt and Kristiansen 2007; Rao 2005). Research in Tanzania shows that constraints are largely physical: access to venues such as Internet cafés and libraries determine extent of use (Furuholt and Kristiansen 2007). As access to smart phones increases, such mundane spatial barriers may become easier to overcome but less overt factors such as language, poverty indicators and perceptions of technology continue to impact (Keniston and Kumar 2004). Demographic and socio-economic factors influence the choice of ICTs (Crang and Graham 2007; Selwyn and Facer 2007). Availability of technology does not guarantee use. Several African studies quoted in Obijiofor (2009) show a predominance of Internet use for e-mail, while web surfing remains low. Social attitudes to computers are associated with social hierarchy and status.

Technology can "foster new practices of being, knowing and doing" (Schech 2002: 21), as illustrated through use by social movements. The Arab Spring illustrates these performative dimensions. In Egypt and Tunisia, social media played an important role in influencing key debates before both uprisings, and assisted in spreading democratic messages beyond the countries' borders, during and after demonstrations (Howard *et al.* 2011). ICT was part of broader heterogeneous networks that included television, radio and built upon existing social and kinship capital (Allagui and Kuebler 2011). The power of the media no longer vests in the state alone, enabling distributed voices and visual content that potentially challenge official discourses.

Technology and space

ICT does not necessarily overcome spatial inequalities; the opposite is in fact true. Drawing on Graham and Marvin's *Splintering Urbanism* thesis (2001), Servon (2002: 226) argues that "IT is deeply rooted in geography". In South Africa, the distribution of cell phones and landline telephony as well as online access in social facilities such as libraries and schools reflect Apartheid urban geographies, and subsequent post-Apartheid investment patterns (Odendaal 2011). Spatial constraints do not simply disappear with broadened technology access but co-exist with cyber-space in a hybrid form (Graham 2008) that is more relational than Cartesian (Odendaal 2014). Some *places* remain isolated and under-serviced; what new technologies offer is access to networked *spaces* that connect. Technology is not enough to access those networked spaces, however; they have to be mediated through livelihood strategies and socio-economic appropriation (Odendaal 2010).

Spatial inefficiency coexists with the creation of new technology-mediated spaces. The location of Internet cafés and mobile phones in Africa reflects access differentiation, and also the creation of new places for social and economic relations (Obijiofor 2009). Where individual contact is constrained by income and bandwidth, public and private Internet cafés become important access points. Tele-centres are in many cases the result of private entrepreneurial activity in urban centres that offer a suite of services that include photocopying, faxing as well as Internet. In some cases, they provide telephony, providing services for small businesses in many cases (Falch 2004). A study of 32 Internet cafes in inner-city Johannesburg, South Africa, found that these facilities are used for gathering information on study and job opportunities and compiling curricula vitae, also for organising social events and staying in touch with friends and family. The study reports a high rate of returning customers and almost half of respondents accessed the Internet on behalf of others (Hobbs and Bristow 2007).

Social acquaintance with technology is ongoing and imbued with values. Different technologies take on different social meanings in particular cultural contexts (Thrift 1996). Among the youth of marginalised communities in Cape Town, for example, it is a personalised and symbolic connection to the world; a means of reinforcing identity while remaining connected to selected global cultural icons (Hammett 2009). New spatial modalities of ICT use in developing countries mitigate cost restrictions. Examples are container telecentres and informal phone shops on sidewalks (Donner 2007, 2008).

This has implications for the experience of space and movement between places. Ambient computing anticipates a spatial dimension where the 'cyber' and 'real' co-produce an experiential dimension typified by seamless flows of information and interaction. A hybrid space is possible at the interface between infrastructure and human experience (Dourish and Bell 2007). While traditional networked infrastructures are tangible and fixed, ubiquitous computing is pervasive, mobile and increasingly footloose due to wireless capacity. The challenge of spatial governance lies in the ability to enhance the opportunities presented by potentially empowering network spaces whilst finding creative ways to address old inequalities.

Planning the smart city

The socio-technical embedding of technology results in diverse spatial configurations, very different from the mainstream representation of 'smart cities'. In India, the Modi regime defines them as "cities that leverage data gathered from smart sensors through a smart grid to create a city that is livable, workable and sustainable" (India Ministry of Urban Development 2016). Multinational technology companies are touted to invest in smart grid infrastructure, while the government of Singapore is claimed to be interested in supporting the Indian government in its construction of 10 smart cities on the Delhi–Mumbai industrial corridor (Sethi 2014). Marketing images project a decontextualised spacelessness, a "fuzzy, New Millen[n]ium fantasy" (Sethi 2014).

The India smart city programme is inherently spatial in its range: retrofitting or redevelopment, greenfields development, or what is termed a 'Pan-city' model focused on upgrading existing infrastructure (India Ministry of Urban Development 2016). The performance criteria speak to a modelling approach to urbanism, reminiscent of Modernist, comprehensive urban planning, or as Datta argues, driving aspirations of modernisation and progress (2015: 53).

The IBM Smarter Cities programme makes three connections to the global South: a celebration of how the last 20 years of technology has enabled improved quality of life: an emphasis on climate change and the fact that vulnerability to floods and storms in over half of the developing world is a reality: and the need for improved governance. Technology is the answer (see www.ibm.com). The central operations centre in Rio de Janeiro, Brazil is one of the most visible examples of IBM/municipal collaboration. Following a flash flood in 2007, the company, together with Cisco and the local authority, developed an integrated disaster response and monitoring system that integrates 32 agencies and services and relies on 400 active cameras in the city for continuous surveillance. As in most socio-economically divided cities, the concern is that only well-off neighbourhoods benefit and that it is an interim measure that detracts from real infrastructural problems.

The Rio Control room has raised concerns about privacy (Singer 2012), one of the manifestations of informational territorialisation (Firmino 2016). CCTV cameras, security electronic applications ('apps') and access to drone technology enable private surveillance of public space. Use of CCTV in Curitiba in Brazil, for example, by private households in the surveillance of public spaces are indicative of a layer of territory that sits between public and private; a form of privately imposed surveillance urbanism tacitly permitted by the state (Firmino and Duarte 2015). This unmediated and 'invisible' intervention in space represents a form of control not dissimilar to more overt exercises of surveillance power. In the post-9/11 world cities have become more militarised in the desire to anticipate danger and exclude perceived disruptive social groups (Graham 2011). In highly unequal societies where fear of crime and desire for local infrastructural governance drives the demand for gated living, technology provides an increasingly sophisticated ally.

When using a socio-technical lens to consider the smart city, many layers of human–material interactions evolve, indicative of spatial practices that bridge traditional binaries of public/private and social/technical.

Representations of the smart city show an eerily place-less visual narrative that seemingly applies everywhere. The contrasts between physical qualities of cities in the global South and this visual language are staggering. If this is indicative of a new form of urbanism, one that is data-driven and essentially accessible to a broad populace, it begs the question how, in the urban areas of the global South, can this translate into transformative planning practice that is cognisant of local place? How do the emerging practices of the 'real' smart city interface with popular conceptions and offer a more contextualised and hopeful alternative? The following section explores these questions with some tentative answers supported by examples.

Technology, space and planning in cities of the global South

Cities in the global South clearly vary in size and character but they do share a number of characteristics: increasing informalisation that is indicative of the failure of the formal economy to provide livelihoods and spatial inequalities as the fallacy of trickle-down economics is displayed in fragmented urban forms and income disparities. The following examples explore these dimensions in relation to ICT.

Rendering the 'invisible' visible

At the World Urban Forum in 2014, in Medellín, Shack/Slum Dwellers International (SDI) convened a session entitled "Smart Cities from the bottom up". Together with the Sante Fé Institute, the organisation is working on uncovering the "science of slums" (Brelsford and Bettercourt 2015), systematically mapping the spatial logics that underpin informal neighbourhoods. The Institute uses geographic information systems (GIS) and other technical tools to analyse the logics of grassroots spatial practices, and how various spatial interventions can assist in improved access. The second part of this initiative relates to what it uncovers, which is of interest to understanding smart practices at a grassroots level. Re-blocking practices (where slum dwellers reorganise their own settlements spatially to enable utility provision) and self-enumeration enable control and generation of data by slum dwellers, empowering them to engage the state.

The acts of documentation and systematic mapping are processes of 'making the invisible visible'. This is the strapline used by the Map Kibera Trust in describing their work in Kenya. In addition to the invisibility of slums on conventional maps and in planning documents, mapping is often outdated as land use and circulation routes change on an ongoing basis to suit local conditions. Technically it therefore makes sense to enable local residents to map and update local conditions, but the actual process of mapping is an act of power.

The MapKibera project initially trained residents in using a range of technologies to map places of interest in Kibera (Hagen 2010). Java editing software was used to map and share this data through OpenStreetMap, a community-driven "wikipedia of maps" that captures local knowledge about places (OpenStreetMap 2015). The project has evolved into three spinoffs: more detailed mapping on prioritised thematic areas, ongoing media development using Kenyan open source tools that enable mapping through use of mobile phones, online video news reporting and SMS monitoring of local issues (Hagen 2010). This learning is now used in two other slums in Nairobi and the website has evolved into a training platform where information and techniques are shared (MapKibera 2015).

Technology-mediated spatial governance

The term 'governance' is often used in relation to smart city discourses, with the promotion of transparency and more integrated decision-making seen as important outcomes. The predominance of social media signifies a shift to a more decentralised form of e-governance where citizens contribute content. But it also reveals a new form of oppositional politics as illustrated by a contemporary case in Cape Town.

The Social Justice Coalition Cape Town (SJC-CT) is a civil society organisation based in Khayelitsha, engaged in monitoring communal sanitation in the area (SJC-CT 2014a). The lack of sufficient maintenance, the limited numbers of facilities and high number of attacks on women at night in communal sanitation areas, together with the fact that many of the toilets provided do not have doors, have caused great and justified embarrassment to the city administration SJC-CT (2014b). SJC-CT uses digital tools to monitor and report on such issues.

The organisation's website includes online petitions, responses to public press releases and links to news articles. One of the drop down menus on the site is entitled 'Imali Yethu', a Xhosa phrase for 'our money'. The menu contains pages with places where service delivery contracts are in place for sanitation, refuse removal and policing in Khayelitsha. The detail of each contract is displayed, with the expected frequency and scope of maintenance tasks, the contact person and detail of the company contracted and time frames. The aim is to empower communities with the necessary facts to monitor and share information on service delivery.

The online dissemination of surveys and reports, as well as links to media from activist organisations, as a counter to municipal evaluations, has proven to be one of the central tasks of the organisation. Using this information to motivate for more rigorous upkeep of communal toilets has resulted in revised service-level agreements between the City of Cape Town and contractors (Mitchell 2014). The fact that these audits on sanitation services are updated monthly is important for ongoing operations, as part of the organisation's 'quiet activism' (Robins 2014).

Negotiating 'new' and 'old' spaces

New spatial configurations have emerged from the interface between ICT and the informal economy that provide opportunities for augmented urban design. Fieldwork I did in Durban, South Africa over the last decade shows interstices on the streets where umbrellas and trolleys combine in physical space to provide telephony to the passer-by, co-existing with containers situated on open spaces often providing community focal points. Telephony extends the repertoire of goods for sale by hawkers as they diversify and agglomerate. Boundaries between the public and private became blurred as a pedestrian takes a moment to have a telephone conversation in a public space. The informal–formal enterprise spectrum incorporates a number of entry points for the telephony trader: from second-hand mobile sales to sidewalk phone booths and sponsored container shops (Odendaal 2014).

Observations show how it extends to community – albeit transient community – as pedestrians go about their business, stopping to make a phone call at a table with an umbrella located on a paved space and then extending to an interchange with the vendor and fellow callers (Odendaal 2014). The space can become private again as booths in shops allow for separation from the bustle of city life. In informal spaces, the smart city is never far away physically, or from our consciousness.

Conclusion

This chapter departs from an infrastructure-led approach to the complexities of cities in the global South. Using examples from the African context I explored a more nuanced reading of the relationship between spatial governance and technology that rejects the notion of the 'smart city' as a standalone, technology-driven solution for complex urban problems. Digital space is on a continuum of urban space that stretches from the physical to the virtual. Newly defined land uses and uses of open space require urban design that is mindful of expanded livelihoods, that accommodates flexibility and acknowledges the fleetingness of exchange. Digitally enabled mobilisation and mapping broaden methodologies of planning, which provide us with clues on the entry points for technology appropriation in urban space.

The relationship between social agency and technology represents an ongoing dance between technical innovation and individual appropriation. Quite when technology evolves from being a connector to an essential part of the social infrastructure people employ to access the city depends on how well it fits in with the day-to-day strategies of the marginalised for whom we plan.

Note

1 Web 2.0 refers to the transition from online presence dominated by connected but static web sites to one where user-generated content and shared operability dominate.

References

Allagui, I. and Kuebler, J. (2011) "The Arab Spring and the role of ICTs: Editorial introduction", *International Journal of Communication* 5: 1435–1442.

Bosch, T. (2011) "Talk radio, democracy and citizenship in (South) Africa". In H. Wassermann (ed.) *Popular Media, Democracy and Development in Africa*, London: Routledge, pp. 75–87.

Brelsford, C. and Bettercourt, L. (2015) "Optimal re-blocking as a practical tool for neighborhood development", New Mexico: Santa Fe Institute. Available: www.santafe.edu/media/working papers/15-10-037.pdf [accessed 3 November 2015].
Cisco website. Available: www.cisco.com [accessed 20 March 2016].
Crang, M. and Graham, S. (2007) "Sentient cities: Ambient intelligence and the politics of urban space", *Information Communication and Society* 10(6): 789–817.
Dada, D. (2006) "The failure of e-government in developing countries: A literature review", *The Electronic Journal of Information Systems in Developing Countries* 26(7): 1–10.
Datta, A. (2015) "The smart entrepreneurial city: Dholera and 100 other utopias in India". In S. Marvin, A. Luque-Ayala and C. McFarlane (eds) *Smart Urbanism: Utopian Vision or False Dawn?* London: Routledge, pp. 52–70.
Davidson, L. (2014) "Emerging markets will lead in smartphone growth next year", *The Telegraph*, 29 September. Available: www.telegraph.co.uk/finance/newsbysector/mediatechnologyandtelecoms/11126997/Emerging-markets-will-lead-smartphone-growth-next-year.html [accessed 10 May 2016].
Donner, J. (2007) "The rules of beeping: Exchanging messages via intentional 'missed calls' on mobile phones", *Journal of Computer-Mediated Communication* 13: 1–14.
Donner, J. (2008) "Research approaches to mobile use in the developing world: A review of the literature", *The Information Society* 24(3): 140–159.
Dourish, P. and Bell, G. (2007) "The infrastructure of experience and the experience of infrastructure: Meaning and structure", *Environment and Planning B: Planning and Design* 34(3): 414–430.
Falch, M. (2004) "Tele-centres in Ghana", *Telematics and Informatics* 21: 103–114.
Firmino, R. (2016) "Connected and controlled", LA+ *Interdisciplinary Journal of Landscape Architecture*: Special Issue on Tyranny, April: 42–47.
Firmino, R. and Duarte, F. (2015) "Private video monitoring of public spaces: The construction of new invisible territories", *Urban Studies* 53(4): 741–754.
Furuholt, B. and Kristiansen, S. (2007) "A rural–urban digital divide? Regional aspects of Internet use in Tanzania", *The Electronic Journal of Information Systems in Developing Countries* 31(6): 1–15.
Graham, M. (2008) "Warped geographies of development: The Internet and theories of economic development", *Geography Compass* 2(3): 771–789.
Graham. S. (2011) *Cities under Siege: The New Military Urbanism*, London: Verso.
Graham, S. and Marvin, S. (2001) *Splintering Urbanism: Networked Infrastructures, Technological Mobilities and the Urban Condition*, London: Routledge.
Hagen, N. (2010) "Putting Nairobi slums on the map", *Development Outreach* (World Bank Institute) 12(1): 41–43.
Hall, P. (2002) *Cities of Tomorrow* (3rd edn), London: Blackwell.
Hammett, D. (2009) "Local beats to global rhythms: Coloured student identity and negotiations of global cultural imports in Cape Town, South Africa", *Social and Cultural Geography* 10(4): 403–419.
Heeks, R. (2003) "Most egovernment-for-development projects fail: How can risks be reduced?" Government Working Paper Series, Vol. 14, Manchester: Institute for Development Policy and Management, University of Manchester. Available: http://unpan1.un.org/intradoc/groups/public/documents/NISPAcee/UNPAN015488.pdf [accessed 1 June 2016].
Heeks, R. (2008) "ICT4D 2.0: The next phase of applying ICT for international development", *Computer: Journal of the IEEE Computer Society* 41(6): 26–33.
Hobbs, J. and Bristow, T. (2007) "Communal computing and shared spaces of usage: A study of Internet cafes in developing contexts", ASIS & TIA Summit, Las Vegas, March, pp. 22–26.
Howard, P.N., Duffy, A., Freelon, D., Hussain, M., Mari, W. and Maziad, M. (2011) "Opening closed regimes: What was the role of social media during the Arab Spring?", Report to the Project on Information Technology and Political Islam. Available: http://papers.ssrn.com/sol3/papers.cfm?abstract_id=2595096 [accessed 18 November 2015].
IBM website. Available: www.ibm.com [accessed 20 March 2016].
India Ministry of Urban Development (2016) India Smart Cities Mission website. Available: http://smartcities.gov.in [accessed 18 March 2016].
The Internet Society (2016) Official website. Available: www.internetsociety.org [accessed 15 May 2016].
Javuru, K. (2013) "New media and the changing public sphere in Uganda: Towards deliberative democracy?". In A.A. Olorunnisola and A. Douai (eds) *New Media Influence on Social and Political Change in Africa*, Hershey, PA: IGI Global, pp. 357–378.
Keniston, K. and Kumar, D. (eds) (2004) *IT Experience in India*, Delhi: Sage.

Kleine, D. (2009) "The ideology behind technology: Chilean microentrepreneurs and public ICT policies", *Geoforum* 40: 171–183.
Krishna, S. and Walsham, G. (2005) "Implementing public information systems in developing countries: Learning from a success story", *Information Technology for Development* 11(2): 123–140.
Luque-Ayala, A. and Marvin, S. (2015) "Developing a critical understanding of Smart Urbanism?", *Urban Studies* 52(12): 2105–2116.
MapKibera (2015) Available: http://mapkibera.org/about/ [accessed 3 November 2015].
Mitchell, H. (2014) "Information and communication technologies and urban transformation in South African township communities", Unpublished MPhil Thesis, University of Cape Town.
Mwesige, P.G. (2009) "The democratic functions and dysfunctions of political talk radio: The case of Uganda", *Journal of African Media Studies* 1(2): 221–245.
Mytton, G. (2000) "From saucepan to dish: Radio and TV in Africa". In R. Fardon and G. Furniss (eds) *African Broadcast Cultures: Radio in Transition*, Oxford: James Currey, pp. 21–41.
Netchaeva, I. (2002) "E-government and e-democracy: A comparison of opportunities in the north and south", *International Communication Gazette* 64(5): 467–477.
Obijiofor, L. (2009) "Mapping theoretical and practical issues in the relationship between ICTs and Africa's socioeconomic development", *Telematics and Informatics* 26(1): 32–43.
Odendaal, N. (2010) "Information and communication technology and urban transformation in South African cities", Unpublished PhD Thesis, University of the Witwatersrand.
Odendaal, N. (2011) "Splintering urbanism or split agendas? Examining the spatial distribution of technology access in relation to ICT policy in Durban, South Africa", *Urban Studies* 48(11): 2375–2397.
Odendaal, N. (2014) "Space matters: The relational power of mobile technologies", *URBE: Brazilian Journal of Urban Management* 6(1): 33–45.
OpenStreetMap (2015) Available: www.openstreetmap.org/about [accessed 18 November 2015].
Paterson, C.A. (1998) "Reform or re-colonisation? The overhaul of African television", *Review of African Political Economy* 78: 571–583.
Rao, S.S. (2005) "Bridging digital divide: Efforts in India", *Telematics and informatics* 22(4): 361–375.
Robins, S. (2014) "Data-driven activism empowering", *Cape Times*, 15 December. Available: www.iol.co.za/capetimes/data-driven-activism-empowering-1.1795657#.Vkw7J9JriM8 [accessed 18 November 2015].
Schech, S. (2002) "Wired for change: The links between ICTs and development discourses", *Journal of International Development* 14: 13–23.
Selwyn, N. and Facer, K. (2007) *Beyond the Digital Divide: Rethinking Digital Inclusion for the 21st Century*, Bristol: Futurelab.
Servon, L. (2002) "Four myths about the digital divide", *Planning Theory and Practice* 3(2): 222–227.
Sethi, S. (2014) "What on earth is a 'smart city'?", *Business Standard*, 18 July, New Delhi. Available: www.business-standard.com/article/opinion/sunil-sethi-what-on-earth-is-a-smart-city-114071801449_1.html [accessed 18 November 2015].
Singer, N. (2012) "Mission control, built for cities: IBM takes 'smarter cities' concept to Rio de Janeiro", *New York Times*, 3 March. Available: www.nytimes.com/2012/03/04/business/ibm-takes-smarter-cities-concept-to-rio-de-janeiro.html?_r=0 [accessed 18 November 2015].
Social Justice Coalition Cape Town (2014a) [accessed December 2014].
Social Justice Coalition Cape Town (2014b) "'Our toilets are dirty'. Report of the social audit into the Janitorial Service for communal flush toilets in Khayelitsha, Cape Town: 14–19 July 2014", Available: http://nu.org.za/wp-content/uploads/2014/09/Social-Audit-report-final.pdf [accessed 18 November 2015].
Thrift, N. (1996) "New urban areas and old technological fears: Reconfiguring the goodwill of electronic things", *Urban Studies* 33(8): 1463–1493.
Townsend, A.M. (2014) *Smart Cities: Big Data, Civic Hackers, and the Quest for a New Utopia*, New York: W.W. Norton & Company.
Warschauer, M. (2004) *Technology and Social Inclusion: Rethinking the Digital Divide*, Cambridge: MIT Press.
Watson, V. (2014) "African urban fantasies: Dreams or nightmares?", *Environment and Urbanization* 26(1): 215–231.

18

Balancing accessibility with aspiration

Challenges in urban transport planning in the global South

Anjali Mahendra

Introduction

Growing urbanisation and economic activity in countries of the global South are placing immense demands on the transport sector. Rising incomes are increasing motorisation at unprecedented rates, with the number of vehicles doubling in some countries every 5–7 years (ADB 2010). This is occurring against a backdrop of deficient or non-existent public transport systems, exclusionary planning that neglects the mobility needs of the disadvantaged and disproportionate investment in road capacity catering to more affluent vehicle owners (Ahmed *et al.* 2008).

Consequently, cities in these countries are facing the brunt of serious challenges related to traffic congestion, air pollution and related health issues, traffic-related injuries and deaths, as well as unsustainable trajectories of energy consumption and climate change. Vehicular emissions account for almost a quarter of global carbon dioxide emissions (ITF 2010) and over 90 per cent of the increase in transport-related carbon dioxide emissions is expected to occur in the global South (UNCSD 2012). Yet, in the absence of other alternatives, a large percentage of people, especially the poor, still rely on walking and bicycling for their mobility needs (Leather *et al.* 2011), locating themselves as close to employment opportunities as possible. These modes constitute about a third of all trips in India, more than a third of all trips in many Latin American cities, almost two-thirds in China and almost 70 per cent of trips in African cities like Dakar (UN-Habitat 2013).

Globally, 1.3 million deaths and 78 million non-fatal injuries requiring medical care were caused by road accidents in 2010 (Bhalla *et al.* 2014). Some 90 per cent of road traffic deaths occur in low- and middle-income countries that comprise only 54 per cent of the registered vehicles in the world (WHO 2015). The premature deaths attributable to transport-related air pollution from particulate matter are estimated at 230,000 per year globally (Bhalla *et al.* 2014). Together, the productivity losses from traffic congestion and economic costs of road accidents and deaths from air pollution can represent 5–10 per cent of the GDP in many cities of the global South (ADB 2010; UN-Habitat 2013; UN News Centre 2012).

These challenges remain unresolved by transport planning in the global South, partly because of how dissociated it is from urban land and economic development planning and partly due to significant institutional and capacity limitations. Achieving accessibility for all people to work, education, health and leisure amenities is a key goal of transport planning; however, this is increasingly undermined as the rapid pace of private motorisation creates strong tensions between the public good and private aspirations.

In addition, the diverse local mobility issues and systems of urban governance across countries mean lessons from one context must be carefully assessed and contextualised before adoption in another. Transport planning in cities of the global South is mired in some of these complications of practice. The purpose of this chapter is to highlight some key common challenges faced by cities as a broad, though not exhaustive, survey of issues. The chapter will conclude with suggestions for a way forward that is befitting to these contexts.

Key urban transport problems and root causes

Transport is one of the planning sectors that has significant economic, environmental and equity implications not just locally but also regionally and globally. The schematic of 'root problems' and global and local 'manifestations' of urban transport challenges shown in Figure 18.1 provides an overview of how these linkages play out.

Over 1 billion motor vehicles operate in the world today, a number projected to double by 2030 (Sperling and Gordon 2008: 1). Despite this growth, about three-fourths of the world's population will still not own motor vehicles in 2020 (Leather *et al.* 2011) and current trends show this is the ignored majority in transport investments in cities of the global South. This has a direct economic outcome as the labour market for this vast section of the population is limited by the lack of accessibility arising from a lack of reliable and affordable transport options. The result is the development of transport systems that are fraught with inequities.

In many cities of the global South, lower income people are disproportionately affected by the negative externalities generated by transport, including road accidents, air pollution and project displacement (Drabo 2013; Vasconcellos 1997). With the incompatible mix of both motorised and non-motorised vehicles traveling at widely different speeds, road conditions are often unsafe for pedestrians and cyclists (Pucher *et al.* 2005). Lower income households living in peripheral areas of cities may lack access to the public transport network, if this even exists. The limited connectivity to employment centres and other amenities creates heavy reliance on informal modes, with multiple modes needed to make a single trip, resulting in high trip costs and long waiting times. Lower income people spend higher shares of their income, about 25–35 per cent on transport (Hook 2005), as compared to the vehicle-owning affluent classes. Figure 18.2 schematically shows how transport choices vary by income in cities of the global South.

This chapter argues that there are four pressing mobility challenges common across cities of the global South.

Unmanaged motorisation, rapid growth of two-wheelers and cars and declining non-motorised accessibility

Among regions seeing high growth in motorisation, Asia's has come at a high price. Asia alone accounts for nearly 60 per cent of the road fatalities that occur every year. Between 36 and 69 per cent of all crashes resulting in injuries in Asian and African countries occur in urban areas, and even these numbers are under-reported (TRL 2000).

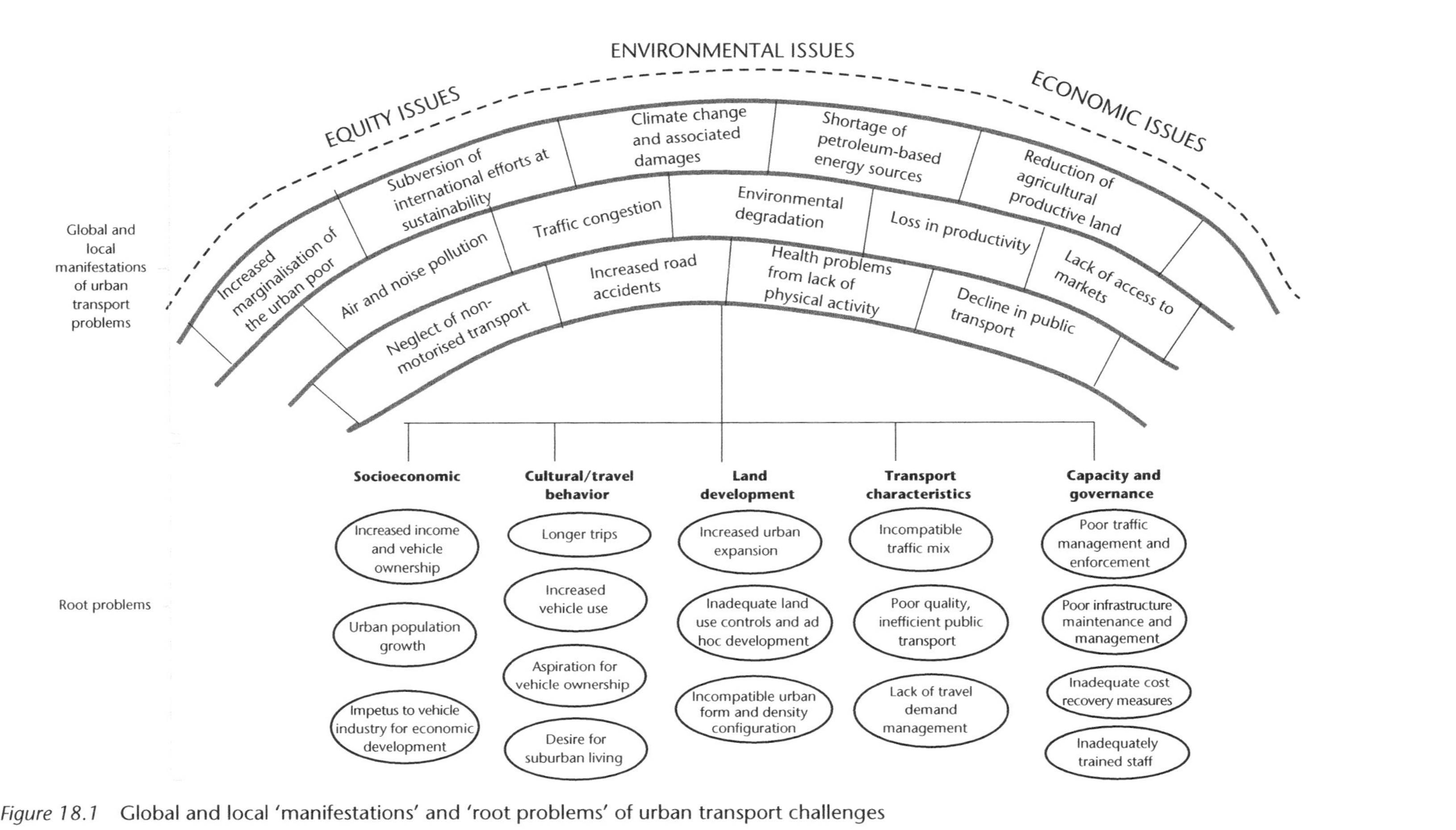

Figure 18.1 Global and local 'manifestations' and 'root problems' of urban transport challenges

Source: adapted from Dimitriou (2011).

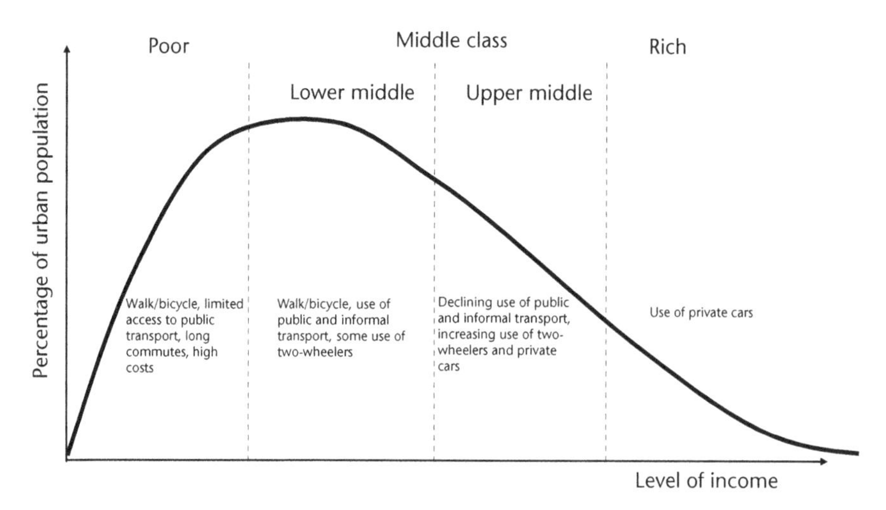

Figure 18.2 Transport choices by income class

Source: adapted from CODATU and AFD (2014).

The motorisation rate or the number of motor vehicles per thousand inhabitants is relatively low in Asian countries, but the rapid growth in cheap, affordable two-wheelers is leading the process of mass motorisation more than cars. In 2008, nearly 80 per cent of the world's 315 million two-wheelers were in Asia. Figure 18.3 shows the increase in motorisation expected in major Asian countries in 2025 and 2035, based on current trends.

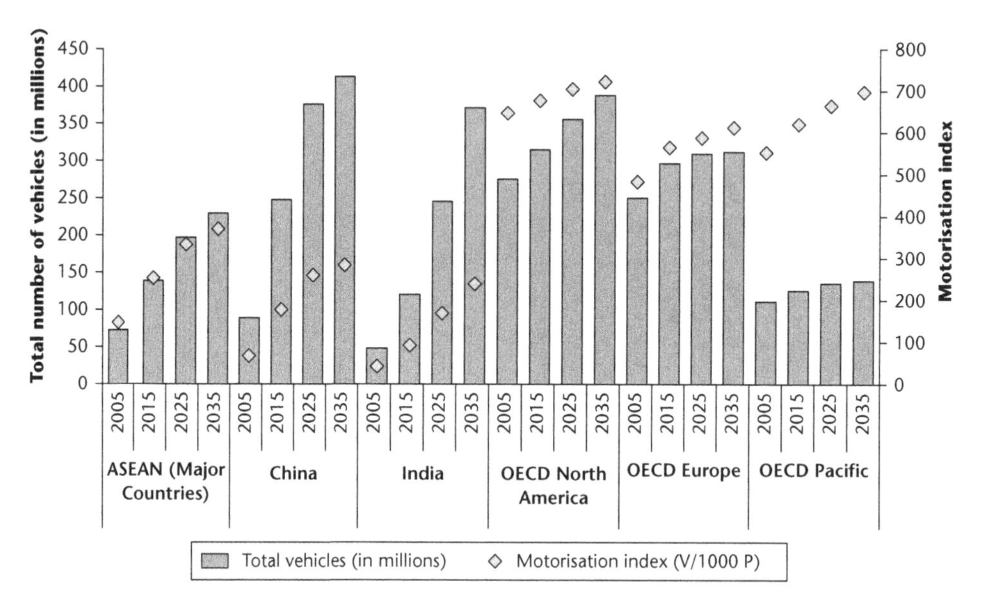

Figure 18.3 Growth in motorisation in Asia compared to OECD countries

Source: Fabian (2012).

Two-wheelers in Asia present a conundrum for planners and policy makers as they pose serious safety concerns, while also engendering a lifelong preference for personal motorised vehicles over other transport modes. Still, they are an important transport mode for a large proportion of people in Asian cities who lack access to good quality, affordable public transport options (Pai *et al.* 2014). Two-wheelers are the predominant mobility mode of the middle class, with sales in countries like India exceeding those of cars by 5 to 1, mostly driven by their competitive price. It is found that when GDP per person reaches about USD5,000 two-wheeler sales begin to decline and sales of private cars begin to increase (Symonds 2008). This threshold is being challenged by car manufacturers wooing the two-wheeler driver with ultra-cheap cars such as the Tata Nano. Termed the "world's cheapest car" (Able 2014) and "The People's Car", it was launched at an entry price of USD2,000 in India in 2009 (Kurczewski 2009) while the typical two-wheeler sold for USD800–1,500 (Symonds 2008).

The immense growth in two-wheeler ownership by the burgeoning middle class in developing countries (Parker 2009) signals an impending crisis with declining public transport patronage and rising dependence on private motorised modes. Public transport here refers to organised, fixed route, fixed fare mass transport services (typically forms of bus or rail services) that are available to all residents.

The rapid growth in two-wheeler users thus has implications for the ridership and use of public transport which loses mode share, primarily because it is similar socio-economic groups that rely on two-wheelers and public transport for their mobility needs. Large cities in India, like Delhi, Kolkata and Mumbai, which have higher public transport supply and mode shares, have lower two-wheeler shares. However, the converse is true in small to mid-sized public transport-deficient cities, which have the highest two-wheeler shares in the country (Pai *et al.* 2014). Women in these cities rely heavily on two-wheelers as their only transport option given the unreliability and safety concerns on the limited public transport that exists.

To accommodate this motorisation growth, most Asian cities continue to sacrifice the needs of pedestrians and cyclists, the majority of road users, through road widening and extension schemes. Dimitriou and Ernst (2001) describe how traffic speeds in Asian cities have continued to decrease despite massive expenditures on road infrastructure, resulting in what they call "an undeliverable vision of motorization" in these countries. They suggest that this trend "looks set to accentuate the 'mobility differential' between the 'haves' and the 'have nots'" in what are already "significantly unequal societies" (Dimitriou and Ernst 2001: 98). It is causing a decline in the already low modal share of public transport, which mostly comprises captive rather than choice riders.

Isolated planning across transport modes and the problem of last-mile access

The lack of an integrated approach to transport planning is ubiquitous across the relatively low-income cities of the global South, particularly in South Asia and sub-Saharan Africa. This includes integration not only across transport modes – public, private, informal, non-motorised and freight – and the infrastructure supporting them, but also crucial integration across the interlinked land use, transport and economic development sectors.

The absence of intermodal integration creates conflicts in the policy agenda that sets priorities and investments for road usage by different modes. Road fatalities are a manifestation of these conflicts. In some countries, while policies that encourage more sustainable and non-motorised transport modes exist, city budgets and infrastructure on the ground do not reflect this, as there are no designated lanes or footpaths to protect vulnerable road users from private motorised vehicles (Wegman 2013; WHO 2013).[1] At a global level too, analyses of funding streams for transport from a range of international sources, shows a minuscule proportion of funding directed towards these objectives (Sakamoto *et al.* 2010).

Integration across modes in terms of infrastructure, travel schedules and fares is crucial for maximising access for the majority of people in cities who do not own private vehicles. The lack of integration, reflected in the problem of last-mile connectivity, is pervasive. For example, in sprawling cities like New Delhi and Mexico City, although good quality metro systems do exist, last-mile access to the stations is severely limited. In most cases, distances to the nearest station are long enough that walking for regular commutes is often not possible. People therefore rely on informal modes such as auto rickshaws and cycle rickshaws in New Delhi, minivans knows as *colectivos* in Mexico City, *matatus* in Nairobi and so on. Like two-wheelers, informal transport modes (alternatively referred to as paratransit or intermediate public transport – IPT) present another conundrum for cities. While widely used, their unregulated operations also impose significant economic and environmental burdens on cities. Yet the informal transport sector is an important source of jobs and income.

The large and growing use of informal modes[2] indicates a crucial gap between transport supply and demand that city agencies are not able to meet. Informal transport poses no costs to public agencies as it does not receive any direct operating subsidies, yet it offers highly flexible, demand-responsive service to passengers in terms of locations served and operating times.

These services are often operated by independent operators who may not be subject to fare regulation. They can charge fares at will, while following unsafe practices or operating inefficient, polluting vehicles. However, informal transport services can play an important role in maximising connectivity as feeder services to public transport operating on trunk routes, instead of competing with key public transport routes. This works well in large cities like Bangkok where higher capacity informal modes like *songtaews*, vans with defined routes, *tuk-tuks* and motorcycle taxis that operate on demand on user-defined routes, offer much needed last-mile access to complement the formal public transit modes (CODATU and AFD 2015a).

Currently, in many cities, passengers pay significantly higher cumulative costs when using a combination of informal transport modes and public transport modes (where the latter exist). Integrating informal transport into urban transport plans and policies can allow efficiencies of fare integration that lower user costs, rationalisation of routes so that they support instead of compete with public transport routes, and better operational practices among operators through regulation and incentives.

Informal transport is a significant source of livelihood in many cities of the global South; for example, researchers estimated that in 2013, auto rickshaw driving created 200,000 jobs in the city of Mumbai (Shlaes and Mani 2013). There are debates in the literature about whether the sector should be formalised or not (Kring and Rothboeck 2014). Recognising the entrepreneurship of individual operators, the challenges they face and the reliance of large majorities of people on these modes, integrating informal transport is a more reasonable approach. Ensuring safe access to public transport stations for not only non-motorised modes but also two-wheelers and informal modes can help solve the last-mile access problem. This is often ignored in cities that are building public transport networks in a manner isolated from planning for other modes.

Misaligned sectoral plans and conflicting government policies

The colonial legacy of many countries in the global South has left behind a patchwork of investments and policies that have determined the future of urban transport in their major cities, for better or worse. For example, in India, the British built the suburban rail network in Mumbai in the 1850s, which has today become one of the busiest commuter rail systems in the world, used by almost 8 million commuters every day (Rao 2000). The system provides the original model of transit-oriented development (TOD) as the entire metropolitan region of Mumbai has expanded around it and derives the benefits of this access. At the same time, land acquisition for urban development in India has been constrained, subject to an antiquated Land Acquisition Act of 1894, also instituted by the British. It was only in 2014 that the Indian government revised this Act amid much controversy. This example highlights the historical context within which today's transport planning challenges in the global South must be understood. Urban transport planning, as it came about in many of these countries as they gained independence, has lacked consideration of land use, land markets and urban density. Transport and land use planning are rarely integrated, with cities missing the benefits of the important interactions and synergies between them.

Gakenheimer (2011) describes the key planning question with respect to land use–transport integration as (i) whether to plan land use in a way that reduces trip-making, reduces trip lengths, and encourages development that is oriented around public transit with transport services planned to complement the land use; or (ii) to let development be steered by market forces, with limited regulation on location, and instead rely on transport sector policies such as vehicle user charges (Mahendra 2008) and restrictions and new mobility services (e.g. Uber) to meet the travel demand generated at these locations. Gakenheimer (2011) argues that the first of these – land use-based urban transport planning – is emerging much more importantly in cities of the global South than in the North. Globally, perspectives have shifted towards the first approach in recent years because of growing concerns about urban sprawl, its resource footprint, climate change, congestion and air pollution.

Transport planning is thus demand driven, responding more to the growing number of private vehicles and estimates of how these will increase, and responding much less to new locations of land development that are primarily in peripheral areas where land is cheaper. Land development led by private developers has been rapid in these locations where city jurisdictions exert limited control. This in turn leads to growth in private vehicle traffic as commuters must travel longer distances to access jobs and other key urban services. Although ideas such as TOD are being advanced in many cities, with some proven examples of success, the reality of land governance and markets and late investments in high-quality public transit systems means that this concept turns on its head and is often manifested as development-oriented transit in the global South.

The disconnect between transport and land use planning is also a reflection of limited technical capacity and traditionally siloed modes of functioning. This is the planning challenge with perhaps the greatest long-term impacts on quality of life in the city. It is directly within the control of city agencies but requires enabling national policies, an ability to enforce plans at the city level and coordination across municipalities in a metropolitan region.

In many countries of the global South, the economic development agenda is also dissociated from spatial realities of how cities are growing. The lack of planning or enforcement where plans do exist, has led to fragmented growth. National and regional investments in special economic zones in remote locations and intercity corridors for purposes of facilitating inter-regional trade and freight movement have ignored connectivity to the rest of the city. This can be avoided by better integration of spatial planning and economic planning at the local, regional and national scales.

In addition, most countries lack a national transport policy and countries such as India and China support the motorised vehicle industry as a key pillar of economic development. Over the last two decades, the automobile-related manufacturing industries have grown rapidly in India and China, now constituting about 6–7 per cent of GDP and employment in both countries (Rode *et al.* 2014). This creates incentives to encourage private vehicle sales for economic reasons that may work against local goals of managing congestion and limiting pollution. Given the heavy economic costs of growing motorisation, it is imperative to pursue an alignment of transport and economic development policies across national, sub-national and local scales.

The financing chasm

Local public transport agencies in many cities of the global South rely heavily on national-level resources and subsidies for their capital investments, operations and maintenance. Even though subsidies bring economic benefits (UN-Habitat 2013), reliance on often unpredictable levels of subsidies makes local transport operations planning problematic. With high upfront capital investments and low revenues, cities are caught in an "underfunding trap" (Ardila-Gomez and Ortegon-Sanchez 2016). Some cases of public–private partnerships in operating city public transport services exist; however, these remain few and far between because of weaker institutions in the global South that are unable to support and derive benefit from private investment. In addition, in the absence of strong national policy frameworks within which cities can design and operate their transport systems, national-level financing may not reach the local level for intended sustainable transport improvements and is often re-categorised for other urban investments or for expansion of roads (Mahendra *et al.* 2013).

With increasing resources spent on road construction, the budget allocation for pedestrian facilities is often in the range of only 0.2–5 per cent of total transport budgets in Asian cities (Leather *et al.* 2011). Although road construction and improvement projects have their place in providing basic accessibility, governments often prioritise urban and inter-urban road widening projects as a solution for congestion. For example, Figure 18.4 shows that in Indian cities, cars and two-wheelers comprise nearly 86 per cent of vehicles, but account for only 29 per cent of trips (IIHS 2011). Under the Indian government's Jawaharlal Nehru National Urban Renewal Mission (JnNURM) programme, about 62 per cent of transport sector funding was allocated

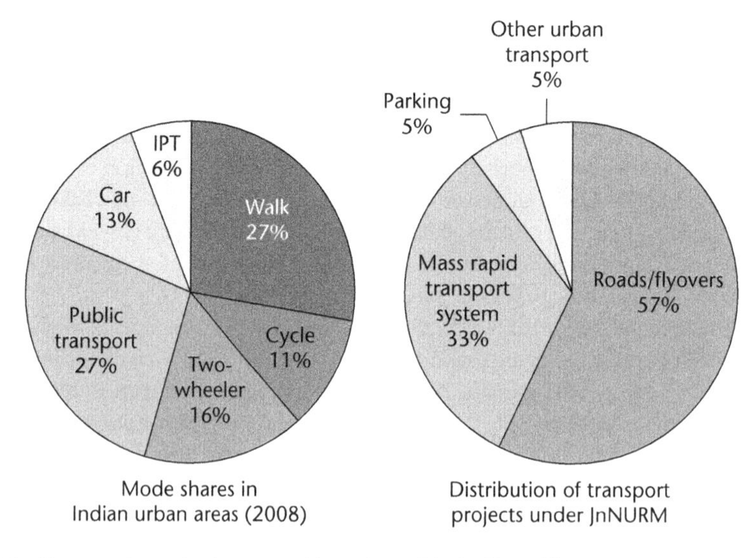

Figure 18.4 Transport mode shares and investment in Indian cities

Source: IIHS (2011).

Note: IPT: Intermediate public transport (auto rickshaws, TATA Magic and similar vehicles).

to roads and flyovers, with only 33 per cent to public transport and 5 per cent to pedestrian and other non-motorised transport modes. This allocation appears questionable, as the Indian Ministry of Urban Development's (MoUD) own study suggests that hardly 30 per cent of roads in most cities have any pedestrian footpaths (MoUD 2008). This mismatch between funding priorities and policy objectives extends to the local level as well.

There is an urgent need to create national policies that develop a coherent vision for sustainable transport investment, create capacity for cities to assess local investment needs and to establish a pipeline of projects that meet policy goals, toward which investment can be directed.

Frequently, perverse incentives in existing financing structures result in consequences opposite to purported goals. For instance, many Chinese cities rely heavily on land concessions to generate revenue for public transport and other local investments, a situation that fosters low-density urban sprawl and instead of supporting public transit, creates the need for increased private motorisation (GIZ, EMBARQ and SLoCaT 2012).

National fuel subsidies are another important cause of market distortions that channel financing away from sustainable modes. Fuel subsidies are very expensive for national governments and economically inefficient. Over 20 per cent of annual global fossil fuel subsidies amounting to USD500 billion is spent by governments in Asia (IMF 2013). Yet, national governments have been slow to reduce fuel subsidies, as this is nearly always politically unfavourable (Mahendra *et al.* 2013).

The challenges described above are common across many countries of the global South, yet emerging good practices in some cities offer crucial lessons in planning specific to this context. Evidence from the global South points to four recommendations towards a strategy for change, discussed in the next section.

Strategy for change involving four key recommendations

Enhance accessibility for all by investing in public, shared and non-motorised transport and manage aspirations by disincentivising private vehicle use

Worldwide, experience in travel demand management and the extremely limited application of user fees or charges to control congestion are evidence of the political difficulty of implementation. Even though car ownership is relatively low in cities of the global South, they often do not have the necessary alternatives available in good quality public transport that would make disincentives to manage vehicle use more publicly acceptable. Singapore, London and Stockholm are examples of cities that have implemented these policies successfully, but none so far in the global South where institutions and city-level authority are typically weaker.

Following the vehicle purchase quotas in Singapore, Shanghai has implemented a system of auctioning a limited number of vehicle license plates each year at a relatively high price, while Beijing in 2011 also implemented a lottery system called the Yaohao (CODATU and AFD 2014). A lack of transparency in license allocation and redistribution of revenues has led to limited acceptance of these measures, but the measures have undoubtedly tempered the growth in the number of vehicles in both cities (CODATU and AFD 2014).

High-quality public transport has the potential to increase the accessibility of low- and middle-income workers to key labour markets, reduce spatial and economic inequalities, and bring public health and environmental benefits. Recent research shows that Bus Rapid Transit (BRT) projects implemented in Latin American and Indian cities resulted in a 52 per cent reduction in fatalities (Carrigan *et al.* 2013). The TransMilenio BRT in Bogotá carries 2.2 million passengers per day and saves almost 40 minutes per trip as compared to driving a private vehicle (CODATU and AFD 2015b). Its impacts are far reaching because it was implemented along with complementary policies including a ban on private vehicles at specific times and days through the *Pico y Placa* scheme (Mahendra 2008), a fuel tax used to fund public transport investments, and the creation of 350 km of bicycle lanes (CODATU and AFD 2015b). Although the BRT network is considered to have limited coverage in locations of the city where the underprivileged live, its environmental and job creation benefits have persisted over time.

Establish national government policy and incentives for integrating urban transport planning across modes, and across land development, economic development and environmental plans

It is crucial to establish metropolitan transport authorities that can plan and coordinate across modes and across economic development, land use and transport plans, for improved mobility and access to jobs beyond municipal borders. Further, informal transit has an important place in providing feeder and last-mile connectivity as part of an organised, safe and integrated public transit system. If implemented well, national policy can play an instrumental role in creating incentives for metropolitan regions to take these actions, while providing greater decision-making authority, national financing and capacity building support to cities (Gakenheimer 2011). It can also create a strong, enabling framework to direct financing appropriately from national and international sources to local governments (Mahendra *et al.* 2013).

Brazil and India are two countries that instituted such national policies. In India, the National Urban Transport Policy established in 2006 stated that the "policy focus is on moving people rather than vehicles" and improving public transport, pedestrian and bicycle amenities in Indian cities (MoUD 2014: i). It called for cities to prepare comprehensive mobility plans integrated with urban development planning. It also required states to establish a Unified Metropolitan Transport Authority (UMTA) in all cities with million plus residents, with the UMTA being backed by state legislation and responsible for integrated citywide transport planning by professional staff. However, few cities have built the technical or financial capacity to do any of this.[3]

In Brazil, the National Law on Urban Mobility in 2012 required over 3,000 municipalities with populations higher than 20,000 to adopt urban mobility plans *(plano mobilidade urbana, PMU)* by 2015. Similar to the Indian policy, the PMU needed to integrate local land use and urban development master plans, emphasising public and non-motorised modes and improving overall accessibility through better integration of transport and land development. As in India, a large share of municipalities lack the capacity to submit coherent plans that can actually be implemented (CODATU and AFD 2015b). But Brazil's mobility law also requires cities to ensure public participation in the development and implementation of their mobility plans, with cities finding innovative ways to share information and engage the public.

These national laws play a crucial role in bringing urban mobility issues on the political agenda and directing national funding to cities to plan and manage urban transport more effectively (CODATU and AFD 2014). City-level capacity, however, has proven to be a bottleneck in implementation.

Tap into national, international and new land-based sources of financing and limit dependence on national transfers or subsidies for operations

To solve the challenge of financing urban transport, an appropriate mix of complementary financing instruments that potentially involves multiple levels of government is necessary

(Ardila-Gomez and Ortegon-Sanchez 2016). Such arrangements will necessarily vary across cities based on the level of local fiscal authority, existing national policies governing urban transport, and the level of participation by the private sector. For instance, Mexico's public transit funding programme PROTRAM involves cost sharing between national, state and municipal governments and the private sector (Hidalgo *et al.* 2012).

Many untapped, viable sources of finance exist to cover the significant financing gaps seen in the transport sector and some have been used successfully in the global South. Economic instruments of transport demand management such as user fees or congestion charging can raise revenues while resolving serious negative externalities of the transport sector. But these policies typically need a strong urban authority for implementation. The use of fuel taxes to fund public transport is one example of a policy that helps correct the incentives in favour of a more balanced transport system. For instance, vehicle drivers in Colombia pay a fuel surcharge generating a funding source that helped finance 20 per cent of the investment for the first three Transmilenio BRT system lines (CODATU and AFD 2014).

A key financing source that has proved effective around the world is the capture of increases in land value that inevitably result from transport improvements. For example, in Mumbai, the extensive Mumbai Urban Transport Project (MUTP) has been significantly financed by the sale of public land to private developers (CODATU and AFD 2014). Similarly in Brazil, the new satellite city of Aguas Claras, located about 20 kilometres outside Brasilia, was connected to the city through an underground metro line, 85 per cent of which was financed through the sale of developed land plots in Aguas Claras.

Leveraging land to finance urban transport has, in the successful cases, certain prerequisites: – the land is either owned or bought by public agencies, the original value of the land is low, and the property market is thriving. In addition, the development and sale of land is also managed jointly with the construction of the transport infrastructure (CODATU and AFD 2014), pointing again to how critical it is for cities to have an integrated approach to land development and transport improvements.

Recognising the key role the transport sector plays in contributing to global carbon emissions, climate finance is emerging as a relatively small, but fast-growing source of international funding. Hanoi in Vietnam has used climate finance from a range of international climate funds such as the Clean Technology Fund and the Global Environmental Facility (GEF), to finance a metro line and improve BRT stations (CODATU and AFD 2014).

Enhance performance of the urban transport system, governance and planning capacity using emerging technologies

Intelligent transportation systems that help public transport agencies provide information to passengers, while managing fleets, routes, and operations from a centralised control centre are increasingly being adopted by more cities in the global South. Fare integration technologies have also become more prevalent. There is now another kind of technological revolution underway in urban planning as cities and their stakeholders realise the potential of the information generated through mobile technologies used increasingly by residents. In transport planning, in particular, cities are now using data from mobile phone use and applications to circumvent the need for time and labour intensive origin-destination surveys that were typically done once in a decade. In addition, satellite images of land development are being used to ascertain the extent of urban activity and adopt a more regional perspective to planning for urban growth and services.

New mobility transport services have emerged that take advantage of mobile phone applications and allow citizens to share vehicles (bicycles and cars), share rides, access taxis, rickshaws and similar intermediate public transport options. Car-sharing companies like ZipCar and the so-called 'transport network companies' (TNCs) such as Uber and Lyft are expanding services rapidly in cities of the global South. While disrupting traditional mobility models, these TNCs are also competing with existing traditional taxi services and informal transport in these cities, with disruptive impacts on the livelihoods of these operators. Many countries of the global South do not have appropriate regulations in place to manage these new modes and deal with critical issues such as pricing or the safety of passengers. Although an interesting phenomenon in itself, the jury is still out as to whether these emerging modes have enough of an impact on car ownership in the global South to temper current motorisation trends. With the limited evidence that exists, there appear to be benefits for cities that integrate new mobility solutions, informal and public transport to enhance mobility choices for citizens, as long as this is complemented by appropriate regulation and a focus on improving accessibility for all classes of people, including those that rely on non-motorised transport.

The use of open data and technology holds much promise to enhance performance and increase transparency, accountability and public participation in transport planning in the global South. For instance, the Florianopolis metropolitan region in Brazil established an online public engagement platform to seek inputs from residents on its regional mobility plan (Worker 2014). Another interesting Brazilian example is São Paulo's Mobilab, a laboratory for urban mobility solutions set up by the city in collaboration with researchers and entrepreneurs to analyse open data (Zottis 2014). Considered the first city in Latin America to open its public transport data to the public, São Paulo's open GPS data from bus systems has resulted in the creation of 60 user-oriented apps that allow people to plan their trips and check bus arrival times. The city is using the large amounts of data generated each year to monitor performance of the public transit system (Zottis 2014).

In a similar way, information and communication technologies have helped spawn an interesting array of user-friendly applications. Some examples are: crowdsourced accident reporting to identify accident hot spots and implement improvements as in New York City, use of radar data to understand origin-destination patterns and vehicle speeds as in São Paulo (Zottis 2014), collaborative mapping and route planning for *matatus* (informal transport) in Nairobi using mobile phone applications (Digital Matatus Project 2016), and GPS and mobile phone-based Dial-a-Rickshaw services for 'Ecocabs' cycle rickshaws in Fazilka and 'G-Auto' auto rickshaws in Rajkot, India (Kring and Rothboeck 2014).

In sum, these new technologies are proving powerful in facilitating essential transport planning tasks, including: conducting annual citywide mobility surveys and user perception surveys of transport needs and challenges; measuring transport system performance and impacts in terms of road safety, air pollution, vehicle speeds, congestion and carbon emissions; and enhancing transparency of operations and public participation in planning.

Conclusions and directions for further research

With much of the transport planning research continuing to be generated from the West, there remain important knowledge gaps in our understanding of transport planning in the global South. First, there is a crucial need for empirical evidence on land use and transport interactions in these cities and to what extent built form factors such as density, diversity or the extent of mixed use, design, distance to transit and accessibility to destinations (the 5 Ds often cited in transport planning studies in the West) (Ewing and Cervero 2010) affect trip-making and vehicle kilometres travelled. These relationships are bound to be different in contexts where large proportions of people walk and cycle out of need, where there is heavy use of two-wheelers and densities are often many orders of magnitude greater than those seen in Western cities.

Second, when project appraisals are done to assess alternatives, the social and economic costs, including the health impacts and costs of motorised transport must be factored in. These more comprehensive assessments of urban transport projects and policies will help balance mobility priorities with other crucial social, economic and environmental goals such as promoting equitable access, reducing congestion, air pollution and health impacts and providing access to jobs for those residing in remote or informal settlements.

Third, more empirical evidence must be gathered on the impacts of two-wheelers and informal transport in cities of the global South. This is important to determine their benefits and challenges and accordingly design policies to manage their growth and integration with other modes as well as incentives to encourage innovation in these modes. So far, these modes remain understudied and therefore largely ignored in urban transport planning.

Finally, it is important to assess the economic costs of conflicting programmes at national and local levels that subvert the best intentions to improve accessibility and quality of life in cities of the global South. Some examples are: local pollution reduction or climate action plans in the presence of national fuel subsidies; or programmes to manage extreme levels of congestion in cities in the face of national support of the automobile industry as a key pillar of economic development; or unregulated private sector-led fragmented land development causing urban sprawl, while national governments invest in public transport projects that mostly serve high densities. Presenting the economic costs of such misaligned policies to decision makers and urban change agents in these cities is a critical first step to correcting the misalignment.

Despite political stalemates, bringing the right information and analysis into the public domain can allow civil society organisations and the public to demand and drive change. This democratisation of transport planning would be a welcome trend, particularly in the global South where inequities in access to high-quality transport services have exacerbated the rise of private motorisation. The task at hand is to ensure that urban transport services fulfil their role as an enabler of access to jobs, education, health care and other amenities for all sections of the population.

Notes

1 For example, India's first National Urban Transport Policy, adopted in 2006, encourages cities to support walking and cycling, but analysis of city budgets in the city of Pune (Menon 2011) shows a very small budget directed toward such infrastructure.
2 Informal modes comprise a heterogeneous range of small- to medium-sized vehicles with an intermediate capacity between automobiles (14 people) and large buses (50 people or more) with fragmented, individualised ownership (CODATU and AFD 2015a). Informal modes typically have a market share ranging from about 10–95 per cent across cities of the developing world, with significantly higher rates – about 70–75 per cent – in many African and some Latin American cities, and lower rates – about 25–50 per cent – in Asian cities where two-wheeler ridership is prevalent (CODATU and AFD 2015a).
3 Ten years on, although there are 53 'million plus' cities in India, only 10 of these have UMTAs and experts suggest the actual number is closer to 5.

References

Able, V. (2014) "Tata Nano: The car that was just too cheap", *The Guardian*, 3 February. Available: www.theguardian.com/commentisfree/2014/feb/03/tata-nano-car-cheap-poor-safety-rating [accessed 11 April 2016].

Ahmed, Q.I., Lu, H. and Ye, S. (2008) "Urban transportation and equity: A case study of Beijing and Karachi", *Transportation Research Part A* 42(1): 125–139.

Ardila-Gomez, A. and Ortegon-Sanchez, A. (2016) *Sustainable Urban Transport Financing from the Sidewalk to the Subway: Capital, Operations, and Maintenance Financing*, Washington, DC: The World Bank. Available: http://hdl.handle.net/10986/23521 [accessed 11 April 2016].

Asian Development Bank (ADB) (2010) "Sustainable transport initiative operational plan". Available: www.adb.org/sites/default/files/institutional-document/31315/sustainable-transport-initiative.pdf [accessed 24 October 2016].

Bhalla, K., Shotten, M., Cohen, A., Brauer, M., Shahraz, S., Burnett, R., Leach-Kemon, K., Freedman, G. and Murray, C.J.L. (2014) *Transport for Health: The Global Burden of Disease from Motorized Road Transport*, Global Road Safety Facility. Washington DC: Institute for Health Metrics and Evaluation and World Bank. Available: http://documents.worldbank.org/curated/en/2014/01/19308007/transport-health-global-burden-disease-motorized-road-transport# [accessed 30 January 2017].

Carrigan, A., King, R., Velasquez, J.M. Raifman, M. and Duduta, N. (2013) "Social, environmental and economic impacts Of BRT systems: Bus Rapid Transit case studies from around the world", EMBARQ, A programme of the World Resources Institute. Available: www.wrirosscities.org/sites/default/files/Social-Environmental-Economic-Impacts-BRT-Bus-Rapid-Transit-EMBARQ.pdf [accessed 30 January 2017].

CODATU and Agence Francaise de Développement (AFD) (2014) *Who Pays What for Urban Transport? A Handbook of Good Practices*. Available: www.unep.org/Transport/ASTF/pdf/whopays_forwhat.pdf [accessed 11 April 2016].

CODATU and AFD (2015a) *Paratransit: A Key Element in a Dual System*. Available: www.codatu.org/wp-content/uploads/transports_collec_artisanal_V03ecran_EN.pdf [accessed 11 April 2016].

CODATU and AFD (2015b) *Urban Mobility: A Source of Solutions Against Climate Change*. Available: www.codatu.org/wp-content/uploads/climat_EN_V_Ecran.pdf [accessed 11 April 2016].

Digital Matatus Project, Collaborative Mapping for Public Transit Everywhere (2016) "About". Available: www.digitalmatatus.com/about.html [accessed 11 April 2016].

Dimitriou, H. (2011) "Transport and city development: Understanding the fundamentals". In H.T. Dimitriou and R. Gakenheimer (eds) *Urban Transport in the Developing World: A Handbook for Policy and Practice*, Washington, DC: Island Press, pp. 8–39.

Dimitriou, H. and Ernst, J. (2001) "The undeliverable vision: Problems and prospects of motorisation in Asia", *Competition and Change* 5: 73–102.

Drabo, A. (2013) "Intra-country health inequalities and air pollution in developing countries", *Oxford Development Studies* 41(4): 455–475.

Ewing, R. and Cervero, R. (2010) "Travel and the built environment", *Journal of the American Planning Association* 76(3): 265–294.

Fabian, B. (2012) "Overview on transport data and MRV potential in Asia", Pasig, Phillipines: Clean Air Initiative for Asian Cities Center.

Gakenheimer, R. (2011) "Land use and transport in rapidly motorizing cities: Contexts of controversy". In H.T. Dimitriou and R. Gakenheimer (eds) *Urban Transport in the Developing World: A Handbook for Policy and Practice*, Washington, DC: Island Press, pp. 40–70.

GIZ, EMBARQ and SLoCaT (2012) "Prospects for national-level programmes and funds for sustainable urban transport in China", Summary of Workshop Results, Expert Workshop 1–3 November, Mutianyu. Available: http://sustainabletransport.org/final-workshop-summary-report-on-financing-sustainable-urban-transport [accessed 14 October 2016].

Hidalgo, D., Pai, M., Carrigan, A. and Bhatt, A. (2012) "National investment in urban transport: Towards people's cities through land use and transport integration", EMBARQ India. Available: www.wrirosscities.org/sites/default/files/National-Investment-Urban-Transport-EMBARQ-India.pdf [accessed 14 October 2016].

Hook, W. (2005) "Urban transport and the Millennium Development Goals". Available: http://siteresources.worldbank.org/INTTSR/Resources/Hook_MDG_and_Transport_Article_final_nov05_no_pictures.pdf [accessed 14 October 2016].

Indian Institute for Human Settlements (IIHS) (2011) "Urban India 2011: Evidence", presented at India Urban Conference, November, Delhi, India. Available: http://iihs.co.in/knowledge-gateway/wp-content/uploads/2015/08/IUC-Book_02-03-12-LOW-RES.pdf [accessed 14 October 2016].

International Monetary Fund (IMF) (2013) "Energy subsidy reform: Lessons and implications", International Monetary Fund. Available: www.imf.org/external/np/pp/eng/2013/012813.pdf [accessed 11 April 2016].

International Transport Forum (ITF) (2010) "Reducing transport greenhouse gas emissions: Trends and Data", OECD. Available: www.internationaltransportforum.org/Pub/pdf/10GHGTrends.pdf [accessed 11 April 2016].

Kring, T. and Rothboeck, S. (2014) "Promoting transition towards formalization: Selected good practices in four sectors", International Labour Organization, ILO DWT for South Asia and Country Office for India, New Delhi. Available: www.ilo.org/wcmsp5/groups/public/---asia/---ro-bangkok/---sro-new_delhi/documents/publication/wcms_344607.pdf [accessed 11 April 2016].

Kurczewski, N. (2009) "Tata Nano launched in Mumbai", *The New York Times*, 23 March. Available: http://wheels.blogs.nytimes.com/2009/03/23/tata-nano-launched-in-mumbai/?_r=0 [accessed 11 April 2016].

Leather, J., Fabian, H., Gota, S. and Mejia, A. (2011) "Walkability and pedestrian facilities in Asian cities: State and issues", ADB Sustainable Development Working Paper Series, Asian Development Bank. Available: www.adb.org/sites/default/files/publication/28679/adb-wp17-walkability-pedestrian-facilities-asian-cities.pdf [accessed 11 April 2016].

Mahendra, A. (2008). "Vehicle restrictions in four Latin American cities: Is congestion pricing possible?" *Transport Reviews*, 28(1): 105–133.

Mahendra, A., Raifman, M. and Dalkmann, H. (2013) "Financing needs for sustainable transport systems for the 21st century", Environmentally Sustainable Transport (EST) Asia Forum, Indonesia. Available: www.wrirosscities.org/sites/default/files/Financing-Needs-for-Sustainable-Transport-Systems-21st-Century.pdf [accessed 14 October 2016].

Menon, R. (2011) "Pune transport budget analysis 2011–2012", Report 9, Parisar. Available: http://admin.indiaenvironmentportal.org.in/files/file/transport%20budget%20analysis%202011-12%5B1%5D.pdf [accessed 14 October 2016].

Ministry of Urban Development (MoUD), India (2008) "Study on traffic and transportation policies and strategies in urban areas in India", final report. Available: https://casi.sas.upenn.edu/sites/casi.sas.upenn.edu/files/iit/GOI%202008%20Traffic%20Study.pdf [accessed 25 May 2016].

Ministry of Urban Development (MoUD), Government of India (2014) "National urban transport policy, 2014". Available: www.itdp.in/wp-content/uploads/2014/11/NUTP-2014.pdf [accessed 24 October 2016].

Pai, M., Gadgil, G., Mahendra, A., Vernekar, S., Heywood, R. and Chanchani, R. (2014) "Motorized two-wheelers in Indian cities: A case study of the city of Pune, India", Working Paper, EMBARQ India, Mumbai. Available: www.wrirosscities.org/sites/default/files/Motorized-Two-Wheelers-Indian-Cities-Pune-EMBARQ-India.pdf [accessed 11 April 2016].

Parker, J. (2009) "Burgeoning bourgeoisie", a special report on the new middle classes in emerging markets, *The Economist*, 12 February. Available: www.economist.com/node/13063298 [accessed 11 April 2016].

Pucher, J., Korattyswaropam, N., Mittal, N. and Ittyerah, N. (2005) "Urban transport crisis in India", *Transport Policy* 12(3): 185–198.

Rao, S. (2000) "How many people take the Mumbai local trains every day?", *Mid-Day*, 3 May 3, 2016. Available: www.mid-day.com/articles/how-many-people-take-the-mumbai-local-trains-every-day/17191155 [accessed 24 October 2016].

Rode, P., Floater, G., Thomopoulos, N., Docherty, J., Schwinger, P., Mahendra, A. and Fang, W. (2014) *Accessibility in Cities: Transport and Urban Form*, NCE Cities Paper 03, LSE Cities, London School of Economics and Political Science. Available: http://newclimateeconomy.report/wp-content/uploads/2014/11/Transport-and-urban-form.pdf [accessed 11 April 2016].

Sakamoto, K., Dalkmann, H., and Palmer, D. (2010) "A paradigm shift towards sustainable low-carbon transport: Financing the vision ASAP", report prepared for the Institute for Transportation and Development Policy. Available: www.policyinnovations.org/ideas/policy_library/data/01592/_res/id=sa_File1/A_Paradigm_Shift_toward_Sustainable_Transport.pdf [accessed 11 April 2016].

Shlaes, E. and Mani, A, (2013) "A case study of the auto-rickshaw sector in Mumbai", EMBARQ India. Available: http://wricitieshub.org/sites/default/files/Mumbai%20auto-rickshaw%20sector_Case%20study_EMBARQ%20India.pdf [accessed 14 October 2016].

Sperling, D. and Gordon, D. (2008) "Two billion cars: Transforming a culture", *Transportation Research (TR) News*, November-December. Available: http://onlinepubs.trb.org/onlinepubs/trnews/trnews259billioncars.pdf [accessed 11 April 2016].

Symonds, M. (2008) "A global love affair", a special report on cars in emerging markets. *The Economist*, 15 November. Available: www.economist.com/node/12544933 [accessed 11 April 2016].

Transport Research Laboratory (TRL) (2000) "Review of road safety in urban areas". Available: http://siteresources.worldbank.org/INTURBANTRANSPORT/Resources/urban_safety_trl_rs_2.pdf [accessed 24 October 2016].

UNCSD (2012) "Sustainable, low carbon transport in emerging and developing economies", *Rio 2012 Issues Briefs*, UNCSD Secretariat and the Partnership for Sustainable, Low Carbon Transport (SLoCaT). Available: https://sustainabledevelopment.un.org/content/documents/403brief13.pdf [accessed 11 April 2016]

UN-Habitat (2013) "Planning and design for sustainable urban mobility: Global report on human settlements 2013". Available: http://mirror.unhabitat.org/pmss/getElectronicVersion.aspx?nr=3503&alt=1 [accessed 11 April 2016].

UN News Centre (2012) "Rio+20: Development banks to invest $175 billion in sustainable transport". Available: www.un.org/apps/news/story.asp?NewsID=42287#.WACrJPmc6ko [accessed 14 October 2016]. Text of Commitment available: www.eib.org/attachments/press/statement_commitment_sustainable_transport_en.pdf [accessed 14 October 2016].

Vasconcellos, E.A. (1997) "The making of the middle-class city: Transportation policy in São Paulo", *Environment and Planning A* 29(2): 293–310.

Wegman, F. (2013) "Road safety in India: A systems approach", presented at the 5th Annual TRIPP Lecture, Delhi: Transportation Research and Injury Prevention Programme (TRIPP), Indian Institute of Technology (IIT) Delhi.

Worker, J. (2014) "Why the public voice matters in urban mobility planning: Lessons from Brazil", The CityFix, WRI Ross Center for Sustainable Cities, World Resources Institute, 22 December. Available: http://thecityfix.com/blog/public-participation-sustainable-urban-mobility-plan-brazil-cities-pac-jesse-worker/ [accessed 11 April 2016].

World Health Organization (2013) *Global Status Report on Road Safety 2013*, Geneva, Switzerland: WHO. Available: www.who.int/iris/bitstream/10665/78256/1/9789241564564_eng.pdf?ua=1 [accessed 11 April 2016].

World Health Organization (2015) *Global Status Report on Road Safety 2015*, Geneva, Switzerland: WHO. Available: www.who.int/violence_injury_prevention/road_safety_status/2015/en/ [accessed 24 May 2017].

Zottis, L. (2014) "How technology can transform Urban Mobility: A Q&A with São Paulo's Ciro Biderman", The CityFix, WRI Ross Center for Sustainable Cities, World resources Institute, 6 October. Available: http://thecityfix.com/blog/technology-transform-urban-mobility-q-a-sao-paulo-brazil-ciro-biderman-mobi-lab-luisa-zottis/ [accessed 11 April 2016].

Part IV

Landscapes of citizenship

19

'Terra nullius' and planning

Land, law and identity in Israel/Palestine

Oren Yiftachel

> We are determined to accelerate our efforts to settle Jews in the Negev. This is necessary to protect state lands against illegal Bedouin invasions that occur daily ... For this end, the government has decided to add five more Jewish settlements. The ministry, under my leadership, will move immediately to plan and build these settlements.
>
> *(Yair Gallant, Israeli Minister for Housing and Construction, 17 November 2015)*

'Planning' in this chapter pertains to the making, content and consequences of public plans governing urban and regional development. As such it covers a wide range of design, legal, regulative economic, ethnic and political decisions that together 'produce' societal space. Theorisation of urban and regional planning has advanced in leaps and bounds during the last three decades. Despite the continuing dominance of functionalist, state- and market-centric analyses, notable critical perspectives have emerged regarding the 'darker sides' of urban and regional policies (Flyvbjerg 1996; Yiftachel 1994), most notably the emergence of 'Southern' or 'South-Eastern' perspectives in recent years (Connell 2014; Roy 2005; Watson 2009; 2014; Yiftachel 2006).

Yet, the surge of critical writing has been (inevitably) partial. One aspect not analysed sufficiently as yet, has been the frequently oppressive consequences of the interaction between law, land and identity (for exceptions see Alfasi 2014; Blomley 2002). As the opening quotation by the Israeli housing ministry shows, planning, land and law are strongly intertwined in several illuminating ways: it is identity that largely determines the allocation of land, implemented through settlement planning. In turn, it is the interconnection between land and planning (settlement) policies that largely determines the plight of identity groups in the Israeli/Palestinian space. This 'triad' creates sets of relations that should be brought to the centre of our conceptual and empirical writing about planning.

Yet, planning and urban theory has given only scant attention to these matters, mainly because, as Watson (2009, 2014) and Connell (2014) illustrate, most planning knowledge, tools and concepts emerge from liberal democracies in which citizenship, governmental liberalism, legal (*de jure*) equality, human rights and the rule of law are basic norms that operate within a hegemonic capitalist framework. To be sure, these norms are never fully observed, and are often abused, even in the global North-West. Yet the impact of concepts that 'run against the grain' of liberalism on planning and urban theories has been too marginal in the theoretical and historical discussion (for exceptions, see Blomley 2002; Porter 2014).

Drawing on these shortcomings, this chapter foregrounds the relations between planning, land, law and identity. It focuses in particular on the concept of 'terra nullius' (TN) – land deemed as 'empty' of rights – as key to understanding these relations. The empirical focus is on the charged context of Israel/Palestine where the settler Jewish state has extensively used legal and planning tools to seize, control and manage contested indigenous Bedouin lands.

Terra nullius

The concept derives from a Roman legal doctrine and evolved during the period of European imperialism to denote a legal setting in which lands are classified as emptied of sovereignty, ownership or possession rights. The concept was widely used – explicitly and implicitly – by European empires and settlers to dispossess, evict and at times genocide indigenous populations, while exploiting the resources of their lands and regions (Daes 2008; Du Plessis 2011; Fisch 1988; Macklem 2008; Wolfe 2006). While the precise conditions for classifying land as 'empty' vary by time and region, a common denominator emerged of annulling indigenous histories, legal systems and property rights. TN represented one of the most effective and notorious hallmarks of the racist colonial period, as noted by Judge Hollinworth of the Australian High Court in the famous 1992 Mabo case:

> Australia was declared to be *terra nullius* ... This strategy enabled the British to class their occupation as 'peaceful settlement' rather than invasion ... no treaties or agreements (were signed) with Indigenous leaders. No compensation or legal recognition of Indigenous property rights was made ... This legal concept enabled generations of indigenous dispossession.
>
> *(High Court of Australia 1992)*

TN is far more than a legal concept. It is a frame of mind typifying colonial and ethnocratic regimes. While the concept rests on legal foundations, its most powerful effect lies well beyond the legal – stripping indigenous peoples of their culture, histories and codes of governance. The concept endows the invading or expanding entity the power and legitimacy to define when and where land is considered 'empty', and hence who is a 'rightful' owner.

Needless to say, not all legal and planning interventions are based on TN principles. In recent decades, the concept of indigenous, multicultural and democratic planning 'from below' have gained considerable traction in research and professional practice (see Alfasi 2014; Sandercock 2003; Yiftachel *et al.* 2016). Yet, TN has been, and still is, an important foundation over which planning and law often construct contested space, with little critical research and theorisation.

In recent decades, serious challenges and indigenous mobilisation have eroded the use and applicability of TN (see Altman 2014; McAulsan 2013; Macklem 2008; Sandercock 2003). It was famously declared null and void in Australia following the 1992 Mabo decision (Porter 2014). The case of Australia has become a symbol for a counter-movement, in which indigenous and other marginalised groups strengthen their resistance to the hegemonic order, using their indigeneity and human rights norms as important tools in the struggle for land and planning rights (see Stavenhagen and Amara 2013; Yiftachel *et al.* 2016).

Importantly, the TN concept is often used implicitly by policy makers, through the discourses, narratives, norms and practices of hegemonic groups that attempt to seize and control ethnic and racial minorities. Moreover, TN is not limited to reconstruction of the legal past and present, as it also reconstructs the future, mainly through land allocation and urban and regional planning. The conceptual framework of 'empty land' has been vital for the imagination, practices and abuses of redeveloping land, both in Israel/Palestine and elsewhere. This has led many (though of course not all) planning professionals to often treat development sites as 'tabula rasa'. In such a manner, TN (explicitly or implicitly) has been vital for the colonisation and planning of contested lands in colonial and postcolonial eras, and must be exposed and unpacked in order to offer better understanding of the use and abuse of this concept, as well as articulate ways to resist and overcome its consequences.

Land, law and planning in southern Israel/Palestine

Let us move to an analysis of a telling case study of the struggle over the Negev region (in Arabic, *Naqab*) lying at the southern half of Israel/Palestine. The region has been inhabited for centuries by Bedouin Arabs until being taken over by Israel in 1948, and consequently thoroughly Judaised. In previous centuries, semi-nomadic Bedouin Arabs gradually settled into permanent localities, combining agriculture with pastoral grazing (Bailey 1980; Falah 1983; Meir 1997). The historic Bedouin region was governed by well-established codes of ownership, partition, sales and conflict resolution (for details see Abu-Sitta 2009; Amara *et al.* 2013; Frantzman *et al.* 2012; Nasasra *et al.* 2014; Yiftachel *et al.* 2012, 2016).

Historically, the legal challenge facing the settling Jewish group was more complex than most other settler societies, because modern land laws were already established by the previous imperial Ottoman and British rulers. Hence, in order to gain statutory control over contested lands, it was necessary for the legal system to construct new 'truths' that erased previous possession, ownership and development rights, already recognised by modern imperial regimes. In the Negev, the state approach has been termed 'the dead Negev doctrine' which became the Israeli version of TN (Yiftachel *et al.* 2012).

The doctrine relied, and manipulated, the Islamic tradition where 'dead' (*mawat*) land was coded into the 1858 Ottoman Land Law as deserted, unpossessed, unused lands, lying at least 2.5 km from the edge of a locality (OLC 1858: Clauses 6, 103). Importantly, and contrary to the use of this concept by Israel more than a century later, the Islamic and Ottoman law sought to encourage development and cultivation, even without prior government approval. It enabled the population to gain property rights through 'vivification' and development of unpossessed, uncultivated, lands.

British rule over Palestine, beginning in 1917, brought several changes to the legal geography of land, some of which under Zionist pressure (Abu-Sitta 2009; Essaid 2013). Notable among the changes was the 1921 'dead land ordinance', which prohibited 'vivification' of deserted land without government approval (Essaid 2013). The new ordinance required possessors of mawat lands to register the land within a two-month period, ending on 16 April 1921. However, the move proved a failure, as reported by the Abramson Land Settlement Commission in late 1921 (Abramson Commission 1921), as only 4.4 per cent of the land in Mandatory Palestine was registered, almost entirely in the urban areas, and only negligible portion as a result of the mawat ordinance (Abramson Commission 1921; Bunton 2007).

The failure prompted the British to launch a comprehensive land registration effort, culminating in a new Settlement of Title Ordinance (1928), which to date forms the legal basis for land settlement between Jordan and Mediterranean Sea. Under the ordinance, registration progressed professionally, mainly along the Mediterranean coast and northern valleys, where Jews held significant tracts of land. The process moved slowly, and by British departure in 1948, only 20 per cent of Mandatory Palestine was surveyed and registered (Forman 2002).

A full analysis of the key role played by the British imperial authorities in the Negev must await another paper (see also Nasasra *et al.* 2014; Kark and Frantzman 2012). Suffice it to say here that the legal framework created by the British authorities proved vital for later Zionist colonisation and the dispossession of the indigenous. The British themselves, however, explicitly protected indigenous Palestinian rights and customs (particularly in the Negev), and even appointed tribal courts to preside over conflicts in the region. Yet, their early legislative steps were used and manipulated years later by the Israeli state. Hence, echoes of global TN were imported by the British to Palestine, and formed the foundation for the Israeli version of TN in the region, and later in the occupied West Bank (see B'tselem 2012).

Following independence in 1948, associated with mass eviction of the Bedouins, Israel adopted the Mandate land settlement legislation almost unchanged. However, its *interpretation* changed radically, borrowing selectively and manipulatively from a range of past legislation. This resulted in the listing of unattainable conditions for proof of ownership by the Bedouins. This interpretation became the main axis for the prolonged land planning conflict in the region.

At the same time, Israel progressed with land registration, reaching the Negev by the early 1970s. The Bedouins filed 220 tribal land claims, covering 1.5 million dunam (roughly a seventh of the Negev region). Hopes among the indigenous population were high, given the recognition of previous regimes, and by the pervasive legalised pre-1948 Jewish land purchase from the Bedouins, amounting to over 160,000 dunams (Kark 2002; Fischbach 2003). Even major Zionist organisations such as the Jewish National Fund (JNF) and the Palestine Development Fund recognised Bedouin ownership by purchasing from them vast tracts of their lands, after registering the land as belonging to the Bedouins before its purchase (Kark 2002). This obvious confirmation of indigenous land ownership was later deemed by Israeli

Figure 19.1 A customary Bedouin land sale document ('sanad') dated 1911, displayed by the head of one of the Araqib tribes

Source: author.

courts as "documentation of transactions" and hence a series of non-binding cases, whose legal logic was "unclear" (see also Kedar *et al.* in press 2017; Supreme Court 2015; see Figure 19.1).

In parallel, Israel has attempted for decades to coerce the Bedouins to urbanise into newly created Bedouin towns. It was steadfast, until the late 1990s, in its refusal to recognise the 46 unrecognised Bedouin-Arab localities, now hosting close to 100,000 people. This refusal, in turn, rests on the denial of any Bedouin property rights on lands inherited from their ancestors of many generations. This denial is achieved by 'emptying' the land of its history, belongings and past legal conventions. This results not only in refusing to recognise their land rights, but denying these communities most basic services, such as roads, public transport, water, electricity, education or health facilities. In recent years, Bedouins also face the highest levels of house demolitions ever recorded in the state's history, reaching 945 demolitions in 2014 – five times higher than in the occupied West Bank (Negev Coexistence Forum 2015). Hence, TN in the Negev bridges past and future, and aggressively impedes the ability of Bedouin communities to subsist, let alone develop and prosper in their traditional localities.

Denial and erasure

The legal basis of the denial is Israel's total refusal to recognise the validity of a pre-state indigenous land system. History, however, tells a different story, in which a well-established indigenous system operated for generations. In 1903, for example, the Ottomans also appointed tribal courts in Beersheba, thereby sanctioning Bedouin indigenous frameworks of authority and justice (Ben-David *et al.* 1991). Capitalist commercialisation began to take effect in the region, and was augmented by the Ottoman 'Tansimat' (government reforms and reorganisation), part of which spawned the abovementioned 1858 Land Code. Under these circumstances, Bedouin agriculture developed rapidly. While Israel denies the existence of pre-Mandate Bedouin agriculture (which means the land was not 'dead' and awards significant rights under the Ottoman system), plenty of evidence shows otherwise (a full account appears in Yiftachel *et al.* 2016). For example, British geologist Edward Hull noted in 1883 on the Negev landscape: "The district is extensively cultivated by the Terabin Arabs . . . The extent of the ground which is cultivated . . . , is immense, and the crops are wheat, barley, and maise must vastly exceed the requirements of the population" (Hull 1886: 138–139).

Against this history and geography, Israel developed 'the Dead Negev Doctrine' (DND) as a sophisticated way to claim indigenous lands. The doctrine puts great emphasis on formal dates, rather than actual history and geography. It represents a highly formalistic interpretation of Ottoman and British legislation, opposed to the recognition granted by these regimes to Bedouin land rights and possession. This was reinforced in 1935, in the answer of the British High Commissioner to Zionist claims of receiving land for settlement in the Negev in 1935: "The cultivable land in the Beersheba sub-district is regarded as belonging to the Bedouin tribes by virtue of possession from time immemorial" (Government of Palestine 1937).

From all accounts, the British land settlement process routinely registered cultivated lands in the names of their holders, if neighbouring land owners approved (Hilleli 1983). It is highly likely that if the British land registration would have arrived at the Negev, at least all cultivated and inhabited lands would be registered on the names of their holders, as was done in all other regions registered by the British (see Forman 2002; Forman and Kedar 2003).

Yet, Israeli courts continued to rule according to the DND, which made it all but impossible for the Bedouins to prove in court their land ownership rights. The doctrine has in effect *emptied the land backward*, that is, it asserts (with no systematic evidence) land and settlement rights based on their own reconfiguration of events and practices in the distant past. By accepting the doctrine, the courts have in effect been telling the Bedouins: "your parents and grandparents did not know, nor were they told by previous regimes, but now we are telling you – they were trespassers, and so are you!".

Consistent court ruling against Bedouin land claims relies on the most important precedent of *Hawashleh*, where the Supreme Court ruled in 1984 that the claimed land was neither settled permanently (because clusters of tents were considered 'encampments' and not 'settlements'), nor systematically cultivated during the nineteenth century. They also nailed down the decision by noting that once the last date for registration had passed (1921), there would be no reprieve. Even if the Bedouins continued to live on their ancestors' land and bequeath it from generation to generation, they would be considered trespassers, and the land should be registered as state property. In the absence of a constitution or any other legal recourse, the Hawashleh Supreme Court ruling became a binding precedent, in effect 'emptying' the entire Bedouin living areas from previous ownership.

In the decades that followed, the courts continued to fully rely on the Hawashleh precedent. This allowed them to systematically ignore a growing body of evidence about pre-state years; Bedouin legal representation was weak, unorganised and seriously lacking research. Many tribes also decided to ban what they considered to be biased courts. By the end of 2014 the state had won over 200 land claim cases, while the Bedouins had won none (Negev Coexistence Forum 2015).

As noted, the customary land system formed the baseline foundation for the Ottoman and British determination of land ownership. Plenty of internal indigenous documentation and verbal testimony demonstrates once more the existence of an active and fully functioning indigenous land system. However, these indigenous documents were deemed by Israeli courts as "lacking legal relevance towards the state . . . the source of rights [on which they draw] is flawed, and the documented land transferred have no relevance for the law, lacking any legal value" (Al-Uqbi *et al.* vs. State of Israel 2012, clause 7).

Naturally, the Israeli TN version has had profound planning implications. By and large, during the last seven decades Negev planning was premised on the 'emptying' of indigenous ownership or residential rights. Within this setting, urban and regional planning became a major player, by establishing 110 Jewish settlements (towns, villages, farms) in the region. During the same period, the state denied, through planning means, any recognition of Bedouin Arab localities; those existing prior to 1948 and those settled by the state in new locations as 'temporary solutions' during the 1950s and 1960s.

The consequences have been grave, as the unrecognised Bedouin localities evolved into slum-like sprawling semi-urbanised developments, lacking basic services such as roads, schools, electricity, water and health clinics. All Bedouin localities are placed at the lowest socioeconomic decile in Israel, and suffer constant state oppression, most notably over unauthorised building. Given the lack of municipal status or outline plans, about 70,000 unauthorised structures exist, and house demolition has become the main 'language' in which the state 'speaks' to its Bedouin citizens (Negev Coexistence Forum 2015).

Bedouin towns and villages around Beersheba have thus evolved into 'gray space', alluding to people, developments, lands and transaction that are neither equally integrated into citizenship, membership or plans, nor evicted or destroyed. 'Gray spacing' is a common process in colonial and neo-colonial settings, as well as in the hyper capitalist metropolis of the twenty-first century. It typically results in a systemic and long-lasting denial or rights and in stratified system of citizenship, leading to a regime described as "creeping urban apartheid" (Tzfadia 2013; Yiftachel 2009).

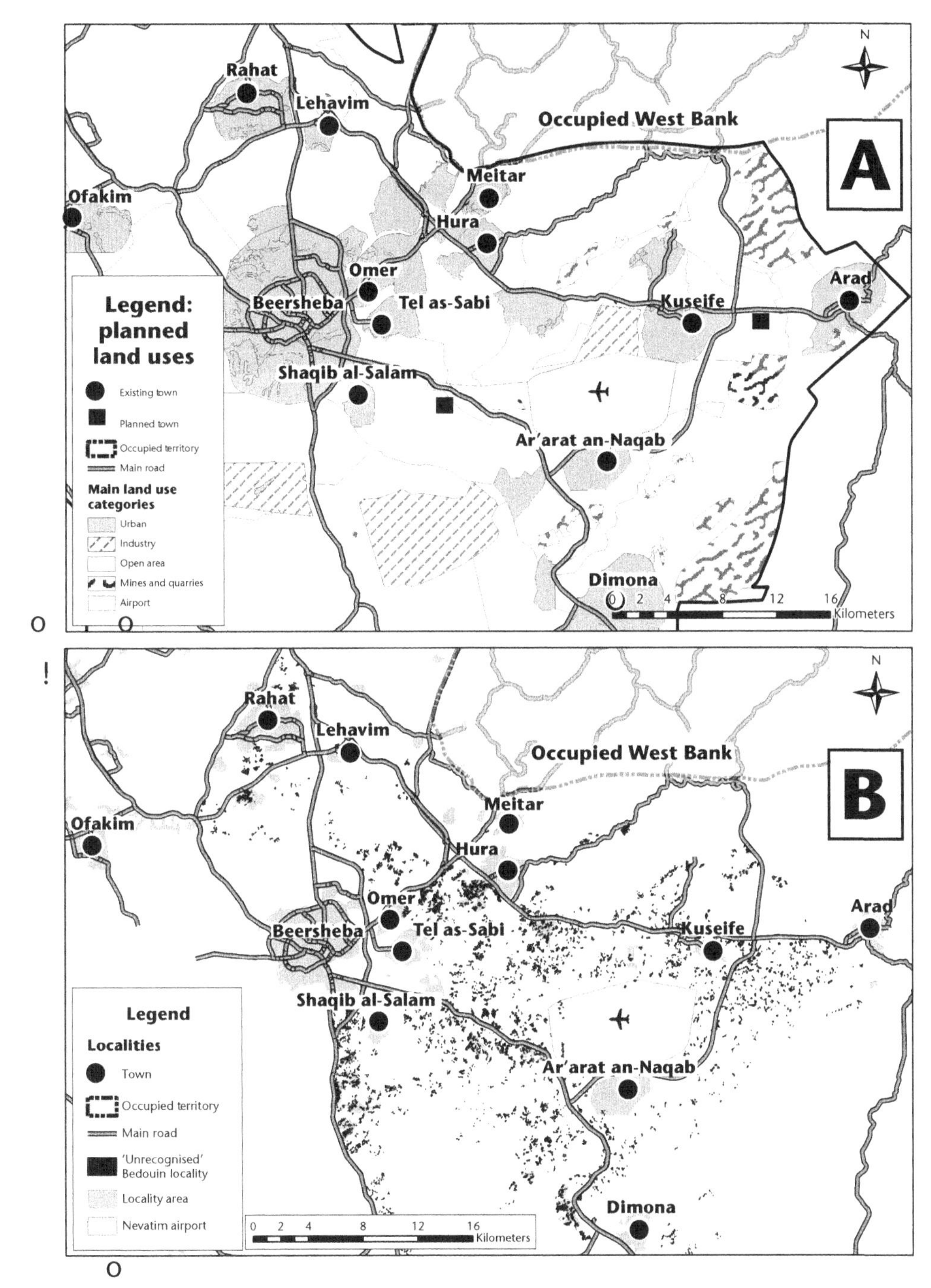

Figure 19.2 Planning as terra nullius: Bedouin localities invisible in Beersheba's Metropolitan Plan. A: Beersheba Metropolitan Plan (approved 2012). B: Spatial distribution of Bedouin localities based on aerial photography, 2012, noting the dozens of unrecognised localities not appearing in the plan

Source: RCUV (2013).

Policy changes and Bedouin resistance

The massive gray spacing of Bedouin development at the outskirts of Beersheba, coupled with persistent and often increasingly organised Bedouin resistance 'from below' caused a rare compromise by the Israeli planning and land systems. During the decade from the mid-1990s, 11 previously unauthorised localities hosting some 30,000 residents were recognised, and the authorities began to create outline plans that would make their future development fully legal. The compromising line reached a peak a decade ago, with the appointment of a special commission of inquiry, known as the Goldberg Commission, which recommended to "recognise the Bedouin localities and give them legal status" (2008: preface). The Goldberg report was met with severe criticism from hard-line Jewish circles, and some scepticism from their Arab counterparts. It was never adopted by the government, nor implemented.

However, the period of compromise was overridden by a new hard-line approach prevalent since 2009. During this phase the authorities returned to rejectionist policy, refusing to recognise the remaining 35 unrecognised Bedouin localities hosting some 70,000–75,000 Bedouins. In parallel, a new legal assault was launched, in which the state places 'counter claims' against Bedouin land claims. More than 45 years after the launching of the original claims and nearly 70 years since independence, the state has 'remembered' to claim the unregistered Bedouin lands (see Figure 19.2; Amara 2013; Negev Coexistence Forum 2015).

The legal assault is central to the enforcement of a TN concept, given the new research and many challenges emerging from the communities (e.g. Amara *et al.* 2013; Nasasra *et al.* 2014; Yiftachel *et al.* 2016). In addition, state attorneys are well aware of the Hawashleh precedent outlined above and are able to fully exploit it. Given the lack of state constitution, and the reliance of the Israeli legal system on precedents, the state is confident to continue and win all cases brought to the courts (Sheehan 2013).

The main new government policy known as the 'Prawer Plan', was released in 2012, with a concerted effort to totally reorganise and urbanise Bedouin indigenous space. The main effort focused on passing new legislation that would offer Bedouins land and housing packages in the existing towns or in partially recognised localities. The Prawer plan offered to 'resolve' Bedouin land claims with the allocation of 15–20 per cent of claimed land to the Bedouins (Azmon 2015). The plan thus threatens the vast majority of Bedouins with massive dispossession, in return for receiving suburban style residential blocks and small compensation in unknown quantity and quality, at unspecified locations (Prawer Report 2012).

The Prawer bill and plan, with a new policy document composed by Benny Begin (the responsible minister until 2014) were adopted by the government in February 2013. It encountered widespread opposition from local communities, human rights organisations and international bodies, including the European Union and the United Nations. This caused Yair Shamir, the responsible minister and Begin's successor, to temporarily shelve it. A new minister was appointed in 2015 – Uri Ariel, a renowned leader of West Bank settlers, who voiced a desire to reintroduce the Prawer plan in the near future.

It is beyond the limits of this chapter to describe the many ways in which Bedouins have resisted government plans, although it is worth mentioning that despite their marginality, the Bedouins have managed to stage a serious challenge to the authorities. The various stages of this growing resistance are documented elsewhere (see Amara and Yiftachel 2014; Karplus and Meir 2013; Meir 2005; Nasasra 2012; Shmueli and Khamaissi 2014; Yiftachel *et al.* 2016).

A notable act of resistance was the preparation by the Regional Council of Unrecognised Villages (RCUV) of an alternative master plan for the Bedouin region. This act of counter-planning produced a 350-page document with in-depth analysis and proposals. It demonstrated in professional and legal terms the viability and desirability of fully recognising all Bedouin communities and their possession of traditional land holdings (RCUV 2013). The alternative plan has become a major tool in the generation-long struggle of communities to hang on to their indigenous lands and lifestyle, and resist their urbanisation (Livnat 2011).

Overall, the structural tensions between, on the one hand, an ethnocratic state regime keen to expand its control and develop the region mainly for Jews; and, on the other, the persistence of indigenous Bedouins keen to protect their land and localities, have suspended Bedouin space and society in limbo (see Shamir 2000; Yiftachel 2009). The long-term 'gray-spacing' and 'creeping urban apartheid' bears in the second decade of the twenty-first century the strong imprint of the colonial TN approach.

A non-final word

As we have seen, TN has been a key tool – used explicitly and implicitly – in the colonisation of contested lands and massive resources from indigenous and minority groups, particularly in the global South-East. More recently, post- and neo-colonial regimes have applied TN approaches, paving the road to spatialising new visions of development as if space and society were 'tabula rasa' to be radically remoulded.

The case of the Negev/Naqab has illustrated this process in some detail. The Bedouins (the traditional Arab owners of southern Palestine) were classified – through a manipulative Israeli legal procedure – as trespassers on their ancestors' lands. Their homeland regions became subject to a radical transformation through Judaisation and privatisation. The DND has been used to dispossess traditional owners from most of their lands, leaving them in impoverished, seriously underdeveloped localities, subject to repeated waves of house demolition and criminalisation. This plight, growing research shows, is rather typical for informally urbanising populations in ethnocratic states of the global South-East (Yiftachel 2006; Watson 2013).

TN has thus fundamentally influenced the planning of space in the majority of world societies. Yet, it is largely absent from leading urban, regional and planning theories, as noted by scholars such as Roy (2009), Watson (2009, 2014), Robinson (2006), Porter (2014) and Yiftachel (2006, 2009). This chapter reinforces their call for a readjustment of the 'camera angle' in planning and urban theories. Instead of universalising theories designated to explain and guide cities and regions of the developed liberal world, theorists should *dare and theorise* from the South-East, in order to *de-colonise planning* thought, and – more importantly – actual cities and regions. Scholars should thus analyse and conceptualise prevalent processes occurring in (the very diverse) sites of the 'non-West' and use them – where and when relevant – to explain and de-colonise spatial processes in comparable settings. In such an endeavour, the role of land, law and identity in general, and the concept of TN in particular, should become leading topics.

References

Abramson Commission (1921) *General Report of the Commission to Enquire into the Conditions of Land Settlement in Palestine* (Chair – Major Abramson), PRO CO 733/18, 174761.

Abu-Sitta, S. (2009) "The denied inheritance: Land ownership in Beer Sheba", paper presented to the International Fact Finding Mission Initiated by RCUV.

Alfasi, N. (2014) "Doomed to informality: Familial vs. modern planning in Arab towns in Israel", *Planning Theory and Practice* 15(2): 170–186.

Altman, J. (2014) "The political ecology and political economy of the Indigenous land titling 'revolution' in Australia", *Maori Law Review*, March: 1–17.

Al-Uqbi, S. *et al.* vs. State of Israel. (2012) C.A. 7161/06. Beersheba District Court 14 May, 2015, Decision of Supreme Court, 15 March.
Amara, A. (2013) "The Israeli policy of counter claims against the Bedouin-Arabs", Coexistence Forum for Civil Equality, Beersheba. Available: www.dukium.org/wp-content/uploads/2011/06/NCF-CounterClaims-Dec10.pdf [accessed 5 February 2016].
Amara, A. and Yiftachel, O. (2014) "Confrontation in the Negev: Israel land policies and the indigenous Bedouin-Arabs", Israel: The Rosa Luxemburg Foundation. Available: www.rosalux.co.il/confrontation_in_the_negev_eng [accessed 5 February 2016].
Amara, A., Abu-Saad, I. and Yiftachel, O. (eds) (2013) *Indigenous (In)justice: Human Rights Law and Bedouin Arabs in the Naqab/Negev*, Cambridge, MA: Harvard University Press.
Azmon, E. (2015) Consultant and expert on Bedouin affairs, Personal communication, 18 April 2015.
Bailey, C. (1980) "The Negev in the nineteenth century: Reconstructing history from Bedouin oral tradition", *Asian and African Studies* 14(1): 35–80.
Ben-David, Y., Kressel, G. and Abu-Rabia, G.N. (1991) "Changes in land usage by the Negev Bedouin since the 19th century", *Nomadic Peoples* 28: 28–55.
Blomley, N. (2002) "Law, property and the geography of violence: The frontier, the survey and the grid", *Annals of the Association of American Geographers* 93(1): 121–141.
B'tselem (2012) *Under the Guise of Legality: Israel's Declaration of State Land in the West Bank*. Available: www.btselem.org/download/201203_under_the_guise_of_legality_eng.pdf [accessed 5 February 2016].
Bunton, M. (2007) *Colonial Land Policies in Palestine, 1917–1936*, Oxford: Oxford University Press.
Connell, R. (2014) "Using Southern theory: Decolonizing social thought in theory, research and application", *Planning Theory* 13: 210–223.
Daes, E.-I.A. (2008) "Standard-setting activities: Evolution of standards concerning the rights of indigenous peoples: On the concept of 'indigenous people'". In C. Erni (ed.) *The Concept of Indigenous Peoples in Asia: A Resource Book*, Copenhagen/Chiang Mai: IWGA and AIPP, pp. 3–78.
Du Plessis, W.J. (2011) "African indigenous land rights in a private ownership paradigm", *Potchefstroom Electronic Law Journal* 14(7): 44–69.
Essaid, A. (2013) *Zionism and Land Tenure in Mandate Palestine*, London: Routledge
Falah, G. (1983) "The development of the 'planned Bedouin settlements' in Israel 1964–1982: Evaluation and characteristics", *Geoforum* 14: 311–323.
Fisch, J. (1988) "Africa as *terra nullius*: The Berlin Conference and International Law". In S. Förster, W.J. Mommsen and R.E. Robinson (eds) *Bismarck, Europe and Africa: The Berlin Africa Conference and the Onset of Partition*, Oxford: Oxford University Press, pp. 347–375.
Fischbach, M. (2003) *Records of Dispossession*, New York: University of Columbia Press.
Flybvjerg, B. (1996) "The dark side of planning: Rationality and 'realrationalität'". In S. Mandelbaum, L. Mazza and R.W. Burchell (eds) *Explorations in Planning Theory*, New Brunswick, NJ: Center for Urban Policy Research Press, pp. 383–394.
Forman, G. (2002) "Settlement of the title in the Galilee; Dowson's colonial guiding principles", *Israel Studies* 7(3): 61–83.
Forman, G. and Kedar, S. (2003) "Colonialism, colonization and land law in mandate Palestine: The Zor al-Zarqa and Barrat Qisarya land disputes in historical perspective", *Theoretical Inquiries in Law* 4(2): article 11.
Frantzman, S., Yahel, H. and Kark, R. (2012) "Contested indigeneity: The development of an indigenous discourse on the Bedouin of the Negev, Israel", *Israel Studies* 17(1): 78–105.
Goldberg, E. (2008) "The Goldberg Commission Report, to propose a policy for regulating Bedouin settlement in the Negev". Available: www.moch.gov.il/SiteCollectionDocuments/odot/doch_goldberg/Dvar_Shofet_Goldberg_ENG.pdf [accessed 5 February 2016].
Government of Palestine (1937) "British response to the Jewish Agency concerning Bedouin land", Government for the British Mandate for Palestine Files, DCF/32–72.
High Court of Australia (1992) "Decision: Mabo and others vs the Government of Queensland". Available: www.austlii.edu.au/au/cases/cth/HCA/1992/23.html [accessed 8 November 2016].
Hilleli, A. (1983) "Landed property rights: General history of the development of property in this land". In B. Shmueli, N. Kliot, and A. Soffer (eds) *The Lands of Galilee*, Haifa: Haifa University Press and Ministry of Defense, vol. 2, pp. 575–611 (in Hebrew).
Hull, E. (1886) *The Survey of Western Palestine: Memoir on the Physical Geology and Geography of Arabia Petraea*, London: The Palestine Exploration Fund.

Kark (2002, new edition) *Frontier Jewish Settlement in the* Negev, *1880–1948*, Tel Aviv: HaKibbutz HaMeuhad (in Hebrew).

Kark, R. and Frantzman, S. (2012) "Empire, state and the Bedouin of the Middle East, past and present: A comparative study of land and settlement policies", *Middle Eastern Studies* 27(2): 1–25.

Karplus, Y. and Meir, A. (2013) "Past and present in the discourse of Negev Bedouin geography: A critical review". In M. Nsasra, R. Ratcliffe, S. Abu-Rabia-Queder and S. Richter-Devroe (eds), *Rethinking the Paradigms: Negev Bedouin Research 2000+*, Abingdon: Routledge, pp. 32–59.

Kedar, S. Amara, A., Yiftachel, O. (in press 2017) *Terra Nullius in Southern Israel/Palestine*, Palo Alto, CA: Stanford University Press.

Livnat, Y. (2011) "The Bedouins: The story of internal colonialism". In H. Katz and E. Tzfadia (eds) *Neglecting State, Controlling State*, Tel–Aviv, Resling, pp. 107–126. (in Hebrew).

McAulsan, P. (2013) *Land Law Reform in Eastern Africa*, London: Routledge.

Macklem, P. (2008) "Indigenous recognition in international law: Theoretical observations", *Michigan International Law Journal* 30(1): 177–184.

Meir, A. (1997) *When Nomadism Ends*, New York: Westview Press.

Meir, A. (2005) "Bedouin, the Israeli State and insurgent planning: Globalization, localization or glocalization?", *Cities* 22(3): 201–215.

Nasasra, M. (2012) "The ongoing Judaisation of the Naqab and the struggle for recognizing the indigenous rights of the Arab Bedouin people", *Settler Colonial Studies* 2(1): 81–107.

Nasasra, M., Richter Devroe, S. Radcliffe, R. and Abu Rabia, S. (eds) (2014) *The Naqab Bedouin and Colonialism: New Perspectives*, London: Routledge.

Negev Coexistence Forum (2015) "Community under attack: The situation of the human rights of Bedouin community in the Negev-Nagab 2015", Beersheba: Negev Coexistence Forum. Available: www.dukium.org/wp-content/uploads/2015/12/HRDR_2015_ENG.pdf [accessed 8 November 2016].

Porter, L. (2014) "Possessory politics and the conceit of procedure: Exposing the cost of rights under conditions of dispossession", *Planning Theory* 13: 387–406.

Ottoman Land Code (OLC) (1858) English translation. Available: http://ra.smixx.de/Links-A-E/Ottoman-Land-Code-1858—1927-.pdf [accessed 5 December 2016].

Prawer Report (2012) "Prawer team plan to implement Goldberg Commission proposal for regulation of the Bedouin settlement in the Negev", 31 May. Available: www.truah.org/images/stories/PrawerReportedTranslatedFINAL.pdf [accessed 8 November 2016].

Regional Council of Unrecognized Bedouin Villages (RCUV) with Bimkom – Planning for Planning Rights, and Sidra – Arab Women in the Negev (2013) "Master plan for recognition in Bedouin localities, Beersheba, RCUV". Available: http://bimkom.org/eng/the-alternative-master-plan-for-bedouin-villages-in-the-negev/ [accessed 5 December 2016] (Hebrew, Arabic and abridged English version).

Robinson, J. (2006) *Ordinary Cities: Between Modernity and Development*, London: Routledge.

Roy, A. (2005) "Urban informality: Toward an epistemology of planning", *Journal of the American Planning Association* 71(2): 147–158.

Roy, A. (2009) "Why India cannot plan its cities: Informality, insurgence, and the idiom of urbanization", *Planning Theory* 8(1): 76–87.

Shamir, R. (2000) "Suspended in space: Bedouins under the law of Israel". In N. Blomley, D. Delaney and R. Ford (eds) *The Legal Geographies Reader: Law, Power and Space*, New York: Wiley, pp. 231–257.

Sandercock, L. (2003) *Cosmopolis II: Mongrel Cities of the 21st Century*, New York: Wiley & Sons.

Sheehan, J. (2013) "Applying an Australian native title framework to Bedouin property". In A. Amara, I. Abu-Saad and O. Yiftachel (eds) *Indigenous (In)justice: Human Rights Law and Bedouin Arabs in the Naqab/Negev*, Cambridge, MA: Harvard University Press, pp. 228–253.

Shmueli, D. and Khamaissi, R. (2014) *Israel's Invisible Negev Bedouin: Issues of Land and Spatial Planning*, New York: Springer.

Stavenhagen, R. and Amara, A. (2013) "International law on indigenous peoples and the Naqab Bedouin Arabs". In A. Amara, I. Abu-Saad and Yiftachel, O. (eds) (2013) *Indigenous (In)justice: Human Rights Law and Bedouin Arabs in the Naqab/Negev*, Cambridge, MA: Harvard University Press, pp. 158–193.

Supreme Court (2015) "al-'Uqbi vs the State of Israel", Civil Appeal C.A. 4220/12.

Tzfadia, E. (2013) "Exceptionalism". In T. Hatuka and T. Fenster (eds) *The Planners*, Tel-Aviv: Resling, pp. 65–75 (in Hebrew).

Watson, V. (2009) "Seeing from the South: Refocusing urban planning on the globe's central urban issues", *Urban Studies* 46(11): 2259–2275.

Watson, V. (2013) "Planning and the 'stubborn realities' of global South-East cities: Some emerging ideas", *Planning Theory* 12(1): 81–100.
Watson, V. (2014) "The case for a Southern perspective in planning theory", *International Journal of E-Planning Research* 3(1): 23–37.
Wolfe, P. (2006) "Settler colonialism and the elimination of the native", *Genocide Studies* 8: 387–409.
Yiftachel, O. (1994) "The dark side of modernism: Planning as control of an ethnic minority". In S. Watson and K. Gibson (eds) *Postmodern Cities and Spaces?* Oxford: Blackwell, pp. 216–234.
Yiftachel, O. (2006) "Re-engaging planning theory", *Planning Theory* 5(3): 211–222.
Yiftachel, O. (2009) "Critical theory and 'gray space': Mobilization of the colonized", *City* 13: 246–263.
Yiftachel, O., Kedar, S. and Amara, A. 2012. 'Rethinking the dead Negev doctrine: Property rights in Bedouin regions', *Law and Government* (Mishpat U-Mimshal) 14(1): 7–147 (in Hebrew).
Yiftachel, O., Roded, B. and Kedar, S. (2016) "Between rights and denials: Bedouin indigeneity in the Negev", *Environment and Planning A* 48(11): 2129–2161.

20

The intent to reside

Residence in the auto-constructed city[1]

Gautam Bhan, Amlanjyoti Goswami and Aromar Revi

In the uncertainty that still surrounds notions of what, if anything, a 'city of the global South' is or could be, one empirical reality holds up to a fair amount of scrutiny. If we take planning to mean deliberate attempts by the state to shape the built environment using law, plans and policy, then many cities of the South have been built in some tension with this notion of planning. Put quite simply: for many reasons, large parts of these cities simply neither look like their plans nor do they fit into neat categories of law, especially in the latter's understanding of ownership and property. Here, the 'large' is significant. Variations from planning are not a Southern phenomenon and certainly evidence of variation is to be found in all cities. The claim that there could be something Southern about a mode of urbanisation is then partly about the extent of the disjuncture and how fundamental or not it is to understand urbanisation itself.

Drawing from a range of cities from São Paulo to Istanbul, Mexico City to New Delhi, Teresa Caldeira has recently offered a "a characterization of modes of the production of space that are different from those that generated the cities of the North Atlantic" (Caldeira 2016: 2). She argues that significant parts of cities of the South are built by residents themselves. These are done in incremental and particular temporalities and, most importantly, with "transversal engagements with official logics of legal property, formal labor, colonial dominance, state regulation, and market capitalism" (Caldeira 2014). She terms this "auto-construction", the basis of a mode of the production of space she describes as "peripheral urbanization" (Caldeira, 2016). One part of her argument is particularly important for us here. Caldeira argues that while "peripheral urbanization unfolds in quite different ways" in different cities, it "is remarkably pervasive, occurring in many cities of the south, regardless of their different histories of urbanization and political specificities" (2016: 2). It is this sense of being "pervasive" that Southern urban theory has sought to examine and possibly build from. Caldeira reminds us that this does not mean an argument that denies variation within the South, but one that suggests that there are still theoretical claims that can be made, which "articulate general features while remaining open and provisional to account for the ways in which the modes of operation [they] characterize vary and constantly transform" (2016: 3).

What does auto-construction mean for planning in cities of the global South? Southern urban theorists have taken different routes to understand auto-construction and its disjunctures with planning, in turn studying informality, illegality, violation, the failure of planning, corruption, the implementation gap and even the meaning of regulation, intervention and policy. In this chapter, we argue that one of the uncertainties that auto-construction introduces into urbanisation is to confound any simple understanding of *urban residence*. In different contexts of peripheral urbanisation, this confounding has particular and locally determined impacts. We follow the impacts in a specific context: the contemporary Indian city. Here, we argue that the spatial illegality that results from auto-construction's "transversal engagements with official logics" has acted as a basis to deny recognition of legitimate presence in the city and thereby block access to rights and resources. We detail this through describing two specific exclusions: the lack of secure tenure, and impediments in access to basic urban services like water and sanitation. Given our challenge to think about planning practice, we then suggest a policy response. We propose a framework that re-frames residence in its legal and bureaucratic instantiations as a response to a history of auto-construction. We call this the Intent to Reside Framework.

One caveat is due here before we proceed. Though we use this chapter to suggest policy and thereby state action, this does not imply that we believe it to hold any greater or lesser power in reaching desired outcomes. As befits an auto-constructed landscape, we see policy and state action as only one of many political fields in which urban outcomes are contested and shaped. Yet we also stand against debates in Southern urban theory that use the presence of auto-construction to argue that thinking about policy and planning is futile, at worst, or irrelevant, at best. Planning does not have to be slotted between being either hegemonic or futile. As one of the authors has argued elsewhere, even if plans and policies do not determine urban outcomes as imagined in a rational-modernist framing, they certainly shape and influence them (Bhan 2016). Suggesting frameworks for how state action should be conceptualised and designed is then an important theoretical and political task, and is the beginning of outlining a Southern urban practice that can take on the tasks that theory has set out for us.

We proceed as follows. The first section describes the contexts of spatial illegality and auto-construction, focusing on the determination of legitimate urban residence. The second then shows how such illegality endangers secure tenure and hinders access to water in urban areas. The third then lays out the Intent to Reside Framework. The fourth concludes.

Residence and insecure tenure

When cities are built in "transversal engagements" with logics of property and planning, two particular risks ensue that are relevant to the arguments of this chapter. The first is that many such localities or neighbourhoods often do not possess secure tenure precisely because of the way they were built and settled. We understand secure tenure to be the set of arrangements, as Payne has phrased it, that ensure "the right of all individuals and groups to effective protection by the state against forced evictions" (2004: 167). Payne further states that "security of tenure derives from the fact that the right of access to and use of the land and property is underwritten by a known set of rules, and that this right is justiciable" (2004: 167–168). What is important is that this "known set of rules" needs not be codified or legal, but includes a vast array of social and political relations and norms, including explicit as well as tacit negotiations with state actors. Security is, then, a matter of degree, and one that varies with time. It is also, critically, not linear. In work on Delhi, Bhan has shown precisely how long-held and seemingly stable arrangements of settling and inhabiting the city were undone in a wave of evictions in the past two decades that have transformed the city's landscape (Bhan 2016). Auto-construction implies that negotiations with law and planning never cease – each step of presence and incremental development is negotiated and uncertain.

This is not a new conversation in talking about Southern urbanism, but the discussion is usually limited to and reduced to 'the slum'. Yet while slums are certainly a key type of auto-constructed neighbourhoods with uncertain tenure, they are not the only form of auto-construction. In cities ranging from São Paulo (Holston 2009), Cairo (Bayat 2000), New Delhi (Bhan 2016) and Jakarta (Simone 2010), there is a range of housing typologies from irregular subdivisions, unauthorised housing and peri-urban and rural landscapes part of expanding city regions, that are all marked by tensions with formal logics of law and planning. The precise nature of insecurity of tenure may be different even among these neighbourhoods and across these cities, but the relationship between such insecurity and the history of auto-construction is what is shared.

Even within auto-constructed neighbourhoods, different households have different degrees of tenure security. Let us take an example from an Indian city. In every *basti*[2] in an Indian city, each household has a different degree of claim to legitimacy, if not legality (see Bhan 2016). The terms of this legitimacy vary: the number of years of residence, the possession of different kinds of formal identifications on paper, belonging to different social identities, etc.

Residence is thus a key vector in determining what happens to individual households *within* a basti, and not just to the basti itself. This becomes starkly clear often when any attempt is made to intervene into an auto-constructed landscape, say to try to develop/redevelop them or offer some form of tenure legalisation. This moment is a double-edged sword for settlements, as multiple lines of exclusions and inclusions emerge.

Let us take an example to make this clear. In Mumbai, the Slum Rehabilitation Scheme offered residents of *bastis* a chance to formalise their residence in the city as their neighbourhood was redeveloped into multi-storey flats in which they would have legal and secure tenure. The caveat is this: only 'legible' residents can be part of the redevelopment. How is legibility determined? Here, legibility hinges on fitting into a 'cut-off' date, a measure of the number of years one has lived in the *basti*. Thus, only households that can prove residence before 1 January 1995 (later moved to 1 January 2000) are eligible to be part of the redevelopment. The cut-off date then illustrates that it is not just that the threat of eviction marks auto-constructed landscapes, but that even the possibility of inclusion, redevelopment and formalisation comes striated along lines of how residence is determined, and how its legitimacy is evaluated. Each of these follows from the particular history of auto-construction.

Access to basic services and entitlements

The second example of how uncertainty in residential status resulting from auto-construction – whether at the household level or the neighbourhood – impacts urban life comes from access to basic services and entitlements. Let us take one kind of each: (a) water, as a basic urban environmental service, and (b) food, as an entitlement provided as part of social protection and welfare policies. Auto-construction can disrupt a safety net in the provision of these basic needs in two distinct ways: direct, *de jure* exclusions by rule or law; or indirect, *de facto* exclusions through the process of trying to access what theoretically one should be able to. The former represents clear obstructions that spatial illegality presents in establishing universal access by obstructing provision to particular households or neighbourhoods because of their "transversal engagement", to bring back Caldeira's words, with law and planning. The latter reminds us that even if we removed all such obstructions, the process of implementing a safety net still brings exclusions in practice that must be addressed.

In cities like Delhi, the exclusion in access to water is clear. The Delhi Jal Board is not obligated to "provide water supply to any premises which have been constructed in contravention of any law".[3] It is important that this does not just exclude the slum – unauthorised colonies have vast networks of private self-provision ranging from tankers to storage units that bypass public supply. Applications for water connections to the Municipal Corporation of Greater Mumbai's Water Department are divided into "slum/non-slum" categories. Within "slum", there is a further division to those before 1 January 2000 and those after 1 January 2000. The application states that no applications can be entertained from "hutments on footpath and roads, non-declared slums on private land, or hutments on land developed for public purpose, or on any land affected by a project" undertaken the government.[4] The exclusion therefore works – like with slum redevelopment – through a cut-off date. A recent judicial challenge to this exclusion in the Bombay High Court has resulted in its reinforcement. Denying the petition filed by the *Pani Haq Samiti*, the Bombay High Court articulated a common fear underlying the denial of water to 'slum' residents – that services would make residents feel entitled to tenure security: "you would not want to move away from that place if you have water".[5] In both cases, a service that residents would otherwise be entitled to cannot be provided to them because of the configuration of their residence in the city.

Insecurity of tenure can also result in *de facto* exclusions through the requirements of process. Even if the Bangalore Water Supply and Sewerage Board (BWSSB) does not have *de jure* exclusions for households without tenure, applying for a water connection requires an application along with "sanctioned plan or Tax Paid Receipt" – a requirement certain to exclude many households, particularly poor households without tenure. Inclusion may also be differentiated with a distinction between what level of amenities can be provided to communities with or without security of tenure. As the BWSSB outlines, it offers "individual household connections" for those with land tenure and "community-level services" such as shared metered connections for communities without security of tenure.[6]

Other *de facto* exclusions stem from a lived consequence of living in auto-constructed landscapes: the inability to exist on paper. Let us take the case of a ration card, the address-linked identification that entitles Indians to subsidised food. While there is no *de jure* exclusion on a household getting a ration card no matter where they live, in practice, it is widely known that getting a ration card living on occupied land – one form of the "transversal" relationship with law – is incredibly difficult. Once attained, any changes through births or deaths in the family are near impossible. Most importantly, in the case of eviction, getting cards in a new location can take years, if they are to happen at all (Bhan and Menon-Sen 2006).

One recent study found that 33 per cent of residents in the slum they studied in Mumbai had no ration cards despite repeated attempts to get one (Subbaraman *et al.* 2012). As the authors argue:

> there is no specific regulation that prevents residents of non-notified slums from getting a ration card – however, the ration card application states that an applicant must supply some form of 'residential proof'. Since residents live in a situation of informality, without official residential documents such as home ownership papers or rental agreements, providing appropriate paperwork is frequently impossible.
>
> *(Subbaraman* et al. *2012: 657)*

The inability to prove residence on paper is a primary barrier to access to almost any entitlements from public programmes. While ad hoc provisions of Voter IDs and Ration Cards continue, the slow and uncertain acquisition of these documents implies that every resident must remain without access to public benefits until they are able to build a residence that is deemed legitimate.

Not only does the fact of auto-construction create specific impacts for residents, it can also create a set of circumstances that shape state action in multiple, often paradoxical, ways. In some cases, utilities that may want to provide water services to all neighbourhoods in the city – for better financial health, ethical or constitutional commitments, or as a response to political pressure from residents – find themselves unable to do so without violating regulations that they are often not part of setting. Should they proceed anyway, they risk finding themselves in court challenges or the subject of audit regulations. On the other hand, it is also possible that public agencies reluctant to provide services to 'illegal' neighbourhoods find the illegality of the neighbourhood a mode of legitimate exclusion, a way to refuse citizens that are entitled to services because they claim them from auto-constructed locations. Auto-construction thus implies that the delivery of services and the resolution of tenure become matters of negotiation and discretion rather than capacity, efficiency or technology. Until the uncertainties of tenure, legality and residence embedded into the history of the production of urban space are resolved, several of the desired outcomes of planning – services and tenure are just two examples – become impossible to attain.

The intent to reside

Can we then re-think how such a legitimate residence can be built? It is here that we propose a new framework of thinking about how to understand and bureaucratically grasp urban residence: the intent to reside (ITR). The ITR approach works on embracing universal (or quasi-universal) entitlements through evidence of an *intention* to reside in the city that includes residents at an early stage of this residence. The ITR approach is, in a sense, the antithesis to a 'cut-off date'. Rather than asking residents to prove that they deserve to be included as urban residents by surviving for years in the city, it includes them from the very beginning. It attempts at being more mindful of errors of exclusion within a context where operationalisation and implementation of services are themselves premised on conditions and modes of residence and spatial illegality.

How does it work? The approach is rooted in legal and bureaucratic precedent in the Indian context. This is pivotal: without such locations it remains a purely hypothetical exercise unable to find rules and procedures to link into, or have any foundations to survive a judicial challenge. In its simplest form, the approach argues that residence must be measured at six months of presence in the city with a multiplicity of forms of identification to be used as proof of such presence.

The ITR draws upon several different existing bureaucratic mechanisms for understanding residence. For example, the National Population Register (NPR 2011) understands the term "usual resident" as follows:

> a usual resident of a local area, for the purpose of NPR, is defined as a person who has stayed in the local area for the past 6 months or a person who intends to stay in the local area for the next six months.

For NPR purposes, there is a Registrar General of India approved "Local Register of Usual Residents" for the purpose of verification. An ITR approach could build on this idea by suggesting that residents, in the context of provisioning of services, could produce evidence of having been in the city for six months and thereby provide evidence of ITR, that could in turn enable service provisioning as enforcement of constitutionally guaranteed fundamental rights.

Such an idea of residence may strike many as untenable but it does draw upon long-standing as well as contemporary legal precedents articulated by the Supreme Court. In *Jagir Kaur and Another vs Jaswant Singh* (1963), the Supreme Court offered a formulation that indicates how judicial precedent could be set for the ITR. We excerpt at length given the importance of the text of the verdict. The judges stated (emphasis added):

> Having regard to the object sought to be achieved, the meaning implicit in the words used, and the construction placed by decided cases thereon, we would define the word 'resides' thus: a person resides in a place if he *through choice* makes it his abode permanently *or even temporarily*; whether a person has chosen to make a particular place his abode depends upon the facts of each case. Some illustrations may make our meaning clear: (1) A, living in a village, goes to a nearby town B to attend a marriage or to make purchases and stays there in a hotel for a day or two. (2) A, a tourist, goes from place to place during his peregrinations and stays for a few days in each of the places he visits. (3) A, a resident of a village, who is suffering from a chronic disease, goes along with his wife to a town for medical treatment, takes a house and lives there for about 6 months. (4) A, a permanent resident of a town, goes to a city for higher education, takes a house and lives there, alone or with his wife, to complete his studies.
>
> In the first two cases, A makes only a flying visit and he has *no intention* to live either permanently *or temporarily* in the places he visits. It cannot, therefore, be said that he 'resides' in the places he visits. In the last two illustrations, though A has a permanent house elsewhere, he has a clear *intention* or *animus manendi* to make the places where he has gone for medical relief in one and studies in the other, his temporary abode or residence. In the last two cases it can be said that though he is not a domicile of those places, he '*resides*' in those places.

More recently, in 2011, the Supreme Court in *Ruchi Majoo vs Sanjeev Majoo* also stated as follows:

> Thus residence is a concept that may also be transitory. Even when qualified by the word ordinarily the word resident would not result in a construction having the effect of a requirement of a person using a particular place for dwelling always or on permanent uninterrupted basis.

Such a formulation allows even a regular and cyclical mobility as long as intent is proven by the people residing. This is an inclusive formulation, moving beyond concern about 'free riders'. All residents are potentially eligible until specific criteria rule them out, as opposed to them needing specific criteria to be included (Drèze and Khera 2010).

There is a second part to the framework. This definition of residence, while removing the need to prove long-term residence in the city, will not address arguments about the difficulty of getting any formal paper documents within the overall context of spatial illegality. How then can proof of ITR be established?

Several current policies in India are innovating on how to think about identity and residence. Such new regimes of identification are seeking processes to ease Voter IDs (recent changes found ways to include the homeless, for example, even without proof of address), suggestions on making Ration Cards barcoded and portable, or even creating a new national identification system altogether such as the Aadhar, the new Unique Identification number being given to all Indian residents (note: residents, not citizens). Each of these has pros and cons, but each is, in principle, compatible with an ITR approach as they seek to expand the ease of inclusion rather than play the role of gatekeepers. For example, the Aadhar card accepts 18 proof of identity and

33 proof of address documents as a nationally valid list of documents. These include documents that are 'letters of introduction' from various actors that theoretically makes inclusion as expansive as possible. An adapted list that could suffice for the purpose of the ITR approach could include *any* of the forms of identification in Table 20.1.

Such a list of documents is indicative, but it is rooted in an approach that seeks to expand the ways and means in which residents – particularly more recent residents – can acquire legitimacy. The wider the form and number of multiple identifications that are acceptable, the less the danger of exclusion. When combined with the imagination of needing to be in the city for only six months, the ITR approach then offers not just a lower burden of proof but also a more open and inclusive processes to get there.

Concluding provocations

The arguments of this chapter have sought to do two things. The first is to locate the possibilities of provision of some basic environmental services and entitlements in auto-constructed cities, marking real challenges that this location makes apparent. The second is to suggest the beginnings of conceptualising urban practice that emerges from a close understanding of specific locations and histories of urbanisation. In that sense, for both theory and practice, the chapter reminds us that place matters, taking seriously one of the foundational premises of Southern urban theory.

Our arguments here suggest that planning – in the way we have understood it here as an assemblage of attempts to shape the built environment – must be taken seriously as a core part of imagining and assembling any apparatus that seeks to universally deliver services or entitlements

Table 20.1 Forms of identification under an expanded Intent to Reside Framework

No.	Form of identification
1	Passport
2	Bank statement/Passbook
3	Post office account statement/Passbook
4	Ration card
5	Voter ID
6	Driving license
7	Government photo ID cards/service photo identity card issued by public sector unit
8	Electricity bill (not older than 3 months)
9	Water bill (not older than 3 months)
10	Telephone landline bill (not older than 3 months)
11	Property tax receipt (not older than 3 months)
12	Credit card statement (not older than 3 months)
13	Insurance policy
14	Signed letter having photo from bank on letterhead
15	Signed letter having photo issued by registered company on letterhead
16	Signed letter having photo issued by recognised educational instruction on letterhead
17	National Rural Employment Guarantee Scheme (NREGS) job card
18	Pensioner card
19	Freedom fighter card
20	Central Government Health Services (CGHS) card
21	Certificate of address having photo issued by MP or MLA or gazetted officer
22	Income tax assessment order
23	Vehicle registration certificate
24	Registered sale/lease/rent agreement
25	Address card having photo issued by Department of Posts
26	Caste and Domicile Certificate having photo issued by state government
27	Disability ID card/Handicapped medical certificate issued by the respective state/UT government administrations
28	Gas connection bill (not older than 3 months)
29	Passport of spouse
30	Passport of parents (in case of minor)
31	Unique Identification Authority's Aadhar card

in cities of the South. Questions on universal access have been limited to ideological debates on the intention of the state to 'provide' or 'withdraw' in different political economies, or on the capacity of the state – either financially or in terms of delivery – to keep to whatever commitments it makes. Another set of debates focus on modes of delivery, arguing between cash transfers and direct provision; or between targeted approaches versus universal coverage. These are important debates but most, if not all, remain markedly aspatial and dislocated from contexts of existing urban settlements and their logics, particularly the character and mode of their urbanisation. We hope that arguments that use residence, for example, as a conceptual and empirical link between spatial histories of urbanisation and the design of service delivery and the provision of entitlements will begin to close this disciplinary gap.

Such discussions can add to the history of thinking about spatialising the provision of different types of social protection entitlements. Smita Srinivas has compellingly described the distribution of entitlements using a triad of work, place and workplace (Srinivas 2010). Historically, she argues, social security entitlements have been understood and delivered as a worker, resident or citizen with entitlements being delivered through the state (place), the employer (workplace), or even through a third agency, but on the basis of either work status (work) or citizenship (place). The arguments of this chapter have sought to extend this frame, noting that 'place' is often marked and undone precisely by its tensions with formal logics of law and planning.

This is not a singular or linear move. Residents of auto-constructed neighbourhoods are, after all, still workers and citizens. They therefore consistently make other claims to rights and entitlements, many of which may be unaffected by spatial illegality. These claims – especially those that are more centred on work and workplace, rather than place – can then also bolster the legitimacy of residents to challenge, mitigate, or temper the impacts of their spatial illegality, for example, the right to education. India's most recent and new fundamental right guarantees access to free education within the geographical proximity of one's home. Local admission then becomes another 'proof of address' that can add legitimacy to a household with uncertain tenure. If rights to education are secured, then can they trump spatial illegality? As Nivedita Menon reminds us, rights are often experienced like this, in the contestation between "diverse discourses of rights" and claims that play off against each other (Menon 1998). As complex political fields negotiate who is entitled in urban areas to what as whom, we argue that spatial illegality is often under appreciated as a vector that shapes substantive access to core development outcomes. A reminder of its resilience is a current case in the Punjab High Court that is hearing a challenge to the right to education. The challenge asks: does a slum present within the appropriate spatial radius from a school count as being 'in the local area'? In other words, is the right to education to be extended to all households in a 5km radius, or all legal households? The question at stake is, in fact, of the salience of the legacy of auto-construction.

In closing, we reiterate the importance of engaging with Southern theory to ask what it implies for planning practice in different locations. Echoing Caldeira, we argue that just as peripheral urbanisation takes specific forms, so must the practices that emerge to engage with them. From the Indian context, we have shown that the ITR framework can be one such response. It is determinedly pragmatic, seeking to be effective in actually existing public and state apparatus, but at the same time it is both normative and historically rooted. Moving forward, we urge more such formulations that lay seeds of new forms of urban practice from the South, to parallel the challenge to epistemes.

The ITR framework has found a first home in the new Karnataka Affordable Housing Policy 2016, where eligibility criteria for access to improved or new housing is defined using a one-year parameter of residence and the expanded set of identification documents (Government of Karnataka 2016). Effectively, this implies that in the southern Indian state of

Karnataka, home to Bangalore, one of India's fastest growing metropolises, the 'cut-off date' is no longer a legal practice. As this policy begins implementation, the real gains and challenges will become more visible on the ground, but in its beginning is the opportunity to study a new vocabulary of Southern urban practice in action.

Notes

1 A version of this paper has been previously published as Bhan *et al.* (2014).
2 The Hindi/Urdu word *basti* (related to *basna*, to settle; plural: *bastis*) means settlement. Colloquially, it is the word most commonly used by residents of urban poor settlements to describe their homes and hence it is the word used here. Colloquially, *bastis* are understood to represent settlements typically marked by some measure of physical, economic and social vulnerability. It is these settlements that are often called 'slums' in English, a use that I argue is insufficient and often reductive.
3 Chapter 3, Section 9.1a of the Delhi Jal Board Act. See: www.delhijalboard.nic.in/djbdocs/about_us/act.htm [accessed 9 February 2017].
4 See application form at aquaptax.mcgm.gov.in.
5 *Pani Haq Samiti vs Bombay Municipal Corporation*. CWP 10 of 2012.
6 From 'Services to the urban poor' on the BWSSB website. Available: http://bwssb.org/services/ [accessed 12 May 2014].

References

Bayat, A. (2000) "From 'dangerous classes' to 'quiet rebels': Politics of the urban subaltern in the global South", *International Sociology* 15(3): 533–557.

Bhan, G. (2016) *In the Public's Interest: Evictions, Citizenship and Inequality in Contemporary Delhi*. Athens, GA: University of Georgia Press; New Delhi: Orient Blackswan.

Bhan and Menon-Sen (2006) *Swept off the Map: Surviving Eviction and Resettlement in Delhi*. New Delhi: Yoda Press.

Bhan, G., Goswami, A. and Revi, A. (2014) 'The intent to reside: Spatial illegality, inclusive planning and urban social security". In O. Mathur (ed.) *State of the Urban Poor Report 2014*. New Delhi: Oxford University Press, pp. 83–94.

Caldeira, T.P.R. (2014) "Peripheral urbanization", paper presented at the LSE Cities Public Lectures, London. Available: https://lsecities.net/media/objects/events/peripheral-urbanisation [accessed 9 February 2017].

Caldeira, T.P.R. (2016) "Peripheral urbanization: Autoconstruction, transversal logics, and politics in cities of the global South", *Environment and Planning D: Society and Space* 35(1): 3–20.

Drèze, J. and Khera, R. (2010) "The BPL census and a possible alternative", *Economic and Political Weekly* 45(9): 54–63.

Government of Karnataka (2016) *Karnataka Affordable Housing Policy 2016–2026*. Bengaluru: Government of Karnataka.

Holston, J. (2009) "Insurgent citizenship in an era of global urban peripheries", *City and Society* 21(2): 245–267.

Jagir Kaur and Another vs Jaswant Singh (1963) AIR 1521, 1964 SCR (2) 73.

Menon, N. (1998) "State/gender/community: Citizenship in contemporary India", *Economic and Political Weekly*, 33(5): PE3–PE10.

National Population Register (NPR) (2011) "Introduction to NPR". Available: http://censusindia.gov.in/2011-Common/IntroductionToNpr.html (accessed 19 May 2017).

Payne, G. (2004) "Land tenure and property rights: An introduction", *Habitat International*, 28(2): 167–179.

Ruchi Majoo vs Sanjeev Majoo (2011) AIR 2011 SC 1952.

Simone, A. (2010) *City Life from Jakarta to Dakar: Movements at the Crossroads*. New York and Oxon: Routledge.

Srinivas, S. (2010) "Industrial welfare and the state: Nation and city reconsidered", *Theory and Society* 39(3–4): 451–470.

Subbaraman, R., O'Brien, J., Shitole, T., Shitole, S., Sawant, K., Bloom, D.E. and Patil-Deshmukh, A. (2012) "Off the map: The health and social implications of being a non-notified slum in India", *Environment and Urbanization* 24(2): 643–663.

21

Living as logistics

Tenuous struggles in the remaking of collective urban life

AbdouMaliq Simone

Make+shift life

If you are one of the scores of millions of residents across much of the urban South who cannot depend upon one specific job to earn your keep, that lacks sufficient documentation to secure a place to live over the long term, or that can't afford to get sick or into any kind of trouble, what is it that you pay attention to in order to know something about what to do? What happens if the people you rely upon for support or information are no longer available or suddenly turn against you? What happens when the skills you have to ply are also those of an increasing number of residents, and competition becomes increasingly fierce for opportunities? The question here is how can you best know what is going on and try to situate yourself in a position where opportunities might 'come your way'; how to be at the right place at the right time when there is no clear map available?

The enormous transformations of the built environment and the enhanced possibilities of consumption that have marked even some of the most marginal of the world's cities should not detract from acknowledging just how dependent the majority of the urban residents of the South are on constantly putting together some workable form of income and inhabitation. The makeshift character of much of what this majority does is quite literally 'make'+'shift'. Whatever they come up with rarely is institutionalised into a fixed set of practices, locales or organisational forms. This doesn't mean that relationships and economic activities do not endure, that people do not find themselves rooted in the same place and set of affiliations over a long period of time. Rather, these stabilities inhere from a constant recalibration of edges, boundaries and interfaces. Whatever appears to be stable largely depends upon its participation in a series of changing relationships with other activities, personnel and sites. Whatever is made then shifts in terms of its availability to specific uses and users, as well as its exposure to new potentials and vulnerabilities.

The objective of this chapter is to consider the various ways in which the notion of the make+shift shows up and operates in urban conditions that have substantially constrained opportunities for the elaboration of residential and economic opportunities on the part of the urban majority in cities of the South. To what extent is the make+shift configuring tentative, problematic, yet potentially generative affiliations with practices usually associated with the very neoliberal processes that otherwise tend to intensify the individuation of residence and livelihood and disentangle long-honed collective collaborations? To what extent are a younger generation of urban residents turning themselves into instruments of logistics in order to continuously reorient themselves to both the exigencies of normalisation as well as to concretely stretch the terms and spaces of a normal life?

While the chapter builds on past work exploring the intersections of labour, informality, development and regulation theory (Kudva and Beneria 2005; Piore and Safford 2006; Unni and Rani 2008) – particularly in the trajectories of ushering young people into the work force – it attempts to hone in on processes whereby flexibility and uncertainty are being thoroughly internalised and transformed into a resource for piecing together survival in urban contexts just as these very same things constitute a locus of incessant vulnerability.

As urban accumulation increasingly turns toward the incorporation of sentiment, experimentation and the elaboration of uncertainty as a means to homogenise space, target and immobilise parts of the urban population and marketise urban effort, are there options for the majority to deploy urban operations in ways that hold open a multiplicity of opportunities? Instead of devolving into chaos, brutal capital accumulation or becoming more proficient copycat imitations of their Northern counterparts, could it be that the big metropolitan areas of the South are concretising new ways of 'being' cities – ones that make broader and more judicious use of varied actors who inhabit them? This possibility is informed by a substantial body of work that examines the ways, albeit often problematic in terms of their aspirations and methods, in which residents did manage to compose and choreograph complicated distributions of space and opportunities (Bayat 2010; Boudreau *et al.* 2016; McFarlane and Silver 2017) Does such judicious use really need the conventional tropes of justice? Is it possible that even as the familiar modes of collaboration, autoconstruction and informality are fading, new formations of collective life, less readily discernible, are coming to the fore, and in ways that alter the social arrangements and subjective experiences of being in the city?

These question stem from the observation of intensifying dissatisfaction about the character of contemporary urban life expressed by a wide range of actors. Even as built environments become more spectacular, service provision more widespread and income levels and consumption rise, there is a growing sense of uncertainty and worry stemming from perceptions that spaces of manoeuvrability are shrinking. Across social classes and sectors, there is a prolific belief that urban life entails either dealing with too many variables or simply being misguided altogether; that attempts to balance and productively intersect divergent forces becomes a complex task immune to available tactics and sensibilities.

Cities have always been places of complaint and contestation, but especially now, even as the poor, the barely and solidly middle class, municipal governments, property and infrastructure developers in most cities are making more sincere and concerted efforts to work with each other, discontent is growing. From gridlocked traffic, the lack of affordable land and housing, the sheer volume of infrastructure requiring repair or adaptation, the dissipation of collaboration among neighbours, or the lack of necessary finance and capacity for needed developments, residents are feeling that the work of managing everyday life is becoming harder.

This dissatisfaction is particularly significant since it was in the realm of the everyday that many residents of these cities sought to make the city work for them, where they could assess the efficacy of their efforts. Now, rampant property markets may be running roughshod over much of the urban landscape, but they are producing a plurality of vacancies, *terroir vague*, and generally uncertain dispositions. As objects of both real and fictitious investment, it is increasingly difficult to predict the outcomes of the impulsive rush to hedge the costs of past projects with speculation on the future. It is difficult to tell when the now normative paradigm of 'just build and occupants will follow' will run its course.

In cities where residents largely built their own districts, economies, regulations and authorities, the perceived constriction of manoeuvrability not only operates against valued histories and practices, but it also intensifies a sense of urgency that is addressed through half-hearted submissions to increasingly formatted modalities of residence and work. For states that often felt they could impose their agendas and operating procedures with little popular consent or consultation, their efforts now to enrol citizen participation are often met with indifference or scepticism. As cities operate in an expanding network of regional and global transactions, the necessity to 'get things right', and the efforts made to elaborate environments that promote growth and attempt to mitigate inequality, often seems to makes things worse.

The conundrum of informalisation

Once thought compartmentalised to production in the global South, economic informalisation is growing across regions. Such informalisation entails occupational multiplicity, falling wage levels in comparison to productivity growth, extensive outsourcing and subcontracting of production, self-employment, shortening of work hours and careers, drastic cutbacks of secondary benefits and the relaxation of regulations and compliance (Breman and van der Linden 2014). Increasing rates of self-employment are substantially composed of hidden wage labour without the contractual obligations of the wage relationship, as workers use supplies, resources or instruments provided by others and are obligated to generate a fixed return for these suppliers at the end of every day regardless of their earnings. While a global working class may continue to grow, their everyday experience is one of contracting opportunities for a more secure and predictable life – uncertain, unstable, and insecure (Vosko 2010). As Lordon (2014) emphasises, in a time of neoliberal affect, individuals do not identify with concrete labour, nor with the pleasures and demands of this specific task, or the craftwork ideal of a job well done, but rather with abstract labour – with the general capacity of being put to work. This, however, may be only a medium-term phenomenon as informal work demonstrates the capacity of what Harriss-White calls hybrid innovation systems capable of generating new multiplier effects among various forms of skill, location and financing that in the past had few concrete sites of articulation (Harriss-White and Prosperi 2014).

This growing informalisation of contemporary economic life stems from long histories of detachment, of not belonging and of people not being affiliated to a constancy that might provide security. The global South experienced enormous levels of inactivity. Particularly since the 1980s, this inactivity amounts to 30–40 per cent of potential "working man hours" (van der Linden 2014). In his examination of the historiography of precarious labour, Marcel van der Linden argues for a fundamental distinction in the nature of precarity between North and South. Precarious labour in advanced capitalist countries has now spread across all sectors, becoming less conjunctural and more an effect of the restructuration of international competition. While substantially whittled down and particularised, many of the residues of state regulation and social protection laws and welfare remain in place.

Perhaps more significantly, as Povinelli (2011) has pointed out, precarity has become the defining character of late liberalism, compensating for crises of legitimacy with the prolific identification of risks from which various scales of polity offer protection in exchange for the normalisation of behaviour. This protection constitutes then the demarcation of viable collective life. Such normalisation exceeds compliance to particular standards or rules but rather entails a process of continuous self-improvement, self-regulation, which then presumes access to viable platforms, affordances and infrastructures from which to conduct it. For those unable or unwilling, they find themselves increasingly on their own (Bear and Mathur 2015; Berlant 2011).

For the global South, the spread of precarious works across sectors is especially structural in nature. It exerts a greater impact through the near absence of social protection regimes or the application of income supplements contingent upon stringent adherence to specific procedures of household organisation. A disposable population is constituted not only to provide flexible labour for mobile and eventually transient capital investment but also to leverage the state's access to enjoining the game of financial speculation.

Here a disposable population is bundled in aggregate as that which can be offered as "wholesale life commodities" offered in advance through a state's compliance with austerity measures, structural adjustment, debt repayment and budget cuts in health, education and social services (Tadiar 2013).

> Put another way, the seemingly limitless resource that is the future (as part of the seeming limitlessness of life itself) is in actuality the lives of people whose own futures are offered up as exchange values extractable in the present.
>
> *(Tadiar 2013: 30)*

In part, this is a continuation of a long-term process whereby households compensate for the stagnation of real wages by financing necessary housing, food, transport, mobility, health and education through credit (debt), with enormous effects and yet more deductions from wages, along with an increase in the social disciplinary mechanisms that come with the need for debt reduction (de la Rocha 2007).

If the vast zones of the urban poor, with their piecemeal and oscillating attachments to the larger metropolitan system, in terms of provisioning of services, citizenship, legality and institutional participation, are no longer 'required' as the living antithesis of that figure of the human to which urban resources are mobilised in support, are these populations completely expendable? If the surplus value of urban life – of its recursive and reflexive symbolic infrastructures – is decreasingly contingent on the dispossession of the poor, on the extraction of their contributions by force, by relegating them to a condition of sheer survival, then what? In line with the notion of precarity being a defining feature of late liberalism, what Tadiar seems to suggest is that the process of rendering people as impoverished does not so much relegate them to a reserve or expendable surplus but to lives that can be used in any way imaginable to elaborate a future for the non-poor.

Such practices run counter to the ways in which cities throughout the global South largely worked, not to the extent to which they constructed a particular kind of person, inhabitant, citizen, but rather the way everyday practices availed spatially and materially heterogeneous environments with densities. These densities not only involved those of bodies, but ways of doing things and a wide range of technical devices that put things into a plurality of different relationships – with different scope, degrees of visibility and duration. The sheer diversity of the overall built environment and the activities that took place within it, and in close proximity to each other, precipitated discussions, compensations, repairs, alliances, trade-offs and short-term pooling of information, contacts and resources that supplemented official income and earnings. At the same time, the composition of the built environment reiterated a sense of separateness among residents, the unavailability of any overarching reference point of easy commonality and, as such, these were localities of fractures that necessitated the constant reworking of lines of articulation (Benjamin 2008; Chattopadhyay 2012; Dovey 2014; Perlman 2010; Sundaram 2009).

As cities everywhere experience a declining manufacturing base, an enlarged labour market of both educated and low-skilled workers who are unemployed or underemployed, and underproduction in relationship to the available material base, a critical question is where work will come from, especially for a growing youthful population. This is a population almost entirely born and raised within these cities, with few connections and references anywhere else. If work once was generated through the very spatial enrichments of urban inhabitation – i.e. the possibilities for residents living in heterogeneous environments of layering different associations among what was available to them, of interconnecting different activities, needs, aspirations and spaces – as these possibilities through residency decline, more intentional mobilisations of effort will have to be generated. In many cities the hardening of religious identification, the intensification of religious practice and devotion itself becomes a form of work, anchoring the organisation of everyday activities and the attainment of well-being.

Rehabilitation and retrofitting of spatial assets, local environmental and community management, and various forms of service provision are areas to be developed. These areas require new forms of knowledge, sociality and individual capacity in order to bring them about. They require an ability to circulate throughout different facets of city life and, indeed, such circulation is increasingly valued by youth and local economies emphasising the provision of short-term residency and providing a concrete platform for it. How is such circulation best facilitated?

Despite growing dissatisfaction and the concomitant uncertainties in everyday life, urban residents exhibit many ways in which the desires to be in and to creatively engage the city are also intensifying. Many of the manifestations of this desire are restless and provisional. They concern less the aspiration to put down roots or to institutionalise a particular kind of presence in the city than a need to find out as much as possible what is taking place in the city, how to get an angle on it, how not to get pinned down, and to find the right kind of niche. While segregations persist, there are also multiple boundary crossings and new determination among residents from different walks of life to find ways of intersecting.

Take the example of Abdul Azis, 22, who grew up in a poor family in Sukabumi, some 300 kilometres from Jakarta. After finishing one year of junior high school, he commenced five years of work in various factories at the periphery of Jakarta, working 18-hour shifts and becoming addicted to dextroamphetamines, which employers would disburse in large quantities. Two years ago a childhood friend, working at a Padang restaurant in central Jakarta, telephoned and informed Abdul Azis of a dishwasher vacancy in the kitchen.

At nights after the restaurant closed, Abdul Azis would experiment in the kitchen, coming up with unique drinks that the owner was willing to try out with some of the customers. After a year, he was moved from the kitchen to the front to handle the drinks, at which point he noticed that the restaurant's customers would bring coffee purchased from the outside. He convinced the boss to try out some inexpensive coffee equipment and then, with the help of YouTube and an informal group of baristas that he located through Twitter, set out to learn the coffee trade and enter some local barista competitions. After doing well in these, the boss decided to allocate part of the restaurant space as a coffee shop, which Abdul Azis now runs.

Many workers have passed through the restaurant in the time that Abdul Azis has been there, including his childhood friend. But they continue to maintain their relationships with each other, borrowing and lending, helping to support each other's small moneymaking schemes. Abdul Azis has been offered more lucrative positions at more up-scale coffee places, but he professes a desire to remain where he is for now because he views his job as the critical node that keeps these relationships among ex-workers going, which in turn enables him to "stay *with* the city" and to make it "knowable without having to always get stuck in traffic" (personal communication, 18 February 2015). While urban residents have long attempted to stitch together networks of care provision, Abdul Azis's efforts reflect the more complex efforts often required to build functional systems of 'kinship' from intensely fragmented histories of family ties, work histories, speeded up migratory circuits and the wariness of debilitating obligations.

The collusions among developers, investment banks, management companies, hedge funds and property investors primarily use the built environment as the 'shell' around which elaborate financial manipulations take place. Even if the actual resultant built development is itself empty of significance, this financial game eats up substantial urban space. If residents of mixed-use districts are to preserve their hard won efforts, this may require collective efforts like those managed by Abdul Azis, where resources put into projects are no longer always visible. It will be important to find ways to upscale and collate hundreds of such initiatives and appropriate various corporate screens and contracts where the labor of a district is hedged across various forms of income generation and ways of exerting influence. Who then makes these decisions; who has sufficient clout and trust to mobilise local effort and resources? What kind of politics is sufficient to this task? Clearly a different sense of planning and prototyping is required, one that takes into account the highly variegated forms of place-making taking place in cities of more intensive circulation of populations, which at the same time are increasingly 'conscripted' into more formatted, standardised modalities of residence that constrain spaces for manipulating the built environment. The conventional forms of subsidiary, participative local-planning, while still potentially important tools for fostering democratisation, also have to be supplemented with the proliferation of coalitions of interest that cut across metropolitan territories, that built a connective tissue of information exchange, advocacy and planning across distinct territories. As a result, local institutions not only must focus on cohering local space, but also identifying and actualising trajectories of involvement across various metropolitan sectors; how to maintain the viability of place-based 'collective bodies' as they move and to engage localities as often highly fragmented and provisional assemblages of intensely individuated agendas, histories and ways of doing things.

The capacity to acquire land, build a house and access services for large numbers of residents often depends on complicated negotiations between land owners, brokers, local customary authorities, bureaucrats working in municipal tax, land registration and infrastructure agencies, political party representatives, religious leaders, unofficial mediators and enforcers – as well as both long periods of waiting *and* a capacity to act quickly at a moment's notice. But the formats of deliberations, calculations of costs, circumvention or creation of rules, and the formulas that dictate specific outcomes are often not clear.

This is not because they are by nature opaque – although the ability of these diverse actors to work together does require large measures of invisibility. Rather, the processes of working things out often take many different twists and turns and generate multiple unanticipated feedback loops. The actors are not simply trying to get their financial cut, but also use their involvement in these processes to open up different spaces of manoeuvrability and reach, cultivate dependencies and obligations, as well carve out new spaces of autonomous action. In other words, while each actor may be seeking to fill their pockets, there are also more cloudy strivings at work. The game here is how to mix and match, how to stitch together various agendas and power bases into dispositions that may merely buy time, but in doing so keep open multiple possibilities for how urban space in any particular location might be used and valued.

Lives of logistics

The conundrums of contemporary informalisation suggest the necessities of re-description as a means of deploying a provisional politics. This is particularly the case when the formations of urban rule increasingly are predicated on economies and vernaculars of risk, of states undertaking to protect residents against a wide range of possible dangers. Here, in order to attain recognition of needs, aspirations and realities, urban residents are compelled to work through the modes of visibility and calculation that increasingly determine who is seen as legitimate and eligible for particular rights, services and opportunities (Zeiderman 2016). It is for this reason, then, that I re-describe these practices of keeping possibilities open as a form of logistics.

Logistics configures new forms of territoriality that both facilitate and curtail particular kinds of circulation. This is done through ports, inter-modal transport hubs, warehouses, container parks, information technology infrastructure, as well as labour that is continuously monitored and assessed in terms of optimal efficiencies The actions of employees are tracked in real time through performance quotas, GIS, radio-frequency identification tags and mechanisms that deliver voice instructions for a continuous sequence of tasks, and the use of what are known as KPIs – software programmes that measure worker and organisational efficiencies. Through the use of such tracking technologies, the temporal delay between execution of tasks and their statistical measure is erased (Rossiter 2014).

For global production networks that require the seamless movement of primary materials and finished commodities, the configuration of such new territory also entails a process of detachment of materials, social relations and bodies from their embeddedness in specific locales. Instead, they are rearranged as elements in an intensely standardised series of manoeuvres and movements (Martin 2012). Logistical knowledge is the means to stabilise interconnections across multiple sites and practices. Interchanges, ports and trade zones promise the organisation of stocks and flows in ways seemingly drained of political interest. Here the ostensible consideration is the seamless circulation of effort and goods accomplished by technical efficacies in spaces completely turned over and neutralised for such observations (Toscano 2014).

In the meantime, the complex negotiations of accords, monitoring procedures, fragmented and multiple sovereignties, security apparatuses and labour regimes that go into creating this promise are obscured and become impenetrable to negotiation. Standardisation is a critical element of logistical functioning in terms of attaining the interoperability of different infrastructures and materials, the exchange of data across multiple platforms and attempts to overcome protocological asymmetries – all of which enable goods to pass through increasingly elongated and textured commodity chains (Rossiter 2014).

Knowledge about the heterogeneous circulations, destinations, diversions, costs and assemblages is distributed across various organisations and scales (Cowen 2014). Resultant asymmetries in the access to interconnection potentially destabilise logistics. Martin (2012) provides the example of the illicit appropriation of existent commodity circulation networks and the appropriation of particular circuits of transit by migrants. Logistics attempts to compose a friction-free circulation of inputs and outputs across various spaces often marked by national boundaries and discrepant regulatory frameworks of all kinds, while relying upon a specific set of standardised equivalences, containers, and calculations.

All of this requires a *diffracted* knowledge where learning and know-how that are intended for forging particular anticipated relationships come to be applied to situations for which that knowledge is seemingly 'not prepared'. Here the process of disembedding particular nodes, transit and processing sites from the specificities of their relationships with particular locales, demographic compositions, social and economic histories and cultural practices requires an open-ended sense of how these sites, now acting as nodes, could be articulated in new and various ways. It entails how they are multiply situated in a plurality of different circulations. This process of *diffraction* reiterates the fundamental *instability of interconnectivity* – as well as a potential space through which disruptions and illicit uses might emerge – and thus requires capacities to anticipate instability and pre-empt interruptions (Martin 2012).

In listening to the voices and observing the practices of youth in Abidjan, Hyderabad, Jakarta, Berlin, Athens and Karachi during the past several years, it is clear that youth are taking on their futures in ways that intersect with logistics. For them, everyday life is wrapped up in the compulsion to act quickly and the diffusion of incessant opportunities that need to be seized. But there is also evidence of failure all around, of risks that compound debt, so that the capacity to wait, even with all its disciplinary and submissive connotations, is a critical tool in terms of cultivating an ability to read and play 'the field'. Yet at the same time, many youth also actively attenuate fears of failure by adamantly 'making their moves' and then 'moving on', taking what they can from any situation without investments over the long term. Other youth surmise particular trajectories of change, and willingly position themselves in situations where nothing seems to be happening at the moment, but where they anticipate being the recipients of particular amalgamations of forces and opportunities that are 'headed their way' and that have the capacity to fruitfully alter their lives if they are willing to wait.

Youth are also *harvesting* bits and pieces of solidarities, cultural memories and discarded materials, and the fuzzy interstices that are created through uneasy juxtapositions in rapidly restructured urban environments become resources for livelihoods. Urban spaces become fields to be passed through and culled. It is important to have a presence in many types of locations. So, in addition to the readily discernible arenas of action – such as households, factories, markets, malls, institutions, stations – youth insert themselves into more ephemeral formations, such as packs, pop-up shops, short-term rooming houses, over and underpasses, backrooms, plazas, stairwells, vacant lots, game rooms, lobbies, interchanges, WhatsApp groups, clubs, corners, enclaves, swarms, chat rooms and hashtags.

As youth are increasingly aware of their own expendability and the seeming arbitrariness entailed in who gets ahead or not, they actively pursue ways of subtracting themselves from what is expected of them. Additionally, the incorporation of cognition into increasingly automated protocols of behavioural enactment and algorithmically determined self-consideration prompts refusals to commit to particular modes of operating in the city. Concrete courses of action taken or imagined often involve individuals spreading themselves across disparate 'projects' or engagements, hedging their bets, pluralising possible destinations and sources of income. In one sense this is the very dissipation of an integrative subjectivity cultivated by neoliberal logics of flexible labour, where value is placed on resilience and a willingness to persist through contradictions. On the other hand, such individual tactical manoeuvres opt for the construction of a self-visibility more effectively insulated from scrutiny. In some instances, detachment from steady contractual labour is a means of forging a relational autonomy, where mobility is linked to sustaining particular experiences of sociality or relations of care (Millar 2014).

While we may think of logistics as a matter of constructing the seamless transmission of commodities across discrete territories, youth are also faced with the exigencies of 'transporting' themselves across intensely differentiated environments and protocols, which require a continuous detachment of their understandings and performances from the contexts in which they may be most familiar and comfortable. Regardless of crucial class, regional and gender differences, there is a sense of commonality among many youth not wanting to or affording to be caught in readily discernible practices of accumulation and self-enactment. The cruel irony is that the mushrooming of identity politics, where youth hang on to narrowly circumscribed, caricature-like depictions of religious faith, national belonging or ethnic identity, may be largely motivated by the desire not to be 'pinned down' by destinies that seem increasingly predetermined, with little recourse to negotiation or change or which entail "cancellation of what in the human psyche is incompatible with abstract domination" (Berardi 2014).

Even when you are not going anywhere, it may be important to always act as if you are. Hendi, 18, finished junior high but did not go to secondary school. He came to Jakarta at age 15 and found work in a restaurant, which was comfortable and paid well, but served pork, and as a devout Muslim the idea and the smell made him sick, so he quit. Through an older brother, he found work in one of Southeast Asia's textile and clothing markets, Tanah Abang in Jakarta, where he has worked for the past two years. Although he expresses a desire to save enough money to open up a shop in his hometown, he largely concedes that for the conceivable future, Tanah Abang is the only place where a young man without a diploma would be able to find work. He circulates through the market, never content with one specific location for very long, always looking for better deals. Many young men are in his situation; they have fairly stable work as sellers in stalls owned by others, but a large number never stay for very long, but they also never leave Tanah Abang. If so many workers are quitting after a short time working at a particular stall, but also never leave the market, taking a job somewhere else within it, what gets produced is always a sense of both vacancy and plenitude. Any particular position is never sufficient, but there is always the possibility to move on to something else that probably won't be sufficient either – and it is this duality that becomes a resource to be harvested.

In order for the bosses to retain labour, deal-making becomes more intricate. Different formulas of compensation are negotiated – daily, weekly, monthly and varying combinations of these – accorded on the basis of flexibility of hours and tasks. This deal-making is not the product of any self-conscious mobilisation on the part of the bottom tier of the market's work force. Still, young labourers talk to each other, and through these interchanges get a sense of the market's atmosphere – the fluctuations of prices, the capacity of owners to respond to new trends and volume, the capacity of owners to bundle their goods with others to offer wholesale prices to buyers coming from all over Indonesia and also pooling together their money.

The young sellers get a reading of who is in debt or who is making a lot of money, and then try to find ways to get closer to the real action. Of course, there are a lot of failures in what the owners do, either collectively or individually, and there is always a sense of urgency to recover from these failures. This happens not only by reducing costs, but taking on energetic labour equipped with new ideas and solid experience. These are situations to be "harvested", as some young labourers put it. Most young labour will never make enough to become owners or ever increase their eventual earnings substantially, but this circulation through the market at least creates the semblance of trajectories, of going somewhere, even if it is not necessarily going forward.

Instability becomes a critical modality through which the power of capital accumulation is recomposed and reiterated. In key ways, the long-term instabilities of urban life strategically maintained by many urban inhabitants, which engendered the continuous updating and remaking of various forms of urban life, have now been 'stolen' by more powerful economic actors. As a result, current trajectories of urban change that emphasise security of tenure, propriety through property, inclusiveness through debt, the equation of affordances with affordability and the enforced promotion of resilience through the attenuation of social contracts have a debilitating impact on the sociability of city life. These technologies of apparent 'stability' disentangle various solidarities, collaboration and economies built on the continuous recalibration and plying of relations among different kinds of actors and activities.

Yet, many of the youth, particularly in the South, demonstrate an unintentional or calculated indifference to this theft. They generate ways of doing things that are not easily included within the constantly shifting frontiers of accumulation and political normalcy, and as such re-order the conventional terms through which life in the city is spoken about and experienced. It is not about finding ways to include what is rendered irrelevant or marginal within the existent modalities of everyday life, but to, perhaps logistically, elaborate new modalities capable of making judicious use of them.

In terms of planning processes, these modalities might best come to the fore in examining how alternative service delivery modalities can provide a basis for an enlarged relational politics that connects the molecular dynamics of place-making and elaborating networks of care with larger citywide questions of distribution of public resources. In addressing this question, the potential power of environmental politics comes to the fore. In almost all infrastructural sectors related to the social reproduction of (poor) households, it is possible to rethink and reconfigure these systems in ways that are labour-intensive, ostensibly favourable for the regeneration of ecosystem services and catalytic for social learning and deepening politicisation (Simone and Pieterse 2017).

References

Bayat, A. (2010) *Life as Politics: How Ordinary People Change the Middle East*, Palo Alto, CA: Stanford University Press.

Bear, L. and Mathur, N. (2015) "Remaking the public good: A new anthropology of bureaucracy", *Cambridge Anthropology* 33: 18–34.

Benjamin, S. (2008) "Occupancy urbanism: Radicalizing politics and economy beyond policy and programs", *International Journal of Urban and Regional Research* 32(3): 719–729.

Berardi, F. (2014) "The neuroplastic dilemma: Consciousness and evolution", Efflux 60. Available: www.e-flux.com/issues/60-december-2014/ [accessed 8 October 2015].

Berlant, L. (2011) *Cruel Optimism*, Durham, NC: Duke University Press.

Boudreau, J., Gilbert, L. and Labbé, D. (2016) "Uneven state formalisation and peri-urban housing production in Hanoi and Mexico City: Comparative reflections from the global South", *Environment and Planning A* 48(12): 2383–2401.

Breman, J. and van der Linden, M. (2014) "Informalizing the economy: The return of the social question at the global level", *Development and Change* 45(5): 920–940.

Chattopadhyay, S. (2012) *Unlearning the City: Infrastructure in a New Optical Field*, Minneapolis, MN and London: University of Minnesota Press.

Cowen, D. (2014) *The Deadly Life of Logistics: Mapping Violence in the Global Trade*, Minneapolis, MN: University of Minnesota Press.

de la Rocha, M.G. (2007) "The construction of the myth of survival", *Development and Change* 38(1): 45–66.

Dovey, K. (2014) "Incremental urbanism: The emergence of informal settlements". In T. Haas and K. Olsson (eds), *Emergent Urbanism: Urban Planning and Design in a Time of Structural and Systemic Change*, Burlington, VT: Ashgate, pp. 45–53.

Harriss-White, B. and Prosperi, V. (2014) "The micro-political gains of unorganised workers in India", *Economic and Political Weekly* 49(9). Available: www.epw.in/journal/2014/9/perspectives/micro-political-economy-gains-unorganised-workers-india.html [accessed 9 December 2015].

Kudva, N. and Beneria, L. (2005) *Rethinking Informalisation: Poverty, Precarious Jobs and Social* Protection, Ithaca, NY: Cornell University Open Access Repository.

Lordon, F. (2014) *Willing Slaves of Capital: Spinoza and Marx on Desire*, London: Verso.

McFarlane, C. and Silver, J. (2017) "Navigating the city: Dialectics of everyday urbanism", *Transactions of the Institute for British Geographers*. DOI: 10.1111/tran.12175.

Martin, C. (2012) "Desperate mobilities: Logistics, security and the extra-logistical knowledge of 'appropriation'", *Geopolitics* 17(2): 355–376.

Millar, K. (2014) "The precarious present: Wageless labor and disrupted life in Rio de Janeiro, Brazil", *Cultural Anthropology* 29(1): 32–53.

Perlman, J.E. (2010) *Favela: Four Decades of Living on the Edge in Rio de Janeiro*, Oxford: Oxford University Press.

Piore, M. and Safford, S. (2006) "Changing regimes of workplace governance, shifting axes of social mobilisation and the challenge to Industrial Relations Theory", *Industrial Relations* 45: 299–325.

Povinelli, E. (2011) *Economies of Abandonment: Social Belonging and Endurance in Late Liberalism*, Durham NC: Duke University Press.

Rossiter, N. (2014) "Logistical worlds", *Cultural Studies Review* 20(1): 53–76.
Simone, A. and Pieterse, E. (2017) *New Urban Worlds: Inhabiting Dissonant Times*, London and New York: Polity.
Sundaram, R. (2009) *Pirate Modernity: Media Urbanism in Delhi*, London: Routledge.
Tadiar, N.X.M. (2013) "Lifetimes of disposability within global neoliberalism", *Social Text* 31(2115): 19–48.
Toscano (2014) "Lineaments of the logistical state", *Viewpoint Magazine*, 28 August. Available: https://viewpointmag.com/2014/09/28/lineaments-of-the-logistical-state/ [accessed 1 June 2015].
Unni, J. and Rani, U. (2008) *Flexibility of Labour in Globalising India: The Challenge of Skills and Technology*, New Delhi: Tulika Books.
van der Linden, M. (2014) "San Precario: A new inspiration for labor historians", *Labor: Studies in Working Class History of the Americas* 11(1): 9–14.
Vosko, L. (2010) *Managing the Margins: Gender, Citizenship and the International Regulation of Precarious Employment*, Oxford: Oxford University Press.
Zeiderman, A. (2016) *Endangered City: The Politics of Security and Risk in Bogotá*, Durham, NC: Duke University Press.

22

Informal worker organising and mobilisation

Linking global with local advocacy

Chris Bonner, Françoise Carré, Martha Alter Chen and Rhonda Douglas

Introduction

In most countries of the global South, as Skinner and Watson (Chapter 11, this volume) have demonstrated, informal workers account for over half of non-agricultural employment. They are a common fixture of employment in cities. Organisations of informal workers have developed as important players in urban policy spaces as they seek to address the issues their members face. Over the past 20 years, these organisations have found that, in order to engage with urban policy institutions, they must combine local organising with transnational, or global, organising and strategies.

Urban policies in general and urban space regulations in particular directly affect the lives of informal workers, their ability to work and do so effectively and safely. Urban spaces – streets, settlement patterns, travel routes and the regulatory and physical infrastructures that undergird all of these – are not only the setting for informal workers' activities but often represent the means of access to markets and earning opportunities.

Informal workers face a major challenge of legitimacy in urban policy. Being seen as a marginal, even illegal, workforce, performing incidental economic activities and forming an obstacle to growth, has worked against their participation and representation in urban policy processes. In recent years, as urbanisation has accelerated, interactions with city authorities, including planners and private interests, have become increasingly fraught. Some urban areas have experienced significant economic growth with the attendant increasing power of real estate interests, and of national and multinational formal businesses. These developments bode ill for informal workers, the majority of whom work in public spaces (which are increasingly being privatised) or in private homes (which often do not receive basic infrastructure services).

These workers have found that local advocacy must be enhanced, and can even be triggered, by transnational, or global, advocacy. This chapter addresses the following questions: Why have informal workers and their organisations brought their concerns to global policy venues and how? How do local and global advocacy strategies interact?

The next section reviews informal worker organisations in urban settings. We explore why and how informal worker organisations have formed transnational networks for global advocacy and, thereafter, we examine how they have combined global and local advocacy to achieve changes in city planning practices in three urban areas: home-based workers in Bangkok, Thailand; street vendors in Delhi, India; and waste pickers in Bogotá, Colombia.

Informal worker organisations in cities

Informal workers have organised and mobilised in urban areas as they sought ways to mitigate frequent conflicts with police or municipality regulatory units (e.g. permit offices and courts). They have also organised for improved access to public amenities in settlements where they live and work.

The chapter focuses on challenges and goals of organisation faced by three groups of workers. Home-based workers, a group that produces goods for domestic as well as global value chains, organise because land use and basic infrastructure policies regarding settlements affect their work activities, transportation policies affect the delivery and sale of their goods, and taxation policies affect the costs of material and equipment. The location of their home and work matters greatly to their ability to produce and access buyers.

Street vendors organise into associations or unions, sometimes incorporating existing market organisations, to regulate their operations in shared space, to limit conflicts and to improve access to customers. They organise to enter urban policy and planning venues (Brown and Lyons 2010) to maintain access to vending space, demand market infrastructure and defend their historical access to prized vending space that is threatened by large, formal retailers and the spread of self-contained shopping areas. To these ends, they seek to create negotiation platforms (Carré *et al.* 2016).

Waste pickers, who collect, sort and reclaim waste for recycling, organise to contend with the rise of city contract awards for garbage collection to corporations ostensibly specialising in recycling. Their goal has been to be recognised as part of cities' waste management systems (Samson 2009).

These and other informal workers also need to organise to access social protection, public health systems and other state-related services and insurance mechanisms that often overlook them because their role as workers is invisible. Hence, cross-sector/cross-worker group organising around shared demands is necessary.

Forms of informal worker organisations

Informal workers organise into associations, unions, cooperatives and self-help groups, and into hybrids of these types (Bonner and Spooner 2011a, 2011b; Carré 2013). Their organisations also form local, national, regional and transnational networks. The forms that informal worker organisations take are varied and are the subjects of existing research (Bonner and Spooner 2010, 2011b; Chen *et al.* 2015, 2007; Schurman *et al.* 2012). Almost all membership-based organisations (MBOs) of informal workers – that is, with elected leaders accountable to members for decisions (Chen *et al.* 2007) – share the following goals with different emphases: representation (voice); improving their economic position; building negotiation and collective bargaining platforms; accessing or providing services; mobilising on specific issues, or for social and political inclusion; and forging worker solidarity.

Strategies used include: negotiation and collective bargaining strategies with local and national governments, and/or with employers and other dominant actors in markets and value chains; economic development strategies to leverage or provide services (e.g. credit); legal strategies to access benefits or markets; and mixed approaches such as those developed by India's

Self Employed Women's Association (SEWA) that combines a struggle approach with an economic development approach (Bonner and Spooner 2010; Carré 2013: 6–9). SEWA, a trade union with over 1.9 million members, has a complex struggle strategy (organising, negotiating and advocating) but it also contains within its family of sister organisations a cooperative bank, a housing finance company, an insurance cooperative and over 100 producer and marketing cooperatives run by its members (Chen 2010).

Global urban advocacy

Informal workers must address, locally and globally, their shared cross-sector issues, as informal workers, and also sector-specific issues relating to their trade or occupation. They are directly concerned with 'commons' (space and resources, including waste, and free access to all with regulated and/or negotiated use) as well as frameworks for social protection, occupational health and safety, solid waste management, and corporate social responsibility that are often addressed in global forums.

Why transnational organising and advocacy

Organisations (and local networks) of informal workers in specific sectors have formed transnational networks with same sector organisations. StreetNet[1] International gathers associations and unions of street vendors. Three regional HomeNets – Eastern Europe, South Asia and South East Asia[2] – are multi-country regional networks of home-based workers associations. There is one regional alliance (Red Lacre) and an informal global alliance of waste pickers, and the International Domestic Workers' Federation[3] gathers domestic worker unions and associations.

Why would informal worker organisations create, or join, networks with a global reach? Global networks of informal worker organisations began forming in the 1990s, inspired in large part by the example and leadership of SEWA. Concerns included the employment and income implications of 'globalisation' in all its manifestations; the globalisation of production and markets but also the fact that ideas, technologies and systems are rapidly transmitted and replicated by business and governments, thus requiring workers' organisations to do the same. To address their marginalisation, informal worker organisations have combined the global with the local; they use transnational activism to support and augment their actions as organisations in local and national settings (Bonner and Carré 2013; Bonner and Spooner 2011b; Carré *et al.* 2016; Mitulah 2010).

Networks of informal worker organisations engage with international organisations (e.g. International Labour Organisation (ILO), other United Nations (UN) agencies) and international debates/forums because both affect regional, national and local policies and practices. In engaging in international venues, informal workers and their organisations aim to secure recognition as workers needing labour standards and social protection. Through the ILO tripartite social dialogue and processes, national governments can be pressured to hold policy discussions on the status of informal workers, and their access to social protection. In low- and middle-income countries in particular, national governments may enact legislation to comply with a convention.[4] International standards are a norm to be invoked in collective advocacy and bargaining in national and local settings.

By its constitution, the ILO aims to undertake consultations, negotiations and eventually to issue directives that engage official representatives of governments, worker groups (almost exclusively unions of workers who are employees) and employer groups. (It has given observer status to some international NGOs.[5]) Tripartite standard setting processes aim to establish labour standards for specific groups of workers and types of work. Domestic workers, a group able to claim employee status, have successfully used ILO tripartite standard setting processes. Until recently, other groups 'officially' defined as self-employed,[6] such as street vendors, have advocated within this tripartite process to be recognised as own-account workers (not employers) deserving of policy attention but not subjects of standard setting discussions. (The 2002 International Labour Conference's stated goal was to reduce the greater decent work deficits of informal workers: in opportunities, rights, protection, and social dialogue (International Labour Conference 2002).) This situation began to change in 2014–2015 with the ILO standard setting processes on "Facilitating transitions from the informal economy to the formal economy", which included the self-employed; the resulting recommendation covers *all* categories of informal workers (International Labour Conference 2015; WIEGO 2014).

Beyond advocacy on labour standards and social protection, informal worker organisations, particularly those of the self-employed, have formed networks in order to engage more effectively in other international forums, be they focused on the environment (e.g. United Nations Conferences on Climate Change), economic development (e.g. World Social Forum) or urban development (e.g. World Urban Forum). International debates offer opportunities for informal worker organisations to gain a foothold in policy discussions with ramifications for their future.

Types of transnational networks of advocacy and mobilisation

Exactly how international forums are used for advocacy and eventually for negotiation varies with each worker group and venue. The form or structure of transnational networks ranges from formalised structures to less formal alliances. StreetNet International, officially launched in 2002, has 51 affiliates in 47 countries. It operates with a *trade union-like structure* based on direct representation by national affiliates. The International Domestic Workers' Federation (launched in 2013) also has a formal *trade union-like structure* with elected representatives from 54 affiliates in 44 countries. The three regional networks of home-based workers all have a *hybrid structure* of informal worker organisations and support NGOs, in part because home-based worker organisations are less developed due to the nature of the work (scattered workers isolated in their homes). However, HomeNet South Asia went through a transition from its NGO-type structure and governance to one that includes democratic decision-making by MBOs of home-based workers (Bonner and Carré 2013; Chen *et al.* 2015). The Latin American Waste Pickers Network ('Red Lacre'), a regional/multi-country alliance of waste pickers, is based on a *social movement philosophy and form.*

From global to local: how global networks interface with local advocacy

While local organising is the bedrock for improving the livelihoods of informal workers, participating in a transnational network of organisations of similar workers plays a part in the activities and impacts of local organisations on urban policies. Overall, active participation in a global or transnational network serves to enhance organisational capacity and increase visibility. The enhancement of capacity for organising and advocacy occurs through the cross-national and cross-organisational sharing of knowledge about strategies, models and lessons learned from experience. Realising that problems are shared across national settings and, thus, that some solutions can likewise be shared opens new avenues for local action, opportunities to experiment and broadens the range of options that a group might try. It also enables local organisations to cull lessons and develop arguments for presenting alternative policy approaches to local governments. Activism at several levels of governance reinforces impact at all levels, as has been found in the human rights field (Keck and Sikking 2014; Sikking 1993).

Transnational exchanges foster the learning and sharing of strategies and lessons between organisations. Informal worker leaders use the information acquired through network activities in engagements with local authorities, including citing success with particular approaches in other countries, or organising visits by local officials to other countries. Exchanges help build solidarity across the informal worker movement. Solidarity develops through interpersonal relationships sprung from learning exchanges, cross national assemblies and grouped representations on joint agendas at the International Labour Conference (ILC) and other global forums. Institutional solidarity (organisational ties) also develops during preparatory meetings for selected ILC discussions. This and the international visibility of an organisation's struggles and achievements break the isolation of local leaders.

Three examples below illustrate several ways in which local advocacy and negotiation were bolstered and their impacts amplified by transnational network activities.

Home-based workers: making the case for infrastructure improvements

In Bangkok, Thailand, HomeNet Thailand (HNT), an affiliate of the regional network HomeNet South East Asia (HNSEA), promoted engagement between informal workers including significant numbers of home-based workers, other residents and several levels of government. HNT sought negotiations to address the needs of informal workers and others displaced to outlying districts due to public and private developments, particularly home-based workers. The immediate issue was the lack of transportation.

HNT has organised informal workers since 1992. By 2013, it had 3,286 members nationally and 529 members (HBWs, some street vendors and others) in districts throughout Bangkok, notably in three outlying districts. (In one of these, Nong Chok, three committee members represent approximately 150 home-based members.) In the relocation areas, the infrastructure as a whole, and public transport in particular, were significant hurdles. Relocated populations had previously lived and worked in central city areas. Over time, resettled people had adopted activities feasible at a distance from the city centre – taxi-driving for men and home-based work for women. But home-based workers (HBWs) still need access to the city centre to buy supplies or deliver finished goods. Public buses were few, irregular, and made for lengthy rides to markets (up to a full day), entailing financial and opportunity costs. Private alternatives (taxis and motorbikes) were unaffordable. Poorly maintained roads were difficult for those with a means of transport. These risks affected all residents, most of whom are poor, but were a particular hindrance to HBWs and others relying on transport for their work (HomeNet Thailand 2013; Tangworamongkon 2015).

HNT and its members faced a complex web of public authorities – national, regional and local – with which to seek negotiation, as well as members' lack of knowledge of this complexity. As it has done for other policy areas, HNT initiated and then facilitated a series of dialogues between the HBWs and other residents of these districts to identify shared transit concerns, build a common line of argument and priority requests, and prepare for meetings with officials. It then convened a 'forum' between city residents, including HNT representatives, and relevant agency officials to achieve cooperation across all districts to address these transportation issues. These forums included representatives of the National Housing Authority (NHA), the Nong Chok District Office, the Bangkok Mass Transit Authority (BMTA) and BMTA's labour union as well as selected HBW representatives and other interested community members and staff of a local foundation.[7] In addition to introducing city and regional officials to the needs of home-based workers, the dialogue process helped workers understand how to navigate bureaucracies at several levels.

As a result of these negotiations, the BMTA approved two additional buses on the route from the district to a main market area and improved the regularity of service. These gains went part of the way toward meeting HNT's goals; the BMTA did not accede to requests to add a more direct route to a nearer market town. The changes improved the access of HBWs to the city centre as well as that of other district residents. The BMTA also committed to processing the request for a pedestrian bridge in Chalong Krung district, infrastructure necessary to negotiate a dangerous road crossing and an improvement useful to all community residents, HBWs among them. The HNT strategy had been to involve representatives of the broader communities, beyond HBWs, in negotiations and to seek changes that would benefit most residents in order to establish broad support.

In a policy environment where deep-seated political conflict and frequent shifts in government over the past decade have made continuous work aimed at policy change challenging, HNT provided training and support so local HBW leaders developed the skills necessary to articulate their demands to officials across levels of government and different agencies. This organising work by HomeNet Thailand was buttressed by the interactions of its leaders with other worker leaders through transnational networks.

The way in which HNT handled these negotiations reflects, in part, the organisation's affiliation with HNSEA, and links with HomeNet South Asia, the transnational networks.[8] The emphasis on dialogue and mediation, imparting information to city authorities and bridging 'silos' of public authorities are approaches fostered by the network and used by member organisations elsewhere. Importantly, belonging to HNSEA also brought HNT to participate in the global *Inclusive Cities for the Urban Working Poor* project, coordinated by the global research-policy-action network WIEGO.[9] Through the project, HNT and the other two networks participated in the Informal Economy Monitoring Study, a 10-city study of dynamics in the urban informal economy. The study found that transport costs represent one-third of business costs for home-based workers in three of the four Asian cities in the study, including Bangkok; and that one-quarter of the home-based workers who incur transport costs operate at a loss.[10] These cross-national findings, particularly those on the transport constraints faced by home-based workers displaced from central Bangkok, played a significant role in legitimating *local* home-based workers' claims. Instrumental to legitimating local efforts, a 2014 regional conference of urban authorities on home-based work was convened by HNT and HomeNet South Asia where the acting Governor of Bangkok was the chief guest and South Asian policy experiences were presented. The ensuing declaration presented a roadmap for city authorities in South and South East Asia to handle home-based worker issues in a participatory manner (Sinha 2014).

Preserving market access for street vendors

The story of activism among organised street vendors in Delhi, India, and its evolution toward a national legislative approach, which includes consultation with street vendor local committees, illustrate the several ways in which local, national and global activism are used by informal worker organisations. The local and national campaigns mobilised research, legal strategies and alliance-building spanning local, national and international dimensions.

In Delhi, India, about 6,000 women street vendors are organised by SEWA Bharat, a branch of SEWA, a founding member of StreetNet International, the global network of street vendor organisations, and of WIEGO. SEWA together with the National Association of Street

Vendors in India (NASVI) participated in a multi-year struggle to secure the national 2014 Street Vendors Act (Street Vendors Protection of Livelihood and Street Vendor Regulation Act). In addition to high transportation costs, fines and harassment, street vendors face a constant threat of eviction from markets and their homes.

The process leading to the passage of this law illustrates the multi-pronged and multi-level approach that informal worker organisations devise. Since 1998, NASVI has dealt with the daily challenges of urbanisation, urban renewal and economic reforms. A first step was to conduct a survey of street vending in six cities of India in 2002, working through local affiliates. The report (*Hawkers in the Urban Informal Sector: A Study of Street Vending in Six Cities*) highlighted the increasing harassment of street vendors by local authorities and their growing exclusion in city plans (Bowmik 2001). It generated public discussion; it was presented at a national workshop where the Minister of Urban Development announced that a National Task Force on Street Vendors would be set up to frame a policy.

The resulting national policy, jointly developed with NASVI, SEWA and other organisations, recommended that state and local governments register street vendors, issue vendor identification cards, and amend legislation and practice to reduce their vulnerabilities (Sinha and Roever 2011). It called for Vending Committees at the town and ward levels with representatives from street vendor organisations. The policy was not widely implemented in large part because local governments are controlled by state governments; few states formulated their own policies based on the national one. In response, NASVI demanded a national *law*; it launched a campaign both on the streets with the vendors and at the highest levels of national government. It argued that the issues of street vendors relate not only to urban planning but also to livelihood, labour, employment and social protection – all subjects under the purview of the national government.

Concurrently, street vendor organisations won several court cases against government entities, including a landmark Supreme Court judgment in October 2010 directing the appropriate government authority to enact a national law by June 2011 to protect the fundamental rights of vendors and hawkers. The campaign and combined advocacy efforts of NASVI, SEWA, and other organisations eventually gained the support of two key ministries (late 2011) (Chen *et al.* 2013). The draft law was formulated by the Ministry of Housing and Urban Poverty Alleviation in consultation with NASVI and other organisations of street vendors and ultimately passed in 2014.

The membership of SEWA and the NASVI in StreetNet International supported the activities of these well-established organisations; it provided examples of policy approaches for the India campaign as well as cross-country solidarity for the member leaders. Cross-national research also played a role; the findings from the 10-city study on the largely negative impact of city policies on street vendors around the world provided credible evidence of the need for the national law to regulate vending and protect street vendors. Also, membership in WIEGO mobilised the resources of legal scholars whose multi-country analyses of street vendor regulations supported the formulation of the law (Transnational Development Clinic 2011).

In this example, national law directs city practices. The 2014 Act mandates that municipalities cannot bypass town vending committees. Representative committees include: 40 per cent from street vendor organisations; 10 per cent from voluntary organisations/NGOs; and 50 per cent from the municipality, transport sector, police, health and sanitation units and banking sector. This legislative success illustrates how a national movement builds, first, from local organised activism to gain national momentum, the latter of which is strengthened by transnational advocacy and research (e.g. Inclusive Cities Project participation and Informal Economy Monitoring Study (IEMS)), and then results in changes in urban policy. In turn, local implementation necessitates further local activism in order to deliver concrete progress to street vendors.

In 2014, NASVI had to take the city of Delhi to court to seek full implementation. As of early 2016, the Act has strengthened the hand of the organisation and its Delhi members. NASVI opposed the closing of the Qutub Road Market (2,000 vendors, 600 of whom are SEWA members),[11] the eviction of the Book Bazaar (200 vendors) and the extortion practices of fictitious 'trade unions'[12] in the Velodrome Road Market (12,000 vendors) (Sankrit 2015). In turn, the national law reverberates internationally. Its mandate that cities cannot bypass negotiations with local vending committees is an example widely used at the ILC where both governments and international labour federations take note. It also has been featured in StreetNet International's outreach activities as a pro-worker approach, expanding the repertoire of policy approaches for which to advocate in other settings.

Amplifying waste picker struggles in Bogotá, Columbia

In Bogotá, Colombia, waste pickers have undertaken a lengthy struggle to establish themselves as key actors in public waste management systems. They acted because of threats to their access to waste; they had lost regulatory and policy battles to private interests that have the resources and political influence to shift policy in their favour. Waste management has been affected by a policy tendency to favour privatisation and by private interests' new perception of waste as a valuable commodity. Bogotá's experience is unique among cities in that a local organisation, the Asociación Cooperativa de Recicladores de Bogotá (ARB), has been active for decades and has a long history of legal struggle on behalf of waste pickers.

Waste pickers have historically been invisible in city waste policy discussions due to their limited organisation, the lack of policy information about their contribution to waste recycling and their vulnerability to corruption in local government. Nevertheless, ARB has organised waste pickers and engaged in public activism to integrate them in public wage-management systems. The organisation, founded in 1990, has over 2,368 members (2016). ARB has pursued four concurrent strategies. The first is the legal defense of waste pickers' labour rights, and seeking court rulings to underpin negotiation of an inclusive model of waste management. The second is the information and mobilisation of waste pickers and allies to influence the policy decision-making process. The third entails building partnerships with national and transnational organisations and networks to strengthen the standing of ARB leaders as negotiators and to gain access to expert knowledge in relevant domains. The fourth is the development of policy proposals to negotiate with authorities and, leveraging worker mobilisation, to overcome the resistance of public officials and/or present evidence of the role of waste pickers in the waste recycling value chain.

These multi-pronged efforts led to a 2012–2013 Constitutional Court decision ordering several authorities in charge of Bogotá's waste to devise a system of payment for waste recovery. ARB presented a research-based paper on the structure of costs for waste pickers, a document later used by the Commission to determine rates. During the process, ARB played an important role to ensure payments were made directly to individual waste pickers and to authorise private warehouses to operate as official weighing centres. The payment system was implemented on March 2013 and now benefits 5,000 waste pickers.

The engagement of the organisation and of other waste picker movements with transnational networks occurred in the later stages of this process. The participation of ARB in the WIEGO-led IEMS provided critical evidence on the status of waste pickers in Bogotá, notably the impact of city policies and practices in the waste management sector. The *Inclusive Cities* project also provided opportunities for waste picker leaders from Bogotá, including Nohra Padilla, to engage in global debates on climate change and the green economy. These engagements served to highlight the contributions of waste pickers to reducing carbon emissions and to raise the profile of the waste picker leaders. It thereby reinforced the local campaign for recognition of waste pickers as environmental agents.

Conclusion

Combined global–local activism encounters challenges, both internal to networks and external, in achieving lasting impacts. Internally, familiar challenges include working across cultures, languages and distinct organising traditions and political histories. They also include ensuring representativeness and sustainability. For representativeness and resilience, networks benefit from the strength of local and regional organisations. For example, associations of street vendors have been in existence for many years and have provided the basis for national and global organising by StreetNet. Still, a thinner base of local IW organisations has not prevented the formation of global networks and their achieving goals (with local impacts) beyond the levels initially expected from such a 'thin' base. Some of the examples discussed in this chapter, particularly that concerning HBWs, are such cases. The challenge for IW organisations is how to maximise the global reach of networks and their limited resources to provide support for growing local organisations and strengthening their capacity for advocacy and negotiation in local and national settings. Along with these challenges, there is the familiar danger that an unrepresentative leadership may become increasingly out of touch with membership concerns. Here again, the danger is lessened where member organisations have strong traditions of representation and where global networks develop representation structures. Also, we find that some local organisations are very dependent upon NGOs for direction and support and take longer to evolve into a representative structure. Yet others remain very local with no regional or transnational links and fight their local battles on their own.

Beyond organisational challenges, global–local advocacy encounters the far greater challenges posed by the rapid evolution of cities, the dominance of private interests over common use access to space, the prioritising of (some) formal businesses in infrastructure development and servicing, and land use policies, as well as the continuing, possibly exacerbated, stance of not recognising informal activities as part of the economy, and penalising them. In these hostile circumstances, informal worker organisations win some battles and lose some; sometimes a victory entails maintaining an historical practice (e.g. access to the central city), while at other times it entails inserting IWs as agents into an activity perceived to be 'modern', such as waste recycling.

In their dealings with national and local entities, IW networks and member organisations operate on two levels: seeking to alter a specific local regulation or practice on the one hand; and pushing for shifts in the mindsets of local regulators, planning officials and police toward informal work, on the other. These opposite approaches are meant to be mutually reinforcing; using a concrete situation to turn widely held ideas 'on their head' and winning allies; and, conversely, using dialogue, research and public debate to foster a more inclusive approach to informal activities, and the inclusion of informal worker representatives in relevant decision-making bodies. Our experience in cases in which WIEGO affiliates have engaged is that it has been possible to achieve changes in regulations and policies. It is a far longer-term process to alter prevailing views in city planning agencies and the profession, let alone find ways to alter the unstable balance between economically and politically powerful private real estate or other business interests and local government and city residents.

Based on the experience of global networks and their local affiliates so far, the most promising examples of shifts in policies and practices towards urban informal workers are in countries where there is an activist supreme, or constitutional, court willing to address informal work and strong IW organisations that engage in advocacy combined with legal struggles. Examples of victories in Colombia and India illustrate the combination of pressure 'from above' and 'from below'. Court decisions, achieved in part because strong local movements followed a legal strategy, are longer lasting and less likely to be reversed than local government policies. Some national contexts with a long tradition of participatory democracy, relative to other countries, and strong associations with a history of advocacy or legal struggles are also more propitious to IWs achieving gains in local planning. Such is the case of some states in Brazil, as with waste picker gains in Minas Gerais. Even strong local IW organisations do better with achieving policy impacts when they belong to, and have the backing of, a national federation with capacity to leverage support in the policy sphere as well as resources for research. The case of the street vendor policy in India illustrates the role of combined action by organisation and federation. A national federation plays an important role where policy (on IW, on planning) can be made at the national level. In turn, the effective implementation of national policy concerning informal workers can only occur with the monitoring and activism of local IW organisations.

Notes

1 www.streetnet.org.za/show.php?id=19 [accessed 20 February 2017].
2 www.homenetsouthasia.net/; http://wiego.org/wiego/homenet-south-east-asia [both accessed 20 February 2017].
3 http://wiego.org/wiego/international-domestic-workers-federation-idwf [accessed 20 February 2017].
4 ILO Recommendations are a standard commanding legitimacy that governments may follow.
5 For the tri-partite ILO standard setting processes, only union national centres are included. Global Union Federations and global networks have NGO observer status along with selected NGOs.
6 Historically, the self-employed have been considered to belong to the entrepreneur category.
7 Previous meetings centered on water and sanitation improvements involved representatives of the Nongchok District Office, the National Housing Authority (NHA), the Metropolitan Waterworks Authority (MWA) and the Bangkok Metropolitan Administration's Health Department.
8 Leaders participated in the International Labour Conference and other forums where informal workers have come together (Global Conference of HBWs, WIEGO General Assembly), all activities contributing to leaders' experience in advocating for change at national and local levels.
9 www.wiego.org [accessed 20 February 2017].
10 The Informal Economy Monitoring study was conducted in 10 cities in Africa, Asia and Latin America.
11 Communication from SEWA Bharat (March 2016).
12 Fictitious 'trade unions' are organisations set up to extort fees from street vendors for using the space.

References

Bonner, C. and Carré, F. (2013) "Global networking: Informal workers build solidarity, power, and representation through networks and alliances", WIEGO Working Paper No. 31. Cambridge, MA and Manchester, UK: WIEGO, September. Available: http://wiego.org/sites/wiego.org/files/publications/files/Bonner-Global-Networking-Informal-Workers-WIEGO-WP31.pdf [accessed 11 January 2016].

Bonner, C. and Spooner, D. (2010) "Work in progress: Organising labour in the informal economy – forms of organisation and relationships", XVII World Congress of Sociology, Gothenberg, Sweden, 11–17 July, pp. 1–30. Available: www.globallabour.info/en/Spooner%20Bonner%20(Organising%20in%20the%20informal%20economy%20(ISA%20RC-44,%202010)).pdf [accessed 11 January 2016].

Bonner, C. and Spooner, D. (2011a) "Organizing in the informal economy: A challenge for trade unions", *International Politics and Society* IPG 2/11: 87–105.

Bonner, C. and Spooner, D. (2011b) "Organizing labour in the informal economy: Institutional forms and relationships", *Labour, Capital and Society* 44(1): 126–152. Available: http://wiego.org/sites/wiego.org/files/publications/files/Bonner_Spooner_Organizing_Labour.pdf [accessed 11 January 2016].

Bowmik, S. (2001) *Hawkers in the Urban Informal Sector: A Study of Street Vending in Six Cities*, Dehli: National Association of Street Vendors, May. Available: http://wiego.org/sites/wiego.org/files/publications/files/Bhowmik-Hawkers-URBAN-INFORMAL-SECTOR.pdf [accessed 20 February 2017].

Brown, A. and Lyons, M. (2010) "Seen but not heard: Urban voice and citizenship for street traders". In I. Lindell (ed.) *Africa's Informal Workers: Collective Agency, Alliances and Transnational Organizing in Urban Africa*, London and New York: Zed Books and The Nordic Africa Institute, pp. 33–64.

Carré, F. (2013) "Defining and categorizing organizations of informal workers", WIEGO Organizing Brief No. 8, September. Available: http://wiego.org/sites/wiego.org/files/publications/files/Carre-Informal-Worker-Organizations-WIEGO-OB8.pdf [accessed 11 January 2016].

Carré, F., Horn, P. and Bonner, C. (2016) "Modes and venues of bargaining for informal worker organisations in the global south", draft presented at Radcliffe Institute for Advanced Studies Workshop Gender, Precarious Work, and Labor Organizing, Cambridge, MA: Harvard University, February 25–26.

Chen, M.A. (2010) "The Self-Employed Women's Association". In T.K. Oommen (ed.) *Social Movements II: Concerns of Equity and Security*, New Delhi: Oxford University Press, pp. 221–236.

Chen, M.A., Bonner, C. and Carré, F. (2015) "Organizing informal workers: Benefits, challenges and successes", background paper for the *Human Development Report* 2015. Available: www.hdr.undp.org/sites/default/files/chen_hdr_2015_final.pdf [accessed 30 May 2016].

Chen, M.A., Jhabvala, R., Kanbur, R. and Richards, C. (2007) *Membership-Based Organizations of the Poor*, New York: Routledge.

Chen, M.A., Bonner, C., Chetty, M., Fernandez, L., Pape, K., Parra, F., Singh, A. and Skinner, C. (2013) "Urban informal workers: Representative voice and economic rights", background report for the World Development Report 2013. Available: http://wiego.org/sites/wiego.org/files/publications/files/WDR2013_Urban_Informal_Workers.pdf [accessed 30 May 2016].

HomeNet Thailand (2013) "Homeworkers in Thailand and their legal rights protection". Available: http://wiego.org/sites/wiego.org/files/publications/files/T05.pdf [accessed 30 May 2016].

International Labour Conference (2002) *Decent Work and the Informal Economy*, Geneva: ILO. Available: www.ilo.org/public/english/standards/relm/ilc/ilc90/pdf/rep-vi.pdf [accessed 30 May 2016].

International Labour Conference (2015) "Recommendation No. 204 concerning the transition from the informal to the formal economy", adopted at 104th Session, Geneva, 12 June. Available: www.ilo.org/ilc/ILCSessions/104/texts-adopted/WCMS_377774/lang--en/index.htm [accessed 30 May 2016].

Keck, M.E. and Sikkink, K. (2014) *Activists Beyond Borders: Advocacy Networks in International Politics*, Ithaca, NY: Cornell University Press.

Mitullah, W. (2010) "Informal workers in Kenya and trans-national organizing: Networking and leveraging resources". In I. Lindell (ed.) *Africa's Informal Workers: Collective Agency, Alliances and Transnational Organizing in Urban Africa*, London and New York: Zed Books and Nordic Africa Institute, pp. 184–201.

Samson, M. (2009) *Refusing to be Cast Aside: Waste Pickers Organising Around the World*, Cambridge, MA: WIEGO.

Sankrit, R. (2015) "SEWA Bharat and street vendors in Delhi", Inclusive Cities Project, Cambridge, MA: WIEGO. Available: www.inclusivecities.org/wp-content/uploads/2015/09/IC-Delhi-SV-Case-Study-Summary.pdf [accessed 30 May 2016].

Schurman, S.J. and Eaton, A.E. with Gumbrell-McCormick, R., Hyman, R., DiLeo, C., Berroterán, G.M., Ryklief, S., Varga, M. and Viajar, V. (2012) "Trade union organizing in the informal economy: A review of the literature on organizing in Africa, Asia, Latin America, North America and Western, Central and Eastern Europe", report to the American Center for International Labor Solidarity, New Brunswick, NJ: Rutgers University. Available: www.solidaritycenter.org/wp-content/uploads/2014/11/Rutgers.Trade-Union-Organizing-in-the-Informal-Economy.pdf [accessed 11 January 2016].

Sikkink, K. (1993) "Human rights, principled issue-networks, and sovereignty in Latin America", *International Organization* 47: 411–441.

Sinha, S. (2014) "Regional conference of City Authorities on Home-Based Workers – Report", Bangkok, 7–9 May. Available: http://wiego.org/sites/wiego.org/files/reports/files/Regional_Conference_HBWs_Report_May_2014.pdf [accessed 30 May 2016].

Sinha, S. and Roever, S. (2011) "India's national policy on urban street vendors", WIEGO Policy Brief (Urban Policies) No. 2, Manchester, UK: WIEGO. Available: http://wiego.org/sites/wiego.org/files/publications/files/Sinha_WIEGO_PB2.pdf [accessed 30 May 2016].

Tangworamongkon, C. (2015) "Home-based workers create city dialogue in Bangkok", 21 December. Available: www.inclusivecities.org/impact/bangkok/ [accessed 30 May 2016].

Transnational Development Clinic (2011) *Developing National Street Vendor Legislation in India: A Comparative Study of Street Vending Regulation*, New Haven, CT: Jerome N. Frank Legal Services Organisation, Yale Law School.

WIEGO (2014) "WIEGO network platform: Transitioning from the informal to the formal economy in the interests of workers in the informal economy", Manchester, UK: WIEGO. Available: http://wiego.org/sites/wiego.org/files/resources/files/WIEGO-Platform-ILO-2014.pdf [accessed 30 May 2016].

23
Is there a typical urban violence?

Fernando M. Carrión and Alexandra Velasco

Introduction

During the last 25 years, violence has become one of the most important issues for Latin American cities due to changes in its nature, its social and economic impacts, the increase in its magnitude and the appearance of new types of violence. Violence extends throughout countries and cities of the region with differing characteristics, causing changes of urbanism (construction of walls in cities), new behaviour in the population (anxiety, helplessness), changes in social interactions (reduced citizen participation and socialisation) and the militarisation of cities (heavy-handed policies, military in the streets), as well as a reduction in the quality of life of the population (homicides, property loss). However, the current model of cities also produces violence that has never been seen before, such as 'non-criminal' violence that arises from fragmentation, exclusion, density, the dispute over public space and access to services. In other words, the problems of transportation, services, environment, poverty and housing have incorporated violence, so much so that they have become an additional urban problem, which requires new urban planning explicitly tied to policies of public safety.

Despite the rise of violence in Latin American cities, it has not been given the importance it deserves, nor has it been incorporated into discussions on the problems of development and the characteristics of urban life. The relationship between cities and violence has been overlooked; this oversight can be attributed to a methodological problem that arises from violence being defined and understood as rooted in certain attributes and not in social relationships. Therefore, every extremity of the relationship has been researched independently, without having been able to create the desired connections. And when there have been attempts to create links among these social relationships, they have been done under an unequivocal determinism of the urban toward violence (aetiology).

How violence is defined remains an unresolved issue and is something that needs to be addressed. Defining the methodological starting point is crucial; in this chapter we pose questions that are relevant to understanding violence and will provide some clarity on the definitional issues. Given the prevalent anti-urban positions, the starting point for understanding violence is to analyse if there is a causal relationship between cities and violence.

Cities are vessels for criminal acts, which supposes an autonomy of cities when facing violence. This is why asking the question if cities are sites of crime is pertinent in as much as the *locus* of social practices has a spatial expression. As violence is plural, is it possible to find specifically urban violence if we understand cities as places of concentration of the highest density of heterogeneity and, therefore, where the rituals of daily life can produce friction, conflicts and contradictions that lead to typical 'city' acts of violence?

From these reference points the structure of the analysis is organised as follows: first, should we address violence from its aetiology? Is the city the cause of violence or is it a relationship? Does the city contain violence or does it generate it? Second, is there a causality of the city toward violence or is violence expressed in it because of its geographic locus? Are the so-called classical causes of violence – inequality, unemployment, poor education and lack of family unity – urban in nature? Third, the chapter examines the transition between public space and private space and how these generate specific types of violence. Finally, in the conclusion there is an attempt to understand how we should address violence in the city: with urban policies, with public safety policies or a combination of both?

Cities and violence

Two misconceptions about their relationship

Violence and cities change constantly because they are historic: however, ignorance about the transformation of the relationship between violence and cities – and vice versa – has created many misconceptions, among which we can point out two. There is a predominant vision caused by the temptation to find the causality of violence in cities, through an anti-urbanism that leads one to believe that urban violence is synonymous with violence and that cities are the origin of violence, through which cities end up being criminalised. For a long time, cities were considered a problem because they were assumed to be a source of chaos and anomie and an artificial product that challenges nature. Today, this opinion is trying to be overcome because there are trends that have begun to see cities as a *solution*, in as much as women become visible and public,[1] poverty is reduced,[2] there's better quality employment and unemployment is lessened (OIT 2003: 16), services are better and the rates of violence are different than in the rural areas. That is to say, it isn't cities in abstract but rather the concrete model that is related to a particular type of violence; in other words, *urban violence* can be reversed under a model that is distinct from that of current urbanism. Good urban policies could be transformed through good safety policies to improve life quality in cities, and to integrate people fostering citizenship.

Due to a lack of studies of the relationship between cities and violence and a deterministic vision of the urban as violence, a number of proposals have been developed that are not based on reality. Two proposals include:

- *Situational prevention*, which seeks to reduce the opportunities for violence (routine activities) and the incentives for the perpetrators (rational choice) in the spaces or sites where crime normally occurs (urban design) (Crawford 1998: 8). Three types of actions have defined situational prevention in public spaces. First, video cameras, community police and rapid response. Second, the expulsion of certain segments of the population (youth, homeless) under the pretext of privatisation and strict right of entry. Third, the provision of services and equipment in certain spaces previously considered dangerous (the classified and stigmatised city of 'others').

- *Broken windows*, which derives from the causal logic established between disorder in the streets and violence. This disorder comes from the neglect or lack of interest in repairing a broken window (something negligible), that later turns into a kind of public waste dump (something important), that brings in gangs, beggars, alcoholics and drug addicts and causes the neighbourhood to demand security (Kelling and Coles 2001: 20).

These kinds of proposals show a misunderstanding of the relationship between cities and violence, as the next section suggests.

Cities and violence: cause or relation?

Without studies that document the reality of this relationship, there is the danger that socially stigmatising images will be created along with policy proposals that circulate independently of the real processes,[3] born in the mass media and/or in the demands of the public.

Violence cannot be understood based on natural (biological) or moral (religious or traditional) causes, nor through exclusive consideration of the legal deviation (anomie). Nor is it understood through the existence of one or several attributing causes (so-called *risk factors*) that define it. Rather, it should be thought of as a particular social relationship of social conflict (Carrión 2008: 24) and, as such, as a *complex social and political construction* (Sozzo 2008: 10) that takes shape in specific territories and times.

This type of affirmation leads us to understand violence as a social condition that has multiple direct and indirect actors that change with time and space; and that a 'before and after' the violence doesn't exist because it is an ongoing continuum. There is not a single cause of violence, nor multiple causes for multiple episodes of violence, because this indisputable determinism does not exist.

Violence is more than a criminal act; there are forms of violence that are not legally classified due to there being a "mass of events" (Sozzo 2008: 19) that are undistinguished from each other. Criminality is a powerful feeling that incites real violence while being part of that violence. Fear is a powerful feeling that guides daily life, be it as a social mechanism (solidarity), an individual response (being armed, personal defence) as well as a factor organising time and space in cities.

Violence is not a social pathology that comes from certain attributes (risk factors or causes). The information that comes from the indicators of this paradigm (risk factors) does not reflect the reality (suicide, gender, contract killings) and the plurality of violence; but the indicators do create stigmas, for example, that Latin America is the most violent continent in the world. If it is viewed according to the definition of violence used by the WHO (2002: 4), one should ask why homicides (violence toward another) and not suicides (violence toward oneself) are the measure that quantifies the degree of violence in a country. In 2000, deaths caused by war were 18.6 per cent, homicides were 31.3 per cent and suicides were 49.1 per cent of violent deaths (WHO 2002: 8). If that's so, why do homicides and not suicides determine which are the most violent neighbourhoods, cities and countries? Rather, it is a specific social relationship of conflict that is plural,[4] historical and relational. In other words, if violence cannot be understood based on aetiology (Carrión 2008: 12), cities should not be considered as the cause of violence. However, this does not mean one should ignore the relationship between violence and cities or cities and violence.

Cities and violence: particular violence or vessels for violence?

If one accepts the statement that there is a historic relationship between cities and violence, one must also ask if there is particular violence that is experienced in cities and/or if they are a vessel for violence. To understand the relationship between violence and cities requires going beyond the methodologies that interpret the relationship as if it were a pathology rooted in certain attributes (factors), when in reality it has to do with social relationships. Thus, we have to take into account:

- *Violence* – as defined by Alvaro Guzman (1994: 170) – should be understood as "a product of a social relationship of conflict that, as a consequence, ties together at least two opponents, actors, individuals or groups that are passive or active in the relationship".
- *Cities* – as understood by one of the classic scholars of modern urbanism, Louis Wirth (1988: 4) – should be seen in terms of a "relatively large, dense and permanent settlement of socially heterogeneous individuals".

If the city brings together the greatest diversity of people in a relatively reduced territory (density), it is possible to think that it gathers an arena of relationships, where, on the one hand, social conflicts arise, some of which can result in acts of violence that are typical of cities (urban violence) and on the other hand, there is a tendency toward a concentration of violent events (geography of violence).

Urban violence dates from time immemorial, it is inherent to cities; however, the violence of the past is distinct from that which exists today because the relationship between violence and cities has a history; and every incident of violence is historical. For example, "during the Garcia era there was more emphasis on the public good and crimes that affect the functioning of the theocratic State, (while) liberalism is more concerned with crimes that threaten the individual and property" (Goetschel 2005: 97). During the last 25 years, the most important change in crime has been the shift from the *traditional* to the *modern*. The former is that developed by a culture of strategies of survival or asymmetries of power; the latter has an explicit predisposition to commit a criminal act and, therefore, is organised, international and technologically developed (Carrión 2008: 26). The statement that violence is historical therefore carries with it the recognition that violence is in a constant process of change, whether through a change in magnitude or a transformation of its characteristics.

The Latin American city suffers a significant transformation caused by decreasing rates of urbanisation (return to the 'constructed' city),[5] internationalisation (global cities), and state reforms that assign more power to the municipality in relation to the national government, albeit less compared to the cities itself since the private sector, in the form of real estate or big transnational companies, have emerged as key actors for privatisation.

While the levels of urbanisation (city size) in Latin America have generally decreased, the rates of homicides have increased.[6] For example, countries with the highest levels of urbanisation are not the most violent: Chile, Argentina or Uruguay – which have high levels of urbanisation – have lower rates of violence compared to countries that have a lower level of urbanisation such as Ecuador, Guatemala and Honduras. In addition, the biggest cities are not those that have the highest rates of violence. In Mexico, Mexico City itself does not have more violence than Guadalajara; in Colombia, Medellín and Cali are more insecure than Bogotá; in Brazil, Rio de Janeiro, despite its high rates of violence, doesn't have higher rates than Recife and São Paulo doesn't have as much insecurity as Rio de Janeiro. In other words, it cannot be said that the greater the city size, the greater the violence.

However, the urbanisation of violence produces a double effect: cities produce a particular type of violence – urban violence – and are also a vessel of violent scenarios in the sense that these are social practices that are developed and expressed in that space. There are growing urban violences compared to those in the rural areas due to a subtle differentiation between both: while in the rural areas violence is against people, the family and traditions, violence in the city is related to public space, property, gangs, labour struggles, vandalism and squatting, linked to social coexistence.

Mutual relationships: cities and violence

Violence has become a component for understanding cities and this instils violence with certain explicit qualities. It is impossible to deny that violence creates a particular type of spatial organisation, for example, through the projection of images of fear, that change into an integral element of cities, or to deny the conflictive nature of cities that produces specific violence types. This supposes that there is a dialectical relationship and not a unidirectional determinism between violence and cities, born out of the so-called risk factors.

From the city to violence

The social production of territory is an important element in the behaviour of certain types of violence. It is clear that there exists a geography of violence that comes out of the social division of space and a particular logic of urbanism.

Land uses have an important significance in the production of some types of violence. The statement is as simple as "they rob banks where there are banks" and this becomes a key element in understanding that certain types of violence have a direct relationship with spatial organisation. For example, car theft occurs near banking and commercial areas, during working days and hours. The most common street crimes have patterns and identified points connected to some collective transportation hubs, urban centralities or public spaces. These facts of violence cannot be disassociated from the variable of time. This observation should lead to the formulation of security policies based on the essential need to tie physical planning with security policies; that is to say, security actions to land use,[7] as well as urban planning that reduces inequalities, constructs multiple centralities and recognises distinct orders. It is necessary to recover the geography of violence, not from socio-territorial stigmatisation, but in order to have geo-referenced information that allows for reality-based decision-making.

Residential segregation creates a symbolic and real violence that ultimately is expressed in exclusionary relationships between the places where those with more economic resources settle versus those with fewer economic resources. The existence of unequal urban space increases insecurity, as well as vandalism, social reprisal, stereotypes and the search outside the market for that which others have legally. Residential segregation creates barriers that drive inequality and violence, but security policies also increase the fragmentation. The city of the 'others' – the city of the poor, the 'savage city' – is criminalised, and as a result, is further 'othered'. For example, "in a socio-economically segregated city like Santiago, the perception of insecurity in public spaces implies a near inexistence of interactions between habitants that belong to distinct social strata" (Rodríguez and Winchester 2004: 132). Cities – essential places of encounter with the other – end up being places of exclusion. The significant polarisation between rich and poor makes it so that the "relationship of dependence, or at least of compassion, that until now underpinned all the forms of inequality, now shows itself as a new *non-place* in world society" (Beck 1998: 122). Residential segregation confirms that while the habitants of poor neighbourhoods find security in the sense of belonging, identity and community participation, on the other hand the inhabitants of the rich neighbourhoods contract private security and create 'bunkers' in order to defend and preserve the neighbourhood's internal homogeneity. Thus, when one talks of the fears of the city, those who experience these fears are those that are integrated more so than those that are excluded (Castells 1999: 63).

Cities have started to change their structure and become 'foreign', in ways suggested by Castells:

- Cities go from the classic *urban segregation*, in which the parts are integrated through the public space and urban centrality, to *urban fragmentation*, producing ruptures of the previous socio-territorial unity to give way to the formation of cities converted into "discontinuous constellations of spatial fragments" (Castells 1999: 438).
- Cities shift from having public space that is a fundamental element for meeting, to a city where mobility represents a shift "from the space of places to the flow of places" (Castells 1999: 422), creating agoraphobia and the return of nomadism.

That is to say, the *city of borders*, with its limits and impassable barriers, where exclusion is a way of being outside of oneself (of the city), is achieved through territorial, generational, ethnic, migratory and gender ruptures. One should keep in mind that borders are synonyms of separation, in which for some their path is regulated while for others it is denied – not all can pass. This causes the city inhabitant to behave like a foreigner or stranger because they don't walk the familiar paths and when they leave their habitual territory they are asked for their identification: a passport or a national identity document.

Nowadays our cities are not of the citizens but of the foreigners because fragmentation leads to the loss of reference space for social construction and the feeling of belonging. As a result, insecurity is perceived in the space of the other and security in the space of oneself; it is for this reason that space is defended from the other and why the other is always outside, excluded, strange: foreign.

Fear has become the strongest urban image in Latin America. The unequal city expresses a symbolic fear, a consequence of the daily risk of living there, although the fear differs according to the social condition of the inhabitant. In seeking safety, closed, single-purpose, autonomous and specialised enclaves are developed and people abandon the public space to seclude themselves in the domestic space that, in many cases, is more violent and dangerous. The images of fear that cities produce are sometimes linked to places of natural origin, embedded in the urban centrality, such as hills and rivers and those of human production like informal markets, transport stations and main streets. One of the noteworthy elements of these cases is that urban policies, brought about for rehabilitation efforts, do not have immediate effect because there exists a social inertia that gives a permanent feeling of insecurity (Silva 2004: 25).

From violence to the city

While cities are not a determining factor of violence, it seems appropriate to engage in a reverse methodological exercise: to think of the impact that violence has on cities. It cannot be ignored that the increase in insecurity leads to violence being one of the most destructive problems to the quality of urban life and that it erodes the condition of the public sphere of cities. The worsening of life conditions is part of the process of urban violence, where the safety reactions of the population bring about a new social behaviour that leads to more violence: self-interest, anxiety, uncertainty, marginalisation, neglect, isolation, mistrust and aggression. From this perspective, important explicit impacts emerge from the violence in cities: citizenship, time and space.

Citizenship is borne historically in the city, due to the population's affiliation or belonging to the political community that grants it rights and responsibilities. Since the city is where citizenship is constituted, we can agree that violence restricts the origin and the source of the quality of citizenship.

> Violence increasingly affects more settings of our social life: work, family, school; and for this reason has become one of the factors that most deteriorates the habitability and quality of life of the city. Social coexistence is one of the most worrying issues for citizens.
>
> *(Corporación Región-Medellín 1993: 13)*

One cannot ignore the effects that violence and the fight against it have on the population: the erosion of citizenship, and the modification of citizens' daily conduct, including changes to daily routine and the paths and spaces travelled daily, the limitation of social relationships because all that is unknown is suspicious, reducing personal vulnerability by getting guns, dogs and alarms or learning personal defence.

There is a *chronology of violence*, borne out of the relationship between time and violence that is expressed through two modalities. On the one hand, some crimes are more common during certain hours, for example, homicides are nocturnal and linked to holidays, juvenile suicide occurs more often at the end of school semesters, and firework accidents more at the end of the year. A chronology of this type produces restrictive human behaviour and focalised urban policies that tend to increase the individual attitudes of the population that, in many cases, turn into the instigating elements for acts of violence or inhibition of social or economic activities. On the other hand, some crimes limit the times that the city is utilised by inhabitants, causing cities to disappear at night.

Cities are losing public and civic space, and private/walled urbanisation is widespread. As a result, cities are becoming increasingly more private and domestic. Proof of this is the significant increase in home life, for example, watching movies, eating and working at home. Today there are parts of the city that the police cannot enter. Objective and subjective violence unfold in the territory in unequal ways, producing different impacts; closed housing developments, closed social and athletic clubs, as well as verticalisation and urban centralities have much to do with violence and the fear that it has created. Fear would seem to have made itself the driving force of the contemporary city. Today, fear is an urban principle.

Public space/private space

In Latin America, public spaces are being consumed by the private sector that privatises everything, leading to segmentation that causes people to never be in the same place at the same time. Also, agoraphobia is expressed most clearly in the quintessential symbolic element: the plaza, but also in the streets, parks and sidewalks. In this context, to mention *situational prevention* is, to say the least, a contradiction because if 'public space is the city', what must be done is not to cover up public space but rather propose a new urbanism based on the construction and production of cities, that is to say, the public space.

The nexus between poverty and wealth has been lost. According to Bauman this is caused by the division of the world population between the globalised rich that dominate the time, and the localised poor that are stuck in the space (cited in Beck 1998: 121). Public space loses its sense of inclusion and becomes a *non-place of the poor*[8] and is subject to a permanent attack (agoraphobia), to the extent that plazas are on the road to extinction. Meanwhile, the rich build their private spaces exclusively for themselves: the social club, the sports club, the supermarket and the closed neighbourhood. The dual city finds its new dimension.

Without quality public spaces, coexistence is not possible nor is urban structure satisfactory and, as such, neither is public safety. This is why land management and construction of public spaces is not a task for the police (repression) nor is it a process for social control (discipline). Public space is the space of dispute for liberty, integration, visibility and representation (Carrión 2007a: 16). From this, we can see three examples of interesting cases that move from the private to the public, though these are not without conflicts.

Gender violence

Until very recently, gender violence wasn't historically recognised as a specific type of violence (crime). The current recognition of gender violence is happening through the lens of public safety, which produces two important changes. First, it is registered as a right, particularly related to the citizenship of women; and second, it is recognised as a way to allow the private-domestic world to become more visible as a part of the public landscape, eliminating the public–private dichotomy (Carrión 2007b: 1). In other words, the recognition of gender violence is linked to democracy, to respect for diversity, to the fulfilment of rights and the creation of a community with an urban sense. Violence isn't defined by the space where it occurs but by the asymmetries of power that exist between the sexes. However, it is correct to recognise that in cities it has been feasible to change the unequal patterns of gender, because while in the rural areas there are traditional cultures that assign distinct and specific roles to women, in the city women become public, visible and their rights are recognised.

Gang violence

Gangs become an institutional substitute for that which prepares young people to be part of daily life. It is a mechanism to face the scarce liberty that they have in the private-domestic space, obligating them to fight for the public space, which is elusive for them because their social and symbolic appropriation has a condition of exclusionary occupation for authority. In this unrest, it becomes a diffuse conflict in terms of the plurality of actors (other gangs, the community, police, the city governments); different to a conflict of polarised actors in which safety policies define gangs as the new enemy. This is the case of the '*Maras*', a gang in El Salvador and Honduras.

Young people separated from their families find in gangs a public collective space of symbolic expression and social integration. "A gang is a type of fraternal organization that gives young men autonomy from adult authority. These young men that lack freedom at home, create their own privacy in urban public spaces where they feel protected from authority" (Goubaud 2008: 36). That is where they feel belonging and authority, and part of an elite recognised by the media and public policies. From this perspective, public space is a defining element of safety and group identity where they construct their street, neighbourhood, urban and even virtual communities. The migratory phenomenon, new communication technologies and the opening of international markets give way to new strategies for the groups of emigrant youth that establish ties between their place of origin and destination as territories that are initially differentiated and later united, thanks to virtual integration. This means that the gang ends up transforming its neighbourhood space into a virtual space.

Private security

This grows with state reforms when the minimal conditions for security privatisation are introduced and there is a noticeable increase in violence. Since the mid-1980s, there has been an expansive increase in the market for private security goods and services. According to Frigo (2003: 2):

> In Latin America, private security is a rapidly expanding economic sector … In the last 15 years, private security has gained a relevant place in the world as well as in our region … The world market for private security has a value of 85 billion dollars, with an average annual growth rate between 7 per cent and 8 per cent. In Latin America the estimated growth was 11 per cent.

This economic condition positions a new political actor on the scene of public safety and does so under a new precept: the profitability of capital investment. In addition, it turns into an actor for control of public order. Today in Latin America, private security companies (more than 4 million) have more than twice the number of enforcement personnel than the police (approximately 2 million) (Frigo 2003: 2). A situation such as the one described introduces two elements that we should highlight: the change of the right to public safety to an object of economic transaction, a process that creates a highly segmented and exclusive market. Paradoxically, the state itself is one of the principal consumers of this service turned commodity and those who have the money can get security and those that lack resources don't have access to this 'service'. Is this private security or deprivation of security?

Conclusion

Today, the central debate about violence and public safety centres on the need to break the unilateralism of dominant views because social phenomena cannot be understood from the natural sciences or the restrictive point of view of the aetiology.

While cities once were conceived as merely a catalogue of problems, they are also places where citizenship is constructed, where better innovations are produced, and are the quintessential places for productivity. Cities are the places where services are provided in the best way, the scenes of growing employment and decreasing poverty, as well as where women are free to be in public, young people can express themselves, and where politics evolve. Such a consideration is fundamental so as not to fall into the prejudice of causality and also to understand that good urban policies can do a great deal from the perspective of 'de-securitising' interpersonal relationships and satisfying public and individual liberties.

If we consider that the crisis of public space is a significant expression of the urban crisis, we can conclude that urban policies and urban planning can truly contribute to the reduction of crime. In other words, the development and construction of new public spaces – such as significant and symbiotic spaces – will only be possible with a new urbanism and new public safety policies.

There is a dialectical relationship between the city and violence that allows for the structuring of two important analytical perspectives. One has a dynamic borne out of the changes in the elements of the equation that seek to alter the relationship; that is, that violence, cities and their relationship are historic. The other perspective is that it is evident that there is violence that is part of the city (urban violence) as well as the view that the city is a space where multiple violent acts occur. It should be also understood that violence creates many problems in the city, therefore there is a need for more work on public safety and urbanism.

Notes

1 "According to Anderson, this growing presence of women in cities reveals social factors and complex social and economic processes, among which can be included that cities provide conditions of 'viability' for single women, women that want to become independent and single mothers, in other words, the so-called 'vitality' of cities" (Arboleda 1999: 24).

2 In all countries, poverty tends to be greater in the rural areas than in the urban areas, and it tends to be less in the biggest cities than in the medium and small cities. On the other hand, in the majority of countries the urban concentration hasn't been a negative factor, as it has allowed for the access to goods and services to a much greater degree than was the case in the time of rural predominance.

(Jordan and Simioni 2002: 15)

3 There are a number of public and private actions that can be mentioned: the privatisation of public space, autonomous closed neighbourhoods, video surveillance, restrictions on alcohol sales, the quadrant plan, lighting, trash collection, urban mobility, which in the long run represent a logistical burden, in the absence of real security strategies.
4 Just as there is gender violence or juvenile violence that takes place, the first in the asymmetries of power and the second in the conflict created by generational borders within age groups, urban violence can also exist.
5 Latin America has 82 per cent of its population living in the cities, which means that the migration from rural to urban areas is decreasing. The demographic pressure in the peripheries of cities will decrease causing a re-urbanisation of what already exists inside the cities.
6 While in 1980 the rate of urbanisation was 1.2 per cent per annum, in 2005 it was 0.5 per cent per annum, which means that it has decreased by half in 25 years (United Nations 2014: 9). During the same period, the homicide rate doubled: "In 1980, the average homicide rate was 12.5 per 100,000 habitants per year. In 2006, it was 25.1, which means delinquency has doubled in the last quarter of a century" (Kliksberg 2008: 5).
7 The planning of urban infrastructures includes the dimension of civil security.
8 "A *non-place* can't be defined as a space of identity nor as relational nor historic" (Augé 1998: 83).

References

Arboleda, M. (1999) *Equidad de género: El campo municipal como espacio de derechos y políticas*, Mimeo.
Augé, M. (ed.) (1998) *Los no Lugares, Espacios del Anonimato*, Barcelona: Gedisa.
Beck, U. (ed.) (1998) *¿Qué es la globalización?* Barcelona: Paidos.
Carrión, F. (2007a) "*De la violencia urbana a la convivencia ciudadana*", Ecuador: Flasco. Available: www.flacso.org.ec/docs/sfsegcarrion.pdf [accessed 20 February 2017].
Carrión, F. (2007b) "El espacio de la violencia de género". In *Revista Ciudad Segura, Vol. 20*, Ecuador: Programa Estudios de la Ciudad, Flacso, pp. 1–3.
Carrión, F. (ed.) (2008) *La Seguridad en su Laberinto*, Quito: Flacso-Ecuador-IMDQ.
Castells, M. (ed.) (1999) *La Era de la Información*, Barcelona: Siglo XXI.
Corporacion Region-Medellin (1993) *Procesos de Urbanización y nuevos Conflictos Sociales*, Seminario: Políticas e Instituciones en el Desarrollo Urbano Futuro de Colombia, Bogotá.
Crawford, A. (1998) *Crime Prevention and Community Safety. Politics, Policies and Practices*, Longman: London.
Frigo, E. (2003) "Hacia un modelo de seguridad privada en América Latina", conference paper, Primer Congreso Latinoamericano de Seguridad Privada, Bogotá, 2–26 September.
Goetschel, A.M. (2005) "La cárcel en el Ecuador, vida cotidiana, relaciones de poder y políticas públicas", unpublished work, Quito: Flacso-Ecuador.
Goubaud, E. (2008) "Maras y pandillas en Centroamérica", Revista URVIO No. 4, Pandillas, Quito: Flacso-Ecuador.
Guzmán, A. (1994) "Observaciones sobre violencia urbana y seguridad ciudadana". In F. Carrión (ed.) *Ciudad y violencias en América Latina*, Cali, Colombia: PGU-Alcaldía Cali, pp. 321–324.
Jordan, R. and Simioni, D. (2002) "Hacia una nueva modalidad de gestión urbana". In *Las Nuevas Funciones Urbanas: Gestión para la Ciudad Sostenible*, Santiago de Chile: Naciones Unidas, Serie Medio ambiente y desarrollo, No. 48, pp. 7–31. Av ailable: www.eclac.org/publicaciones/xml/9/10559/lcl1692e_1.pdf [accessed 15 February 2016].
Kelling, G. and Coles, C. (eds) (2001) *No más Ventanas Rotas*, Mexico: Instituto Cultural Ludwig von Mises.
Kliksberg, B. (2008) "¿Cómo enfrentar la inseguridad en América Latina?", *Revista Nueva Sociedad*, 215: 4–16, Friedrich Ebert Stiftung: Buenos Aires, Argentina.
Organización Internacional del Trabajo (OIT) (2003) *Panorama Laboral 2013*. Lima: OIT/Oficina Regional para América Latina y el Caribe.
Rodríguez, A. and Winchester, L. (2004) "Santiago de Chile: Una ciudad fragmentada". In C. De Mattos, M.E. Ducci, A. Rodríguez and G. Yáñez Warner (eds) *Santiago en la Globalización: ¿Una Nueva Ciudad?*, Santiago de Chile: Ediciones SUR-EURE Libros, pp. 115–136.
Silva, A. (2004) *Imaginarios Urbanos (metodología)*, Bogotá, Colombia: UNAL.
Sozzo, M. (ed.) (2008) *Inseguridad, Prevención y Policía*, Quito, Ecuador: Flacso-Ecuador-IMDQ.

United Nations, Department of Economic and Social Affairs, Population Division (2014). *World Urbanization Prospects: The 2014 Revision, Highlights*. Available: http://esa.un.org/unpd/wup/highlights/wup2014-highlights.pdf [accessed 15 February 2016].

Wirth, L. (ed.) (1988) *El Urbanismo Como Modo de Vida*, México: Universidad Nacional Autónoma de México (UNAM).

WHO (2002) *Informe Mundial sobre la Violencia y la Salud. Resumen*. Washington, DC: Organización Panamericana de la Salud, Oficina Regional para las Américas de la Organización Mundial de la Salud. Available: www.who.int/violence_injury_prevention/violence/world_report/es/summary_es.pdf [accessed 5 September 2016].

24

Urban upgrading to reduce violence in informal settlements

The case of violence prevention through urban upgrading (VPUU) in Monwabisi Park, Cape Town, South Africa

Mercy Brown-Luthango and Elena Reyes

Introduction

Urban violence is a serious development concern, especially in cities in the South where rapid urbanisation and high rates of poverty and inequality result in very bad living conditions for the majority of urban residents. Violence and violent crime have reached disturbingly high levels in countries like South Africa, Colombia and Brazil and in cities like Nairobi and Lagos (Kessides 2005; Mcllwaine 1999). It is argued that

> when a large population group in cities is afflicted by malnutrition, impoverishment, social exclusion and discrimination, ill health and poor conditions as well as restricted access to land and basic infrastructure, increasing levels of criminal violence, lack of safety and general fear in the use of public space are often observed.
>
> *(Bauer 2010: 4)*

High levels of violence and violent crime coincide with equally high levels of poverty and income disparities. Moser and McIlwaine (2014: 331) argue that it is a well-known fact that violence, poverty and inequality are "inextricably interconnected". In fact, the authors argue that research over the last 10 years has shown that violence cannot be completely eradicated, but might indeed be an integral component of the development process (Moser and McIlwaine 2014).

This does not mean that violence only occurs where there is poverty and inequality or that poverty and inequality would inevitably lead to violent behaviour. Violence is a complex phenomenon, and its causes and consequences are directly related to its context, the actors involved and the level at which it happens (Winton 2004). But from the urban perspective, there are some conditions highly interlinked with the occurrence of violence. The first one is accentuated inequality (Morenoff *et al.* 2001). This happens every day in cities, where informal settlements – with inappropriate basic services and scarce health, education and training or employment opportunities – are found in the proximity of other more privileged areas. Impoverishment also triggers urban violence and crime (Briceño-León 2002). With limited opportunities and concentrated poverty, the population is more vulnerable to shocks (Shaw and Carli 2011). They have to resort to insecure, poorly paid jobs that easily disappear in the event of economic decline. Then, the lack of opportunities is accentuated, and there is a perpetuated strain that the individual cannot cope with (Ügur 2014). In some instances, crime becomes the way to obtain what people cannot get through formal means, and violence is the response to the inability to manage the constant strain. Violence is becoming more frequent and lethal (Briceño-León 2002), especially with easy access to arms and illicit drugs, and when there are gangs and organised crime groups, an environment of impunity and corruption, where police presence is limited or ineffective and the existing institutions cannot be relied upon for justice (Briceño-León 2002; Moser and Winton 2002, cited in the World Bank 2003).

From the spatial perspective, the discussion has been focused on how the conditions of the built environment can, on the one hand, create opportunities for violence and crime. The World Bank, in its report *Violence in the City* (2011), mentions several examples. The lack of public spaces makes it impossible for the community to gather in a safe space. Inadequate infrastructure such as narrow alleyways or lack of street lighting creates opportunities for violent assault or other crimes, without perpetrators being seen or pursued. The absence of infrastructure such as water or sanitation increases vulnerability to crime, for instance, when people have to walk long distances to fetch water or use public toilets, especially at night. Thus, the potential of the built environment to reduce opportunities for violence has been incorporated into different theories, such as Defensible Space and Crime Prevention through Environmental Design. These approaches will be described in the next section.

Concepts behind urban upgrading for violence prevention and reduction

One of the most influential planning approaches to violence prevention and reduction is the 'Defensible Space' theory, whose main proponent was Oscar Newman, an architect and urban planner. The basic argument of Defensible Space is that the physical design and layout of urban living environments have an important bearing on safety and vulnerability to crime (Newman 1995). Through the modification of the built environment, residents are able to influence the behaviour of would-be criminals and deter criminal activity (Brunson *et al.* 2001; Reynald and Elffers 2009). The theory has three main principles. First, *territoriality*, or the physical demarcation of private, semi-private and public spaces. Second, *natural surveillance*, which refers to the capacity of physical design to provide opportunities for natural surveillance, e.g. through the orientation of buildings or residents as "policing agents" who take ownership of space and provide "eyes on the street" (Mawby 1977). And the third is *image/milieu*, or the perception of an area's uniqueness (Brunson *et al.* 2001; Mawby 1977; Reynald and Elffers 2009)

Defensible Space theory has dominated much of the discussion around the impact of design on the reduction of opportunities for crime and violence, but there has also been a fair amount of critique of Newman's theory. These relate to:

a) *Methodological issues*: some argue that Defensible Space theory is not empirically robust, that it is merely a "fashionable consensus rather than a set of empirically robust concepts that effectively prevents crime" (Shu 2000, cited in Reynald and Elffers 2009: 27). Newman has also been accused of oversimplifying the nature of crime and qualities of defensible space (Mawby 1977).

b) *Conceptual fuzziness*: it is a common critique that Newman's theory "leaves a great deal of room for various subjective interpretations of what defensible space and its various components actually are" (Reynald and Elffers 2009: 32).
c) *Contradictions*: another criticism is that the elements of design that might prevent crime could simultaneously facilitate crime (e.g. fences and high walls) and Newman failed to recognise these contradictions (Brunson *et al.* 2001; Mawby 1977; Reynald and Elffers 2009).
d) *Physical determinism*: criminologists in particular have criticised Newman's Defensible Space theory for not paying enough attention to social control as well as the impact of structural and neighbourhood factors in contributing to crime and a lack of safety. Chavis and Wandersman (1990) argue that a strong community spirit can encourage residents to work together in order to assert collective control over their public spaces and in so doing prevent crime. Without this sense of community spirit and co-operation, territoriality breaks down (Chavis and Wandersman 1990, cited in Reynald and Elffers 2009). In other words, "environmental design can translate into 'defensible space' only when the social conditions are optimal" (Reynald and Elffers 2009: 31).

Cisneros (1996), while acknowledging the potential of Defensible Space theory, holds that physical design techniques should be a component of more multi-faceted programmes that address crime. Effective solutions to crime problems depend on targeted interventions to address the structural factors that drive high rates of crime and violence (Cisneros 1996: 16). A simpler approach focusing on only one aspect of violence often has little effect on it. For instance, traditional responses to violence and crime, such as increased policing and tougher penalties, have also failed. None of these actions change the conditions in which violence and crime are produced (Acero *et al.* 2012). Therefore, there has been a shift to strategies to prevent, rather than control crime, focusing on generating protective factors that help individuals manage stress without resorting to violence (World Bank 2011) and complemented by the reduction of opportunities for violence (e.g. modifications in the built environment). This is a more comprehensive approach that includes increased participation of the community to foster development by strengthening social capital and improving security and infrastructure.

The above elements are partially incorporated into the Crime Prevention through Environmental Design (CPTED) approach. The term was coined in 1971 by the criminologist C. Ray Jeffery, who intended to establish a link between the environment and the offender (Crowe and Zahm 1994). CPTED is, however, more often used to describe strategies similar to those of the Defensible Space theory instead of Jeffery's approach. Like the Defensible Space theory, CPTED is built on the ideas and principles originally developed by Jane Jacobs (Mawby 1977), who advocated for a rethink of planning policies and more creative and innovative urban planning.

The more comprehensive approach of CPTED is included in its Second Generation that operates with three principles similar to those of Defensible Space: natural access control, natural surveillance and territorial behaviour (Crowe and Zahm 1994). But to change the focus from the built environment to community participation, three other principles are added: activity support, space management and target hardening (Cozens 2008). Activity support and space management aim to create and maintain a positive image and to generate a sense of 'ownership' of space. Target hardening aims at limiting access to a crime target, and is sometimes not considered a part of CPTED. It has resulted in the "fortress mentality", where walls and fences are built for protection and decrease the opportunity for the "eyes on the street" (Cozens 2008: 13). Also similar to the Defensible Space theory, one of the main limitations of CPTED is that it does not address the socio-economic factors that lead to crime.

The application of the principles may lead to positive or negative results, depending on the dynamics of the community (Cozens *et al.* 2005).

Another programme that has been hailed by many as a successful model for urban transformation and the improvement of urban safety is social urbanism (Brand 2010; Brand and Davila 2011; Hernandez-Garcia 2013; Scruggs 2014; Turok 2014; UN Habitat 2011). It adopts an integrated approach, incorporating interventions that cover preventive and control-oriented activities (UN Habitat 2011). It is more comprehensive than Defensible Space, for example, yet the primary focus is on physical improvements in the built environment. It was introduced in Medellín, Colombia in the early 2000s to bring about transformation in this city, notorious for high rates of crime and violence linked to the drug trade. During the 1980s and 1990s Medellín had the highest murder rate in Colombia and one of the highest in the world with 381 murders for every 100,000 of the population in 1991 (Brand 2010). Social urbanism was pioneered by Mayor Sergio Fajardo during his term in office from 2003 to 2007 and earned Medellín the accolade of "the most innovative city in the world in 2013" (Turok 2014).

The defining features of social urbanism are investment in architecturally appealing buildings, improvements in public spaces and the provision of social services and infrastructure such as schools, libraries, and recreational and cultural facilities. These are located in poorer neighbourhoods in order to create a sense of place, local identity and promote spatial equality in the city. Another important component of the programme is substantial investment in safe, high-quality public transport and the construction of new housing projects around stations. These contribute to spatial integration by connecting badly located, inaccessible areas to the rest of the city (Turok 2014). Medellín's aerial cable car, the first of its kind, lowered transport costs and improved the mobility of poor residents (Brand and Davila 2011).

The success of social urbanism is underpinned by an awareness of the root social causes of crime and violence and recognition of the need for more holistic planning to address these (UN Habitat 2011). It is argued that with the introduction of social urbanism in Medellín "an explicit confluence of safety policies began, with generation of public spaces, urban renewal, and socio-cultural programmes" (UN Habitat 2011: 13). Good governance and a focus on transparency, community participation and citizen ownership are also important elements of social urbanism. Other good governance principles highlighted by Turok (2014) as contributing to the success of this model are mature political leadership, extensive dialogue and civil society engagement, devolved responsibilities to all levels of government and institutional capacity building of various sectors of society.

The approach has, however, seen a fair number of challenges and shortcomings. Brand and Davila (2011) argue that the greatest achievement of social urbanism is its symbolic value by contributing towards social inclusion and the creation of a sense of belonging among socially, economically and spatially marginalised communities. But, concerning economic development and improving the material conditions of poor residents, the impact has been minimal (Brand 2010; Fukuyama and Kolby 2011; Scruggs 2014). This, according to Brand and Davila (2011), is mainly because the programme did not expressly set out to address structural conditions like unemployment, poverty and inequality. This again highlights the limitations of physical interventions and their long-term impact in terms of improving people's socio-economic prospects. Sustainability and institutionalisation of the model have also been a concern (UN Habitat 2011). Hernandez-Garcia (2013) points out that community participation, prominent during earlier stages, is not as strong as before. Despite these limitations, similar approaches have been developed around the world, for instance, the Violence Prevention through Urban Upgrading (VPUU) in Cape Town, South Africa. This programme and its implementation in informal settlements will be discussed in the following sections.

Violence prevention programmes and informal settlements: the VPUU in Monwabisi Park, Cape Town

The VPUU programme is being implemented in different areas of the Western Cape province in South Africa. Its methodology takes the principles of CPTED as the basis for the physical interventions, but these are accompanied by a set of projects and services provided in the communities. The aim of these interventions is expressed in the VPUU strategy:

a) *Prevention*, by creating healthy and less violent communities through interventions in the whole human life cycle.
b) *Cohesion*, generating social capital and the appropriation of the interventions by the community.
c) *Protection*, reducing opportunities for crime, with interventions in the built environment and the creation or strengthening of systems to control crime or violence.
d) *Research*, for knowledge management and sharing, to develop and replicate processes.

There are three main partners in the process: VPUU as intermediary between the local government, the community and the donors; the local government as facilitator of processes and provider of infrastructure and other main services; and the community, who is expected to actively participate in the formulation and implementation of the projects. Other local organisations become partners in specific interventions, being mainly responsible for the provision of services. The implementation process is designed to foster the participation and appropriation of the projects by the community to ensure the sustainability of the interventions (Cooke *et al.* 2014). Community participation is one of the core principles of the VPUU methodology with the aim of fostering a "people-centred development approach" where communities are equal partners and are actively involved in the decision-making at each level and stage of the upgrading process and eventually take over and manage the infrastructure and other projects delivered (Cooke *et al.* 2014: 146). In reality, VPUU and the other partners still manage the different projects and a strategy for their sustainability is yet to be established.

The programme started in Monwabisi Park, Cape Town in 2009. This has long been considered as one of the areas with more concentrated crime (Poswa and Levy 2006) and higher levels of poverty and inequality in the city. Monwabisi Park, or Endlovini, is located on the border of Khayelitsha and consists of more than 6,000 dwellings and 17,800 inhabitants (Frith 2012; VPUU 2013). It was founded in 1996 when Khayelitsha became the point of arrival of the Black African population migrating from the Eastern Cape (Bliss and Zagst 2011). Eighty-four per cent of the population in Monwabisi Park lived below the poverty line (set at ZAR3,200 – approximately USD220 – by the City of Cape Town) in 2011 (Frith 2012).

The baseline survey from VPUU (SUN Development 2009) showed that in 2009 more than half of the population felt that crime and violence were intolerable. People felt especially vulnerable at night (even in their homes) and on their way to transport; half of them had actually suffered from robbery and break-ins. Akin to the figures presented by Lemanski (2004), rape, murder, stabbing and house breaking were considered to take place during the evening and night; but other types of crime were likely to happen any time of the day.

The physical conditions of the settlement make the population more vulnerable to crime. Dwellings are easily broken into because they are made of corrugated iron. They are located very close to one another; thus the only access points are narrow pathways and a few streets. This makes surveillance by neighbours very difficult and access for the police almost impossible. The only service that all houses have is electricity, which was provided in 2014 and has to some extent contributed to an increased perception of safety. But, according to VPUU (SUN Development 2009), people had to walk 5–10 minutes to reach shared taps (one per every 42 dwellings) and shared toilets (one per every 23 dwellings). Around 60 per cent of interviewees mentioned that toilets are frequently broken or locked by a neighbour, so people use the bushes instead. These are on the boundary of the settlement and a National Park, one of the areas in which the population feels more vulnerable since it has poor surveillance and was used in the past to commit crimes, including murder.

One of the first interventions from VPUU was the creation of a local committee, the Safe Node Area Committee (SNAC), which acts as the main link between the organisation and the community. Members of the SNAC are elected from existing structures in the community like civil society organisations, ward councillors and local civic groups. There is a plan for the interventions in the area called the Community Action Plan (CAP), which was approved by this committee and signed by the city mayor in 2010. The CAP is derived from a baseline survey conducted in the community, and community profile workshops in which members of the SNAC, local government and VPUU participate. This culminates in the CAP workshop where results of the baseline survey and the outcomes of the community profile workshops are compared for the prioritisation of the different project activities. By 2014 VPUU and its partners were working on 26 of the 37 projects included in the CAP (see Table 24.1).

The CAP, SNAC and a full-time facilitator based in the community are some of the main structures for fostering participation and are meant to ensure consistent and effective communication between the project team and the community. Communication tools include reference group meetings between the members of the SNAC, the VPUU project team and local government officials as well as diamond model meetings to ensure alignment between VPUU's work and local government strategies and priorities as well as that of the community. The diamond model refers to the two points where at the top level the VPUU priorities and work streams are aligned with local, provincial and national strategies and priorities and at a local level with community priorities as contained in the CAP (Cooke *et al.* 2014; 160). The SNAC team meets with the rest of the community on a quarterly basis in order to provide feedback on progress as well as challenges with implementation of the CAP activities (Cooke *et al.* 2014).

As shown in Table 24.1, there is a wide range in the projects planned for Monwabisi Park. In terms of urban upgrading, the focus of VPUU is mainly on the provision of community infrastructure in the form of pedestrian walkways, open spaces and multifunctional buildings. In the case of informal settlements, the methodology also aims at the incremental improvement of the living environment, the provision of tenure, essential services (toilets, water taps, storm water infrastructure and access tracks) and reconfiguration plans; but the main interventions so far have been the Emthonjeni. The word means 'a place by the water/river/well' and VPUU uses it to refer to small public spaces built around the shared water taps. These places are meant to be safe meeting points for the community as well as spaces for the delivery of some services (e.g. early childhood development (ECD) sessions with children). There is also a community facility, the 'Bulelani Precinct', which includes a couple of small fields. Figure 24.1 shows the location of these interventions in Monwabisi Park. Aside from electricity, bigger interventions such as the provision of water and sanitation per household are still pending. These are interventions which the City of Cape Town is responsible for, and it is not clear when they will be provided.

The authors conducted research in Monwabisi Park in 2015 as part of a project looking at the impact of informal settlement upgrading on the nature and extent of violence. In Monwabisi Park, a number of interviews and focus group discussions were held with residents, members of

Table 24.1 Interventions in Monwabisi Park*

Institutional crime prevention		*Social crime prevention*		*Situational crime prevention*	*Knowledge management*
Institutional	Economic	Cultural/social	Safety and security	Infrastructure	Knowledge management
Leadership training	Business support	SDF (youth, women, sports development)	Neighbourhood watch	Community facilities (container facility, Kick-About, Emthonjeni)	Enumeration
Community facilitator's office in MP	Diversification (LED) strategy	ECD activities	Local safety station	Operation and maintenance of existing taps and toilets	GIS system for taps and toilets
Spatial reconfiguration plan		Gender-based violence prevention		Crossing over Mew Way	Household surveys
Tenure certificates		Victim/legal support		Public facilities on Luleka Land	Monument photos
Equitable access to community facilities and schools		Youth development ('24' project, after-school support, Chrysalis Graduates)		Bulk infrastructure (electrification, storm water management, disaster risk management, road system, taps and toilets)	Community newsletters
		Promote healthy lifestyles		Youth/neighbourhood centre	Quality of life indicators
		Poverty and food security			CAP review

Source: adapted by authors from AHT Group AG (2014), Cooke *et al.* (2014) and VPUU (2013).

*Items in grey font were projects not yet implemented by December of 2014.

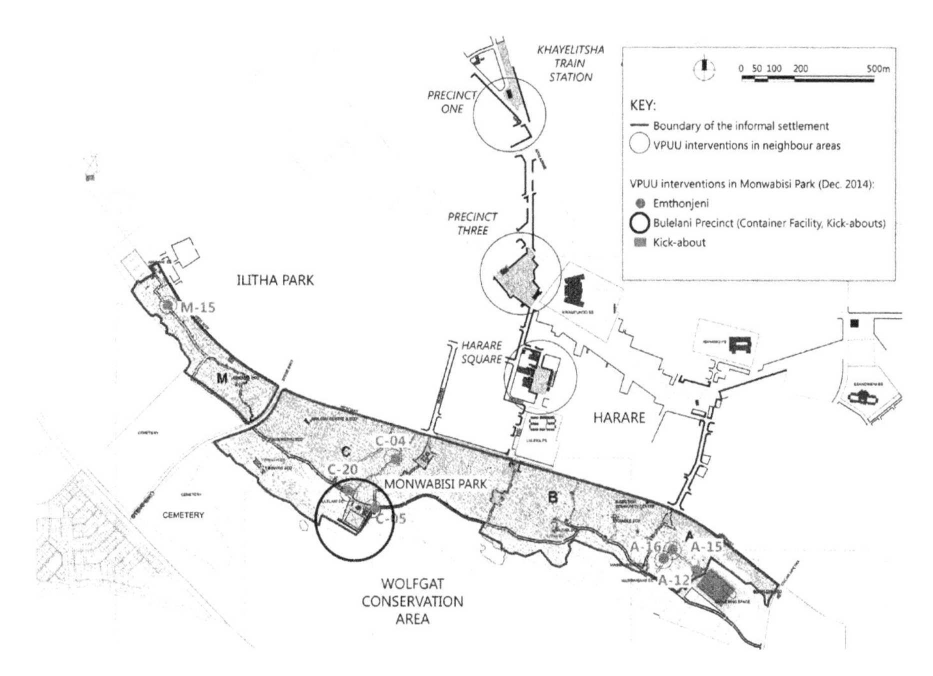

Figure 24.1 The location of the interventions in Monwabisi Park

Source: VPUU (2013), modified by author.

the VPUU team, partner organisations, neighbourhood watch volunteers, participants of youth development programmes and members of the SNAC. A crime-mapping workshop was conducted in which hotspots for crime and violence, before and after upgrading, were pointed out by members of the community on satellite images of the settlement.

The majority of respondents perceived an improvement in the settlement after upgrading, but expressed concern over outstanding issues such as toilets, streets, houses and the unemployment situation. Thirty-five out of the 60 residents who were interviewed as part of a research project from the African Centre for Cities in 2014 (Reyes 2014) felt that there was an improvement in safety since the VPUU interventions, and those participating in specific projects also considered those projects in particular to have had a positive impact. The main intervention impacting on safety is electricity, but other projects such as parks and playgrounds, ECD and sport activities are also perceived to help keep young people and children safer. With the exception of the Neighbourhood Watch, many of the projects seem to be contributing to the generation of skills or conditions to better cope with the existing violence and crime even though many of them have not yet reached the majority of the community. The Tenure Certificates (administrative recognition from the city government) were also regarded as projects that helped to improve the conditions in the settlement. On the other hand, the neighbourhood watch was an example of a project where the expectations of the community didn't match the project's objectives. This led to many misunderstandings between the organisation and the participants and a limited implementation of the watch.

However, the conditions that contribute to crime and violence in Monwabisi Park are perceived as similar before and after the interventions. Robbery and house breaking were and are the main concerns, and the conditions that make people feel vulnerable have not changed. Robbery still occurs in empty fields, on paths on the way to work, or by the bushes.

This also relates to one of the main challenges of the programme: a divergence between the priorities of the community and the objectives and implementation process. For example, streets and toilets are included in the action plan, but may only be implemented in the long term. Employment and housing were also pointed out as priorities for the community, but are not part of the scope of the programme. This in some instances has affected community participation, creating despondency and the feeling of failed delivery.

Part of the challenge lies in the fact that community participation happens mainly through the members of the SNAC, who may or may not have the ear of the majority of the community. This means that the most marginalised members of the community, those who do not participate in the organised structures that the SNAC members are drawn from, might be excluded. This presents several potential issues. First, these members of the community might not have such intimate knowledge of the whole project planning process and the way in which certain project goals and priorities are arrived at. Second, their needs and aspirations may not be considered in the CAP prioritisation process; and third, they might not be able to access some of the services, because the information does not reach them. Another challenge is that in a context of such extreme and concentrated levels of deprivation it is often difficult to reconcile individual needs with broader community interests. It should also be noted that the actual impact of the interventions may only be seen in the long term. And, similar to other approaches, such as social urbanism, the interventions in Monwabisi Park do not address structural problems that contribute to violence and crime.

Discussion

The impact of urban upgrading interventions on violence prevention is hard to prove as there are many factors that contribute to violence or crime. Urban design and planning, and the resulting interventions in the built environment, have a role to play in reducing the opportunity for crime and violence. Frameworks and approaches such as CPTED, Defensible Space and Social Urbanism have made substantial contributions in this regard. However, they have also received a fair amount of critique since they may have negative unintended consequences such as the displacement of violence, enhancing the "fortress mentality" and generating "offensible space". More importantly, they may not be sufficient in contexts where social capital and economic resources are in decline (Cozens *et al.* 2005).

Community participation in upgrading programmes is important, but often limited in reality. Given the size of the settlement, it is especially difficult to balance pragmatism and expediency with an approach which ensures that all voices are heard and represented in the different processes. Community involvement is also hard to sustain particularly when there is a delay in the implementation of certain infrastructure and services. This creates tension between the community and the intermediary, VPUU in this instance, and can result in despondency and a decline in community involvement in the project.

There is not much available literature on Defensible Space theory and its application in informal areas. Since this chapter is primarily concerned with the application of informal settlement upgrading as a violence prevention tool, one has to think about Defensible Space theory in relation to informal settlements. The first issue of concern is that of 'ownership', even symbolic, and residential appropriation of space and how to accomplish this in a context of tenure insecurity or temporality, and the absence of a feeling of belonging, which is very characteristic of informal areas. In Colombia, the real value of the social urbanism programme was in creating a sense of belonging, inclusion and connection to the city, facilitated by substantial investment in transport infrastructure. In Monwabisi Park, the issuing of tenure certificates may lead to an increased perception of safety among residents. This highlights the importance of fostering inclusion and belonging in the city through interventions in the built environment as part of a strategy to reduce violence and crime. Moreover, interventions such as the tenure certificates, which aside from strengthening the sense of ownership of space, grant administrative recognition to the inhabitants, who without the fear of eviction may start to invest in improving their built environment.

Another issue, specifically in relation to Defensive Space theory, is how highly insecure/violent spaces impact on residential appropriation of neighbourhoods. This begs the question: is a safe environment a pre-condition for residential appropriation or does it work the other way around? Newman's Defensible Space theory does not adequately address this. The literature on Defensible Space theory is also silent on what can be referred to as the dark side of territoriality, which is often expressed in mob justice. In a context where residents feel neglected or ignored by police and where the police face real constraints because of an inability to access informal settlements due to high densities and a lack of clearly demarcated roads and pathways, residents often resort to extreme measures in order to feel safe.

A further important consideration is the long-term sustainability of physical interventions in the built environment. In informal settlements, the provision of public spaces and other interventions may create a positive impact on the perception of safety in the short term. A case in point is the provision of electricity in Monwabisi Park. However, violence is multi-faceted in its causes and manifestations. It therefore requires holistic and integrated responses and multi-scale interventions from the individual to the family and community, not neglecting the broader societal level. The VPUU programme, in contrast to some of the other programmes and frameworks discussed in this chapter, does adopt a more holistic approach to violence prevention and includes social crime prevention and institutional crime prevention measures in addition to physical interventions. However, as can be seen from the Monwabisi Park case, in a context of abject poverty and high levels of deprivation where residents face a daily struggle to have their basic needs met, there might be impatience on the side of some members of the community as the longer term benefits of social crime prevention strategies, ECD programmes, for example, might not be immediately apparent. In the long term, in order to have a real impact on high levels of violence and crime and interrupt the intergenerational cycle of violence, overarching structural conditions such as poor education and skills development, unemployment, poverty, inequality and inappropriate housing, need to be addressed.

Conclusion

There is no doubt that urban planning and the design of the built environment play a role and can make a significant contribution to crime and violence prevention. However, what is needed are urban planners and designers who have a deep understanding of the social dimensions and consequences of urban infrastructure. Urban planning for violence prevention should be considered only as a tool to complement broader, more comprehensive programmes that address structural causes of violence. The Monwabisi Park case discussed in the chapter points to the limitations of approaches that rely heavily on physical interventions without due cognisance of and attention to the socio-economic factors that contribute to crime and violence. Informal areas especially might present particular challenges, which some of the models and theories discussed in this chapter do not respond to. Furthermore, it points to the need of a different set of planning skills, in which the work of the urban planner should be combined with the work of professionals from other disciplines.

The case also points to the importance of taking place and context into account when designing interventions to deal with crime and violence in an area rather than applying generalised models in an uncritical manner. For instance, community dynamics should always be taken into account in order to prevent the so-called 'offensible spaces' or, as it may happen in participatory processes, to prevent excluding the most vulnerable members of society. The Monwabisi Park case also highlights some of the challenges of community participation. Whereas there is no argument about the importance of community involvement and ownership of any development intervention, in reality it is often very difficult to design and implement a process where every resident feels that their voice is heard and that their needs and aspirations are adequately represented. This is especially true given the size of the area and the divergent needs of the population.

References

Acero, H., Miraglia, P. and Pazinato, E. (2012) *Estrategias locales de seguridad ciudadana*, URBAL, Recife PE: Provisual.

AHT Group (2014) *VPUU Phase 3, Fifth Semi-Annual Progress Report, July–December 2013*, Cape Town: VPUU.

Bauer, B. (2010) "Violence prevention through urban upgrading: Experience from financial cooperation", German Federal Ministry for Economic Cooperation and Development (BMZ), Frankfurt: Government Printer.

Bliss, F. and Zagst, L. (2011) "Investing in people: Violence prevention and empowerment through urban upgrading in the township of Khayelitsha, South Africa". In POVNET Task Team on Empowerment (eds) *Storyline South Africa 2011*, Berlin: KfW/BMZ.

Brand, P. (2010) "Governing inequality in the South through the Barcelona Model: Social urbanism in Medellín, Colombia", *Interrogating Urban Crisis: Governance, Contestation, Critique*, De Montfort University, 9–11 September. Available: www.dmu.ac.uk/documents/business-and-law-documents/research/lgru/peterbrand.pdf [accessed 1 October 2015].

Brand, P. and Davila, J.D. (2011) "Mobility innovation at the urban margins", *City: Analysis of Urban Trends, Culture, Theory, Policy, Action* 15(6): 647–661.

Briceño-León, R. (2002) "La nueva violencia urbana de América Latina", *Sociologías* (8): 34–51.

Brunson, L., Kuo, F.E. and Sullivan, W.C. (2001) "Resident appropriation of defensible space in public housing: Implications for safety and community", *Environment and Behaviour* 33: 626–652.

Chavis, D.M. and Wandersman, A. (1990) "Sense of community in the urban environment: A catalyst for participation and community development", *American Journal of Community Psychology* 18: 55–81.

Cisneros, H.G. (1996) "Defensible space: Deterring crime and building community", *Cityscape* 15–33.

Cooke, J., Giles, C., Krause, M., Lange, U., Shay, D., Smith, E. and Taani, I. (2014) *Violence Prevention Through Urban Upgrading: A Manual for Safety as a Public Good*, Cape Town: VPUU.

Cozens, P. (2008) "Crime prevention through environmental design". In R. Wortley and L. Mazerolle (eds) *Environmental Criminology and Crime Analysis*, Devon: Willan, pp. 153–177

Cozens, P.M., Saville, G. and Hillier, G. (2005) "Crime prevention through environmental design (CPTED): A review and modern bibliography", *Journal of Property Management* 23(5): 328–356.

Crowe, T. and Zahm, D. (1994) "Crime prevention through environmental design", *NAHB Land Development Magazine*, Fall issue: 22–27.

Frith, A. (2012) Census 2011. Available http://census2011.adrianfrith.com/ [accessed November 7, 2014].

Fukuyama, F. and Kolby, S., (2011) "Half a miracle", *Foreign Policy*. Available: http://foreignpolicy.com/2011/04/25/half-a-miracle/ [accessed 7 September 2015].

Hernandez-Garcia, J. (2013) "Slum tourism, city branding and social urbanism: The case of Medellín, Colombia", *Journal of Place Management and Development* 6(1): 43–51.

Kessides, C. (2005) "The urban transition in sub-Saharan Africa: Implications for economic growth and poverty reduction, Africa region", Working Paper Series No. 97, Urban Development Unit, Transport and Urban Development Department, The World Bank. Available: www.worldbank.org/afr/wps/wp97.pdf [accessed 1 October 2015].

Lemanski, C. (2004) "A new apartheid? The spatial implications of fear of crime in Cape Town, South Africa", *Environment and Urbanization* 16: 101–112.

Mcllwaine, C. (1999) "Geography and development: Violence and crime as development issues", *Progress in Human Geography* 23: 453–463.

Mawby, R.I. (1977) "Defensible space: A theoretical and empirical appraisal", *Urban Studies* 14: 169–179.

Morenoff, J.D., Sampson, R.J. and Raudenbush, S.W. (2001) "Neighbourhood inequality, collective efficacy, and the spatial dynamics of urban violence", Research Report No. 00-451, Population Studies Centre, Institute for Social Research, Ann Arbor, MI: University of Michigan. Available: www.psc.isr.umich.edu/pubs/pdf/rr00-451.pdf [accessed 1 October 2015].
Moser, C. and Mcilwaine, C. (2014) "Editorial: New frontiers in twenty-first-century urban conflict and violence", *Environment and Urbanisation* 26(2): 331–344.
Moser, C. and Winton, A. (2002) "Violence in the Central American region: Towards an integrated framework for violence reduction", Working Paper 171, Overseas Development Institute, London: ODI. Available: www.odi.org/sites/odi.org.uk/files/odi-assets/publications-opinion-files/1826.pdf [accessed 2 February 2016].
Newman, O. (1995) "Defensible space: A new physical planning tool for urban revitalization", *Journal of the American Planning Association* 61(2): 149–155.
Poswa, N. and Levy, R. (2006) "Migration study in Monwabisi Park (Endlovini), Khayelitsha", Strategic Development Information and GIS Department, City of Cape Town, Cape Town. Available: www.capetown.gov.za/en/stats/CityReports/Documents/Other%20City%20Reports/Migration_Study_in_Monwabisi_Park_712200614556_359.pdf [accessed 1 October 2015].
Reyes, E. (2014) "Mainstreaming urban safety and inclusion in South Africa: The case of VPUU in Monwabisi Park, Cape Town", Cape Town: African Centre for Cities.
Reynald, D.M. and Elffers, H. (2009) "The future of Newman's defensible space theory – Linking defensible space and the routine activities of place", *European Journal of Criminology* 6(1): 26–46.
Scruggs, G. (2014) "Latin America's new superstar: How gritty, crime-ridden Medellín became a model for twenty-first-century urbanism". Available: https://nextcity.org/features/view/medellins-eternal-spring-social-urbanism-transforms-latin-america [accessed 1 October 2015].
Shaw, M. and Carli, V. (eds) (2011) "Practical approaches to urban crime prevention", Proceedings of the Workshop held at the 12th UN Congress on Crime Prevention and Criminal Justice, Salvador, Brazil, 12–19 April, 2010. Available: www.unodc.org/pdf/criminal_justice/Practical_Approaches_to_Urban_Crime_Prevention.pdf [accessed 2 February 2016].
Shu, S. (2000). Housing layout and crime vulnerability. *Urban Design International 5*, 177–188.
SUN Development (2009) "Monwabisi Park in-situ upgrade: Baseline survey", Cape Town: SUN Development.
VPUU (2013) "Enumeration survey Monwabisi Park", internal database (unpublished), Cape Town: VPUU.
The World Bank (2003) "A resource guide for municipalities: Community-based crime and prevention in urban Latin America", The World Bank. Available: www-wds.worldbank.org/external/default/WDSContentServer/WDSP/IB/2005/04/21/000012009_20050421125813/Rendered/PDF/320640ENGLISH01eGuide1urbanviolence.pdf [accessed 2 February 2016].
The World Bank (2011) "Violence in the city: Understanding and supporting community responses to urban violence", Washington DC: Social Development Department, The World Bank. Available: www-wds.worldbank.org/external/default/WDSContentServer/WDSP/IB/2011/08/18/000356161_20110818035635/Rendered/PDF/638880WP0Viole00BOX361532B00public0.pdf [accessed 1 October 2015].
Turok, I. (2014) "Medellín's 'social urbanism' a model for city transformation", *Mail and Guardian*. Available: http://mg.co.za/article/2014-05-15-citys-social-urbanism-offers-a-model [accessed 2 February 2016].
Ügur, L. (2014) "Beyond the pilot project, towards broad-based integrated violence prevention in South Africa", PhD thesis, Technische Universität Darmstadt. Available: http://tuprints.ulb.tu-darmstadt.de/4128/ [accessed 1 October 2015].
UN Habitat (2011) "Building urban safety through slum upgrading", UN Habitat, Nairobi. Available: http://mirror.unhabitat.org/pmss/listItemDetails.aspx?publicationID=3222&AspxAutoDetectCookieSupport=1 [accessed 2 February 2016].
Winton, A. (2004) "Urban violence: A guide to the literature", *Environment and Urbanization* 16: 164–184.

25

Starting from here

Challenges in planning for better health care in Tanzania

Maureen Mackintosh and Paula Tibandebage

Introduction: the health sector as a social and economic institution

Health sectors are social and economic institutions. They reflect the broader political economy of their country, and contain within them the economic characteristics, power relations and hierarchies of the broader society – while also, strikingly, providing a location for redistributing resources, responding to need and contesting disadvantage and exclusion (Freedman 2005; Mackintosh 2001).

In Tanzanian health care, the public health sector reflects this low-income economy's acute fiscal pressure, and also society's gendered social hierarchies of employment, training, income and status (Mackintosh and Tibandebage 2006). The private providers are increasingly bifurcated into hospitals and clinics relying on the (limited) insurance market, and lower level facilities operating as small businesses, on a cash-in-hand basis that reflects the business structure characteristics of much of the wider economy (Kida 2009; Mackintosh and Tibandebage 2007). The faith-based health providers reflect the diversity of faiths in Tanzania, while their financial fragility reflects widespread poverty. The role of donors is very large in health, reflecting the broader history of donor conditionality in Tanzania that stretches back to the 1980s.

The Tanzanian government and donors fund salaries and inputs for a public health sector that was initially universalist and inclusive in intention, well reflected in its efforts to develop the health infrastructure in rural areas where the majority of the population live. By 1992 about 72 per cent of the population lived within 5 kilometres of a health facility. Indeed following the 1967 Arusha Declaration, Tanzania chose a socialist path under which its health policy aimed at providing free health care to the entire population. The result was a pyramidal health services infrastructure that continues today: 86 per cent of all facilities in 2014 were dispensaries, and currently the Primary Health Services Development Programme aims to have a dispensary in every village, and a health centre in every ward (URT 2007).

However, the universalistic intentions have been compromised by broader changes in Tanzania's political economy, starting with liberalisation policies under the Structural Adjustment Programme (SAP). Health sector reforms under liberalisation included introduction of user fees in public health facilities and re-introduction of for-profit private practice in the early 1990s. Exemption mechanisms for those unable to pay remain ineffective, and the majority of the population have no health insurance. The universalistic intentions also continue to be compromised by inadequate funding of health care that disproportionately affects the poor. As a result, most attempts to access care now require payment (Mackintosh *et al.* 2013; Tibandebage and Mackintosh 2005; Tibandebage *et al.* 2013). Furthermore, most material inputs to health care, medicines and other essential supplies, are externally produced and imported, reflecting the country's low level of industrialisation, liberalised import regime and import-dependence (Tibandebage *et al.* 2016b).

For Tanzanian health planners aiming to move towards universal health coverage, this context is highly challenging. Those planning health care in Tanzania include both national and local government authorities, since primary health care delivery is decentralised to Local Government Authorities (LGAs) in line with the broader Decentralisation by Devolution (D by D) approach in public sector management (another reflection in the health sector of broader policy shifts). Planning, budgeting, implementation and monitoring are decentralised to the District Medical Officer's office and Council Health Management Teams, with institutional and administrative bodies working at lower levels. This chapter addresses these different levels, in examining some of the challenges. It aims to contribute to building understandings of the scope for developmental planning for better health that is "context- and place-specific in strategy" while focused on practical problem-solving (Srinivas 2016: 190).

The challenge of institutional redesign

The current levels of exclusion and poor quality of care in Tanzanian health care are very serious, and the challenge involved in moving towards universal health coverage is enormous. We argue in this chapter that planning for improvement is a process of institutional (re)design, by which we mean the conscious encouragement of institutions that display 'goodness of fit' with "some larger objectives than those narrow ones embodied in the internal goals of the institution and its immediate environment" (Goodin 1996: 39). Here, the larger objective is universal health coverage, countering social and economic disadvantage. Institutions evolve from the here-and-now, so design must make sense 'from within' and be pursued in good part by insiders (Mackintosh and Tibandebage 2002).

Better institutional design in Tanzanian health care requires planners to tackle two closely interlocking problems in political economy: the perverse incentives inherent in the current market structures, and the lack of political and economic priority assigned to the health needs of the low-income and disadvantaged population. This in turn, we have argued at more length elsewhere (Mackintosh and Tibandebage 2002), requires a rethink of the concepts of planning and regulation. In its most general terms, the health crisis requires explicit moves towards a new social settlement in health care between the better off and the acutely disadvantaged (Mackintosh and Tibandebage 2004; Mackintosh 2001), a settlement that embeds increasingly redistributive health care institutions within a framework of improving health care access for all, including the Tanzanian middle classes who are also struggling to fund and find reliable health care. This challenge in turn requires the rethinking of the public/private interface in Tanzanian health care (Tibandebage *et al.* 2013).

Conceptually, institutional change in health sectors is generally path-dependent, evolving from current organisational structures and culture (the 'here and now'). Planners may divert that evolution into different paths, and occasionally engineer real breaks in trends, but they cannot start from scratch. In investigating scope for planned improvement, we employ concepts familiar to institutional theorists: social norms, incentives (market and non-market), and institutional feedback loops. Norms can be understood as patterns of behaviour that are widespread, are generally tolerated or accepted as proper, are reinforced by responses of others and are quite hard for individuals to resist even if they run against what is felt to be right (Tibandebage and Mackintosh 2002). They "form a web of beliefs and practices whose different strands mutually reinforce each other" (Sen *et al.* 2002), forming feedback loops that may be reinforced by material incentives and financial constraint.

To make these arguments we employ two contrasting lenses to illuminate the institutional characteristics of the current health sector, and the redesign challenge. The next section uses a gender lens to identify some perverse consequences of embedded gender inequalities for maternal care. The following section uses a market lens, exploring perverse incentives embedded in the Tanzanian health care market structure and incentives that are currently undermining health care quality and access. The final section argues for the need for, and possibility of, a more imaginative and flexible regulatory structure, with more emphasis on building compatible incentives and collaborative behaviour, to improve health care in the medium term.

Maternal mortality: a gender lens on health sector inequity

Tanzania continues to suffer a huge burden of mortality and morbidity related to childbearing. World Health Organization data (WHO *et al.* 2015) estimate the Tanzanian maternal mortality ratio (MMR), the number of maternal deaths per 100,000 live births, at 398 in 2015, with a wide uncertainty interval, and a woman's lifetime chance of dying of maternity-related causes at 1 in 36. (Compare the UK, not the best-performing high-income country, with an MMR of 9 and a lifetime risk of 1 in 5,800.) In Tanzania, without good civil registration of deaths and where around half of women give birth at home (Table 25.1), these estimates are highly uncertain (Tibandebage and Mackintosh 2010). What is not in doubt is that these data represent a crisis requiring urgent attention. The causes provide a "gender lens" (Sen *et al.* 2002) that illuminates the discriminatory gendered structure of the health sector, identifying four interlocking institutional patterns of gendered discrimination that underlie the crisis.

Health care in Tanzania, as in many countries, reflects the huge, gendered social inequalities of the wider society, including gendered hierarchies among staff, and extremely unequal quality and access between social classes (CSDH 2008). Inequalities in access to maternal care during delivery by area of residence (rural vs. urban), by education and by wealth status, clearly suggest that the MMR will be higher among poor, less-educated rural women (Table 25.1). Better-off women and women with more education are also more likely to be assisted by qualified staff: 90 per cent of women in the highest quintile had skilled assistance at birth, as compared to 33 per cent in the lowest quintile (NBS 2011: 137).

As Table 25.1 shows, women at all income levels rely mainly on public health facilities for skilled care at birth. Obstetric complications including haemorrhage and sepsis are leading causes of maternal death, so access to emergency obstetric treatment is essential. Yet, most women rely on health centres and dispensaries, which often lack basic supplies, medicines and equipment, and have severe shortages of staff with the skills to handle complications during pregnancy and delivery. For example, in 2012–2013, out of 25 health centres and dispensaries surveyed, 38 per cent did not have oxytocin injectable for treating post-partum bleeding. Even hospital delivery does not assure safety. Essential supplies are often lacking there too, and skilled staff are frequently under great stress. As one hospital director said in 2011: "Women get there [to hospital], *then* they die" (Tibandebage *et al.* 2016a).

Four interlocking structural patterns of disadvantage have sustained a cycle of gendered discrimination underlying the mortality crisis. First, the government hospitals are too few, too

Table 25.1 Tanzania: per cent distribution of live births in the five years preceding 2010 according to background characteristics (most recent birth only)

Health facility				
Background characteristic	*Public*	*Voluntary/ religious*	*Private*	*Home*
Residence				
Urban	67.6	10.2	4.6	16.8
Rural	34.2	6.8	0.9	56.1
Mother's education				
No education	27.7	5.0	1.1	64.0
Primary incomplete	35.3	6.3	1.1	56.0
Primary complete	46.0	8.4	1.6	42.2
Secondary+	66.6	12.7	5.3	15.3
Wealth quintile				
Lowest	28.3	4.5	0.3	65.5
Second	28.6	6.2	1.3	61.7
Middle	36.8	8.3	0.7	51.6
Fourth	50.8	10.4	1.3	36.4
Highest	74.6	9.0	6.0	9.5
Total	41.0	7.5	1.6	48.1

Source: adapted from National Bureau of Statistics (2011: 134–135).

far from most people's homes, while the health centres still generally lack the ability to deal with obstetric emergencies. The inherited health system structure at Independence in 1961 was focused on a few large urban hospitals rather than a dispersed 'cottage hospital' network of the kind that has helped to address maternal mortality in Sri Lanka. Though efforts are now being made to build up health centres' capabilities, the backlog is huge, and has long been reinforced by donors' insistent focus on primary care. The outcome is gender-discriminatory: to reduce maternal deaths requires skilled obstetric intervention, while emergency hospital care is less crucial for other aspects of adult health care, so a bias against the needs of women of child-bearing age is built into the public health sector structure.

Second, improvements in access to supplies for maternal care have been slow in coming, in contrast to donors' major efforts to improve access to treatment for HIV/AIDS, malaria and tuberculosis. Maternity – however lethal – is not a disease, and its priority has not matched the three major diseases within donor funding. Access to medicines in Tanzania is now mainly funded by donors or by out-of-pocket payment (Mackintosh *et al.* 2016a). Government and donors have failed to prioritise supplies for maternal care, both antenatal care and care at birth, including test kits, antibiotics and emergency care supplies.

Nor, third, can pregnant women afford to fill the gap. Maternal care is officially provided free of charge in public facilities in Tanzania. However, the level of commercialisation is quite substantial, especially in urban areas. Of 240 women interviewed in a 2011 study, just 7 per cent had paid nothing for maternal care, and the percentage was lower in urban areas. Table 25.2 shows the mean payments at most recent births recorded in two urban districts (away from the commercial hubs of Dar es Salaam and Arusha). The payments include informal charges and also the money women or their families spent buying supplies in private shops to bring to the facilities. These charges can have a disproportionately severe impoverishing effect on women of childbearing age, who may need to borrow from others

Table 25.2 Mean payments reported by women for most recent birth, by type of facility, two urban districts 2011 (US dollars)

Payment for:	*Urban 1*			*Urban 2*	
	FBO hospital	*Public hospital*	*Public health centres*	*Public hospital*	*Public health centres*
Supplies	1.9	1.5	1.8	6.5	3.8
All non-transport	19.1	2.6	3.0	7.0	4.7
Transport	8.1	4.6	0.8	5.1	7.0
Total payments	27.1	7.2	3.8	12.0	11.7

Source: Mackintosh *et al.* (2013); TShs converted to USD at exchange rate mid-December 2011.

in the wider family, draw on tiny savings or sell assets such as farm animals to pay maternity charges. For a low-income household, finding cash for charges may be impossible and can cause conflict.

Finally, public hospitals display a sharp gender hierarchy of working conditions and status, with most nurse-midwives being female, and many doctors male. Public hospital delivery wards are frequently understaffed and over-crowded. Midwives feel overwhelmed, and often disempowered (Tibandebage *et al.* 2016a), and many complain of lack of support from doctors and managers. Working conditions are poor, and incentives frequently lacking. The outcome can be deteriorating relations between the midwives and the women they serve: "Midwives are not respected. This is because this job is a very hard job. The community has a wrong perception of us, they say we have hard hearts and bad language towards the women we are attending" (Nurse-midwife, public hospital, 2011). Maternal care centrally involves women caring for women, and the gendered power relations of the wider society undermine the quality of those relationships.

Charging for care, paying for imports: health care commercialisation, polarisation and externalisation

A second lens on the institutions of the health sector is provided by close attention to its market structure and the patterns of competition and embedded incentives. Charging for care – through formal charges for consultation, tests and medicines, informal changes, and sending patients to buy medicines in shops – is very widespread in the public sector, and nearly universal in the faith-based and private sectors of Tanzanian health care. Exceptions include some of the 'vertical programmes' that support and supply free-of-charge HIV/AIDS and TB treatment. Charging is almost entirely on-the-spot, out-of-pocket payment, and that charging system influences the behaviour of health care facilities in all three sectors.

Public facilities that can retain their formal fees for facility use – this includes most hospitals and some lower level facilities – rely on that funding for filling gaps in medicines and supplies. The government pays salaries and provides public health facilities with a budget for medicines and supplies at the public wholesaler, but funds for supplies are insufficient and often late. The cash from charges forms an essential resource to help to fill the gaps. Dispensaries and health centres often have to deposit the income from fees and charges with the local authority, leaving them with no cash-in-hand whatsoever; the result can be facilities without soap or disinfectant, while patients are sent to buy not only medicines but also supplies such as gloves and even syringes. This was a frequent comment in a 2012/13 study: "Yes, we have a problem of shortage of supplies because of shortages of financial resources. What we collect [in fees and charges] is not enough [to fill gaps] so we have constant supply shortages" (Hospital pharmacist, public hospital).

There are few studies of the business behaviour of the faith-based and private facilities. Our research in the late 1990s included investigation of the finances of a sample of health facilities, public, private and faith-based (Mackintosh and Tibandebage 2007; Tibandebage and Mackintosh 2002). The finances of almost all the faith-based and private business were fragile, as might be expected in this very low-income market for health services, and all the facilities relied on margins from medicines sales for income to pay staff. The health care and medicines markets are highly price-competitive (Mackintosh and Mujinja 2010), and the private dispensaries and health centres in particular closely watched and responded to competitors' pricing (Mackintosh and Tibandebage 2007). Businesses were mainly small scale, often reliant on family and informally hired staff, and poorly regulated: it is hard to regulate a market of this type with rapid turnover of businesses and very limited regulatory resources. Faith-based facilities struggled unless subsidised by parent religious or NGO bodies, and varied widely in charges and quality (Tibandebage and Mackintosh 2002).

The implications of these market incentives for quality of health services in the non-governmental sectors were severe. Private businesses attempting to staff health centres or dispensaries with qualified people were rapidly undercut by competitors using less skilled, informal and family employees (Mackintosh and Tibandebage 2007). The reliance on medicine sales in conditions of generalised poverty generated a culture of selling part-doses of medicines, with potentially severe effects in terms of outcomes and resistance (Mackintosh and Mujinja 2010; Mujinja *et al.* 2013). The market for treatment and medicines was already polarising in the late 1990s between a higher price sector for a small layer of the better-off, and deteriorating quality for the very low-income majority. In 2006, this polarisation, fed by feedback loops between urban poverty and market behaviour, was confirmed for a socially mixed area of urban Dar es Salaam (Kida 2009). In urban areas, public facilities are fewer and reliance on non-governmental facilities and shops is higher, and Kida's research showed that, where public facilities were experienced as of lower quality, reliance on the private suppliers was greater, and the lower priced, low-quality facilities were the busiest. One well attended example, an illegally operating (unregistered) private dispensary in a squatter area, was run single-handedly by an assistant laboratory technician, offering a wide range of treatments including wound dressings, injections, admission of patients, without appropriate premises, running water or a refrigerator (Kida 2009).

As noted above, the problems of medicine supply to the public sector is one of the drivers of patients' reliance on private facilities and shops. By the late 2000s, international donors had begun to focus large amounts of finance on providing medicines for HIV/AIDS, TB and malaria. The benefits of this funding, especially for HIV and TB, are very substantial. However, the impact on access to reliable medicines more generally has not been strong, and there have been some negative side-effects of the policy. The case of malaria medication illustrates some of these effects. The donors procure medicines through large-scale tenders and the market-entry requirement of product-by-product WHO pre-qualification effectively shuts out local pharmaceutical producers. In 2006 about 90 per cent of the then first-line treatment for malaria, sulphadoxine-pyrimethamine (SP), was sourced from local manufacturers. From 2007, Tanzania shifted to the more expensive first-line combined medication, artemisinin-lumefantrine (AL). Two local manufacturers developed AL formulations but concluded that pre-qualification (costing an estimated USD150,000) was unlikely to provide market access given the scale and pricing power of Asian competitors. One local firm lost an estimated third of its turnover; others also suffered substantial losses (Tibandebage *et al.* 2016b).

Pharmaceutical manufacturing has a long history in Sub-Saharan Africa, and Tanzanian manufacturing began in the 1960s and 1970s, mainly by public sector firms (Banda *et al.* 2016).

The industry is largely African-owned, and serves the domestic and regional markets. Linkages between health sector procurement and local suppliers can provide mutual developmental benefits (Mackintosh *et al.* 2016b). The industry saw a wave of bankruptcies in the 'structural adjustment' years of the 1980s, but revived in the 1990s. It currently faces several challenges, some characteristic of the wider Tanzanian industrial sector, some specific to pharmaceuticals. In addition to the widespread industrial problems of expensive and unreliable energy, poor ports and roads infrastructure, and difficulty recruiting skilled technicians, the pharmaceutical firms face very high levels of price-based competition from Indian exporters benefitting from subsidies and sometimes engaging in dumping; a tax and import duty regime that can disadvantage manufacturing relative to importing; rising technological and regulatory requirements; and rising international barriers to market entry (Tibandebage *et al.* 2016b). The result has been a falling local share of the domestic medicines market, estimated to have declined from 33 per cent in 2006 to 12 per cent by 2012, while imports from outside East Africa rose from 53 per cent to 78 per cent over the same period (Wangwe *et al.* 2014), and threatened loss of relatively skilled industrial sector employment. This process of externalisation, and increasing reliance on imports, to which health donors have contributed, mirrors the wider deindustrialisation that has periodically threatened this very low-income economy.

Institutional design for improving health care: planning from the here and now

How can those concerned to move the health sector towards universal access to care tackle these interlocking institutional blockages? We summarise four starting points for the required rethinking about institutional design.

The first is to recognise that the current health care market structures are indeed, in identifiable ways, perverse in terms of health and development outcomes. Therefore, intervention – including government intervention, civic action and alliances among health and industrial policy makers, health service managers and other stakeholders – is required to reorient cultures and incentives into more appropriate forms. Second, culture and hierarchy, including gendered disadvantage, profoundly influence organisational structure and outcomes and need addressing directly. Third, planners should recognise that private spending – mainly out-of-pocket expenditure by those struggling to get by – is not currently providing value for money in health care (Tibandebage *et al.* 2013), so the role of private funding in the sector has to change. And finally, a wider and more activist definition of health sector regulation is required, that includes the regulatory role of public sector provision itself, and the design of a package of measures to associate better private market incentives with more effective use of public and private funds.

A functioning, inclusive and cost-effective health system relies – worldwide – on primary care that is routinely available, physically and financially accessible, of decent and trustworthy quality, with trained staff and effective preventative and clinical care, and working referral. There are, however, *no* market incentives pushing Tanzanian non-governmental (private and non-profit) provision in this direction, quite the contrary. Planners should therefore seek to work towards a situation where the private sector is not the first resort of the very poor. The private sector should be pushed up-market, to serve as a competent alternative for those with higher ability to pay rather than a very low-quality resort of the urban poor.

To do this requires, unequivocally, a very low-cost or free – that is, very heavily subsidised – service for the poor (Tibandebage *et al.* 2013). It is possible that this could be provided by contracting with competent private providers if they are willing to serve those on low incomes using government funding. It is more likely that this will need to be done predominantly by the public and (subsidised) faith-based sectors, which have, in principle, the organisational competence and cultural commitments to achieve it. The aim should be 'beneficial competition', meaning the active use of accessible public sector, or publically funded, provision to drive out poor quality at the lower end of the market (Mackintosh and Tibandebage 2002). Better public sector dispensaries in the squatter settlements of Dar es Salaam would, on Kida's (2009) evidence, reduce exploitative recourse to unqualified and dangerous private provision at the bottom of the income scale. This effect is now well established in South Asian contexts such as Kerala and Sri Lanka, where competent and accessible, universalist public sector provision has pushed the private sector up-market and underpinned its competence (Mackintosh 2007; Rannan-Eliya *et al.* 2003)

Accessible competent care has also to extend to maternal care including obstetric emergencies. An effort is underway to improve public health centre capability, including theatres in health centres; also required are sustainable solutions to the chronic shortages of life-saving drugs for maternal health care including handling obstetric emergencies such as haemorrhage. Having a dispensary in every village and a health centre in every ward will not help the majority of the poor women in rural areas if these primary level facilities continue to face severe shortages of health workers, medicines and other essential supplies, a consequence of low social priority of maternal care. Disempowerment of nurse midwives and other maternal health care professionals contributes to poor quality maternal health care, and therefore to poor health outcomes (Tibandebage *et al.* 2016a). Alongside shortages of staff and supplies and poor infrastructure, the system of hierarchy and management actively de-motivates maternal health workers, who are mainly women. We showed that at the health facility level, supportive, communicative and participatory management really could make a difference to culture and outcomes; management training can create more flexible and interactive management hierarchies.

One potential avenue for tackling the critical recurrent shortage of medicines and supplies in the public sector, exacerbated by long import supply chains and a low political priority for medicines funding evidenced by funding delays and a build-up of government debt to the public medicines wholesaler, is to actively strengthen local medicines production. This in turn requires a more active industrial policy and supportive public sector procurement: the aim is a medium-term effect in improving supplies and creating a stronger local industrial lobby for better public procurement (Mackintosh *et al.* 2016b).

More effective use of public subsidy is essential, to build up desirable and complementary non-governmental provision, while effective regulatory intervention is only possible in Tanzania if the resource constraint on inspection and enforcement can largely be side-stepped. This requires collaboration between government and non-government actors to build on identified desirable norms of behaviour: to value and strengthen providers who successfully serve the health care needs of the poor; to achieve legitimacy for formal regulations via negotiation; to strengthen the legitimate claims of low-income patients; and to find synergy between supervision and support. Concepts of regulation, still generally understood as arms-length rule-setting and impartial enforcement in the health sector, would need to shift to a more interactive and collaborative concept of the regulator's role, more characteristic of the interactive behaviour of industry regulators in Tanzania, who become involved in supporting local manufacturers to upgrade to meet required standards, and indeed to involve manufacturers in standard setting. The much-invoked 'level playing field' is a casualty of this rethink, being replaced by selective support for good provision.

It follows that the concept of 'regulation' needs to be broadened and more context-appropriate. We follow the socio-legal literature in understanding the organisation of regulatory activity and the issues it addresses – the 'regulatory space' – as historically and contextually constructed, and open to

Table 25.3 Principles of collaborative regulation

Principle	*Objectives*
1 Promote beneficial competition	1.1 To improve the outcomes of competition for patients and for good providers in all sectors 1.2 To ensure that competition works to undermine poor providers
2 Promote beneficial collaboration	2.1 To create and support collaborative professional and institutional cultures 2.2 To improve provision of public goods and shape markets to work in more desirable ways
3 Promote negotiation	3.1 To create negotiated processes that legitimise rules and ensure that incentive structures they embody make sense to providers 3.2 To shape the government's own regulatory culture into more open and collaborative forms

Source: Mackintosh and Tibandebage (2002) reprinted by permission of Taylor & Francis Ltd.

active redesign (Hancher and Moran 1998). Table 25.3 outlines our principles of a more collaborative approach to regulation. Examples include: accreditation schemes for genuinely charitable faith-based providers; rewarding and supporting preventative care by private providers; strengthening community and individual confidence and scope for complaints and claims; improving the status and working conditions of nurses and strengthening nurse management.

Finally, how resources are allocated and utilised is also important. Studies and reviews in Tanzania have documented governance problems including lack of transparency and accountability, and corruption, for example, in the use of public resources, procurement of health supplies and health workers' absences from work (Kida and Mackintosh 2005; Mamdani and Bangser 2004; World Bank 2013). Addressing these governance problems requires, among other things, strengthening accountability by promoting community involvement not only in identifying needs, but also in monitoring use of public resources, and improving logistics and information systems between and within different levels of governance.

Conclusion

We have argued in this chapter for a rethinking of the nature of the planning process for health care. To move towards universal access to health care, from the current crisis situation for much of the Tanzanian population, requires a process of institutional diagnosis of the 'here and now' in institutional terms, and then a project of active institutional redesign, to realign culture, incentives and objectives. In the process, health planning, while led by government health officials, is necessarily a collaborative process among many stakeholders in public, private and faith-based sectors, and across services, manufacturing and trade. While regulating health markets is highly problematic, perhaps the biggest challenges are cultural and organisational: to move towards a situation where the health sector organisation and culture actively challenges and compensates for some of the worst social inequities in the wider society.

References

Banda, G., Wangwe, S. and Mackinstosh, M. (2016) "Making medicines in Africa: An historical political economy overview". In M. Mackintosh, M.G. Banda, P. Tibandebage and W. Wamae (eds) *Making Medicines in Africa: The Political Economy of Industrializing for Local Health*, Basingstoke: Palgrave Macmillan, pp. 7–24. Available: www.palgraveconnect.com/pc/doifinder/10.1057/9781137546470 [accessed 19 January 2016].

Commission on the Social Determinants of Health (CSDH) (2008) *Closing the Gap in a Generation*, Geneva: World Health Organization. Available: www.who.int/social_determinants/final_report/csdh_finalreport_2008.pdf [accessed 19 January 2016].

Freedman, L.P. (2005) "Achieving the MDGs: Health systems as core social institutions", *Development* 48: 19–24.

Goodin, R. (1996) "Institutions and their design". In R. Goodin (ed.) *The Theory of Institutional Design*, Cambridge: Cambridge University Press, pp. 1–53.

Hancher, L. and Moran, M. (1998) "Organising regulatory space". In R. Baldwin, C. Scott and C. Hood (eds) *A Reader on Regulation*, Oxford: Oxford University Press, pp. 148–172.

Kida, T.M. (2009) *The Systemic Interaction of Health Care Market and Urban Poverty in Tanzania*, Maastricht: Shaker Publishing BV.

Kida, T.M. and Mackintosh, M. (2005) "Public expenditure allocation and incidence under health care market liberalization: A Tanzanian case study". In M. Mackintosh and M. Koivusalo (eds) *Commercialization of Health Care: Global and Local Dynamics and Policy Responses*, UNRISD Social Policy in a Development Context Series, Basingstoke: Palgrave Macmillan, pp. 267–284.

Mackintosh, M. (2001) "Do health systems contribute to inequalities?". In D. Leon and G. Walt (eds) *Poverty, Inequality and Health: An International Perspective*, Oxford: Oxford University Press, pp. 175–193.

Mackintosh, M. (2007) "Planning and regulation: Strengths, weaknesses and interactions in the provision of less inequitable and better quality health care", background paper for the Health Systems Knowledge Network of the WHO Commission on the Social Determinants of Health. Available: www.who.int/social_determinants/resources/csdh_media/planning_market_regulation_2007_en.pdf [accessed 2 March 2016].

Mackintosh, M. and Mujinja, P.G.M. (2010) "Markets and policy challenges in access to essential medicines for endemic disease", *Journal of African Economies* 19 (Supp. 3): 166–200.

Mackintosh, M. and Tibandebage, P. (2002) "Inclusion by design: Rethinking regulatory intervention in Tanzanian health care", *Journal of Development Studies* 39(1): 1–20.

Mackintosh, M. and Tibandebage, P. (2004) "Inequality and redistribution in health care: Analytical issues for developmental social policy". In T. Mkandawire (ed.) *Social Policy and Development*, Basingstoke: Palgrave Macmillan, pp. 143–174.

Mackintosh, M. and Tibandebage, P. (2006) "Gender and health sector reform: Analytical perspectives on African experience". In S. Razavi and S. Hassim (eds) *Gender and Social Policy in a Global Context: Uncovering the Gendered Structure of 'the Social'*, Basingstoke: Palgrave Macmillan, pp. 237–257.

Mackintosh, M. and Tibandebage, P. (2007) "Competitive and organisational constraints on innovation, investment and quality of care in a liberalised low income health system: Evidence from Tanzania", *European Journal of Development Research* 19(1): 81–99.

Mackintosh, M., Banda, G., Tibandebage, P. and Wamae, W. (eds) (2016b) *Making Medicines in Africa: The Political Economy of Industrializing for Local Health*, Basingstoke: Palgrave Macmillan. Available: www.palgraveconnect.com/pc/doifinder/10.1057/9781137546470 [accessed 19 January 2016].

Mackintosh M., Kida T., Tibandebage, P., Ikingura, J. and Jahari, C. (2013) "Payments for maternal care and women's experiences of giving birth: Evidence from four districts in Tanzania", REPOA Working Paper 13/3, REPOA: Dar es Salaam. Available: www.repoa.or.tz/documents/REPOA_WORKING_PAPER_13.3.pdf [accessed 19 January 2016].

Mackintosh, M., Tibandebage, P., Kariuki, J., Karimi Njeru, M. and Israel, C. (2016a) "Health systems as industrial policy: Building collaborative capabilities in the Tanzanian and Kenyan health sectors and their local suppliers". In M. Mackintosh, M.G. Banda, P. Tibandebage and W. Wamae (eds) *Making Medicines in Africa: The Political Economy of Industrializing for Local Health*, Basingstoke: Palgrave Macmillan, pp. 147–165. Available: www.palgraveconnect.com/pc/doifinder/10.1057/9781137546470 [accessed 19 January 2016].

Mujinja, P.G.M., Koivusalo, M., Mackintosh, M. and Chaudhuri, S. (2013) "NGOs, consumer rights and access to essential medicines: Non-governmental public action in a low-income market context". In J. Howell (ed.) *Non-Governmental Public Action and Social Justice*, Basingstoke: Palgrave Macmillan, pp. 112–135.

Mamdani, M. and Bangser, M. (2004) "Poor people's experience of health services in Tanzania: A literature review", *Reproductive Health Matters* 12(24): 138–153.

National Bureau of Statistics (NBS) (2011) *Tanzania Demographic and Health Survey 2010*, Dar es Salaam.

Rannan-Eliya, R.P., Jayawardhane, P. and Karunaratne, L. (2003) "Private primary care practitioners in Sri Lanka". In A. Yazbeck and D.H. Peters (eds) *Health Policy Research in South Asia: Building Capacity for Reform*, Washington, DC: The World Bank, pp. 195–225.

Sen, G., George, A. and Östlin, P. (2002) "Engendering health equity: A review of research and policy". In G. Sen, A. George and P. Östlin (eds) *Engendering International Health*, Cambridge, MA: MIT Press.

Srinivas, S. (2016) "Healthy industries and unhealthy populations: Lessons from Indian problem-solving". In M. Mackintosh, M.G. Banda, P. Tibandebage and W. Wamae (eds) *Making Medicines in Africa: The Political Economy of Industrializing for Local Health*, Basingstoke: Palgrave Macmillan, pp. 183–199. Available: www.palgraveconnect.com/pc/doifinder/10.1057/9781137546470 [accessed 19 January 2016].

Tibandebage, P. and Mackintosh, M. (2002) "Institutional cultures and regulatory relationships in a liberalising health care system: A Tanzanian case study". In J. Heyer, F. Stewart and R. Thorp (eds) *Group Behaviour and Development: Is the Market Destroying Cooperation?* Oxford: Oxford University Press, pp. 271–289.

Tibandebage, P. and Mackintosh, M. (2005) "The market shaping of charges, trust and abuse: Health care transactions in Tanzania", *Social Science and Medicine* 61: 1385–1395.

Tibandebage, P. and Mackintosh, M. (2010) "Maternal mortality in Africa: A gendered lens on health system failure". In L. Ranitch and C. Leys (eds) *Socialist Register 2010*, London: The Merlin Press, pp. 168–183.

Tibandebage P., Mackintosh, M. and Kida, T. (2013) "The public–private interface in public services reforms: Analysis and illustrative evidence from the Tanzanian health sector", paper presented at the 12th Annual Research Workshop, REPOA, Tanzania. Available: www.open.ac.uk/ikd/sites/www.open.ac.uk.ikd/files/files/working-papers/ikd-working-paper-66.pdf [accessed 19 January 2016].

Tibandebage, P., Kida, T., Mackintosh, M. and Ikingura, J. (2016a) "Can managers empower nurse-midwives to improve maternal health care? A comparison of two resource-poor hospitals in Tanzania", *International Journal of Health Planning and Management* 31(4): 379–395.

Tibandebage, P., Wangwe, S., Mackintosh, M. and Mujinja, P.G.M. (2016b) "Pharmaceutical manufacturing decline in Tanzania, how possible is a turnaround to growth?". In M. Mackintosh, M.G. Banda, P. Tibandebage and W. Wamae (eds) *Making Medicines in Africa: The Political Economy of Industrializing for Local Health*, Basingstoke: Palgrave Macmillan, pp. 45–64. Available: www.palgraveconnect.com/pc/doifinder/10.1057/9781137546470 [accessed 19 January 2016].

United Republic of Tanzania (URT) (2007) *Primary Health Services Development Programme: MMAM 2007–2017*, Dar es Salaam, Ministry of Health and Social Welfare.

Wangwe, S., Tibandebage, P., Mhede, E., Israel, C., Mujinja, P. and Mackintosh, M. (2014) "Reversing pharmaceutical manufacturing decline in Tanzania: Policy options and constraints", REPOA Brief No. 43, Dar es Salaam. Available: www.repoa.or.tz/documents/REPOA_BRIEF_43.pdf [accessed 19 January 2016].

World Bank (2013) "Service delivery indicators Tanzania". Available: www.sdindicators.org/tanzania/ [accessed 19 January 2016].

WHO, UNICEF, UNFPA, World Bank Group, United Nations Population Division (2015) *Trends in Maternal Mortality 1990–2015*. Available: www.who.int/reproductivehealth/publications/monitoring/maternal-mortality-2015/en/ [accessed 19 January 2016].

Part V
Planning pedagogies

26

Learning from the city

A politics of urban learning in planning

Colin McFarlane

Introduction

We all know that learning matters. It is a fundamental – perhaps the fundamental – process of scholarship, and it is vital to urban planning and to practice. But how often do we step back and actually think about what learning *is*? How often do we think through what learning *does*, and whether different forms of learning might help support different ways of doing things?

While the history of urban research is predicated upon the question of how we might come to *know* the city, the question of learning itself has remained black-boxed. This is likely to be due in part to a sense that learning is a straightforward and therefore unremarkable process, that the creation and transformation of knowledge and perception is a background story to the central drivers of both urban change and the urban condition. It is also likely to be partly a result of a lingering sense that learning, if not quite *a*political, is a step removed from the formation of political struggle and the practice of urban contention.

My starting point is that such a position masks the many ways in which learning is caught up in the production of urban knowledge, policy, planning, ways of seeing and forms of action. Learning involves new forms of knowledge and, sometimes, new ways of seeing the world. Learning has the potential to challenge and transform the ways of knowing and seeing the city that we inherit, sometimes take for granted, and routinely put to work. In this sense, learning is dialectical: it is about both gaining knowledge and what Swati Chattopadhyay (2012) has called a kind of 'defamiliarisation'. As Steve Pile (2001: 263) has put it: "Knowingness and unknowingness are constitutive of the city: each clads buildings in layers of visibility and invisibility, familiarity and surprise". Learning is a dialogue between what we know and don't know, a dialectic of learning and *un*learning that has the potential to be both transformatory or mere reconfirmation of the status quo.

For urban planners interested in more collaborative forms of planning between different groups in the city (e.g. Healey 2006; Schön 1983), the question of 'whose knowledge counts?' is an old and familiar one, but it remains politically vital. Learning plays an important role here. As a form of wayfinding (Ingold 2000) and struggle, the form and politics of learning in the city cannot simply be restricted to the domain of specialist and expertise knowledge and their effects, as important as these are. We need alongside this to repeatedly ask who 'we' – critical urban researchers, planners and so on – learn from, with, for what ends and under what conditions of power and inclusion. We can understand planning as a particular kind of urban archive of knowledge about how the city might best be envisioned and organised, where learning is vital to their production.

A more collaborative form of planning involves two forms of learning: learning about the city, and learning between different knowledges of the city. Alternative urban archives aim to pull together critical pedagogies of ways of knowing and being in the city – pedagogies of writing, talking, seeing, walking, telling, hearing, drawing, making (Mbembe and Nuttall 2004), which reflect the grounded realities of cities in motion. This is a pedagogy *of* and *with* the city, entailing a form of urban planning committed to dialogic exchange in the context of often profoundly unequal power relations between planners, policymakers, researchers, residents, civil society groups and others.

Given that the unequal circuits of urban planning tend to create truncated space, when they create space at all, for marginalised knowledges of the city, the ethical and political challenges here are vast, and cannot be stepped around or wished away. A key research challenge is to better understand how marginal groups learn the city, and our role as critical urban researchers in bringing those learning processes to urban planning.

In what follows, I consider how learning has been conceived in urban and cognate debates, develop a conception of learning as produced through distributed assemblages that become political in different kinds of ways, explore examples of urban learning in practice, and outline a critical urbanism of learning that can feed into the planning process. I will draw on illustrative examples throughout, including from India and Brazil. My purpose is not to provide an exhaustive survey of different ways of conceiving learning in planning or to argue for a model of learning in planning, but to spotlight some of the key concerns and lessons for thinking about learning as a central political problem in the generation of new archives of urban planning.

I have in mind, then, critical urban researchers and planners, and my argument is that it is these groups in particular who would benefit from reflecting further both on how learning takes place in planning and among other groups in the city, and how residents and activists might play a role in this. But this does not mean that the learning I have in mind here is restricted to urban researchers and planners alone – part of my argument is that learning proceeds through forms of co-production of knowledge, and can generate new ways of knowing and seeing urban challenges and possibilities. I am not arguing, then, for a particular model of learning for urban planning, but for a conceptualisation of learning useful for different planning situations, an expansive conception of learning that can act as one guide for ways of building more inclusive learning processes in planning in order to realise socially progressive goals.

Conceptualising learning

Learning has been an important part of debates on planning. Think, for instance, of John Friedmann's (1987) *Planning in the Public Domain*, or on how meaning is formed in planning debates (e.g. Wagenaar 2011), or Patsy Healey's *Collaborative Planning* (2006), or Argyris and Schön's (1978) work on organisational learning. The constitutive role of learning in processes of urban planning, models, change and politics has been identified in debates on urban planning and policy transfer, from Anthony Sutcliffe's *Towards the Planned City* (1981), Ian Masser and Richard Williams' (1986) collection *Learning from Other Countries: The Cross-National Dimension in Urban Policy-Making*, and Anthony King's (e.g. 2004) surveys of colonial urbanism, to Joe Nasr and Mercedes Volait's (2003) collection *Urbanism: Imported or Exported?* and McCann and Ward's (2011) collection, *Mobile Urbanism*. Literature on mobile planning and policy (e.g. Harris and Moore 2013; McCann 2007; Peck and Theodore 2010, 2015; Roy and Ong 2011; Temenos and McCann 2012) is one important example here. This disparate work has considered, for instance, how certain cities learn from particular policy discourses, such as discourses of 'knowledge cities', 'creative cities', or neoliberal, revanchist and punitive ideologies of urban development (e.g. Florida 2005; Peck 2005; Ward 2006). Despite the surge in critical literature on travelling urban knowledge, policy and planning, there has been relatively little attempt to consider how learning itself might be conceptualised.

If these debates are varied and distinct, all of them contain two central claims or assumptions about learning. The first is that learning is a process of potential transformation. All forms of learning have at least some sense of purpose, whether that purpose is to realise a strongly normative planning position or to better understand an urban process or even just to find out more about a city as a visitor. There are, however, differences to these forms of learning. Learning for fun or research purposes might, for instance, be less politically motivated and perhaps more open to different positions than learning to instantiate a particular planning vision. In the latter, learning can be highly selective, preconditioned by ideological visions, and closed to difference (e.g. Peck and Theodore 2010, 2015). If learning is often about transformation, then, the nature and extent of that transformation varies tremendously across urban domains. Part of my argument in this chapter is that learning, and here I have in mind planners in particular, is more likely to shift positions if there is a genuine commitment to listening to different positions and to being prepared to shift your own position in order to realise more progressive ends.

The second claim, more implicit, is that learning, even where it is explicitly described as uncertain – as in, for instance, strands of organisational theory that emphasise creativity and invention (e.g. Amin and Cohendet 2004) – refers to a process involving particular constituencies and discursive constructions, entails a range of inclusions and exclusions of people and epistemologies, and produces a means of going on through a set of guidelines, tactics or opportunities. As a process and outcome, learning is actively involved in changing or bringing into being particular assemblages of people-sources-knowledges. Learning about and in cities, occurs not simply through formal, linear and cognitive processes, but in practical immersion in urban space-times. It involves not simply the absorption of codified information – plans, policies, maps, data-sets and so on – but a myriad set of processes that stretch beyond the individual subject to the social and material processes that co-constitute the city. Learning occurs through an assemblage of people, things, practices, rhythms and spaces, and is part of the architecture through which we come to perceive urbanism, whether we are policy makers, researchers, residents or activists.

The spatialities of learning matter a great deal to both these claims about learning. Learning is translocal in its geographies, and involves an ongoing labour in forging and developing connections between different sources, routes and actors. The shifting relations of knowledge and space have been key questions at stake in recent influential accounts of the translocal nature of urban policy and planning (Soderstrom 2014), including – and this is to mention just a few – a series of edited collections such as Roy and Ong's (2011) *Worlding Cities*, McCann and Ward's (2011) *Mobile Urbanism*, Harris and Moore's (2013) collection on planning histories and circulating knowledge, Bunnell *et al.*'s (2012) collection on global urban frontiers, Cochrane and Ward's (2012) collection on researching the geographies of urban policy mobility and McFarlane and Robinson's (2012) collection on comparative urbanism (and see Robinson 2006).

In *Learning the City: Knowledge and Translocal Assemblage*, I argued that learning emerges through the interaction of three processes: translation, coordination and dwelling (McFarlane 2011). Translation refers to the distribution and comparison of knowledges, ideas and resources across multiple space-times, from activists sharing ideas about how to protest against the state to planners and policy makers seeking to learn from different cities. Coordination refers to the building of functional systems that anchors these multiple translations, and may include discourses or maps or policy frameworks or traffic lights systems, that attempt to deal with complexities. Finally, dwelling refers to how learning is lived, and how over time we attune or educate ourselves through learning assemblages in different ways of seeing and inhabiting cities. These three processes shift and entangle in learning assemblages, through slow processes of bricolage, sudden improvisations that may nonetheless draw upon historical repertoires through which people get a feel for how things work and might get done, or in tactical learning through which activists might produce ideas or approaches that disturb, resist, or initiate alternative ways of seeing the urban condition.

The key point here is to see learning as a process of translating experience and information that is then organised in different ways according to context. The contexts of, say, planners and residents, are often (though not always) quite different, because context is shaped by what people are trying to learn and to what ends. For example, a planner looking to learn about 'best practice' in international approaches to social housing is likely to be learning through different sources from a group of residents looking to learn new ways of ensuring community facilities or safety in their housing development. Moreover, the differential power relations between the learning and priorities of planners as opposed to most residents is a vital part of what kind of learning is privileged in the planning context. Everyone learns cities, but the ways in which people learn, the resources with which they can learn, the routes through which that learning takes place and the political framings and objectives that organise that learning, are profoundly diverse and often deeply contested. In other words, it is vital to attend to what learning *is*, *how* different groups learn, and what kinds of learning are *prioritised*.

I am not claiming that this conception of learning through translation, coordination and dwelling is entirely distinct from other debates on learning. There are, for example, resonances with Healey's (2006) conception of collaborative planning, or long-standing interventions in organisational learning (e.g. Schön 1983), or with how urban geographers like Eugene McCann, among others, foreground the spatialities and mundane practices of learning (e.g. see McCann 2007; McCann and Ward 2011; Temenos and McCann 2012). But, in offering a more expansive approach to learning than most other accounts of how learning operates in cities, my approach to learning here allows us to see the diverse steps, contexts and routes through which learning occurs in practice. It allows us to develop a conception of learning for urban planning that, from the start, recognises that learning takes place not just in offices and over data but through informal relationships and encounters, that learning is often messy and happenstance, that it often includes a wide range of people, things and places, and that planners can learn not just from different groups from the city but from the ways in which those groups learn.

In short, this is a conception of learning that takes place 'out there' in the city, and that therefore might be constituted not just by the learning of data and policy but through the ways in which economically marginal groups themselves learn the city. Building from this conception of learning, in the next section consider I examine a particular form of learning: *incremental learning*. I consider incremental learning in relation to informal settlements, and explore how incremental learning might itself become a useful source of learning for urban planners interested in pursuing improvements to life in informal settlements. Before doing so, a word on what incremental is and why it matters.

Incrementalism is central to how people learn the city, and is common to a whole range of urban processes and forms, from housing and policy to infrastructure and culture. As Simone (2008: 28) has argued, cities,

> no matter how depleted and fragmented, still constitute platforms for trajectories of incrementalism. Houses and limited infrastructure are added onto bit by bit; the mobilization of family labor buys time for a small business to grow; migration is used as an instrument to pool together savings in order to start a new economic activity; mobile work crews are formed to dig wells, help with construction, or deliver goods until they make enough contacts to specialize on one particular activity.

What kind of learning is incrementalism? It is not a process of linear addition, but instead involves translation. As different economists, planners, urbanists and geographers have long shown, learning often occurs in practice, through improvisation as the best laid plans begin to unravel in the heterogeneities of the city, and often in small incremental adjustments (e.g. Argyris and Schön 1978; Hirschman 1967; Nasr and Mercedes Volait 2003; Schön 1983). My focus here, however, is on how residents learn incrementally, and how planners might learn not just with residents, but through the ways in which residents learn. For example, residents tweaking and tinkering with urban infrastructure in informal settlements translate the use of materials between contexts, for example, in using discarded objects to help secure makeshift toilets or housing at different points (McFarlane *et al.* 2014). Similarly, residents working in markets or searching for employment can use existing networks in new ways to help open opportunities, although the capacity to do so is very different depending on the situations people find themselves in (Simone 2014). Some groups, Simone (2008: 28) continues, "are able to organize labor, money, and contacts to finish roads, complete water reticulation projects, or electrify their compounds and neighborhoods, while others languish". Incremental learning, then, involves translation and coordination, but also emerges through dwelling in the city and attuning, as much as one can, to the changing conditions of urban life.

Incremental learning and urban planning

As an often continuous effort, we see incremental learning in the makeshift urbanism that constitutes much of the daily life for people within marginalised informal settlements, especially women who tend to take on the majority of household construction and maintenance. As Neuwirth (2006: 21–22) has written: "With makeshift materials, they are building a future in a society that has always viewed them as people without a future. In this very concrete way, they are asserting their own being". This makeshift urbanism, what De Certeau (1984: xv) may have called bricolage, refers to "ways of 'making do'", a process that is forced on residents partly as a result of exclusions from the formal planning process of housing and infrastructure. Incrementalism is a central process of urban life, an atunement of perception to what urbanism – in conditions of often extreme inequality – might enable and delimit, and how best to negotiate it.

What do these incremental learning processes mean for urban planning? One of the challenges for urban planning today is to create mechanisms that support the incremental learning of the urban poor. In many cities, this would require a commitment on the part of planners, as a first step, to a different kind of politics: one in which demolishing informal settlements gives way to a politics of provisioning, support and consolidation. For example, in work that Renu Desai, Steve Graham and I conducted on sanitation infrastructure in informal settlements in Mumbai, we found that there was a very significant variation in the extent to which incremental strategies are supported by formal planning mechanisms in different informal settlements. For instance, in the neighbourhood of Khotwadi in west Mumbai, the links that many residents and civil society groups have to the dominant political party in the city – the

Shiv Sena – helped ensure that incremental learning strategies around maintaining toilets, drains and water connections were actively supported by the local state. In contrast, a few miles to the east the much poorer, 'illegal' (in the eyes of the state) and largely Muslim neighbourhood of Rafinagar received little and often no such support. Here, incremental learning strategies ranging from self-built latrines and drainage systems to housing and water and solid waste-management systems are left largely adrift, outside of a few relatively well-connected or better off individuals.

As a first step, then, there is a need simply for basic political, legal and ethical recognition of all informal settlements as a necessary basis for supporting urban life (for more information on the Mumbai sanitation research, see Desai *et al.* 2015; Graham *et al.* 2013; McFarlane *et al.* 2014). Here, learning entails the planning process first learning new ways of seeing informal settlements – and this seeing is a political and ethical commitment – and second moving towards a position of learning with residents to support incremental learning processes.

There is, of course, a rich world of urban social movements that take up this very question. In Mumbai, for instance, one example is the 'Right to Pee' movement, which has a commitment to basic sanitation as a human and constitutional right for all residents in the city. Other social movements operating in informal settlements actively draw upon and promote incremental learning as part of their politics and advocacy work on urban planning. The most high-profile example here is Slum/Shack Dwellers International (SDI).

SDI is a collection of non-governmental (NGO) and community-based organisations (CBOs) working with urban poverty, particularly housing and sanitation, which operates throughout Asia, Africa and Latin America. It is a translocal experiment in building a new form of urban sociality; a learning movement based around a structure of what its leaders call 'horizontal exchanges' involving small groups of the urban poor travelling between neighbourhoods to learn from one another. The movement espouses a range of tactics that its leaders describe as indispensable to a development process driven by the urban poor. These include daily savings schemes, exhibitions of model house and toilet blocks, the enumeration of poor people's settlements, training programmes of exchanges, and a variety of other tactics. In seeking grassroots participation and horizontal exchange, SDI seeks to place urban learning at the centre of social and political relations. Its insistence upon learning from and between the urban poor emerges from the context of a failure – deliberate or otherwise – of the state to ensure collective provision of urban infrastructure, services and housing.

In SDI, learning is first and foremost a practice of working in groups (McFarlane 2011). It draws in the incremental learning of the urban poor and seeks to lever that learning into a new kind of incremental learning with the state, one based on partnership to build housing and infrastructure. We may be tempted to describe this form of incremental learning as 'radical incrementalism', to use Edgar Pieterse's (2008) phrase, but we need to be cautious here. SDI's work is not radical in the sense of challenging the structural conditions of inequality in the city. The learning politics here is one that accommodates the existing system and seeks to find space within it. The emphasis on empowering the urban poor as part of this learning politics could be described as radical, but the political form itself is not radical. It would be more accurate to describe this model of learning as a form of partnership urban planning that seeks to create greater space for the urban poor, where incrementalism is both a form of learning and, to use Arjun Appadurai's (2002) apt phrase, a "politics of patience". In the next section I explore the question of learning and more radical urban change through discussion of a critical urbanism of learning, as a basis for a more collaborative form of learning in urban planning.

Critical learning and urban planning

As the example of incrementalism begins to suggest, attending to how people differently learn cities can open out possibilities for a critical urban planning. Peter Marcuse (2009) usefully outlines three steps here. First, *evaluating* urban knowledges that are presented to us as inevitable by planning elites and in the form of the dominant ideologies of urban planning; second, *democratising* urban planning learning processes by examining who is and who isn't included in them; and third, *proposing* alternative sets of urban knowledges oriented to more socially and ecologically just forms of urban planning. This schema of evaluating–democratising–proposing is useful and important, but the key point here is less to follow a particular strategy or model and more to expand discussion around the different approaches that might facilitate learning for more collaborative, and ultimately more socially just forms of urban planning.

We might think, for example, of the experience with participatory budgeting in Porto Alegre, Brazil, which has been well documented. In post-dictatorship urban Brazil, there has been a tradition of participatory democracy that has been particularly associated with its vibrant social movements and the *Partido dos Trabalhadores* (PT, the Workers Party), especially in the high-profile successes in Porto Alegre. As an experiment in planning through urban democratisation and learning, Porto Alegre stands apart in Latin America for its scope and sustained commitment and was one of the most successful cases of municipal socialism on the continent (Abers 2000; Baiocchi 2001; Goldfrank and Schrank 2009; Sintomer *et al.* 2008). Learning through participatory budgeting is coordinated by a strict and clear set of procedures for organisation, representation and participation. These efforts have not, of course, been without their critics. As Baiocchi (2001) argues, a central criticism of these sorts of participatory planning forums is that they can reproduce class hierarchies, giving increased influence to local elites, and that they reproduce power relations around the political competence of 'experts' against non-experts.

However, in contrast to what we might expect to see if more powerful groups were manipulating the process, the vast majority of investment in early participatory budgeting in Porto Alegre went to poorer areas of the city (Baiocchi 2001). Sintomer *et al.* (2008: 166–167) write:

> [Participatory budgeting] has provided for a reversal of priorities: primary health care was set up in the living areas of the poor, the number of schools and nursery schools was extended, and in the meantime the streets were asphalted and most of the households have access to water supply and waste water systems ... [revised budgeting formula ensure] that districts with a deficient infrastructure receive more funds that areas with a high quality of life.

Learning was central to the process, and included ordinary residents learning about how planning operates and how budgets are managed, or how services operate and might be improved. There was a commitment by planners and residents to the spirit of the evaluating–democratising–proposing schema outlined by Marcuse. The success of urban participatory learning forums in Porto Alegre is, to be sure, partly a function of the city's relative wealth compared to other cities in Brazil. But it is about more than this. The municipal government, in the form of the PT, was committed to experimenting with participatory democracy and – crucially – was aware that this had to mean allowing autonomous civil society debates and institutions to flourish. The fact that participatory budgeting in the city was later to become watered down and lost its radical edge only serves to illustrate the need for political commitment to collaborative learning on the part of the state and the planning apparatus.

For those concerned with more collaborative forms of learning for urban planning, the larger question here is what sorts of criteria and procedures might function as practical tools? Writing in the context of controversies around science and the environment, Callon *et al.* (2009) highlight several potential formats, with different potentials to lead to change: *focus groups*, which the authors view as useful in identifying priorities, but that are episodic and generally do not lead to changing relations between experts and laypersons; *public inquiries*, which Callon *et al.* (2009) claim succeed only where there is genuine commitment to involve the public; *consensus conferences*, expert-layperson forums that examine on particular issues by raising awareness, stimulating debate and leading to the production of citizen reports – a meaningful start, but not often a sustained collaboration; and *citizens' panels and juries*, which often privilege local points of view but that are rarely about dialogue. If none of these procedures are themselves satisfactory – and they take us some way from the more radical agenda set out by Marcuse (2009) with which I began this section – specific procedures will be more or less relevant for particular issues and at particular times and spaces. There is no reason why, for instance, local focus groups could not be used to learn about and identify priorities in relation to, say, the construction of community facilities in a poor neighbourhood, neither is there any reason to ignore the possibility of urban consensus conferences whereby citizen reports would be produced on issues such as the locating of urban garbage grounds, perhaps generating media attention. Exploring learning for collaborative urban planning will entail different kinds of forums, then, which are contingent on the issues at hand, and driven by an experimental ethos that is committed to collective learning. Moreover, rather than an appendage or bureaucratic procedure within urban planning, if they are to succeed they should be central to the very conception and nature of urban planning itself.

While the forms of learning forums can and should vary according to the issue and the goals in mind, there is of course a general question of principle at stake. More radical forms of urban learning forums, such as those around participatory budgeting in Porto Alegre, are concerned not with participation per se, but with an ethos of *coproduction*. As Beebeejaun *et al.* (2015) argue, coproduction is explicitly concerned not just with involving different groups to include their perspectives, but with genuine collaboration that challenges the relations of power and that is reflected in the knowledge produced. Diana Mitlin (2008) shows in her work on SDI that coproduction can be an explicit political strategy used by activists to develop partnerships with the state in which the knowledge of the urban poor – on housing, infrastructure and services, for example – drives the knowledge used and strategies put in place for urban development. In some cases in the SDI movement, coproduction has meant shared responsibilities for managing budgets and has developed strong relationships between low- and mid-level urban officials and the leaders of NGOs and CBOs, through which civil society activists have learnt how planning and policy processes operate and might be challenged in ways that sharpen their own political competencies. As Satterthwaite and Mitlin (2014: 215) argue, more than participation, coproduction is about challenging visions of what urban development is and how the city might be planned, by both acting to challenge and reset the priorities of urban development, while at the same time remaining accountable to the neighbourhoods and communities in which civil society groups operate.

Conclusion

To live in cities is to come up regularly against the unknown as well as to the limit points of how different groups can and cannot learn in contexts of ever deepening urban inequalities. Planners, residents, activists and of course researchers, approach the unknown not with a blank slate but with what they already know and in the context of their own lives, resources, plans, hopes, fears, etc. The critical purchase of conceptualisations of urban learning lies not in a straightforward call to know *more* of cities, but to expose, evaluate and democratise the politics of learning cities by placing learning explicitly at the heart of urban planning debate.

This is not a straightforward process. While I have attempted to show the potential of learning for informing a more socially just form of urban planning, I have also identified the limits of learning. Especially important here is the question of political will. Unless there is a political commitment in city (and quite often central state) planning to develop new forms of *learning with* different groups – particularly those groups that are most often excluded (informal settlements in particular, but we might also think for instance of street hawkers and homeless groups) – to *learn from* those marginal groups, and to put those forms of learning into *practice*, then learning is likely to be of secondary importance in the struggle for urban change.

In many cities, the sources of learning appear to be ever-more exclusive, and the agendas that are prioritised often out of touch with more social needs. Current debates about the 'smart city' for example, value particular exclusive groups, spaces and forms of urban development, particularly around well-educated elites living in premium residential spaces and working in high-end service economies, including in science, technology, research, media and finance (Marvin *et al.* 2016). The referent points for these debates is very often the technical and largely positivist renderings of the city produced by corporations like IBM and Cisco, and less the knowledge forms and needs of, say, the residents of informal settlements (which are often conspicuously absent from visions of the smart city). For critical researchers, there is a struggle ahead in both critically exposing whose learning is prioritised in depictions of the urban future, and in developing alternative urban visions foregrounded in more socially and ecologically just concerns and learning.

However daunting the struggle of proposing more socially just approaches of urban learning for planning is, cities always provide possibilities and cause for hope. Examples such as those of participatory budgeting, or elements of the work of movements like SDI, are of course important sources of such hope. As David Harvey (2008: 33) argues in relation to urban social movements, there is the potential to "reshape the city in a different image from that put forward by the developers". Cities are, and have always been, places of unexpected encounters, progressive ideas, forms of knowledge and activism, and can generate not only inventive ways of perceiving and acting in urban space but new forms of urban learning, planning and possibility.

References

Abers, R. (2000) *Inventing Local Democracy: Grassroots Politics in Brazil*, Boulder, CO: Lynne Rienner.

Amin, A. and Cohendet, P. (2004) *Architectures of Knowledge: Firms, Capabilities and Communities*, Oxford: Oxford University Press.

Appadurai, A. (2002) "Deep democracy: Urban governmentality and the horizon of politics", *Public Culture* 14(1): 21–47.

Argyris, C. and Schön, D. (1978) *Organizational Learning: A Theory of Action Perspective*, Reading, MA: Addison-Wesley

Baiocchi, G. (2001) "Participation, activism, and politics: The Porto Alegre experiment and deliberative democratic theory", *Politics and Society* 29: 43–72.

Beebeejaun, Y., Durose, C., Rees, J., Richardson, J. and Richardson, L. (2015) "Public harm or public value? Towards coproduction in research with communities", *Environment and Planning C* 33: 552–565.

Bunnell, T., Goh, D., Lai, C.-K. and Pow, C.P. (2012) "Global urban frontiers: Asian cities in theory, practice and imagination", *Urban Studies* 43(13): 2785–2793.

Callon, M., Lascoumes, P. and Barthe, Y. (2009) *Acting in an Uncertain World: An Essay in Technical Democracy*, Cambridge: MIT Press.

Chattopadhyay, S. (2012) *Unlearning the City: Infrastructure in a New Optical Field*, Minneapolis, MN: University of Minnesota Press.

Cochrane, A. and Ward, K. (2012) "Researching the geographies of policy mobility: Confronting the methodological challenges", *Environment and Planning A* 44(1): 5–12.
De Certeau, M. (1984) *The Practice of Everyday Life*, Berkeley, CA: University of California Press.
Desai, R., McFarlane, C. and Graham, S. (2015) "The politics of open defecation: informality, body and infrastructure in Mumbai", *Antipode* 47(1): 98–120.
Florida, R. (2005) *Cities and the Creative Class*, New York: Routledge.
Friedmann, J. (1987) *Planning in the Public Domain*, Princeton, NJ: Princeton University Press.
Goldfrank, B. and Schrank, A. (2009) "Municipal neoliberalism and municipal socialism: Urban political economy in Latin America", *International Journal of Urban and Regional Research* 33(2): 443–462.
Graham, S., McFarlane, C. and Desai, R. (2013) "Water wars in Mumbai", *Public Culture* 25: 115–141.
Harris, A. and Moore, S. (2013) "Planning histories and practices of circulating urban knowledge", *International Journal of Urban and Regional Research* 37(5): 1499–1509.
Harvey, D. (2008) "The right to the city", *New Left Review* 53: 23–40.
Healey, P. (2006) *Collaborative Planning: Shaping Places in Fragmented Societies* (2nd edn), London: Palgrave Macmillan.
Hirschman, A.O. (1967) *Development Projects Observed*, Washington DC: Brookings Institution
Ingold, T. (2000) *The Perception of the Environment: Essays in Livelihood, Dwelling and Skill*, London: Routledge.
King, A.D. (2004) *Spaces of Global Cultures: Architecture, Urbanism, Identity*, London: Routledge.
McCann, E.J. (2007) "Expertise, truth, and urban policy mobilities: Global circuits of knowledge in the development of Vancouver, Canada's 'four pillar' drug strategy", *Environment and Planning A* 40(4): 805–904.
McCann, E.J. and Ward, K. (eds) (2011) *Mobile Urbanism: Cities and Policymaking in a Global Age*, Minneapolis, MN: University of Minnesota Press.
McFarlane, C. (2011) *Learning the City: Knowledge and Translocal Assemblage*, Oxford: Wiley-Blackwell.
McFarlane, C. and Robinson, J. (2012) "Experiments in comparative urbanism", *Urban Geography* 33(6): 765–773.
McFarlane, C., Desai, R. and Graham, S. (2014) "Informal urban sanitation: Everyday life, comparison and poverty", *Annals of the Association of American Geographers* 104: 989–1011.
Marcuse, P. (2009) "From critical urbanism to right to the city", *City* 13(2): 185–197.
Marvin, S., Luque, A.L. and McFarlane, C. (2016) (eds) *Smart Urbanism: Utopian Vision or False Dawn?*, London: Routledge.
Masser, I. and Williams, R.H. (1986) *Learning from Other Countries: The Cross-National. Dimension in Urban Policy-Making*, Norwich: Geobooks/Elsevier.
Mbembe, A. and Nuttall, S. (2004) "Writing the world from an African metropolis", *Public Culture* 16(3): 347–372.
Mitlin, D. (2008) "With and beyond the state: Co-production as a route to political influence, power and transformation for grassroots organizations", *Environment and Urbanization* 20(2): 339–360.
Nasr, J. and Volait, M. (2003) (eds) *Urbanism: Imported or Exported?*, Oxford: Wiley Blackwell.
Neuwirth, R. (2006) *Shadow Cities: A Billion Squatters Now, A New Urban World*, New York: Routledge.
Peck, J. (2005) "Struggling with the creative class", *International Journal of Urban and Regional Research* 29(4): 740–770.
Peck, J. and Theodore, N. (2010) "Mobilizing policy: Models, methods and mutations", *Geoforum* 41(2): 169–174.
Peck, J. and Theodore, N. (2015) *Fast Policy: Experimental Statecraft at the Thresholds of Neoliberalism*, Minneapolis, MN: Minnesota University Press.
Pieterse, E. (2008) *City Futures: Confronting the Crisis of Urban Development*, London: Zed Books.
Pile, S. (2001) "The un(known)city … or, an urban geography of what lies buried below the surface". In I. Borden, J. Kerr, J. Rendell and A. Pivaro (eds) *The Unknown City: Contesting Architecture and Social Space*, Cambridge: MIT Press, pp. 262–279.
Robinson, J. (2006) *Ordinary Cities: Between Modernity and Development*, London: Routledge.
Roy, A. and Ong, A. (eds) (2011) *Worlding Cities: Asian Experiments and the Art of Being Global*, Oxford: Wiley Blackwell.
Satterthwaite, D. and Mitlin, D. (2014) *Reducing Urban Poverty in the Global South*, London: Routledge.
Schön, D. (1983) *The Reflective Practitioner: How Professionals Think in Action*, London: Temple Smith.
Simone, A. (2008) "Emergency democracy and the 'governing composite'", *Social Text* 26(2): 13–33.

Simone, A. (2014) *Jakarta, Drawing the City Near*, Minneapolis, MN and London: University of Minnesota Press.

Sintomer, Y., Herzberg, C. and Rocke, A. (2008) "Participatory budgeting in Europe: Potentials and challenges", *International Journal of Urban and Regional Research* 32(1): 164–178.

Soderstrom, O. (2014) *Cities in Relations: Trajectories of Urban Development in Hanoi and Ouagadougou*, Oxford: Wiley-Blackwell.

Sutcliffe, A. (1981) *Towards the Planned City, Germany, Britain, the United States and France, 1780–1914*, Oxford: Basil Blackwell.

Temenos, C. and McCann, E. (2012) "The local politics of policy mobility: Learning, persuasion, and the production of a municipal sustainability fix", *Environment and Planning A* 44(6): 1389–1406.

Wagenaar, H. (2011) *Meaning in Action: Interpretation and Dialogue in Policy Analysis*, New York: M.E. Sharpe.

Ward, K. (2006) "Policies in motion, urban management and state restructuring: The trans-local expansion of Business Improvement Districts", *International Journal of Urban and Regional Research* 30(1): 54–75.

27

Campus in Camps

Knowledge production and urban interventions in refugee camps

Alessandro Petti

Introduction

When we think about refugee camps, one of the most common images that comes to our mind is an aggregation of tents. However, after more than 60 years since their establishment, Palestinian refugee camps are constituted today by a completely different materiality. Tents were

Figure 27.1 Campus in Camps is a space for communal learning and production of knowledge grounded in lived experience and connected to communities in Dheisheh Refugee Camp, Bethlehem Palestine

Source: Campus in Camps/Anna Sara.

first reinforced and readapted with vertical walls, later substituted with shelters, and subsequently new houses made of concrete have been built, making camps dense and solid urban spaces.

There is therefore a gap between the image that we have in our mind when we think and talk about refugee camps and the actual materiality of camps today. This challenges us to find meanings in a reality that is in front of our eyes, but we can hardly understand. Camps are no longer made of fragile structures. Yet, at the same time, they are not cities either. Cities have a series of public institutions that organise, manage and control the lives of inhabitants. Today, the United Nations Relief and Works Agency for Palestine Refugees in the Near East (UNRWA), despite its role as a purely humanitarian agency being challenged by the refugee community, does not govern the camp. The camp, as we know, has developed its own form of social and political life. We lack the right vocabulary to describe this new condition as the prolonged exceptionality of its condition has produced different social, spatial and political structures.

In 2012 in an effort to intervene in such an unstable and socially and politically charged urbanity of exile we[1] founded 'Campus in Camps' as a means to address the numerous needed spatial and social interventions in Palestinian refugee camps. Campus in Camps was created as an experimental educational programme with Al-Quds University (Al-Quds/Bard Partnership) and hosted by the Phoenix Center in Dheisheh Refugee Camp in Bethlehem. It was implemented with the support of the GIZ Regional Social and Cultural Fund for Palestinian Refugees and Gaza Population on behalf of the German Federal Ministry for Economic Cooperation and Development (BMZ), in cooperation with UNRWA Camp Improvement Program.

Figure 27.2 The project of the concrete tent in Dheisheh refugee camp deals with this paradox of a permanent temporality that petrifies a mobile tent into a concrete house. The result is a hybrid between a tent and a concrete house, temporality and permanency, soft and hard, movement and stillness

Source: Campus in Camps/Anna Sara.

Taking Campus in Camps as a case study in the following pages I'll try to answer some of the following questions: what is the role of the university in the greater transformation of society? How can the knowledge that is produced inside its walls be relevant and useful for students that live in marginalised communities? What kinds of structures or institutions are required for the accommodation of interests and subjects born from the interaction between students, teachers and the broader social context? How can the attention of educational institutions move from the production of knowledge – based on information and skills – to processes of learning – based on shifts in perception, critical approaches, visions and governing principles? And how to reconcile theory with action, and combine a rigorous understanding of the problems with pragmatic and effective urban interventions?

Neighbours' schools

In 1987, in an attempt to suppress the *intifada* (the Palestinian civil protests against its military occupation), the Israeli government banned people from gathering together and closed all schools and universities. As a reaction, Palestinian civil society grew through the organisation of an underground network of schools and universities in private houses, garages and shops. Universities were no longer confined within walls or campuses and teachers and students began using different learning environments in cities and villages.

These gatherings and assemblies reinforced the social and cultural life among Palestinian communities. Learning was not limited to the hours spent sitting in classrooms; mathematics, science, literature and geography were subjects that could be imparted among friends, family members and neighbours.

In order to resist the long periods of curfew imposed by the Israeli army, these self-organised spaces for learning included self-sufficiency activities such as growing fruits and vegetables and raising animals. Theoretical knowledge was combined with knowledge that emerges from action and experimentation. Learning became a crucial tool for gaining freedom and autonomy. People discovered that they could share knowledge and be in charge of what and how to study.

The classical structure, in which 'expert teachers' transmit knowledge and students are mere recipients to be filled with information, was substituted by a blurred distinction between the two. A group dynamic opened this new learning environment to issues of social justice, inequality and democracy. The First Intifada was, in fact, a non-violent movement that not only aimed at changing the system of colonial occupation but also at creating new spaces for social change. For example, youth and women now had the opportunity to challenge traditional and patriarchal sectors of Palestinian society. Within these processes, education was perceived as an essential tool for liberation and emancipation. The knowledge produced within the group structure was no longer distant and alienating, but rather grounded in the present political struggle for justice and equality.

At the beginning of the 1990s, this open and community-based system of learning was not considered by the newly established Palestinian Authority (PA). The national Palestinian educational curriculum continued to be drawn on the basis of the Jordanian national system, ignoring these challenging and rich experiences.

However, most of the leaders of this underground network became key figures in the Palestinian non-governmental sector. For many, the state-building process of the last years became centralised, bureaucratised and, in some cases, authoritarian. The non-governmental sector is the space where these experimental practices in health, environment, human rights and education have continued developing.

In Palestine today, most NGOs, much like the PA, are internationally funded. Although donors are operating in support of the local population, they are in fact not accountable to the people, often pursuing the cultural and political agendas of the donor states. Philanthropy has thus become one of the main vehicles for Western intervention in the politics and culture of Palestine.

Bearing these dangers in mind, the network of NGOs still seems to be an important tool for developing different policies. In particular, non-governmental spaces are able to react more efficiently to the needs of marginalised sectors of society that are not represented by state policies. A new type of common space has thus emerged through NGO culture, one that has not yet been adequately understood and theorised.

Critical learning environments

In this context, Campus in Camps originated from a collective cumulative thought that aimed at bringing together theory and action, learning in a contextual environment and project-based interventions in refugee camps. The desire for such a programme maturated in an ongoing dialogue that started in 2007 between the UNRWA Camp Improvement Program, directed by Sandi Hilal, and the Refugee Camp Communities of the Southern West Bank. From this ongoing dialogue emerged the urgency from the communities to explore and produce new forms of representation of camps and refugees beyond the static and traditional symbols of passivity and poverty.

In three years of teaching at Al-Quds Bard University, a liberal arts college based in Abu Dis, I was convinced that the university can play a decisive role in creating a space for critical and grounded knowledge production connected to greater transformations and the democratisation of society. In particular, I became convinced that 'moving' the campus to more marginalised geographical areas and sectors of society could create a truly engaged and committed university. The university campus and the refugee camp are both 'extraterritorial islands', of different sorts of course: one utopian and one dystopian. Both are removed from the rest of the city. Campus in Camps aimed to transgress the borders between the 'island of knowledge' and the island of 'social marginalisation'. In conversation with Al-Quds Bard students from refugee camps, I realised that their narrations, ideas and discourses were able to flourish in a protected space, such as the university, but needed to be grounded in context and connected with the community. Reciprocally, by moving to camps, the university was able to open its doors to other forms of knowledge, to an experimental and communal learning able to combine critical reflection with action.

The programme engages young participants in dealing with new forms of visual and cultural representations of refugee camps after more than 66 years of displacement. The aim is to provide young Palestinian refugees who are interested in engaging their community with the intellectual space and necessary infrastructure to facilitate these debates and translate them into practical community-driven projects that will incarnate representational practices and make them visible in the camps. The group of participants in the programme was formed during a long three-month process of interviews, consultations with the community and public announcements in newspapers and mosques. There was not a real selection: instead, a series of meetings allowed both us and the applicants to understand if we shared a mutual interest in embarking on such an experimental project. However, one thing the participants have in common is their engagement with the community. Most of them volunteered in organisations or have been involved in community-based projects.

Campus in Camps' pedagogical approach is fundamentally based on the strict relation between knowledge production (the *Collective Dictionary*)[2] and urban interventions (the initiatives).[3] The *Collective Dictionary* is a series of publications containing definitions of concepts considered fundamental for the understanding of the contemporary condition of Palestinian refugee camps. Written reflections on personal experiences, interviews, excursions and photographic investigations constitute the starting point for the formulation of more structured thoughts and serve to explore each term. Multiple participants developed each publication, suggesting a new form of collective learning and knowledge production. The *Collective Dictionary* aims to establish a common language and a common approach among the participants. This was achieved through education cycles, seminars and lectures. First months of the programme are dedicated to a process that we called 'unlearning', healing from pre-packaged alienating knowledge, knowledge that is not linked with life. We involved professors from universities and community members for lectures and seminars. Based on these first encounters, the participants, together with the project team, discussed the opportunity to involve the guests in a cycle,[4] which was usually structured as biweekly meetings for a minimum of one month. The decision to involve a guest is based on the relevance of the subject in relation to the interest of the group. For this reason, the structure of Campus in Camps is constantly being reshaped to accommodate the interests and subjects born from the interactions between the participants and the social context at large. Over the course of the first year, over a dozen seminars and/or lectures were held in addition to these cycles that gave participants further exposure to experts in a variety of fields.[5] These areas of interest included citizenship, refugee studies, humanitarianism, gender, mapping and research methodologies. Many of these events are open to the public and are the mechanism to connect with members of the camp community as well as university students. The first year culminated in an open public presentation of two days in which more than 100 people from the local community participated.

During the second year, we put more emphasis on the kind of knowledge that emerges from actions. Gatherings, walks, events and urban actions are meant to engage more directly with the camp condition. What is at stake in these interventions is the possibility for the participants to realise projects in the camps without normalising their exceptional conditions and without blending them into the surrounding cities. After over 66 years of exile, the camp is no longer made up of tents. The prolonged exceptional temporality of this site has paradoxically created the condition for its transformation: from a pure humanitarian space to an active political space, it has become an embodiment and an expression of the right of return. The initiatives bear the names of this urbanity of exile: the garden, the pathways, the municipality, the suburb, the pool, the stadium, the square, the unbuilt and the bridge. The very existence of these common places within refugee camps suggests new spatial and social formations beyond the idea of the camp as a site of marginalisation, poverty and political subjugation.

Campus in Camps today is made of two essential and interdependent pillars. The first pillar is constituted by self-organised courses established according to the participants' interests and the camp's requirements. The second pillar is the Consortium formed by local and international universities: Goldsmiths University (London – United Kingdom), Mardin Artuklu University (Mardin – Turkey), Leuven University (Leuven – Belgium), Birzeit University (Birzeit), International Art Academy (Ramallah), Dar El Kalima (Bethlehem) are offering courses, seminars and workshops in Campus in Camps for refugee and non-refugee students. These activities are in constant dialogue with the Popular Committees of Southern West Bank Refugee Camps and UNRWA. These strategic partnerships have already been tested in the past years and have, for the first time, brought together institutions and organisations that rarely work together. At the end-of-year public presentations in 2013, the Director of UNRWA

Operations in the West Bank, Felipe Sanchez, described Campus in Camps as inspirational. "We hope to replicate this effort across the West Bank", he said, "Campus in Camps has connected people to people, institutions to institutions and camps to other camps".

Campus in Camps continues to work as an educational platform that connects young generations in the camp and other generations in the surrounding cities and universities, in order to break the isolation surrounding refugee communities and to offer refugees a platform to engage in a positive and productive way with the rest of the world. Moreover, Campus in Camps aims to create a generational leadership able to introduce new ideas and initiatives in camps that can challenge stereotypes and dominant power relations.

Campus in Camps does not follow or propose itself as a model but rather as public space in formation. Among the several urban interventions realised within Campus in Camps, the construction of the Concrete Tent in Dheisheh refugee camp maybe condenses the limits, paradox and potentiality of our pedagogical approach.

The camp as a heritage site

In December 2013, as part of a collective investigation about *The Unbuilt*[6] in the camp, Campus in Camps' participants found a plot of land in Dheisheh refugee camp called the 'three shelters'. The three shelters' site consisted of three original 1950s UNRWA-built structures (three rooms, one communal toilet and a water reservoir) that were still standing. The plot, no longer in use and closed behind a gate, narrates the camp's foundation and its history. How are we to reconcile this condition with the fact that the camp is always understood and described as a temporary situation of the present with no past, as something that has been established in order to be quickly dismantled and destroyed? Camps are built on the destruction that began in 1948, and for this reason they are 'historical sites' that are constantly destroyed and rebuilt. Refugee camps are also a reconstruction of the demolished villages, re-assemblages of people and social relations. Camps are the embodiment of the Palestinian struggle to exist. Yet it seems that we consider their importance only when they are demolished. Only when they cease to exist.

For instance, when Nahr el-Bared refugee camp in Lebanon was destroyed during the battle between the Lebanese army and the Islamic militias, Palestinian refugees promptly demanded its reconstruction. And they did so not by asking for tents, but by demanding the exact reconstruction of their concrete houses that were built through several years of sacrifice. The same happened after the 2002 invasion of Jenin refugee camp. Here, the significance of the camp and the rebuilding of its exact structures only began to surface once it was lost through military violence.

Further, how do we make sense of the demands of Palestinian refugees in Lebanon to 'Return to Nahr el-Bared camp'? Or, in the case of Syria, what do Palestinian refugees mean when demanding the 'Return to Yarmouk camp'? What does it mean to demand to return to a space never intended for permanence and without a history?

But how can a space that is supposed to be dismantled and disappear actually even have a history?

At this point we understood that claiming that the camp has a history, and a history that needs to be preserved for its cultural, political and social values, was the best way to try to answer the question of 'what is a refugee camp today?'.

What appeared to us as a historical heritage in need of preservation was not just the architecture of the three shelters, it was also the immaterial culture and the meaning of a communal life that people experienced when living in these structures. In fact, we can argue that the entire camp embodies a unique form of a communal form of life against the humanitarian ideology that reduces refugees to numbers and statistics.

After surveying the project site, and in discussion with the local inhabitants, a collaborative design process about the possible preservation and transformation of the plot unfolded among the Campus in Camps' participants. Considering the value of the architectural structures that are anchored to the collective memory of the residents, a non-intrusive approach was selected for the preservation of the site as well as for bringing new uses to the space and, by extension, to the whole camp.

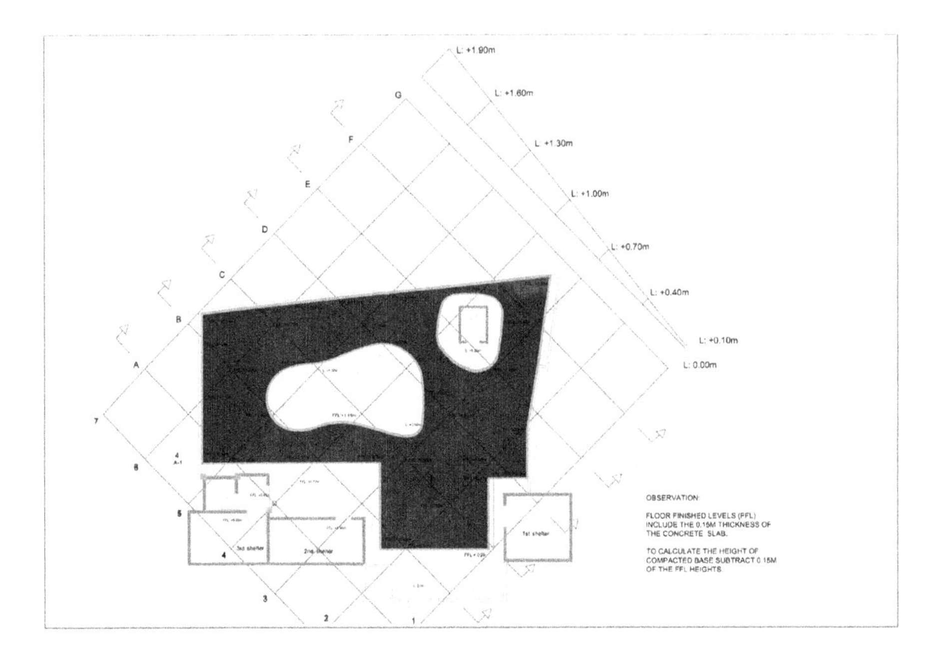

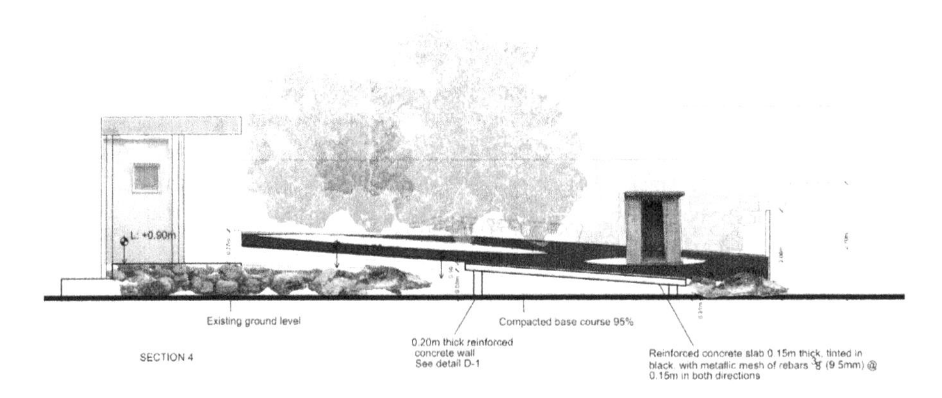

Figures 27.3 and 27.4 Plans for the conservation of the 'three shelters' in Dheisheh Refugee Camp

Source: DAAR.

The project was materialised as a sort of black frame surrounding the historical structures, a 15-cm-thick reinforced concrete platform, seemingly suspended but resting on a compact base course.

This was meant to leave the existing shelters as well as the communal toilet, water reservoir and the olive trees intact as a sign of respect for the past in this new beginning. The platform had a surface area of 120m^2 and the capacity to host activities with more than a hundred people. The black platform was like a theatre stage, ready to host community gatherings, music performances and collective rituals.

The participants of Campus in Camps spent several months in dialogue with the neighbour, and the owners of this site. Together they not only discussed the aim of this project but, with their consent, they began hosting activities such as concerts and screenings. It was crucial for the participants to involve the neighbour in the project. From there an agreement was signed between the popular committee and the owners.

Construction plans began with the excavation for the foundation of the project. After 10 days, one member of a large family prevented the labourers from working on the site. The family, the popular committee and leaders of the camp spent several weeks trying to find a solution. However, this family member stated that, despite the initial agreement to guarantee the collective use of the land for the two coming years, he had now decided to sell it realising that new attention was brought to this abandoned land. In a single night all the shelters were demolished.

Needless to say, this was an extremely frustrating moment for members of the refugee community who witnessed the destruction of this historical site, and a great disappointment for the youth of Campus in Camps, who lost the opportunity to see the materialisation of their new discourse around what constitutes a camp today.

That said, this incident also created a collective awareness of the importance of preserving the camp and its history. This incident brought a new understanding of the camp, no longer as a place without history, but rather as a place full of stories that can be narrated through its urban fabric. These stories have been repressed for fear of normalisation. This moment pushed us to start thinking about how the notion of preservation in a refugee camp is key in order to give meaning and historical importance to a life in exile. And conversely, it prompted us to think aloud about the concept of preservation and cultural heritage to question ways in which systems of values are decided and represented.

Claiming that life in exile is historically meaningful is a way to understand refugeehood not only as a passive production of an absolute form of state violence, but also as a way to recognise refugees as subjects of history, as makers of history and not simply victims of it. Claiming the camp as a heritage site is a way to avoid the trap of being stuck either in the commemoration of the past or in a projection into an abstract messianic future that is constantly postponed and presented as salvation. This perspective offers instead the possibility for the camp to be an historical political subject of the present, and to see the achievements of the present not as an impediment to the right of return, but on the contrary, as a step toward it. Claiming history in the camp is a way to start recognising the camp's present condition, and actually articulate the right of return.

Architecture is able to register various transformations that make the camp a heritage site. And in camps every single architectural transformation is a political statement. Therefore, architecture registers political changes.

When refugees decided to replace the tent with concrete walls, they were forced by the first rigid winters in the early 1950s to confront the necessity to protect their family from the adverse conditions and provide more decent living conditions. They were forced to accept the risk of making life in exile more stable and permanent.

To force people to live in miserable conditions does not bring them closer to return. To negate their right to a life in dignity today is just another form of violence imposed on the most vulnerable segments of Palestinian refugees. Here we need to seriously consider why it is that the right of return should negate the existence of the camp or call for its destruction. In other words, how can we articulate the right of return from the point of view of the condition of the camp?

After the destruction of the 'three shelters' site, the popular committee of Dheisheh offered to build the new structure in a plot of land inside the Garden of Al Finiq, a community centre entirely built by the refugee community. The garden is constantly being reshaped and transformed in an ongoing design process, an architectural laboratory for the camp.[7] The centre is named after the Finiq (the phoenix), from the legendary bird reborn from the ashes; in the same way the refugee community seeks to rebuild their culture on the ashes of destroyed villages.

Campus in Camps' participants saw in this new occasion to build a gathering space in the garden of Al Finiq, a possibility to materialise, to give architectural form to narrations and representations of camps and refugees beyond the idea of poverty, marginalisation and victimisation.

The project tried to inhabit the paradox of how to preserve the very idea of the tent as of symbolic and historical value. Because of the degradability of the material of the tents, these structures simply do not exist anymore. And so, the re-creation of a tent made of concrete today is an attempt to preserve the cultural and symbolic importance of this archetype for the narration of the Nakba, but at the same time, an attempt to engage the present political condition of exile.

The Concrete Tent is today a gathering space for communal learning. It hosts cultural activities, a working area and an open space for social meetings. We are aware of the danger of monumentalisation and oversymbolism, but we decided to take the risk in order to make architecture that engages with social and political problems that concern the refugee community that

Figure 27.5 Concrete Tent (2015). The tent is the architectural structure that over the years has been used as a basic element for the construction of refugee camps

Source: Campus in Camps/Anna Sara.

Figure 27.6 Concrete Tent (2015). For refugees the tent is the material manifestation of their temporary status in the camp. It is a form of architecture that has an expiry date, and guarantees the refugees their right of return. However, more and more refugee camps are no longer constituted of tents

Source: Campus in Camps/Anna Sara.

we work with. Too often architecture in our context is seen simply as an economic asset with no social and political value. Too often architecture has been humiliated in void formalism, to look green or sustainable or efficient, apolitical answers to political problems. Too often within the humanitarian industry, architecture has been reduced to answering to the so-called 'needs of the community'. Rarely has architecture been used for its power to give form to social and political problems and to challenge dominant narrations and assumptions.

The Concrete Tent deals with the paradox of a permanent temporality. It solidifies a mobile tent into a concrete house. The result is a hybrid between a tent and a concrete house, temporality and permanency, soft and hard, movement and stillness. Importantly, the Concrete Tent does not offer a solution. Rather, it embraces the contradiction of an architectural form emerged from a life in exile.

Conclusion

To conclude, in this article I aimed at broadening the investigation on how knowledge can be grounded in action and be embedded and visible in urban interventions in refugee camps. Based on experiences of Campus in Camps educational programme, I aimed to show and reflect on the ways in which universities can strengthen their social role by paying attention to forms of knowledge production and urban interventions that often remain undetected by academic knowledge. I would like to think of Campus in Camps not as an isolated educational experiment but rather as part of a long path that had stations in the schools influenced by Khalil

Al-Sakakini, where walks were considered a form of knowledge, or to the informal and clandestine learning environment established during the First Intifada in which people were learning from each other and in context.

Recognising the importance of connecting the Palestinian reality to global urgencies in comparative contexts such as the informal settlements in South America, India and the Eastern Mediterranean countries, the Campus in Camps team has started a series of collaborations with other groups and universities, calling these environments tree schools. A tree is, in fact, the minimal element to form a school, a gathering place for people who share similar urgencies. The tree, with its characteristics and history, is the device that creates a physical and symbolic common territory where ideas and actions can emerge through critical and independent discussion among participants. The first tree school was established in Bahia, southern Brazil, on the occasion of the São Paulo Biennale. It joined together activists, artists, quilombola intellectuals, landless movements and Palestinian refugees in discussions of forms of life beyond the idea of the nation-state and the meaning of knowledge production within marginalised sectors of society. After the Bahia experience, we have gone on to activate other tree schools in Shufat refugee camp in Jerusalem, in Cuernavaca, Mexico and in Curitiba, Brazil. Over the coming years, we would like to continue to activate similar environments in other contexts and with other groups who have already expressed interest: among these, a network of teachers and students from Beirut and Turkey, and a group of architects in Bogotà, Manama and Medellin, who have already proposed similar learning environments in slums.

Notes

1 Sandi Hilal and myself are the founding members and initiators of Campus in Camps, an educational programme based in Dheisheh Refugee Camp, with offshoots in other Camps in the West Bank. Our practice moves between art, architecture and pedagogy. Beside our interest in radical pedagogies, we are also the co-directors of DAAR, an architectural office and artistic residency programme that combines conceptual speculations and architectural interventions (www.decolonizing.ps).

2 The *Collective Dictionary* is made by the Campus in Camps' participants Marwa Allaham, Qussay Abu Aker, Alaa Al Homouz, Saleh Khannah, Shadi Ramadan, Ahmad Lahham, Aysar Dawoud, Bisan Al Jaffarri, Nedaa Hamouz, Naba Al Assi, Mohammed Abu Alia, Ibrahim Jawabreh, Isshaq Al Barbary, Ayat Al Turshan, Murad Owdah in dialogue with community members, associations and collaborators (see www.campusincamps.ps/projects/common-1/ [accessed 22 February 2017]).

3 The initiatives have been inspired through dialogue with Alessandro Petti, Sandi Hilal, Mmunir Fasheh, and activated with Tamara abu Laban, Bravenewalps, Ayman Khalifah, Matteo Guidi, Sara Pellegrini, Giuliana Racco, Diego Segatto, Dena Qaddumi (see www.campusincamps.ps/projects/02-the-square/ [accessed 22 February 2017]).

4 Guest professors include artists, architects, theoreticians, lawyers, scholars and policy experts such as: David Harvey, Michel Agier, Ruba Saleh, Basel Abbas, Ruanne Abou-Rhame, Wilfried Graf, Tariq Dana, Felicity D. Scott, Mohammed Jabali, Moukhtar Kocache, Hanan Toukan, Shadi Chaleshtoori, Jeffrey Champlin, Manuel Herz, C.K. Raju, Fernando Rampérez, Emilio Dabed, Samer Abdelnour, Sari Hanafi, Michael Buroway, Gudrun Kramer, Sandi Hilal, Muhammed Jabali, Munir Fasheh, Aaron Cezar, Pelin Tan, Thomas Keenan, Shuruq Harb, Umar Al-Ghubari, Khaldun Bshara, Jawad Al Mahal, Ayman Kalifah (see www.campusincamps.ps/projects/the-house-of-wisdom/ [accessed 22 February 2017]).

5 www.campusincamps.ps/projects/reading-fanon-in-palestine-today/ [accessed 22 February 2017].

6 Isshaq Al Barbary, Ahmad Al Lahham, Aysar Al Saifi, Qussay Abu Aker (2013), *The Unbuilt: Regenerating Spaces*, Campus in Camps, Dheisheh Refugee Camp (see www.campusincamps.ps/projects/09-the-unbuilt/ [accessed 22 February 2017]).

7 Qussay Abu Aker, Naba' Al Assi, Aysar Al Saifi, Murad Odeh (2013), *The Garden: Making Place*, Campus in Camps, Dheisheh Refugee Camp (see www.campusincamps.ps/projects/01-the-garden/ [accessed 22 February 2017]).

28

At the coalface, *take 3*[1]

Re-imagining community–university engagements *from here*

Tanja Winkler

Introduction

We learn how to cut down trees by cutting them down.

(An African proverb)

Community–university engagements expose students to 'real world' complexities by allowing them to explore *the here*: "A world students will actually work in [as opposed to] a hypothetical world" (Connell 2009: 225). Engagements also expose students to a range of skills and values that cannot be acquired through academic study alone (Pain *et al.* 2013). Above all else, engagements allow for the co-production of context-specific knowledge. And when engagements are purposefully conceptualised as transformative initiatives that encompass values of democracy, reciprocity, power sharing and social justice, they can become sites of empowerment for community partners and students alike (Boyer 1996). Thus, for those of us who facilitate community–university engagements through our studio-based or other courses, we do so because we hope to transform our teaching and learning endeavours through collaborative praxes that challenge hierarchical modes of knowledge production (Boyer 1996; cf. also Duminy *et al.* 2014; Greenwood 2008; Saltmarsh *et al.* 2009).

For all of these reasons, Ernest Boyer's (1996) "scholarship of engagement" is receiving much attention in higher education policy circles across various world regions. In the South African context, in particular, engaged scholarships are identified as "a means to redress past inequalities" (RSA 1997: 7). And in this situated *here*, the idea of engaged scholarship has become intrinsically linked to the *social purpose* of higher education institutions, which is "to actively contribute to the forging of a critical and democratic citizenship" (Badat 2013: 1).

However, "engagements without reciprocity and democratic purposes have a tendency to become simultaneously fashionable and disengaged" (Greenwood 2008: 332). Such engagements also tend to reflect the dominant culture of higher education that is characterised by scientific, rationalised, objectified or technocratic knowledge, where approaches to problem-solving are shaped by specialised expertise alone. By relying solely on the expertise of the academy, solutions to identified problems are applied *to* or *on* community partners (Reardon 2000; Saltmarsh *et al.* 2009). These concerns immediately alert us to the fact that there are different types of community–university engagements. For analytical purposes, engagements might then be placed along a continuum. At one end of the continuum, we find what I identify as more *transformative* types of engagements, whereas at the opposite end we find *instrumental engagements* that remain rooted in more traditional and hierarchical modes of knowledge production.[2] But, a narrow focus on instrumentalism tends to negate possibilities for *real* change through democratic, reciprocal, equitable and socially just outcomes.

Transformative engagements, in contrast to *instrumental engagements*, embody the idea of learning and doing "in the company of others" (Saltmarsh *et al.* 2009: 7). This is not to suggest that technical or scientific knowledge has no value. Rather, such engagements allow for a destabilisation of dominant knowledge claims and exclusionary positions by valuing other forms of context-specific knowledge. Here, knowledge is explicitly co-created via processes that are inclusive and respectful, while the political dimension of knowledge production is simultaneously addressed. And by valuing multiple knowledge claims in situated contexts, we might inspire students to become empathic and reflective practitioners who are capable of examining their own professional values when learning *from* and *with* community partners. Overt shifts from *instrumental* to *transformative engagements* then necessitate collaborative efforts to facilitate a public culture of democracy. This has epistemological, curricular and pedagogical implications, for the only way to develop values of democracy is to practice democracy as part of one's education (McLaren 1989; Saltmarsh and Hartley 2011). Arguably, values of democracy are much needed in contemporary African settings where socio-spatial, political and economic inequalities remain part of our everyday realities.

Yet, in order to facilitate *transformative engagements* a number of concerns necessitate thoughtful consideration. These are additional concerns to those more often discussed in the engagement literature, like, for example: Access to community partners; time to build trust with community partners; ensuring transparency and accountability throughout an engagement; and sustaining engagements beyond university timetabled projects. In this chapter I identify at least four additional concerns for *transformative engagements*. These are derived from ongoing critical reflections of our well-intentioned collaborations with community leaders from Europe, an informal settlement located in Gugulethu, Cape Town. They include the role of teaching while simultaneously hoping to enable transformative outcomes in Southern contexts; an inadvertent 'slippage' between facilitating *instrumental* and *transformative engagements*; students' anxieties arising from engagements with community partners; and the timing of engaged scholarship in the curriculum. Before discussing these concerns, a brief explanation of the project's history is required.

Our story 'from here'

In December of 2010, a local NGO – the Community Organisation Resource Centre (CORC) – approached faculty in the Masters of City and Regional Planning Programme at the University of Cape Town to assist them and community leaders in crafting *in situ* upgrading proposals for a state owned parcel of land which had become home to 7,462 residents. The informal settlement of Europe, which emerged in the early 1990s, was established on top of a former waste disposal site that had been used by the municipality for the disposal of solid waste between 1956 and 1987. In 1987, the landfill site was closed by the municipality, but it was neither sealed nor capped. As a result, the entire site is contaminated by methane and other noxious gases. Despite these environmental hazards, residents' priority was to obtain security of tenure so that they may continue to live in Europe without fear of being evicted from the site by the City of Cape Town.

When CORC first approached us, it was this priority that was presented as residents' main concern and not the underlying contamination of the site. By December 2010, CORC had established a rapport with community leaders, and together they had completed self-enumerations of Europe. CORC was thus able to assist us with some of the challenges often discussed in the engagement literature. Collectively, we also imagined a true sharing of the processes and outcomes of our engagements, so that our 16-week studio project would avoid becoming a mere teaching tool for students alone.

In the words of one student:

> It was clearly communicated to students from the onset that collaborative planning was integral to the learning objectives, and that community partners needed to be actively involved during each stage of the process: From identifying the issues under study to collaborating on desired outcomes; from data collection and analyses to the development of proposals. Essentially, the project aimed to expose students to the values of learning and working *with*, as opposed to *for*, community partners. This was seen as a valuable process toward integrating planning theory with practice.
>
> *(Bassa, cited in Bassa* et al. *2015: 423)*

At the first meeting with community leaders in 2010 – and prior to the involvement of the first cohort of students – I, as the studio facilitator, cautioned that the students who would be involved in the project were yet-to-be-trained practitioners with no prior experience of developing suitable, let alone implementable, proposals. In other words, our collaborative project would be students' first exposure not only to democratic planning practices, but also to the field of planning in general. By means of this caution I hoped to avoid raising unattainable expectations. Regardless of my cautions, many of our 2011 efforts failed to meet community leaders' security of tenure expectations, because leaders rejected students' proposals to rehabilitate the land prior to the implementation of engineering services and permanent housing structures (cf. Winkler 2013). It thus took two years before we were invited by leaders to return to Europe, and to continue working on some of the planning proposals we began in 2011. In the interim, we were asked by CORC and community leaders of Langrug (another informal settlement located within the municipal boundaries of Stellenbosch) to engage in similar processes to those initiated in Europe.

In contrast to the Europe experience, the Langrug project was deemed to be a success by community leaders, CORC, municipal officials and students alike. Various reasons led to this success (cf. Winkler 2017). Due to a word limitation, however, I need to confine these to two pertinent aspects. First, the Langrug project was initiated and led by a group of proactive and self-empowered leaders who embraced ideas of reciprocal learning and the co-production of knowledge, and who were committed to the transformative potentials of the project from the outset. Second, we explicitly sought to work *with* municipal officials and local politicians as equal stakeholders of the project. This, in itself, was a radical departure for community leaders, as informal settlement residents in South Africa have become accustomed to more insurgent and oppositional forms of action in response to failed state interventions. Yet, by working *with* officials and politically powerful groups – who, in turn, were able to guide our actions through existing legislative and political structures – the Langrug project gained the necessary political support for transformative outcomes that included, for example, securing residents' tenure to remain in Langrug. Explicitly stated, officials from the Municipality of Stellenbosch possessed *established capacities* that were invaluable to the project, because these capacities encompassed some sense of assurance for the types of actions proposed. Moreover, assurance for what we were doing served to empower all participants of the project to take responsibility for the different tasks and actions that were assigned to them at the start of the project. But such *established capacities* were sorely missing in Europe.

Let us now turn to discussions on our second attempt, in 2014, to facilitate democratic, reciprocal, equitable and socially just engagements in Europe. To initiate discussions, I begin with an excerpt from a student's reflections that captures both students' and community leaders' overall sentiments.

> During the first few days of this two-year degree programme, I remember seeing images of the planning proposals that the 'Class of 2012' developed with the residents of Langrug. These were images of a self-actualised community who were demonstrably taking a lead in decisions that would shape the future of Langrug. I recall determining inwardly that our 'Class of 2014' would aim to match those achievements. However, what transpired in Europe was a ragged and rather dispiriting process of community engagement that seemed to produce more friction than understanding. In my view, it contributed little of value to the community or to our learning about how to facilitate participatory initiatives in a context of severe hardship and spatial injustice.
>
> *(Petzer, cited in Bassa* et al. *2015: 424)*

Deeper assessments of Brett Petzer's sentiments reveal at least four concerns that facilitators might need to grapple with if *transformative engagements* are sought.[3] The first concern spotlights the role and nature of teaching in contexts of everyday hardships. Here, I am referring to the fact that in many Southern contexts a sense of urgency to establish implementable actions tends to overshadow possibilities for experimentations. Yet, teaching and learning need to make allowances for experimentations. Let me explain by focusing on this identified concern before discussing 'inadvertent slippages', 'student anxieties' and 'the timing of engaged scholarships'.

The role of teaching in contexts of severe hardships and socio-spatial injustices

"Teaching is about the *creation* of capacities. [It] is a process that *creates* social reality" (Connell 2009: 225, my emphasis). And *created* capacities and social realities are not only based on *lived realities*. Rather, teaching also involves "the fabricated, invented and imagined nature of human realities that are not given, but that are, like fictions, made up" (Conquergood 1989: 83). As such, teaching tends to be a performative activity that accommodates kinetic understandings of imagined and real experiences (Lamm Pineau 1994). And while teaching might include both imagined and real experiences, these experiences are, nevertheless, set apart from the responsibilities and culpabilities associated with lived realities. This may be argued even if the lines separating performed experiences from lived realities are blurred.

My intention here is not to relegate teaching to something that is fictional or irrelevant to everyday realities. Rather, I am acknowledging that "effective teachers often tell stories that do not mirror the world" (Lamm Pineau 1994: 11). We do so because stories "hold out the promise of reimagining the world" (Conquergood 1989: 83). Through such re-imaginings we hope to inspire students to become critical thinkers who are capacitated to take up transformative positions for socio-spatial justice, equality and democracy. Essentially, teaching is an enactment of possibilities that necessitate experimentation, improvisation, innovation, reflexivity, agitation, irony, parody, jest and learning from mistakes (McLaren 1989). It "temporarily releases, but does not disconnect, us from workaday realities by opening-up spaces for deconstruction and reconstruction" (Conquergood 1989: 83). And by temporarily removing us from the responsibilities and culpabilities of everyday life, teaching provides a valuable medium – within the relative safety of the classroom – for confronting complex and difficult topics concerning poverty, socio-economic and spatial injustices, racism, sexism and homophobia found in contemporary societies (Lamm Pineau 1994).

Yet, as one student reminds us, community–university engagements "plunge [us] into the complicated reality that what we do, decide, omit or suggest have [significant] consequences for people other than ourselves" (Finn, cited in Pain *et al.* 2013: 36). They remove us from the relative safety of the classroom where everyday complexities can be explored. And in contexts of severe hardships and injustices – where actions necessitate some degree of certainty and where actors need to take responsibility for their actions – there is arguably no room for fiction, improvisation, trial and error or mistakes. For this reason, "experimentations and creative thinking beyond a certain point were impossible" in the Europe projects (Petzer, cited in Bassa *et al.* 2015: 425). But this concern is ignored in the engagement literature. Instead, scholars speak enthusiastically and uncritically of community settings as being "sites for experiential learning and growth" (Johnston 1997: 12) without taking cognisance of the fact that teaching necessitates the *creation of capacities*, whereas engagements in contexts like Europe necessitate some sense of assurance and *established capacities* if real change is the desired goal. So, while engagements might enable rich learning experiences for students, they do not necessarily, or always, hold out the same promise for community partners (Netshandama 2010; Winkler 2013). Notably, the presence of established capacities during the Langrug project created a space for experimentations (cf. Winkler 2017). By contrast, a sense of urgency to establish implementable actions without political support and established capacities, resulted in the facilitation of *instrumental engagements* in the Europe case.

An inadvertent 'slippage' between facilitating instrumental and transformative engagements

> Our class was to conduct a spatial planning exercise within Europe where residents live under the most parlous conditions imaginable. Even by the standards of informal settlements in Cape Town, Europe is a precarious and noxious environment for human habitation, because it is sited on a former rubbish dump from which methane and other pollutants seep into the Cape Flats Aquifer. This seepage is not only a threat to residents' health and safety, it is also a threat to the municipality's future water source. In fact, we discovered that the Cape Flats Aquifer could potentially provide 15 per cent of the Cape Metropolitan Area's future water needs, which is significant in a water-scarce context.
>
> *(Petzer, cited in Bassa* et al. *2015: 424)*

I now realise that the nature of the problem under study was too complex and politically fraught to yield *real* opportunities for transformation. Unlike the Langrug project, our collaborative efforts in Europe lacked political support from the City of Cape Town. Furthermore, the leaders with whom CORC had established a rapport were not representative of the many different community groupings found in Europe. There was thus little desire among diverse community groups to work with the leaders who had approached CORC in the first place. These leaders were also less interested in the transformative potentials of the project. Rather, from their standpoint, our project was perceived as a means to enhance their political credibility during an election year.[4] To make matters worse, and in the absence of established capacities from municipal officials, we resorted to the technical expertise found within the university in order to address the negative city-wide impacts and multiple complexities arising from the contaminated land.[5] This created not only a division between knowledge producers (the university) and knowledge recipients ('the community'), but also in a division between 'the language of the expert' and other forms of communication (Sullivan 2000).

> Our final planning proposals were presented to the community in an academic register of Standard English. This served to further heighten an already uncomfortable power dynamic between middle-class students and a socio-economically marginalised group of mostly Xhosa-speaking women and a few men.
>
> *(Petzer, cited in Bassa* et al. *2015: 425)*

Rather than exploring other forms of expression and communication, "academic knowledge flowed in one direction: From inside the boundaries of the university to the community" (Saltmarsh *et al.* 2009: 8). And, rather than becoming a site for the co-production of knowledge, our engagements became an instrumental end-in-itself that inadvertently marginalised opportunities for collective decision-making and power-sharing. Attempts at workshopping leaders' proposals on how to address problems of land contamination were met with denial of the problem by Europe's leadership. Despite an explicit desire to facilitate transformative actions, our project resembled a mere *instrumental engagement* with a narrow focus on 'production' alone.

> The challenge of producing rigorous responses to how residents and students could conceptualise future possibilities for Europe was curtailed by the fact that our project ran somewhat like an office, with a focus on production. There was no time or leeway to learn from critical reflections of the processes we adopted, and there was no time to explore alternative approaches of eliciting inputs from the community.
>
> *(Petzer, cited in Bassa* et al. *2015: 425)*

It was, therefore, "hard to distinguish our total Europe experience from that which is offered through *volunteer-tourism*" (Petzer, cited in Bassa *et al.* 2015: 426). This self-awareness among students served only to heighten their anxieties.

Students' anxieties resulting from engagements

Teaching and learning involve a great deal of emotional work that is exacerbated by fears of failure (Connell 2009). And even if engaged scholars embrace the idea that "the most important learning comes from critically reflecting on failure", for students 'failure' is more problematic given that "it is the one thing above all they are encouraged not to do in the rest of the curriculum" (Pain *et al.* 2013: 41). A credit-heavy studio project that was designed to include 'learning from failure' (however defined), proved to be extremely challenging for students, regardless of the fact that no student failed the course. Students are accustomed to an academic system in which assessments are pre-defined, uniform and predictable. The undefined and unpredictable nature of learning through engagements thus became a source of student anxieties, as the end product – and not the actual participatory process – remained their main focus of concern. In their minds, this end product was reduced to one grade that mattered to them, because academia is entrenched in a grading system where financial scholarships, bursaries and academic achievements are awarded based on grades. Hence, and in the words of a student, Matt Finn (cited in Pain *et al.* 2013: 37):

> The struggle and uncertainty about whether we would be able to do the planned project in time, the negotiation of differences and the friendships formed, the nervousness before meetings with community partners, and the sadness and relief of the last [meeting], all these experiences [were] reduced to one grade, and it is a number that feels like it matters.

Related to the problem of assigning grades to engaged scholarship, Finn (cited in Pain *et al.* 2013: 36) also draws our attention to the fact that students may, at times, feel emotionally incapacitated to participate in projects with community partners.

> It seems important to me that students should not feel compelled, by the group or grading structure, to participate to a level which would never be required of all community participants. The principle of ongoing, informed consent – which allows participants' choices about the level at which they wish to engage – should [also] apply to students. Sometimes we did not want to engage emotionally with something or say how it affected us personally. We would have preferred to remain detached and not have to own a position.

In the Europe project, students' fears of failure, the reduction of their experiences to grades, and the emotionally taxing nature of engagements were further compounded by a sense that "the community leaders with whom we were interacting seemed little invested in the project" (Petzer, cited in Bassa *et al.* 2015: 425). A perceived lack of investment in the project by community partners, as well as a lack of guidance from CORC and studio facilitators during community meetings, created additional tensions and feelings of being inessential to the wider *in situ* upgrading project.

> The lack of guidance we received from CORC during the more difficult phases of the project simply added to our frustrations. While CORC representatives were supposed to assist us during our interactions with the community, they were frequently absent or hours late for meetings. This resulted in tense meetings at which our concerns were dismissed as part of 'the territory' of community engagement. We were told by CORC members that community engagements are difficult, and that we need to learn how to navigate these difficulties. But we were never guided to do so. It felt as if we, in the eyes of the NGO, were yet another inessential constituency that needed to be managed in the most energy-efficient way possible.
>
> *(Petzer, cited in Bassa* et al. *2015: 425)*

Of equal concern, our 'outsider' position in Europe added to students' woes, as some community leaders had difficulties in treating students as equals. This is not to suggest that students were disrespected, but rather that some leaders requested financial assistance from students (thereby representing a carbon copy of our 2011 experiences). However, requests for financial assistance served only to shift the 'balance of power' during engagements. Instead of asserting power over the project, leaders' requests invariably placed them in vulnerable positions where they assumed that if they were to challenge students' proposals this would jeopardise opportunities for financial assistance. While *transformative engagements* are purposefully designed to destabilise traditional hierarchies between community and university partners, our 'outsider' position undermined community leaders' power over desired outcomes. Arguably, engagements confer illusions of democratically shared spaces, since our performing bodies inevitably carry inscriptions of differentiated status (Lamm Pineau 1994).

> The central failure seems to be that we expected people living close to, or below, the breadline to offer us extended unpaid assistance in rounding out our own middle-class education. We had nothing, whatsoever, to offer in return, apart from the tangential hope that representing Europe's challenges on paper might eventually hasten some form of state intervention.
>
> *(Petzer, cited in Bassa* et al. *2015: 426)*

Petzer's concerns about rounding out his own middle-class education stem from an innate awareness of our 'outsider' status. "This is where questions of privilege in education arise" (Connell 2009: 225). Such questions are fundamental concerns for engaged scholarships in Southern contexts where we cannot afford to underplay or ignore inscriptions of differentiated status if we hope to learn *with* community partners. This brings me to the last concern discussed in this chapter.

The timing of engaged scholarship in the curriculum

> We were led into Europe as a group of Masters' students whose total exposure to planning theory and basic planning skills amounted to less than a week of formal lectures. Our technical skills-base – namely: How to represent space visually; how to use essential software; and how to employ GIS – resided, essentially, in those students with undergraduate degrees in architectural and geographical studies. These skills were to be acquired and tested concurrently in the very same moment as our cursory lessons on site analysis and representational techniques took place. Combined with the pressures of community interactions itself, the result of this timing was that we, as individuals, failed to build up a general skill set. There simply wasn't enough time to 'master' basic planning skills while simultaneously hoping to meet community leaders' and CORC's (often different) expectations. There was also little time for mutual-learning and knowledge sharing amongst students, let alone knowledge sharing with the community.
>
> *(Petzer, cited in Bassa* et al. *2015: 425)*

Ironically, engagements were purposefully conceptualised to take place in the first semester of the first year of study precisely because I hoped students would approach the project without predetermined, expert-driven knowledge. I imagined that a more 'naive' and hopeful understanding of planning would encourage students to draw extensively on local knowledge for the purpose of enabling reciprocal learning (as was the case in Langrug). But students' critical reflections suggest otherwise. The timing of our engagements was undoubtedly problematic, as students needed more time not only to acquire basic planning skills, but also to learn about the values of *transformative engagements.*

Conclusion

The aim of this chapter is to reimagine engagements by critically reflecting on "mistakes [that] lay the foundation for future social learning" (Reardon 2000: 72). One approach to this end is to purposefully shift pedagogical practices from facilitating mere *instrumental engagements* to facilitating democracy, reciprocity, power sharing and social justice through *transformative engagements.* Thus, *from here,* I identify at least four concerns for *transformative engagements* in addition to the challenges more often discussed in the engagement literature. These concerns include paying attention to the performative role of teaching, and the difficulty of enabling experimentation, improvisation and learning from mistakes in contexts of severe hardships. Such attentions also necessitate being cognisant of the types of problems under study in order to avoid inadvertent 'slippages' between facilitating instrumental and transformative outcomes. More successful engagements, from a transformative standpoint, tend to be those with established capacities and political support for creative ideas. Students' manifold anxieties that arise from collaborative projects, and the timing of engaged scholarship in the curriculum also surface as concerns that necessitate attention. An awareness of these concerns might allow us to conceptualise and implement *real* transformative endeavours: Namely, the kinds of endeavours desired *for here*.

I began this chapter with an African proverb that encapsulates a pedagogy of learning-by-doing. As a final statement, I end with another proverb that inspires ongoing quests to improve transformative learning despite 'failure'.

> *To get lost is to learn the way.*
>
> *(An African proverb)*

Notes

1 Inspired by cinematography where *takes* refer to successive filming shots, 'take 3' in the title denotes a third iteration of critical reflections on community–university engagements. See Winkler (2013) and Bassa *et al*. (2015) for previous 'takes' and critiques.

2 Instrumental engagements are often reduced to public relations initiatives and marketing ploys. It is therefore argued by some that global competitiveness – from which universities seem unable to escape – necessitates higher world rankings, and a demonstration of a university's 'civic purpose' has become one means by which universities advance their rankings (Porter 2015).

3 At the end of the project, when I invited students to co-publish a reflective article of their learning experience in Europe, the class turned to Brett Petzer to capture their sentiments.

4 The National Government elections of 2014 took place nine weeks after we began the project in Europe. Interestingly (and strategically from leaders' standpoints), we were initially invited by the same leaders to engage with them during the Municipal elections of 2011.

5 Municipal officials were repeatedly invited to participate in the project. They, however, turned down our invitations.

References

Badat, S. (2013) "Eleven theses on community engagement at universities", The Social Responsibility of Universities: Community and Civic Engagement, *4th International Exhibition and Conference on Higher Education* (4-IECHE), Riyadh, Saudi Arabia, 16–19 April. Available: http://contentpro.seals.ac.za/iii/cpro/DigitalItemViewPage.external?lang=eng&sp=1015946&sp=T&suite=def [accessed 12 January 2016].

Bassa, F., Petzer, B. and Winkler, T. (2015) "At the coalface, take 2: Lessons from students' critical reflections", *Interface: Planning Theory and Practice* 16(3): 409–434.

Boyer, E. (1996) "The scholarship of engagement", *Journal of Public Service and Outreach* 1(1): 11–20.

Connell, R. (2009) "Good teachers on dangerous ground: Towards a new view of teacher quality and professionalism", *Critical Studies in Education* 50(3): 213–229.

Conquergood, D. (1989) "Poetics, play, process and power: The performative turn in Anthropology", *Text and Performance Quarterly* 9(1): 82–88.

Duminy, J., Andreasen, J., Lerise, F., Odendaal, N. and Watson, V. (2014) *Planning and the Case Study Method in Africa*, Basingstoke and New York: Palgrave Macmillan.

Greenwood, D. (2008) "Theoretical research, applied research, and action research: The deinstitutionalization of activist research". In C. Hale (ed.) *Engaging Contradictions: Theory, Politics, and Methods of Activist Scholarship*, Berkeley, CA: University of California Press, pp. 319–340.

Johnston. M. (1997) *Contradictions in Collaboration: New Thinking on School/University Partnerships*, New York: Teachers College Press.

Lamm Pineau, E. (1994) "Teaching is performance: Reconceptualizing a problematic metaphor", *American Educational Research Journal* 31(1): 3–25.

McLaren, P. (1989) *Life in Schools: An Introduction to Critical Pedagogy in the Foundations of Education*, New York: Longman.

Netshandama, V. (2010) "Quality partnerships: The community stakeholders' view", *Gateways: International Journal of Community Research and Engagement* 3: 70–87.

Pain, R., Finn, M., Bouveng, R. and Ngobe, G. (2013) "Productive tensions: Engaging geography students in participatory action research with communities", *Journal of Geography in Higher Education* 37(1): 28–43.

Porter, L. (ed.) (2015) "Partnerships of learning for planning education: Who is learning what from whom? The beautiful messiness of learning partnerships", *Interface: Planning Theory and Practice* 16(3): 409–434.

Reardon, K. (2000) "An experiential approach to creating and effective community–university partnership: The East St Louis action research project", *Cityscape* 5: 59–74.

Republic of South Africa (RSA) (1997) *Education White Paper 3: A Programme for the Transformation of Higher Education*, Department of Education, Pretoria: Government Gazette, 15 August.

Saltmarsh, J. and Hartley, M. (eds) (2011) *To Serve a Larger Purpose: Engagement for Democracy and the Transformation of Higher Education*, Philadelphia, PA: Temple University Press.

Saltmarsh, J., Hartley, M. and Clayton, P. (2009) *Democratic Engagement White Paper*, Boston, MA: New England Resource Center for Higher Education (NERCHE).

Sullivan, W. (2000) "Institutional identity and social responsibility in higher education". In T. Ehrlich (ed.) *Civic Responsibility and Higher Education*, Phoenix, AZ: Oryx Press, pp. 19–36.

Winkler, T. (2013) "At the coalface: Community–university engagements and planning education", *Journal of Planning Education and Research* 33(2): 215–227.

Winkler, T. (2017) "The 'radical' practice of teaching, learning, and doing in the informal settlement of Langrug, South Africa". In H. Rangan and M. Kam (eds), *Insurgencies and Revolutions*, RTPI Library Series, London: Routledge, pp. 219–228.

29

Co-learning the city

Towards a pedagogy of poly-learning and planning praxis

Adriana Allen, Rita Lambert and Christopher Yap

Introduction

> "From now on, I'll describe the cities to you", the Khan had said, "in your journeys you will see if they exist". But the cities visited by Marco Polo were always different from those thought of by the emperor. "And yet, I have constructed in my mind a model city from which all possible cities can be deduced", Marco answered. "It is a city made only of exceptions, exclusions, incongruities, contradictions. If such a city is the most improbable, by reducing the number of abnormal elements, we increase the probability that the city really exists. So I have only to subtract exceptions from my model, and whatever direction I proceed, I will arrive at one of the cities which, always as an exception, exists. But I cannot force my operation beyond a certain limit: I would achieve cities too probable to be real".
>
> *(Calvino 1972: 44)*

In *Invisible Cities*, Calvino reimagines the conversations between the Venetian merchant Marco Polo and Chinese Emperor Kublai Khan. Exploring the incongruences that emerge from contrasting the exceptions that make each city unique rather than universally apprehendable, Calvino produces an imaginary and invisible city that is all cities.

Literary philosophers like Calvino help us to grasp the possibility of plural learnings through miniature narratives of the city, constructed by characters whose learning draws on complex bodies of knowledge, experiences and imaginations to explain how cities work, how they change, and why. Calvino never claims universalising explanations, but rather invites the reader to interrogate the junctions between these micro-narratives and the social construction of the urban world.

How do we learn the city? How does such learning relate to our learning of other cities? How can learning produce actionable knowledge that can impact upon urban planning, without claiming to be universal? How can learning activate and consolidate new ways of linking urban theory and planning praxis, and counteract the historical hegemony of global North academia?

Much has been written in recent years about the need to decentre urban theory and, with it, the importance of shifting from the production of academic knowledge 'about' the urban global

South to its co-production. But what does this mean in the actual assembling of urban planning pedagogies? With a few exceptions (McFarlane 2011; McFarlane and Robinson 2012; Watson and Odendaal 2012, among others), the question of how we approach the pedagogy of learning urban change has not yet been analysed in a systematic way, particularly in planning circles. We argue that when seeking to constitute new modes of learning the city, what matters the most is not where learning happens but how learning with others engages with questions of 'where', 'with whom', 'how' and 'why'.

The chapter draws on the pedagogy developed by the first two authors through the four-year journey of 'Learning Lima', a co-learning alliance established by the Bartlett Development Planning Unit (DPU), University College London in 2012, which brings together researchers and international students from the Practice Module of the MSc in Environment and Sustainable Development (ESD)[1] with various institutions and collectives of the urban poor in Lima. This alliance emerged in response to a request by a group of Lima-based NGOs that had become aware of the DPU's work in previous learning platforms in Accra and Mumbai, who felt they could benefit from taking a wider and more radical perspective towards issues of social and environmental inequality by exploring how such issues were being addressed in academic and policy debates elsewhere.

A series of informal exchanges led us to a first unofficial visit to the city, where we met several of these NGOs and visited the neighbourhoods where they work. These exchanges led to the decision to focus our common endeavour on learning how unjust water trajectories manifest in different parts of Lima; where, for whom, and why. Since then, Learning Lima has included four cohorts of ESD students – approximately 200 students – interns from Peruvian academic institutions, and staff from four NGO partners: Foro Ciudades para la Vida, Instituto de Desarrollo Urbano (CENCA), Centro de Investigación, Documentación y Asesoría Poblacional (CIDAP), and Servicios Educativos El Agustino (SEA), in addition to hundreds of local residents and civil society groups from seven case study areas. Together these areas capture the full diversity of Lima, both in terms of ecological structures and the multiple everyday and institutional practices that drive urban change across Metropolitan Lima.

In this sense, Learning Lima constitutes not just a co-learning alliance but also a means for trans-local learning, understood as the set of unique practical and epistemic opportunities that arise from immersing ourselves in exploring the "plurality of connections between different places established both by traceable routes . . . and by immaterial and less traceable ones" (Soderstrom and Geertman 2013: 257). However, weaving such routes into a collective learning process that combines a high diversity of learners and geographies is neither an easy nor straightforward process. We approach such a pedagogical undertaking as a fundamentally political process that opens, in our view, multiple opportunities to explore new ways of conceiving, perceiving and living the city; to contrast and interrogate preconceptions and ultimately, to oxygenate the ways in which we connect urban theory and planning praxis, within a world made of differences.

Both London- and overseas-based members of Learning Lima have been involved in several aspects of teaching, research, planning and advocacy relating to pro-poor development and environmental justice in the city, often working in various capacities simultaneously. The convergence of their multiple engagements with the city makes the co-learning alliance dynamic, flexible, responsive and open. Moreover, the co-learning is sustained through time as teaching, action research and planning are designed to be cyclical and overlapping. Leadership shifts between the DPU, NGO partner organisations and civil society groups. The alliance places great emphasis on documenting the learning process and outcomes; building an incremental legacy that grows organically through inputs from its multiple constituent parts.

While the initial focus of Learning Lima was on water injustices, it soon become evident that some parts of the city, and its social fabric, were rendered invisible in official maps and were thus excluded from public policy and programmes. Similarly, an exploration of unjust water trajectories soon transformed into an investigation of everyday risks. Thus, over time, Learning Lima expanded its scope, particularly through two action-research projects led by the DPU in collaboration with local partners. The first, 'ReMapLima', used drones to generate up-to-date images of otherwise invisibilised settlements in the historic centre and steep slopes of the periphery. These images were used as a cartographic base for participatory mapping and community-led diagnosis of spatial patterns of urban risk. The second project, 'cLIMA sin Riesgo', explored the conditions that produce and reproduce risk accumulation cycles, how and where they materialise, and with what consequences for marginalised women and men, as well as their capacity to act individually, collectively and with state organisations.

This chapter offers some critical reflections on the methodological and epistemic conditions underpinning the pedagogical approach developed through the journey of Learning Lima, with specific reference to a number of interrelated learning pathways: learning translocally, co-learning; learning spatially; learning through individual and collective critical reflection; and the significance of practice-oriented learning for the production of actionable knowledge. The discussion draws on reflexive portfolios produced by ESD students and from observations and interviews conducted by the third author with members of Learning Lima in London and overseas.

From urban knowledge to urban learnings: situating co-learning in discourse and practice

The plurality of ways in which urban knowledge(s) and theory are produced has been the focus of much debate among urban theorists in recent years. In the work of Edgar Pieterse, AbdouMaliq Simone, Colin McFarlane, Jenny Robinson, Gautam Bhan, Sue Parnell and Vanessa Watson, among others, a common denominator is the search for more plural modes of knowledge production and a critical engagement with our theoretical constructs wherever we are. However, such a search for plurality requires an engagement not only with multiple 'wheres' but also 'whos', opening the reframing of urban knowledge through the assemblage of co-learning networks.

The notion of co-learning is rooted in a variety of concepts and pedagogical approaches, including but not limited to: 'communities of practice', defined by Wenger as "social leaning systems" (1998, 2000); "co-production" (Ostrom 1996); participatory action research; and critical pedagogy (Freire 1993). In recent years the term has become ubiquitous across a number of disciplines. Invariably positive, the use of the term correlates closely with the proliferation of 'partnerships' that characterise contemporary research agendas; across the global North–South, across disciplines, and across academic and non-academic institutions.

Despite the diversity of ways in which co-learning can be understood, it is possible to identify some common principles. First, co-learning challenges the individualist epistemic notion of knowledge as 'justified, true beliefs'. By appealing to social epistemology, knowledge is understood to be socially constructed, and so, laden with socio-political and historical presuppositions and biases (Goldman 2010).

Second, co-learning challenges the dichotomy between *codified* and *tacit* knowledge. Within a co-learning framework, tacit knowledge, including social and inherited practices, are valued equally with codified, 'scientific' knowledge, and to some extent the two should be considered inseparable (Polanyi 1966). This combination of knowledges is itself relational. As McFarlane (2011: 3) explains:

> If knowledge is the sense that people make of information, that sense is a practice that is distributed through relations between people, objects and environment … learning as much about developing perceptions through engagement with the city as it is about creating knowledge.

Broadly, a co-learning network can be characterised by a plurality of knowledges; a reciprocity of knowledge production and communication; a shared community of interest and practice; and a real-world engagement that extends beyond academic discourse. Co-learning supports a process of knowledge production within and across groups and organisations, creates spaces to challenge and contrast existing narratives and holds particular significance for the production of actionable knowledge.

Co-learning actors and alliances

The last decade has seen an increased recognition of the need to challenge global North-centred urban planning and urban knowledge production paradigms; characterised by the reconceptualisation of urban citizens as agents of their own development, rather than beneficiaries of interventions or service consumers. Foremost this requires a reformulation of the relationship and terms of engagement with universities in the global North. The co-development of a research agenda that produces knowledge by, with and for citizens in the global South requires a critical examination not only of the social-political-geographical 'spheres' of all 'learners' involved, but also of the relations between them. So, what do trans-local learning engagements mean for a Northern academic institution like the DPU? Do we embody the ways of a Northern institution or do we revert back to the experiences lived through our repeated engagements with different elsewheres? Does working in trans-local groups, with learners are often neither 'here' nor 'there', a different dynamic might emerge capable of disrupting the differentiation between 'observers' and 'observed' that often prevails in overseas study tours?

For the DPU, the development of learning alliances allows the establishment of fruitful connections across our teaching, research and planning praxis. It also offers a continuous reality check in our attempts to respond to our vision and mission while operating within a Northern knowledge production framework. For over 60 years the DPU has dedicated its work to challenging mainstream theorisations of urban change and the role of planning, as well as the hegemony of Anglophone knowledge and its self-proclaimed prevailing authority to produce urban theory. In practical terms, DPU does this by actively engaging in partnerships of equality with multiple centres of knowledge production and planning practice across the urban global South. Moreover, its international membership – both in terms of students and staff – feeds from and into highly diverse translocal trajectories for experiencing, understanding and working across the world.

The departing points chosen to interrogate both existing theorisations and empirical analyses of a particular city are fundamentally important in addressing the above questions. We therefore depart from the deconstruction and reconstruction of existing bodies of knowledge around specific issues and geographies from an environmental justice perspective, as a means to tease out what transformative change actually means in concrete circumstances. This involves interrogating urban processes not only through the maldistribution of 'goods' and 'bads' that characterises highly unequal cities, but also their roots. This perspective takes on a hyper-specificity in each context to the point that the diagnosis and strategies to disrupt unjust urban trajectories cannot exist anywhere else.

The nature of the alliances developed through the DPU's engagement with Lima (and with other cities in the past) is central to the process of learning. These alliances are sustained through an engagement that lasts not only through time but also through a number of distinct mechanisms, as exemplified by the intensive use of social media and the Learning Lima online platform, which is collaboratively developed and managed from London and Lima, as well as being open to contributions from other learners from other cities (Figure 29.1).

It is also important to reflect on the dynamic composition of Learning Lima. Through the platform, links are made across residents and their collectives situated across Lima spatially, as well as with activists, civil servants, local NGOs, academic researchers and planning practitioners, among others. Engagement with those coming from London creates new spaces for participation and critical engagement for heterogeneous urban actors. Indeed, in restoring old or developing new public fora, this process explicitly aims to challenge the socio-political structures that hinder genuine multi-stakeholder urban development planning. In supporting the exchange of ideas and bodies through the city, local and visiting learners navigate and make sense together of the complexity of the city and its parts through transect walks, mapping, workshops, discussion panels and the production of shared outputs such as bilingual online story maps, video documentaries, policy briefs, newsletters and so on.

While all participants are equal learners in this process, the risks, responsibilities and consequences of engagement are different. This means that the learning process weaves together a variety of relationships, some continuous, others sporadic. Some based on the tacit commonalities of living in the same city, others on the opportunity to explore contrasting experiences. Each of these relationships brings different challenges that need to be carefully addressed. For instance, a must in our pedagogic approach is that the process should deliver 'positive' impacts and gains for all, and so expectations about what constitutes such gains need to be discussed. This happens

Figure 29.1 From the desk to the field: ESD participants' first encounter with Lima

Source: Teresa Belkow.

through a careful discussion of the learning objectives that inform the Terms of References (ToR) adopted each year, which tackles the following questions: What will be the focus of the work and why? How does this build upon past work? What are the learning objectives for local communities, partners and participants in the MSc? What will be the outputs? And how will these be produced and shared?

Thus, the ToR are co-designed by local partner organisations, residents in Lima and academic staff and students in London, through ex-ante discussions and the evaluation of the outcomes achieved each year, primarily through face-to-face presentations and debates. Through these engagements we identify specific questions and themes that could drive the next phase of work, knowledge gaps, why these gaps are relevant and to whom.

While the ToR frames the work each year, through the actual process of co-learning, the rights, risks, responsibilities and gains for each participant are continuously negotiated. Thus, for instance, the right of local dwellers to engage in the fieldwork on their own terms becomes a responsibility for the students to manage and in turn an opportunity to learn how to avoid an extractive approach and engage sensitively in the field. Does someone want to be interviewed, photographed or filmed? Is a meeting or a transect agreed with respect to local inhabitants' daily routines? Are certain questions making people uncomfortable? Above all, are we all on the same page with regards to both the content of the issues explored and the manner in which this is done?

Similarly, people might want to share stories or denounce processes that would put them at risk if these are publicly disseminated and it becomes our responsibility to minimise and avoid such risks by taking a cautious approach to the information that is recorded and disseminated. Furthermore, local dwellers might push for an agenda that talks to individual interests but not necessarily to collective ones, and this needs to be interrogated and challenged beforehand but also through every step of the fieldwork.

In short, weaving so many different participants into a common learning arena brings multiple challenges that need to be constantly read and addressed. Such challenges are connected to avoiding purely instrumental and extractive relationships, the conflation of personal expectations and benefits with collective ones, or even the well-intended confusion between sensitive engagement and uncritical sympathy.

Power relations are constantly at play in any co-learning engagement and work in all directions. The only way they can be addressed is by learning to recognise and evaluate them to start with. Community leaders might first interpret the possibility of engaging with members of a British university as a means to obtain financial aid, academic staff might unconsciously approach their engagement on the ground as an opportunity to illustrate an intellectual personal pursuit, local NGOs might initially seek engagement with a Northern organisation as a means to raise their profile in political and policy fora, and the list goes on.

These possible misconstructions can only be avoided by spending considerable time and effort to comb all expectations in the actual design of the work to be done together, and addressed in more subtle ways by securing a repetitive and prolonged engagement that combines academic teaching, action research, planning and advocacy.

Formal discussions are complemented over time by tacit exchanges of knowledge, spoken, written, electronic and face-to-face; visual, scientific and poetic. As well as these links, a critical culture of reflexivity developed between diverse institutions and groups supports the transformation of planners and researchers from experts to learners and of local residents from spectators or beneficiaries, to active agents in the co-production of knowledge and praxis. Learning collectively and across scales plays a key role in such transformation. As expressed by one of the local founding members of Learning Lima:

> Without working collectively, it would be impossible to understand the city. Working with people most of the time and the different ways of seeing the cities for each of them … allows me to understand the different scales of the city, metropolitan Lima, the neighbourhoods, the blocks. It is only collectively that one can make sense of new understandings across scales.
>
> *(Interview with the Director of Foro Ciudades para la Vida, Lima, May 2014)*

Learning spatially: from 'here' to 'there'

> History has a geography, both actual and imaginary, real and figurative. Just as with modernity, which was about a geopolitics of knowledge and a putative teleology of history that localised all truly modern knowledge in Europe and that thus centered the historical vanguard of progress in the heartland of Europe, globalisation also traces geopolitical maps.
>
> *(Mendieta 2007: 18)*

Like geography, urban planning relies on all sorts of spatial distinctions, conceptualisations and assumptions to construct its knowledge. However, how we go about this when approaching the pedagogy of learning urban change is rarely discussed, either in geography or planning circles (Allen *et al.* 2015b).

Our contention is that the co-production of planning knowledge – and indeed of understanding how social and spatial relations coproduce one another – is intimately linked to the activation of co-learning through space and in place. Pedagogically, Learning Lima approaches this enterprise through the use of maps and mapping, not as mere representations of what is 'there' or 'here', but as a means to provoke new framings of urban change, linking epistemological, ontological and methodological questions on how we learn the city.

Figure 29.2 Reading neighbourhood maps together with inhabitants during a transect walk

Source: Luise Fishcer; PROLIMA (2014).

The spatial interrogation of the city offers an entry point to understanding urban injustices, and where and how these can potentially be disrupted. As spatial tools, maps can reveal what is otherwise invisible and the myriad relations that constitute the city. They help to analyse and communicate, but also to perform, because they support the actualisation of ideas (Corner 1999; Dodge *et al.* 2009).

Through Learning Lima we adopt three interrelated sites in the mapping process: the reading, writing and audiencing of maps (Allen and Lambert 2015). The strategic inclusion of various actors in the city within these sites brings to the fore a plurality of knowledges and contributes to reframing previous diagnoses and to identify strategic opportunities for transformative change. As one participant reflects: "I have learnt that the construction of a map is part of the social construction of knowledge . . . this for me leads to the power to get things done" (Interview with ESD student, Lima, May 2015).

Within the 'Site of Reading', we interrogate hegemonic representations of the city and the role these often play in fostering exclusionary planning processes. The overlaying and critical examination of existing maps becomes a means to capture the spatiality of injustice, making evident the socio-environmental power struggles driving these representations, as well as helping to identify who and what is left 'off the map' and why.

The 'Site of Writing' focuses on the collective decision of what to map, how to map and toward what end. It also encompasses the actual process of data gathering in the field and its representation. One important starting point is an open discussion of 'why' and 'what to map' together with local partners and residents.

Mapping is hereby used to foster critical reflection and awareness, helping to grasp the spatiality of problems at various scales and denounce otherwise invisible processes. As observed by one of the community mappers engaged in using free apps to survey her settlement with the support of smartphones:

> People often don't know what is happening at the back of their own settlement ... Working with this technology has meant that a lot of information was gathered about the risk areas. With the drone images, the leaders realised that new roads were being opened and they started to pay attention to the matter, raising awareness of their community and promoting the planning and safeguarding of open spaces.
>
> *(Interview with a female resident and community mapper from José Carlos Mariátegui, Lima, May 2015)*

Beyond the moment of collecting spatialised information, the community mappers together with the supporting NGOs, CENCA and CIDAP, take it upon themselves to deepen the data collection and analysis and devise strategies, exploring legal avenues and various coordinated efforts needed to disrupt the unwanted processes that increment risk for the inhabitants.

Finally, the 'Site of Audiencing' of maps involves making collective decisions on who should see the maps, where they should be displayed, and how to frame new interpretations emanating from the contrasting of existing and newly written maps. In the audiencing, the inclusion of a wide range of actors is sought to provide a scalar jump in the possibilities for advocacy and action. But how are these different forms of spatialised knowledge articulated to travel from the neighbourhood to the city? Towards this aim, we produce online story maps that capture quantitative and qualitative data in a single platform that can be further populated over time.

The three sites of the mapping process – the reading, writing and audiencing – all play a key role in facilitating the co-learning and co-production of knowledge for action if strategically designed and sensitively managed as a fundamentally political process.

Learning as an open-ended reflexive exercise

Recognising the significance of individual subjective insights as a form of learning, the ESD programme encourages students to identify and reflect upon what it terms 'ah-ha' moments; moments of understanding that significantly shift one's perspective. These moments are evocative of Kierkegaard's 'leap' and commitment to subjective truth as the highest truth. For Kierkegaard (2009) the highest truths were arrived at through individual epistemic 'leaps', so called because they are irrational; they cannot be deduced and cannot be taught. These insights can only be arrived at through individual reflection, and as such have unique epistemological significance, offering insights on our positionality vis-à-vis our learning.

Opportunities for critical reflection are instigated at various stops throughout the journey and reflection is pursued in a number of ways, starting with an exploration of the students' assumptions about urban socio-environmental transformation and the skills they require to become active players in pursuing transformative planning (Figure 29.3).

To support this process, students develop two personal portfolios that can take many formats, ranging from videos and posters, to written essays, blogs and diaries. The first portfolio allows students to reflect on their own ways of learning the city and the links they establish between familiar and unfamiliar contexts; cities they 'know' and cities that they are discovering. The second portfolio is produced after the actual fieldtrip overseas, and captures a significantly different learning mode, one that is shaped by a multiplicity of concrete encounters. As observed by one of the ESD students:

> As a citizen you look at Google maps aerial view and street view you see garbage piling up and you can see there is no water and you think "if they just brought the truck more to take the garbage, if there was just a little more water that would do". But when you look at it as a system you see there are topographical, geographical limitations on how water comes into the city and where garbage can be put, it becomes less about criticising and more about critiquing.
>
> *(Interview with a female ESD student, Lima, 2014)*

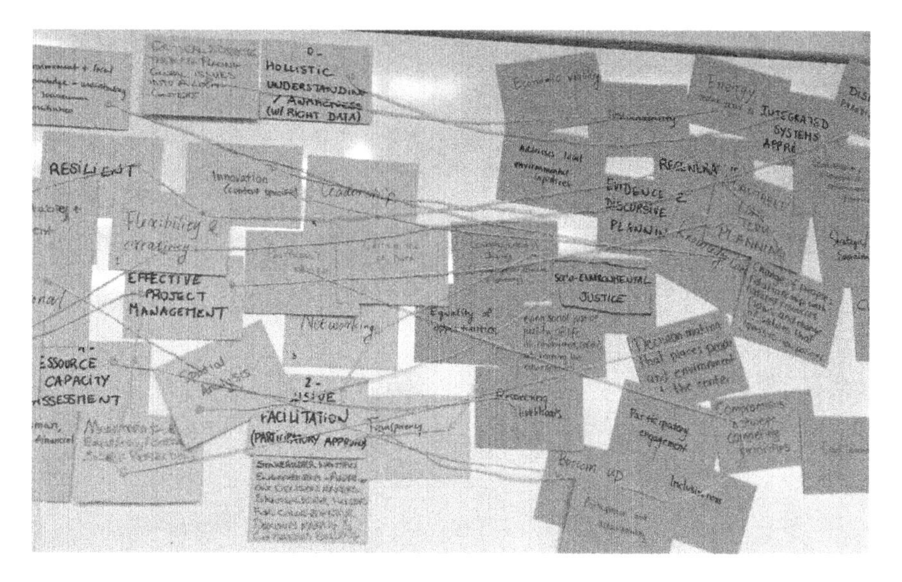

Figure 29.3 Mapping learning expectations on the first day of the ESD practice module

Source: Rita Lambert.

These reflexive portfolios first enable participants to become aware that issues of socio-environmental injustice are rarely, if ever, simple, and second, to realise that complex problems often require complex solutions that stretch well beyond the normative boundaries of statutory urban planning.

Moving from this individual way of learning we approach 'public learning' as a means to reframe the way the city is perceived, experienced and addressed, as a relational process by which ordinary citizens make sense of cities collectively. Such a process has the capacity not only to make the invisible visible and to raise public awareness, but also to foster spaces for collective dialogue and advocacy, to trigger imaginaries of how the city could be more just and resilient, to set precedents for transformative action and to consolidate institutional commitment towards that aim.

This has been done primarily through three simultaneous processes concurring within Learning Lima. The first concerns the active participation of those marginalised in the neighbourhoods studied (Lambert and Poblet 2016). The process allows those taking part in the fieldwork – both as interviewers and interviewees – to make sense of common and diverse experiences (Figure 29.4).

The second process refers to promoting forums for dialogue and using the data collected in the field to guide conversation towards concerted action. The third process relies on drawing public attention to a reality often ignored, offering reframed diagnoses through online story maps, the dissemination of video outputs and travelling public exhibitions that bring together both the co-learning process and outputs (Figure 29.5).

Figure 29.4 Community mappers of Barrios Altos together with cLIMA sin Riesgo surveyors walking into every compound to interview their neighbours

Source: Rita Lambert.

Figure 29.5 cLIMA sin Riesgo and ReMap Lima exhibition opening in November 2015

Source: cLIMA sin Riesgo.

From trans-local co-learning to the production of actionable knowledge

Planning, and indeed its search for knowledge to solve the problems faced by a city, can be understood as the ultimate act of taming the city; but "in this regard, discourses of planning are so bound up with hegemonic power structures that the power at work often becomes hidden or naturalised" (Allen *et al.* 2015a: 11). Whether such taming narratives are linked to the 'smart city' agenda or the assumption that building urban resilience to climate change should be an overriding priority, every attempt to contribute to urban planning theory and praxis is inevitably confronted by the simultaneous challenges of deconstructing the diagnoses from which it departs, and identifying strategies to transform urban injustices. Following the learning pathways discussed here constitutes one specific response to this challenge. We believe that the pedagogical power of the approach lies not just in the outputs produced, but more fundamentally in the learning outcomes achieved.

First, trans-local learning helps us to pause and think about the consequences of where we learn from and how. As Brazilian scholar Marcelo de Souza reminds us: "[n]o contribution in the field of social sciences, urban studies included, is free of 'accent', since every piece of knowledge directly related to social life is both culturally embedded and historically-geographically situated" (2012: 315).

As such, learning the city trans-locally requires us not only to be reflexive and explicit about our accents geographically, but also methodologically, intellectually and politically. The ESD approach to co-learning encourages reflection not only on individual and collective positionalities but also on the dialectic between theory and practice, in so far as they variously support and legitimise, or challenge and undermine one another.

Second, certain modes of learning the city are better suited to enabling actionable knowledge than others. For example, spatial learning and the process of mapping can be impactful as a mode of investigation and communication; enabling learners to engage with, rather than reduce, complexity. Additionally, the process of knowledge production within a co-learning network can itself mobilise and consolidate various organisations and actors.

Third, developing an ethics of practice is a key learning outcome of the ESD programme, both in terms of defining an ethical engagement and reflecting on the role of an urban development planner. Through engagement in a co-learning network, and through critical reflection, students develop an awareness of their own position in the context of international urban development. As observed by a former student: "The planner's role is not to bring knowledge into the system from externally; it is to facilitate the distribution of knowledge throughout the system such that solutions emerge from within it" (Female ESD student portfolio, 2014).

Conclusion

This chapter aimed to reflect on the emerging potentials of dynamic, translocal, university–practitioner alliances and a poly-learning pedagogy. The ESD Practice Module's pedagogical assumptions lead to a particular arrangement of learning pathways and learning outcomes. Central to its approach is the activation of trans-local learning and subsequent commitment to developing co-learning alliances, spatialised learning and critical reflection. The assemblage of learning pathways presented here is not exhaustive, but is intended to promote further reflection and discussion on the nature of poly-learning pathways, and the way they are socially, geographically and historically embedded within wider processes.

The diversity of learning modes adopted – reflexive, spatial, policy-oriented and narrative-based, among others – not only allows for the triangulation or enhancement of different knowledges and modes of learning, but also creates new spaces for critical reflection. It engenders a critical consciousness of the complexities and tensions between urban theories and planning practices, equipping learners with an open mind-set and concrete skills to activate refreshing perspectives on how cities are produced and on the role that planning can play in a more inclusive and progressive practice. A practice that does not seek homogenising explanations and solutions but rather an incisive and continuous commitment to learning urban change as a messy process and to be responsive to difference.

As the quote from Calvino in the introduction to this chapter reminds us, the process of learning the city emerges from our engagement with exclusions, incongruences and contradictions; no matter how far or close we travel, cities only exist through their exceptions.

Note

1 The ESD Practice Module is part of a full-time Masters planning degree. It runs over three academic terms, culminating in a period of overseas fieldwork in a city in the global South. Previous learning alliances have been established in Cairo, Mumbai and Accra.

References

Allen, A. and Lambert, R. (2015) "Learning through mapping". In B. Campkin and R. Ross (eds) *Urban Pamphleteer*, London: UCL Urban Laboratory, pp. 40–42.

Allen, A., Lampis, A. and Swilling, M. (2015a) *Untamed Urbanisms*, London: Routledge.

Allen, A., Lambert, R., Apsan Frediani, A. and Ome, T. (2015b) "Can participatory mapping activate spatial and political practices? Mapping popular resistance and dwelling practices in Bogotá eastern hills", *Area* 47(3): 261–271.

Calvino, I. (1972) *Le Città Invisibili*, Torino: Giulio Einaure Editore. First published in English in 1974 as Calvino, I. (1974) *Invisible Cities*. Translation by William Weaver, Orlando: Harcourt.
Corner, J. (1999) "The agency of mapping: Speculation, critique and invention". In D. Cosgrove (ed.) *Mappings*, Vol. 2, London: Reaktion Books, pp. 213–252.
Dodge, M., Kitchin, R. and Perkins, C. (2009) *Rethinking Maps: New Frontiers in Cartographic Theory*, New York: Routledge.
Freire, P. (1993) *Pedagogy of the Oppressed*, New York: Continuum.
Goldman, A. (2010) "Social epistemology". In *The Stanford Encyclopaedia of Philosophy*. Available: http://plato.stanford.edu/entries/epistemology-social/ [accessed 8 April 2015].
Kierkegaard, S. (2009) *Concluding Unscientific Postscript*, Cambridge: Cambridge University Press.
Lambert, R. and Poblet, R. (2016) "Policy brief No 2: Mapping to reduce urban risk", cLIMA sin Riesgo, The Bartlett Development Planning Unit, University College London. Available in English and Spanish: www.climasinriesgo.net [accessed 11 August 2016].
McFarlane, C. (2011) *Learning the City: Knowledge and Translocal Assemblage*, Oxford: Wiley-Blackwell.
McFarlane, C. and Robinson, J. (2012) "Introduction: Experiments in comparative urbanism", *Urban Geography* 33(6): 765–773.
Mendieta, E. (2007) *Global Fragments: Globalizations, Latinoamericanisms and Critical Theory*, Albany, NY: State University of New York Press.
Polanyi, M. (1966) *The Tacit Dimension*, New York: Doubleday.
PROLIMA (2014). *Plan Maestro del Centro Historico de Lima al 2025*. Lima.
Ostrom, E. (1996) "Crossing the great divide: Coproduction, synergy, and development", *World Development* 24(6): 1073–1087.
Soderstrom, O. and Geertman, S. (2013) "The translocal making of public space policy in Hanoi", *Singapore Journal of Tropical Geography* 34(2): 244–260.
Souza, M. L. de (2012) "Marxists, Libertarians and the city", *City* 16(3): 3015–3031.
Watson, V. and Odendaal, N. (2012) "Changing planning education in Africa: The role of the Association of African Planning Schools", *Journal of Planning Education and Research* 33(1): 96–107.
Wenger, E. (1998) "Communities of practice: Learning as a social system", *Systems Thinker* 9(5): 2–3.
Wenger, E. (2000) "Communities of practice and social learning systems", *Organization* 7(2): 225–246.

30

Learning to learn again

Restoring relevance to development experiments through a whole systems approach

Jigar Bhatt

Introduction

A planner's ability to know whether s/he is 'doing good and being right' requires a set of tools for judging what is 'good' and 'right'. In the past decade and a half one type of evaluative tool – the randomised control trial (RCT) – has come to dominate the way development projects for poverty alleviation are conceived and understood. Thus, paradoxically, while the space for development experimentation in the global South has considerably expanded, the space of creative knowledge production, or how to learn from such experiments, has drastically shrunk. This chapter briefly outlines the rise of the RCT as the experimental and evaluative 'gold standard' in development. It then takes issue with the internal-versus-external validity debate raging between RCT boosters and critics, suggesting that those serious about development and poverty should be focused on the more salient issue of model validity. There is an urgent need to reclaim the space that RCTs have recently monopolised with more relevant intervention and learning models to restore balance in development thinking and practice. But the question of what forms such models should take remains. A proposal presented here borrows from a whole systems approach that has a long tradition in the global South. The goal is not to dispense entirely with RCTs but to explore what a sincerely innovative, 'Southern' inspired approach to development projects' design and evaluation might look like.

The space of development experimentation

Since the Washington Consensus' crisis of legitimacy in the late 1990s, changes in the institutional landscape of development and global balance of power have altered the space for development experimentation. While development continues to unfold in the shadow of the Washington Consensus, a phase of 'post-paradigmatic' development unencumbered by 'pure' notions of central planning or unhindered free markets has allowed for new configurations of state, markets and civil society organisations (Watts 1993). Within this context, old relationships are being reworked and rescaled. A plurality of institutions, particularly non-governmental and civil society organisations, in both the global North and South, design and implement development and poverty alleviation projects (Richey and Ponte 2014).

The North–South development relationship following the Washington Consensus period is governed, if not completely in practice, at least in letter and spirit, by partnership, cooperation and the so called global South's ownership of its own development (Sen 2002). More importantly, these nations' increased wealth and administrative capacity for resource collection and mobilisation has allowed a marshalling of domestic resources for development (Savoy 2014). These trends have resulted, albeit in limited ways, in Southern planners' greater autonomy in initiating and designing development and poverty reduction interventions such as micro-finance and social protection programmes (Yunus 1999; Ferguson 2015).

The rapid rise of RCTs

If the space for development experimentation in the global South has expanded, the space of creative knowledge production, or how to learn from such experiments, has shrunk. Intervention heterogeneity is being met with methodological homogeneity in determining 'what works'. Over the last decade and a half, there has been an explosion of a particular epistemology – RCTs – to learn from development interventions in the global South. The RCT is an epistemic method that estimates the effect of an intervention on a treatment group relative to that of a control group, or counterfactual, which does not receive the intervention. The treatment and control groups are established through a process of random selection that aims to rule out any causes for observed outcomes other than the intervention itself. According to one development planning expert, RCTs have crowded out an appreciation for other methods and 'messy data' in Indian economics programmes (Seddon 2013). At the heart of RCTs' rise is the Massachusetts Institute of Technology's (MIT) Abdul-Jamaal Lateef Poverty Action Lab (J-PAL). Founded in 2003, J-PAL describes its mission as "to reduce poverty by ensuring that policy is based on scientific evidence" (Poor Economics 2016).

By early 2016, J-PAL had initiated 732 RCTs on five continents. It has already completed 485 of those (J-PAL 2016). J-PAL now has six global offices and has spurred partners committed to its mission, namely Innovations for Poverty Action (IPA) at Yale University and the Center of Evaluation for Global Action at the University of California Berkeley. These research centres regularly initiate their own research while also providing training and consulting services to major development, philanthropic and academic institutions across the globe.

This renaissance in RCTs has had a major impact on the practices of multilateral development institutions such as the World Bank, philanthropic foundations such as the Bill and

Table 30.1 Distribution of RCTs by world region

Region	*RCTs in IER 1994–2014 (%)*	*World's population (%)*	*World's poor (%)*
Sub-Saharan Africa	36	12	32
Latin America and the Caribbean	20	9	4
South Asia	19	23	35
Asia and the Pacific	17	32	26
Middle East and North Africa	5	5	1
Europe and CIS	3	13	3

Source: 3ie's Impact Evaluation Repository (IER) and World Bank's World Development Indicators.

Note: Based on the author's calculations of 1,553 RCTs conducted between 1994 and 2014 where a single region could be identified. CIS stands for Commonwealth of Independent States. Demographic figures represent averages between 2002 and 2012. Poverty measurement based on poverty headcount ratio at $1.90 a day (2011 Purchasing Power Parity figures).

Melinda Gates Foundation, bilateral donors such as the German aid agency GIZ and international nongovernmental organisations. For example, by 2010 the World Bank had initiated over 150 RCTs (Legovini 2010). This trend, however, has not been confined to international institutions. RCTs have certainly taken hold among local institutions of the global South. The International Initiative for Impact Evaluation (3ie), a global promoter of learning from RCTs, boasts over 30 members from the global South. Members range from a Rwandan policy institute and Pakistani university to Colombia's National Planning Department (3ie 2015).

Development methods and reflective practice

Methods are important to planners since they wish to know if they are 'doing good' and 'being right' (Hoch 1984). They serve as the bridge between knowledge and action. Many planners help identify a problem, such as lack of access to urban basic services, design a novel intervention, often through a pilot project, and use methods, like RCTs, to learn from that intervention for future expansion or scaling-up. If an intervention is not going as intended, planners may choose to apply lessons learned and adjust course.

This process of learning is tied to reflective practice. Central to reflective practice is the planner's ability to reflect *in* action, thinking what one is doing while one is doing it, and to reflect *on* action, or examine critically what one did after the fact (Schön 1983). Reflective practice is directly connected to the tools or methods used in judging what is 'good' and 'right'. In RCTs we have seen a return to a techno-rational approach of learning rooted in a linear, reductionist and scientistic approach to the world (Clarence 2002). In learning from development projects for poverty reduction, RCTs have become the dominant, 'gold standard' method against which all other learning approaches are judged (Harrison 2013).

Beginning in the 1980s advocates won hard-fought battles to bring participatory, constructivist and social justice-based approaches to development learning. Robert Chambers (1981) encouraged planners to 'put the last first' and Egon Guba and Yvonna Lincoln (1989) ushered in a 'fourth generation' of evaluation rooted in naturalistic knowledge and constructivist methodologies. Participatory rural appraisals and action research linked advocacy planning and cooperative learning (Chambers 1992). Qualitative methods began to share space with quantitative indicators and measures (c.f. Tendler and World Bank Operations Evaluation Dept. 1993). Reflective practice was iterative, rather than a one-off exercise, and knowledge was co-produced through a variety of methods.

These were all in response to a policy-analytic movement that had dominated development since the 1950s. Policy analysis is rooted in the application of microeconomic models of rationality to planning practices and cost-benefit approaches to evaluation research (Friedmann 1987). This approach, however, has never succeeded in resolving several of its central problems, namely, pinpointing a programme's single, specific purpose (Schön and Rein 1995). As one expert at the time pointed out: "Most sophisticated programme evaluations are not able to address the pivotal policy questions 'Are the programmes working, and if not why not and what could be done?'" (Schön and Rein 1995: 13). At the turn of the century, as development learning returned to a techno-rational approach, it endeavoured to resolve at least one of these fundamental flaws – to answer if programmes are working.

The RCT validity debate

The key reason RCTs have become so popular is that direct, linear causality has become the *sina qua non* of learning in development (Bhatt 2011). Structural reasons for this include the

New Public Management movement, which aims to graft private sector accountability regimes onto public sector institutions, and the evidence-based agenda of large global philanthropies (Moynihan 2009). Still, to understand the RCT's allure in development we must pay attention to causality and the RCT's implicit theory of change.

Economics has long enjoyed dominance in development practice and increasingly has come to dominate social science inquiry (Fine and Milonakis 2009). The chief discipline of economics concerned with issues of the global South is development economics. Randomised controlled trials have a rich history in agriculture and psychology (Forsetlund *et al.* 2007), but RCTs in development draw inspiration from their use and history in Western biomedicine. "You can put social innovation to the same rigorous, scientific tests that we use for drugs . . . These economics I'm proposing, it's like twentieth-century medicine" (Duflo 2010).

The rise and dominance of RCTs in development economics has not been without controversy – a vigorous debate has emerged between RCT proponents and opponents. The RCT proponents have enthusiastically advocated for their approach while the opponents have cried foul (Banerjee and Duflo 2008; Bardhan 2005; Deaton 2010; Duflo *et al.* 2008; Ravallion 2009). To understand the debate surrounding this paradigm shift one needs to look at how proponents are claiming to advance learning. They do not claim to have discovered some new determinant of growth, such as a particular policy, technology or institutional innovation. In keeping with the empirical turn in economics, they claim that their specific experimental approach will allow for internally valid, causal estimates that will provide policy makers the evidence they need; something previous econometric studies and practical policy models failed to do (Cohen and Easterly 2010; Milberg 2009). This unsettles many development economists, particularly those who believe their work is based upon a theory of causal mechanisms.

These causal claims allow for something statisticians call internal validity. This is achieved by randomly assigning the intervention to specified units, such as individuals, communities, schools, etc. A properly implemented RCT might allow a researcher to make the following claim: "Compared to the control group, the introduction of an intervention increased the development outcome(s) of those in the treatment group by a statistically significant 7 per cent". The key advantage of an RCT is the implementation of an intervention after the randomisation process is complete (Imai *et al.* 2013).

Opponents have retorted that despite the RCT's claim to eliminate bias, the methodology is not without its trade-offs: RCTs are too dependent on their environment, in other words, they may have internal validity but they suffer from the lack of external validity. Thus, there is no guarantee that a successful outcome as measured by the RCT will have a similar effect if implemented elsewhere. The concerns are whether, first, the treatment effect on a particular development outcome can be translated to other environments (e.g. a different city or neighbourhood), and second, the treatment effect may differ when the implementer is different (Banerjee and Duflo 2008; Deaton 2010).

Model validity

Internal, external or model validity?

The debate between internal and external validity has thus far has been an impoverished one. It has altogether ignored the fact that while RCT proponents claim to be atheoretical, inductive empiricists, they employ an implicit biomedical model of how development and planning works. Thus, the debate over RCTs in development has less to do with internal or external validity and much more to do with 'model validity'. Model validity refers to a 'paradigm fit', or the congruence between a philosophical or theoretical approach to change and the methods used for learning within it (Verhoef *et al.* 2002). Here assumptions about the theory of change, which includes assumptions about problem identification, intervention design and the planning environment and process take on added importance. Development interventions and learning based on RCTs are ultimately based on a biomedical model that involves thinking *symptomatically* rather than thinking *developmentally*.

Thinking symptomatically

As the prized methodology of Western biomedicine, the RCT brings some specific features of reductionist and linear thinking to address issues of poverty and development. In biomedicine a curative model infers a simple, direct causal pathway for the presence of symptoms and focuses on interventions that eliminate them (Jonas *et al.* 2006). For example, a painful rash is treated by prescribing a soothing ointment. This curative approach deals with the immediate cause of pain, thereby 'curing' the symptom while leaving the underlying causes ignored and unaddressed. Similarly, planners may identify a part of the city with a disproportionately high level of waterborne illness in children. An RCT proponent thinking symptomatically may suggest a pilot programme to install chlorine dispensers near water standpipes or market chlorine-based water treatment kits to households and test each intervention's impact on illness (Kremer *et al.* 2011). The waterborne illnesses could be successfully addressed but the root causes for why some parts of the city receive poorer quality water are not investigated and solutions for this imbalance or inequity go unexplored.

Thinking symptomatically can often involve confusing cause and effect. The Urban Services Initiative at J-PAL seeks to partner with institutions in the global South to find solutions to improved water and sanitation (W&S) access among the urban poor. They cite a lack of demand, or willingness to pay, and lack of property rights, or tenure, as causes leading to the effect of low W&S access among the urban poor. This low access to W&S services in turn causes poor health, which can eventually lead to poverty (Duflo *et al.* 2012). In this linear, unidirectional causal model, households' low incomes are the effect of a lack of W&S services resulting from low demand and tenure insecurity. In fact, lack of demand and tenure are not causes at all. These are effects, or symptoms, of underlying structural features of urban poverty (Briggs and Mwamfupe 2000). One does not hear of the middle class lacking demand for services or the wealthy lacking land tenure. Erroneously treating such symptoms of poverty as causes can result in discounting poverty's deeper, more fundamental origins.

Thinking symptomatically also involves examining only one part of a system, such as the recipients of interventions or street-level bureaucrats, through the lens of a microscope – detail is gained at the expense of a broader view and understanding. Currently leaders of the RCT movement have their microscopes honed on poor people's behaviour. While this is partly influenced by the contemporary behavioural turn in microeconomics, Western social scientists have a long heritage in being interested in poor people's decision-making (O'Connor 2009). Currently there is strong interest in learning about why the poor in the Global South, among other things, do not eat more nutritiously, limit childbearing, save more money, become entrepreneurs, and use things that are 'good' for them (Banerjee and Duflo 2012). Magnifying the behaviour of just one subset of individuals comes at the expense of understanding the role of others' behaviour, like those of central or investment bankers, in resolving issues of poverty or development, let alone in creating them.

It is not mere coincidence that the rise of RCTs has gone hand in hand with the rise of individualised micro-interventions such as micro-credit, micro-finance, micro-savings and cash transfers. Intervening in a very local (i.e. individual) way to get at specific material effects (e.g. increases in income) is also part of thinking symptomatically. There is greater model validity between highly individualised, private interventions and the measureable impacts in an RCT. This is not the same as saying that these are the appropriate interventions to deal with development challenges or that RCTs capture all the possible effects from such programmes; they epistemologically do not and cannot (Cartwright 2007). However, RCTs are far more amenable to testing interventions based on private goods, even if they are publically provided, than public goods. An RCT's need to establish a control group that cannot be exposed to an intervention for fear of 'contamination' or spill-over effects complicates its ability to experiment with public goods that are by nature non-excludable and non-rival. Thus, it is far easier to measure if a bed-net distribution programme rather than a neighbourhood storm-water drainage system 'works' at lowering malaria rates. This bias towards highly individualised private goods is evident from an *AidGrade* database that identifies the top 20 represented development programmes among 587 completed experimental and quasi-experimental studies consisting largely of RCTs.

RCTs can only achieve model validity by making heroic assumptions about the planning environment and process. Drawing on Newtonian and Cartesian philosophy, the biomedical model implicitly views the world as a machine with subsystem parts that are independent of the whole and externally controlled (Koithan *et al.* 2012). In fact, assumed independence of parts, including actors in a system, is at the heart of the RCT method. Built into any RCT design is the stable unit treatment value assumption, or SUTVA. The essential idea of SUTVA is that potential outcomes for any unit of an experiment are independent of the treatment assignment of any other unit under study (Sampson 2010). In other words, any learning from an RCT is premised on the assumption that those receiving the intervention (treatment group) had no social interaction with those not receiving the intervention (control group).

Randomised controlled trials also assume that planners and those subject to interventions are separate. Planners only relate to individuals as an intervention's administrators. Planners function as mechanics that *act on* individuals rather than *act with* them. This is considered an engineering approach to development implementation where once knowledge is generated it is put to use in a linear and 'scientific' way (Reddy 2012). This is most evident in the RCT movement's leaders' use of the metaphor of the lever: "It is not easy to escape from poverty, but a sense of

Table 30.2 Top 20 development programmes covered in the *AidGrade* database of experiments

2012	*2013*
Conditional cash transfers	Contract teachers
Deworming	Financial literacy training
Improved stoves	HIV education
Insecticide-treated bed nets	Irrigation
Microfinance	Micro health insurance
Safe water storage	Micronutrient supplementation
Scholarships	Mobile phone-based reminders
School meals	Performance pay
Unconditional cash transfers	Rural electrification
Water treatment	Women's empowerment programs

Source: Vivalt (2014).

possibility and . . . [a] push on the right lever can make a huge difference" (Banerjee and Duflo 2012: x). Thinking symptomatically means imagining the world as a machine.

In addition to biomedicine, the features of thinking symptomatically – focusing on symptoms rather than root causes, defining problems narrowly, intervening in highly localised ways to attain specified effects and assuming independence among parts of a system – have important theoretical and intellectual affinities with rational planning as problem-solving and policy analysis (Friedmann 1987). Randomised controlled trials are an effort at taming what are essentially the 'wicked' problems of poverty and development, for which there is no definitive formulation of the problem and no ultimate test of the solution (Rittel and Webber 1973: 167). It is for this reason that RCTs are often referred to as 'useful fictions'.

Thinking developmentally

If thinking symptomatically and the biomedical model's assumptions of the planning process are inappropriate for wicked problems, what would thinking developmentally, particularly from the global South, look like? There is a rich history of whole systems thinking through Southern healing models of Ayurveda and Chinese Traditional Medicine as well as philosophies of interconnectedness in Ubuntu (Southern Africa) and Buen Vivir (Andean South America). Whole systems approaches

> share a basic worldview that embraces interconnectedness; emergent, non-linear outcomes to treatment; a contextual view of the human being that is inseparable from and responsive to their environment; and interventions that are complex, synergistic, and interdependent. These fundamental beliefs and principles run counter to the assumptions of reductionism and conventional biomedical research methods that dismantle and individually test various interventions.
>
> *(Koithan* et al. *2012: 7)*

Thinking developmentally means focusing on diagnosing root causes rather than symptoms. Such approaches ask different sets of questions. Randomised controlled trials are less focused on questions such as "Why does symptom Y exist?" or "Is Y caused by biological, environmental, emotional or social factors?" (Borgerson 2005). Such questions seek to address the underlying reasons for why certain problems, such as urban poverty, persist. They do not always think of solutions of how to eliminate a symptom but inquire why such symptoms exist in the first place.

Thinking developmentally also means devising interventions that address systemic imbalances. This has the advantage of designing interventions that address *inequality*, something that has been sorely neglected in development since the rise of poverty discourse and analysis (Kabeer 2006). Rather than seeking growth along a single vector, such as per-capita income or even life expectancy, which could actually lead to more imbalance, a whole systems approach seeks to ameliorate root imbalances in power, knowledge and access to productive resources, among others, whose resolution could reduce inequality without favouring any one indicator of welfare over another. According to Traditional Chinese Medicine, "if you reorganise an existing pattern of disharmony into a harmonic pattern of relationships, the original cause will disappear because the conditions in which it was rooted cease to exist" (Beinfield and Korn 2013: 36).

The global and multidimensional worldviews of whole systems "differ from the implicitly reductionist and materialistic worldview of much of mainstream medicine as currently practiced, in which diagnosis and treatment focus on local body parts in isolation from one another" (Bell and Koithan 2006: 293). Thinking developmentally means understanding the individual as one system 'nested' within other social, environmental and political systems where feedback loops exist between all parts of a broader system (Paina and Peters 2012; Rioux 2012). Similarly, instead of assuming an independence of various parts, thinking developmentally treats an environment as interactive, indivisible and whole where individual components are inseparable and interrelated (Bell and Koithan 2006; Koithan *et al.* 2012). This treats *where* the planner chooses to intervene in a system as importantly as the intervention itself. Such an approach has antecedents in development planning in Hirschman's theory of forward and backward linkages where investment in 'leading' sectors would stimulate investment in that sector's input providing and output purchasing firms (Hirschman 1980).

Thinking developmentally also means different assumptions about the planning environment. A planner does not act separately from the implementation environment; s/he is part of the network and system and *acts with* rather than *acts on* subjects. Denying this reality means missing crucial phenomena, such as experimenter–implementer relationships and adaptive agents' reactions to interventions, as possible vehicles of change. More importantly, the planner is self-conscious of her power and position of privilege within the development process. This is part of *acting with* and conscious reflective practice.

Finally, thinking developmentally means a different understanding of whether an intervention 'worked'. Overall effectiveness is determined through global patterns rather than localised 'impact' and complex, often indirect, outcomes are just as acceptable as specified and predetermined ones (Bell *et al.* 2012; Rioux 2012). One aspect of RCTs that is regularly deemphasised is the high risk of a Type II error (i.e. a false negative). As such, the probability of an intervention tested *via* a narrow RCT framing that does not account for contextual factors, like the environment, is more likely to show up as 'failing' to have an impact than not (Bell *et al.* 2012). Rather than testing narrow hypotheses, a whole systems approach asks broadly whether the intervention has led to the discovery of something new. As an empirical approach, it uses more multi-level, thick case observational data to monitor interventions, individuals and their environments iteratively. This is in contrast to the one-off baseline and endline surveys called for by RCT advocates. Also recommended is the use of qualitative data, agent-based modelling and network analysis in addition to quantitative statistics (Paina and Peters 2012). Most importantly, perhaps, is the planner's appetite for uncertainty since thinking developmentally means understanding that an intervention's impacts might be probable rather than certain.

Thinking developmentally in practice

To imagine what a whole systems approach might look like in practice we can assess RCT advocates' most replicated development intervention – school-based deworming. An influential experiment in rural Kenya run between 1998 and 2001 found internally valid impacts of oral deworming drugs (anthelmintics) on school participation and worm burdens, providing evidence of a causal link between health and education (Miguel and Kremer 2004). Since then, over 95 million children have been reached by oral deworming drugs: Kenya scaled up the programme to the national level; Bihar State in India treated 11 million children in 2011 alone; and the World Food Programme has incorporated deworming drugs into its school feeding programmes (J-PAL 2012).

The deworming effort bears all the hallmarks of thinking symptomatically – it identified a symptom of poverty, intestinal worms (helminths), as a cause, tied that symptom to a linear causal model of how decreased worm incidence and school absence reduces poverty, intervened in a localised and highly specified way through the administration of individual deworming pills in select schools, and assumed the independence of participating schools from their wider environment.

How might thinking developmentally through a whole systems approach have addressed this issue differently? It would have viewed intestinal worms within the context of rural Kenya and not mistaken it as a cause, but rather a symptom, of poverty: "Soil-transmitted helminth infections . . . occur wherever there is poverty" (WHO 2015). Meanwhile, in the words of RCT proponents, "rapid reinfection means that the drugs must be taken every 6–12 months to keep worm infections at bay" (J-PAL 2012: 1). It would try to intervene in a way that addressed this chronic reinfection. First, rural development planners might invest heavily in sanitation infrastructure, which can interrupt the intestinal worm infection cycle. This helps restore balance by addressing urban bias and bringing rural sanitation levels closer to that of towns (Bezemer and Heady 2008). Second, since some infections would unfortunately continue, health planners would use locally available resources to treat remaining infections. The papaya fruit has been used for centuries as a natural deworming medicine (Dyce 1865). Studies have shown papaya killing helminths *in vitro* and in animals and humans (Ming and Moore 2013). A recent study in Nigeria showed that dried papaya seed killed intestinal helminths in over 75 per cent of children treated (Okeniyi *et al.* 2007). Third, rural development planners would tie the need for papaya seeds to agricultural development through horticulture programmes. "It is conceivable that *C. papaya*, which thrives effortlessly, can be much more readily available and affordable if commercial and subsistence farming of the papaya plant was encouraged in tropical communities" (Okeniyi *et al.* 2007: 196). The same children who are being dewormed are part of often struggling rural households who rely on agriculture for income. Fourth, economic development planners would provide technical assistance to natural medicine manufacturers in the global South like *Paxherbals*, which began in a wooden hut but now employs over 100 people, to prepare, store and deliver papaya as medicine (Hitchen 2015). Finally, together they would help develop markets to link papaya growers, natural medicine manufacturers and schools.

A whole systems approach like the one above tries to address helminth infections' root causes of inadequate sanitation infrastructure and existing poverty among rural farm households. Planners acknowledge interrelationships and therefore the intervention focuses on multiple actors and levels in a system – sanitation contractors, rural farm households, subsistence and commercial horticulture farmers, natural medicine manufacturers and schools, who become part of a development network. It calls on planners to act with these different participants, actively linking them together, rather than acting on one part of the system in isolation. It conceives of causal relationships as multi-directional and complex, with actors of an intervention potentially engaging and benefitting each other in unexpected ways. Thus, learning must be based on general rather than pre-specified outcomes of interest – a planner will be able to know if she is 'doing good' but not necessarily if she 'was right'.

Conclusion

Thinking developmentally is difficult in today's planning environment. The power of seemingly simple causal relationships propped up by rhetorical and methodological appeals to biomedicine thrive in planning environments ruled by private sector inspired accountability regimes. For a whole systems approach to succeed it is necessary to build a political effort for thinking developmentally that does not rest on rationality alone. There is a distinctly political element to development experimentation and learning that should not be ignored. Development planners need the administrative space that allows for longer experimental time horizons, complex causality, and an appetite for risk. This will only occur when development planners step outside of their technocratic comfort zone to build support for a whole systems approach through alliances and coalitions both inside and outside their immediate institutions and above and below their position of authority.

References

Banerjee, A. and Duflo, E. (2008) "The experimental approach to development economics", Working Paper 14467, Cambridge, MA: National Bureau of Economic Research.

Banerjee, A. and Duflo, E. (2012) *Poor Economics: A Radical Rethinking of the Way to Fight Global Poverty*, New York: Public Affairs.

Bardhan, P. (2005) "Theory or empirics in development economics", *Economic and Political Weekly*, 1 October, pp. 4333–4335.

Beinfield, H. and Korn, E. (2013) *Between Heaven and Earth*, New York: Random House Publishing Group.

Bell, I.R. and Koithan, M. (2006) "Models for the study of whole systems", *Integrative Cancer Therapies* 5(4): 293–307.

Bell, I.R., Koithan, M. and Pincus, D. (2012) "Methodological implications of nonlinear dynamical systems models for whole systems of complementary and alternative medicine", *Forschende Komplementärmedizin [Research in Complementary Medicine]* 19(s1): 15–21.

Bezemer, D. and Headey, D. (2008) "Agriculture, development, and urban bias". *World Development* 36(8): 1342–1364.

Bhatt, J. (2011) "Causality and the experimental turn in development economics", *New School Economic Review* 5(1): 50–67.

Borgerson, K. (2005) "Evidence-based alternative medicine?", *Perspectives in Biology and Medicine* 48(4): 502–515.

Briggs, J. and Mwamfupe, D. (2000) "Peri-urban development in an era of structural adjustment in Africa: The city of Dar es Salaam, Tanzania", *Urban Studies* 37(4): 797–809.

Cartwright, N. (2007) "Are RCTs the gold standard?", *BioSocieties* 2(1): 11–20.

Chambers, R. (1981) "Rural poverty unperceived: Problems and remedies", *World Development* 9(1): 1–19.

Chambers, R. (1992) "Rural appraisal: Rapid, relaxed and participatory", IDS Discussion Paper 311, Brighton: Institute of Development Studies. Available www.ids.ac.uk/files/Dp311.pdf [accessed 1 October 2015].

Clarence, E. (2002) "Technocracy reinvented: The new evidence based policy movement", *Public Policy and Administration* 17(3): 1–11.

Cohen, J. and Easterly, W. (2010) *What Works in Development?: Thinking Big and Thinking Small*. Washington DC: Brookings Institution Press.

Deaton, A. (2010) "Instruments, randomization, and learning about development", *Journal of Economic Literature* 48(2): 424–455.

Duflo, E. (2010) "Social experiments to fight poverty". Available: www.ted.com/talks/esther_duflo_social_experiments_to_fight_poverty?language=en [accessed 4 April 2016].

Duflo, E., Galiani, S. and Mobarak, M. (2012) "Improving access to urban services for the poor: Open issues and a framework for a future research agenda", Cambridge, MA: Abdul Latif Jameel Poverty Action Lab. Available: www.povertyactionlab.org/publication/improving-access-urban-services-poor [accessed 1 October 2015].

Duflo, E., Glennerster, R. and Kremer, M. (2008) "Using randomization in development economics research: A toolkit". In T. Schultz and J. Strauss (eds) *Handbook of Development Economics Volume 4*, Amsterdam and Oxford: North-Holland, pp. 3895–3962.

Dyce, R. (1865) "Carica Papaya: A vermifuge", *British Medical Journal* 1(230): 548–549.

Ferguson, J. (2015) *Give a Man a Fish: Reflections on the New Politics of Distribution*, Durham, NC and London: Duke University Press Books.

Fine, B. and Milonakis, D. (2009) *From Economics Imperialism to Freakonomics: The Shifting Boundaries between Economics and Other Social Sciences*, Oxon and New York: Routledge.

Forsetlund, L., Chalmers, I. and Bjørndal, A. (2007) "When was random allocation first used to generate comparison groups in experiments to assess the effects of social interventions?", *Economics of Innovation and New Technology* 16(5): 371–384.

Friedmann, J. (1987) *Planning in the Public Domain: From Knowledge to Action*, Princeton, NJ: Princeton University Press.

Guba, E.G. and Lincoln, Y.S. (1989) *Fourth Generation Evaluation*. Thousand Oaks, CA: Sage.
Harrison, G.W. (2013) "Field experiments and methodological intolerance", *Journal of Economic Methodology* 20(2): 103–117.
Hirschman, A.O. (1980) *The Strategy of Economic Development*, New York: W.W. Norton & Co.
Hitchen, J. (2015) *Modern African Remedies: Herbal Medicine and Community Development in Nigeria*, London: African Research Institute.
Hoch, C. (1984) "Doing good and being right: The pragmatic connection in planning theory", *Journal of the American Planning Association* 50(3): 335–345.
Imai, K., Tingley, D. and Yamamoto, T. (2013) "Experimental designs for identifying causal mechanisms", *Journal of the Royal Statistical Society: Series A (Statistics in Society)* 176(1): 5–51.
International Initiative for Impact Evaluation (3ie) (2015) "3ie members". Available: www.3ieimpact.org/en/about/3ie-affiliates/3ie-members/ [accessed 1 October 2015].
J-PAL (2012) "Deworming: A best buy for development", Cambridge, MA: Abdul Latif Jameel Poverty Action Lab. Available: www.povertyactionlab.org/publication/deworming-best-buy-development [accessed 1 October 2015].
J-PAL (2016) "Evaluations". Available: www.povertyactionlab.org/evaluations [accessed 16 March 2016].
Jonas, W.B., Beckner, W. and Coulter, I. (2006) "Proposal for an integrated evaluation model for the study of whole systems health care in cancer", *Integrative Cancer Therapies* 5(4): 315–319.
Kabeer, N. (2006) "Poverty, social exclusion and the MDGs: The challenge of 'durable inequalities' in the Asian context", *IDS Bulletin* 37(3): 64–78.
Koithan, M., Bell, I.R., Niemeyer, K. and Pincus, D. (2012) "A complex systems science perspective for whole systems of complementary and alternative medicine research", *Forschende Komplementärmedizin [Research in Complementary Medicine]* 19(s1): 7–14.
Kremer, M., Miguel, E., Mullainathan, S., Null, C. and Zwane, A.P. (2011) "Social engineering: Evidence from a suite of take-up experiments in Kenya", Working Paper, Berkeley, CA: University of California. Available: www.poverty-action.org/publication/social-engineering-evidence-suite-take-experiments-kenya [accessed 4 April 2016].
Legovini, A. (2010) "Development impact evaluation initiative: A World Bank-wide strategic approach to enhance development effectiveness", draft report to the Operational Vice Presidents, Washington, DC: The World Bank. Available: http://documents.worldbank.org/curated/en/2010/06/14595981/development-impact-evaluation-initiative-world-bank-wide-strategic-approach-enhance-developmental-effectiveness [accessed 4 April 2016].
Miguel, E. and Kremer, M. (2004) "Worms: Identifying impacts on education and health in the presence of treatment externalities", *Econometrica* 72(1): 159–217.
Milberg, W. (2009) "The new social science imperialism and the problem of knowledge in contemporary economics". In S. Gudeman (ed.) *Economic Persuasions*, New York and Oxford: Berghan, pp. 43–61.
Ming, R. and Moore, P.H. (2013) *Genetics and Genomics of Papaya*. New York: Springer Science and Business Media.
Moynihan, D.P. (2009) "Through a glass, darkly: Understanding the effects of performance regimes", *Public Performance and Management Review* 32(4): 592–603.
O'Connor, A. (2009) *Poverty Knowledge: Social Science, Social Policy, and the Poor in Twentieth-Century U.S. History*, Princeton, NJ: Princeton University Press.
Okeniyi, J.A.O., Ongulesi, T.A., Oyelami, O.A. and Adeyemi, L.A. (2007) "Effectiveness of dried carica papaya seeds against human intestinal parasitosis: A pilot study", *Journal of Medicinal Food* 10(1): 194–196.
Paina, L. and Peters, D.H. (2012) "Understanding pathways for scaling up health services through the lens of complex adaptive systems", *Health Policy and Planning* 27(5): 365–373.
Poor Economics (2016) "What can you do?". Available: www.pooreconomics.com/what-can-you-do [accessed 16 March 2016].
Ravallion, M. (2009) "Should the randomistas rule?", *The Economists' Voice* 6(2): 1–5.
Reddy, S. (2012) "Randomise this! On poor economics", *Review of Agrarian Studies* 2(2): 60–73.
Richey, L.A. and Ponte, S. (2014) "New actors and alliances in development", *Third World Quarterly* 35(1): 1–21.
Rioux, J. (2012) "A complex, nonlinear dynamic systems perspective on Ayurveda and Ayurveda research", *The Journal of Alternative and Complementary Medicine* 18(7): 709–718.
Rittel, H.W.J. and Webber, M.W. (1973) "Dilemmas in a general theory of planning", *Policy Sciences* 4(2): 155–169.

Sampson, R.J. (2010) "Gold standard myths: Observations on the experimental turn in quantitative criminology", *Journal of Quantitative Criminology* 26(4): 489–500.
Savoy, C.M. (2014) *Taxes and Development: The Promise of Domestic Resource Mobilization*, Lanham, MD: Rowman & Littlefield.
Schön, D.A. (1983) *The Reflective Practitioner: How Professionals Think in Action*, New York: Basic Books.
Schön, D.A. and Rein, M. (1995) *Frame Reflection: Toward the Resolution of Intractable Policy Controversies*, New York: Basic Books.
Seddon, J. (2013) Personal Communication, 9 July, Indian Institute of Human Settlements, Bangalore, India.
Sen, A. (2002) *Delivering the Monterrey Consensus: Which Consensus?*, London: Commonwealth Secretariat.
Tendler, J. and World Bank Operations Evaluation Dept. (1993) *New Lessons from Old Projects: The Workings of Rural Development in Northeast Brazil*, Washington DC: Operations Evaluation Department, The World Bank.
Verhoef, M.J., Casebeer, A.L. and Hilsden, R.J. (2002) "Assessing efficacy of complementary medicine: Adding qualitative research methods to the 'gold standard'", *The Journal of Alternative and Complementary Medicine* 8(3): 275–281.
Vivalt, E. (2014) "How much can we generalize from impact evaluations?", unpublished manuscript, New York: New York University.
Watts, M.J. (1993) "Development I: Power, knowledge, discursive practice", *Progress in Human Geography* 17(2): 257–272.
World Health Organisation (2015) "Epidemiology". Available online: www.who.int/intestinal_worms/epidemiology/en/ [accessed 29 September 2015].
Yunus, M. (1999) *Banker to the Poor: Micro-Lending and the Battle Against World Poverty*, New York: Public Affairs.

Index